AF478224

Human Olfaction

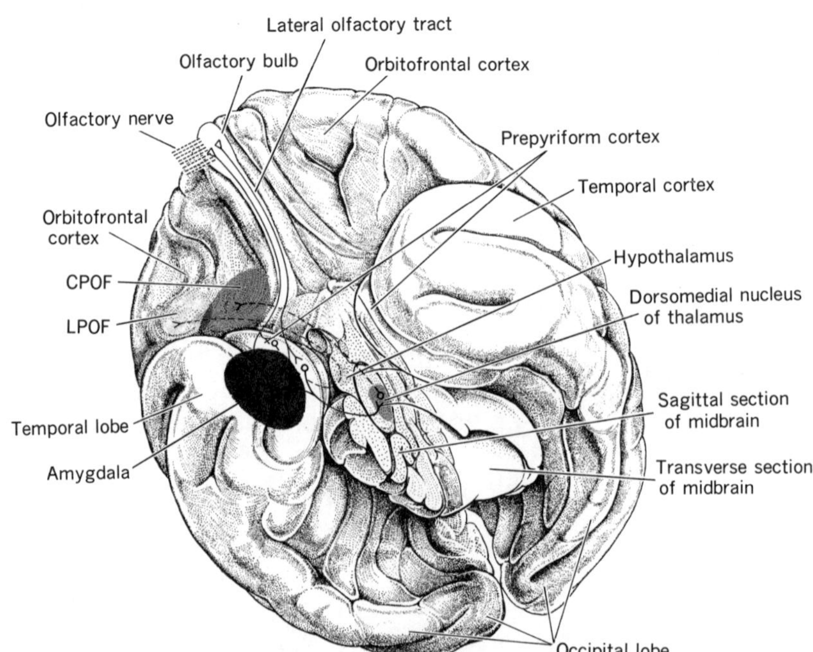

Probable dual olfactory pathways to neocortical olfactory areas in humans. (see pp. 334–351)

Human Olfaction

Sadayuki F. Takagi

Professor Emeritus, Gunma University,
Maebashi, Japan, and
Consultant to Takasago Corporation,
Tokyo, Japan

UNIVERSITY OF TOKYO PRESS

To my two fathers
the late Dr. Katsuyuki Takagi, pediatrician
the late Dr. Satoru Miyamoto, surgeon

This publication is partly supported by The Naito Foundation.

©UNIVERSITY OF TOKYO PRESS, 1989
ISBN 4–13–068148–6
ISBN 0–86008–434–5

Printed in Japan
All rights reserved. No part of this publication may be reproduced or transmitted
in any form or by any means, electronic or mechanical, including photocopy,
recording, or any information storage and retrieval system, without permission
in writing from the publisher.

Contents

Foreword

I am honored that Professor Takagi asked me to write a foreword to his thorough, scholarly book on human olfaction, a field on which he has spent a lifetime of research and to which he has made many notable contributions. I first had the pleasure of meeting Professor Takagi at the International Symposium on Olfaction and Taste (ISOT) held in Tokyo, September 1965. Since then we have met over the years at subsequent ISOT and other meetings around the world. Science is clearly an international enterprise, but the organization of societies for the study of the chemical senses has lagged behind many other fields of biological and medical science.

The early classical works in the chemical senses, olfaction and taste, as in many fields of science, derived from European laboratories. There is the 1895 classic on "The Physiology of Smell" by Hendrik Zwaardemaker, whose contributions included an early classification of odors and the invention of an olfactometer for measuring individual differences of olfactory sensitivity in people. Other early classics are "Der Geruch" by Hans Henning (2nd edition 1924, Barth, Leipzig) and "Die Physiologie des Geruchs- und Geschmackssinnes" in "Handbuch der Physiologie der Niederen Sinnes" by Emil von Skramlik (1926, Thieme, Leipzig).

Study of the chemical senses, taste and smell has intensified and accelerated in recent times signalled by organization of the International Symposium of Olfaction and Taste (ISOT) which the late Prof. Yngve Zotterman, Prof. Lloyd Beidler and I played a role in planning. Its first meeting chaired by Prof. Zotterman was held at the Wenner-Gren Centre in Stockholm, 1962. ISOT has convened since then in connection with each triennial International Congress of Physiology. Chemosensory science in Europe has been signalled by the founding of the European Chemoreception Research Organization (ECRO) in 1980. The Japanese Association for the Study of Taste and Smell was organized in 1967 and the Association for Chemoreception Sciences (AChemS) in the U.S. in 1978.

Olfaction, compared with its fellow chemical sense, gustation, elicits a much richer domain of sensory experience. It may indeed be one of the most varied sensoria of multi-qualities, reflecting the multi-compositions of odor-stimulating chemicals, air-borne or water-borne, not only due to naturally occurring chemicals from plant and animal life, to the ingenuity of the chemist, to the variety and imagination of the perfumer's skill and art, but also to the consequences of industrial and environmental contaminations and pollution. From the fragrant to the foul, odors are rarely neutral; many, if not most, have an hedonic quality, not all the same for all organisms. In animals numerous instances of odor and vomeronasal effects on hormones, sexual and other behavioral processes are well known, due to the joint operation with the vomeronasal system. The latter is not present in old world monkeys, higher primates and humans, yet the variety of effects due to olfaction per se is an intriguing and fascinating subject, as Takagi so thoroughly documents. Hedonic responses to odor may be intrinsic to the odor quality as experienced but also may reflect learned and associative consequences.

On another level, I should like to express my appreciation to Sadayuki Takagi for having introduced my wife and me to the artistry and tradition of the ancient cult of incense, *kōdō*, which he describes in his Appendix: "The Art of Smell." He arranged for such a ceremony on the occasion of another scientific meeting being held at Fukuoka. In deference to the comfort of many westerners in our group, we sat in chairs around a circular table in a "tastefully decorated peaceful room." Sadayuki acted as our master of the ceremony with grace and charm. The fact that olfaction can provide a sensory domain of rich yet subtle character was clearly evident in the course of our ceremony.

One current development is the increased attention and support for the clinical aspects of the chemical senses as reflected in the U.S. National Institutes of Health support for Clinical Chemosensory Research Centers. In this context, Professor Takagi's section on the Clinical Aspects of Olfaction, as well as explications of procedures and treatment developed and employed in Japan, is an informative contribution for western readers.

The wealth of research findings, publications and books on contemporary chemosensory science has been documented by reviews and numerous compendia, most often of multiple authorship. What distinguishes this new work on Human Olfaction is the fact of its single authorship by Sadayuki F. Takagi, making it a fit companion to the early classics of Zwaardemaker, Henning and von Skramlik. There is now mu chmore information from researchers all over the world, an abundance to examine and to analyze. Professor Takagi has not only given us a masterful review, but has been a significant contributor by his own research contributions.

April 1, 1989

CARL PFAFFMANN
Vincent and Brooke Astor Professor Emeritus
The Rockefeller University

Preface

A Question from Dr. Ralph W. Gerard

On October 1, 1954, I arrived at the Neuropsychiatric Institute of the University of Illinois's Chicago campus, at the invitation of Dr. Ralph Waldo Gerard. In the late 1940s, Dr. Gerard had devised a glass capillary microelectrode and recorded the resting membrane potential of a muscle fiber for the first time (Graham and Gerard, 1946; Ling and Gerard, 1949), and led the way in opening a new era in the field of neurophysiology.

Before undertaking these pioneering studies, he had, working with Dr. B. Libet, found a very interesting phenomenon in the brain wave of the frog: one drop of 0.5% nicotine solution produced a striking change in the brain wave of the olfactory bulb (OB) in the same way that the betawave pattern of human EEG can instantly change into alpha-waves. They had published this and related findings in several journals (Libet and Gerard, 1938, 1939, 1941). The regular 6/sec wave elicited by the nicotine solution attracted the attention of many neurophysiologists.

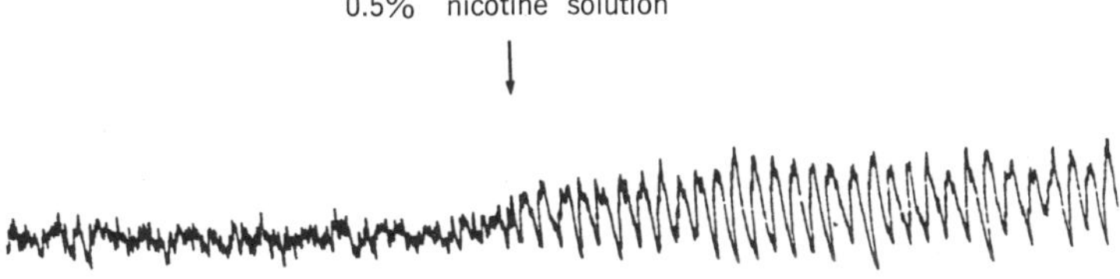

On arriving in Chicago, I found that Dr. Y. Oomura (who was to become professor of physiology at the Kyushu University School of Medicine) had also been studying with Dr. Gerard. The problem that Dr. Gerard gave us was to elucidate the generative mechanism of this regular nicotine wave, employing his glass microelectrode. We looked for a possible relationship between the regular brain waves on the surface of the OB and the spike potentials of neurons inside it in the frog. We managed to find some degree of correspondence between them, but before we could accumulate enough data to publish a paper, I had to leave Chicago to take a post as professor at Gunma University in Maebashi city, Japan.

On the way back, I went through Europe and visited various laboratories in many countries. Having had many fruitful experiences in the United States and Europe, I set out on the return trip to Japan from London aboard a freighter which set sail on January 10, 1957. During the 50 days' voyage, I gave a lot of thought to the choice of a subject or research problem to grapple with in the new laboratory at Gunma University. I felt that this shipboard time should be used to determine the subject of my life's work, so I read many books that I had bought in London. I reflected on my own character and limited abilities. I also considered the fact that I had already begun to study one of the olfactory areas in the brain and that very few investigators, if any, were working on this theme anywhere in the world at that time. Eventually, before landing in Kobe, I decided to take "neurophysiological research on the olfactory nervous system" as my life's work. Thus, the encounter with Dr. Gerard proved to be the impetus for my choice of a research topic. I concentrated my studies solely and consistently on olfaction from October of 1954 for thirty years until April of 1984, when I retired from Gunma University.

It is a great pleasure for me that I finally can send this monograph on olfaction out into the world, based upon my thirty years' research.

In all this time, in the course of studying various aspects of neurophysiology on olfaction, many of which are detailed in this book, I have thought about Dr. Gerard's problem: elucidation of the generative mechanism of the regular nicotine wave of the olfactory bulb. My answer to Dr. Gerard's question will be stated in the Postscript.

June, 1989

Sadayuki F. TAKAGI

Acknowledgements

I would like to express my sincere thanks to my former colleagues in my laboratory of the Gunma University and many friends in other universities as listed below. Without their close collaboration, it would have been impossible for me to write this book.

The core of this book is mainly the data accumulated in the Department of Physiology of the Gunma University School of Medicine over thirty years. For the first part of these years, electrical activities of the olfactory epithelium and the olfactory bulb were studied mainly in the frog (Chapters IV, V, and VI). These chapters owe very much to the earnest efforts of Dr. T. Shibuya (Tsukuba University), Dr. S. Higashino (Tokai University), Dr. N. Ai (Tokyo Gakugei University), Dr. K. Aoki (Sophia University), Dr. M. Okano (Nihon University), Dr. H. Takeuchi (Takasaki National Hospital) and Dr. K. Omura (Tokyo Dental College). Then electrical activities of the olfactory bulb in the rabbit became the subject of research (Chapter VII). Many new data on the olfactory bulb are mainly due to the skillful study of Dr. K. Mori (Osaka Bio-Science Institute) in collaboration with Dr. M. Nakajima (Kyushu University), Dr. K. Kishi (Toho University), Dr. H. Ojima (The Institute of Physical and Chemical Sciences), Dr. F. C. Fujita (Mitsubishi Kasei Institute of Life Sciences) in Japan and Dr. G. M. Shepherd, Dr. M. C. Nowycky, Dr. C. Greer, and Dr. J. Kauer at Yale University. The higher olfactory areas, the anterior olfactory nucleus, the piriform cortex, the olfactory tubercle, the lateral hypothalamus, the thalamus and other areas were also studied in the rabbit (Chapters VIII and X, A). The fruitful results reported in these chapters are entirely owing to the energetic studies by Dr. M. Satou (Tokyo University), Drs. K. Mori and K. Imamura (Osaka Bio-Science Institute), Dr. S. Kogure (Soka University), and Drs. N. Onoda and M. Iino (Gunma University) in the rabbit and the dog. The neocortical olfactory areas were explored and clarified in these mammals by the strenuous efforts of Drs. N. Onoda and K. Imamura, Dr. K. Ariki (Hiroshima University) and Dr. E. Obata (Kagoshima University) and in the cat by the endeavors of Dr. F. Motokizawa (Nara University) (Chapter X, A).

For the last part of my studies, I wanted to clarify the human olfactory nervous system, especially the previously unidentified neocortical olfactory area. For this purpose, I decided to use old world monkeys. As a result, two olfactory areas were discovered in the orbitofrontal cortex, and nerve pathways to them were also demonstrated. In addition, a subcortical olfactory pathway to the lateral hypothalamic area was proven. The new results in Chapter IX were obtained solely by the painstaking efforts of Dr. T. Tanabe (Tokai University), Drs. H. Yarita and M. Iino (Gunma University), Dr. Y. Tazawa (Nihon Medical College), Ms. Y. Ooshima in collaboration with Dr. K. Kawamura (Keio University) and Dr. J. Naito (Iwate University).

During these 30 years, I became engaged in the study of foul odors for over ten years at the government's request as part of the effort to establish an offensive odor control law (Law No. 91 of 1971, Environment Agency). In the course of the project, I felt it urgent and essential to manufacture a standardized olfactometer. With the collaboration of many otorhinolaryngologists from a dozen universities, we examined the olfactory sensitivities of the Japanese people. A standardized "T & T Olfactometer" was manufactured in 1975. Since then, it has been used in universities, hospitals and clinics all over Japan for the measurement of the grades of olfactory sensitivities of normal subjects and patients. Simultaneously, methods to treat olfactory disorders have also been earnestly studied (Chapters II and III). These studies, very probably the world's first on such a large scale, were undertaken by many doctors and subjects; special mention must be made of Drs. B. Toyota and R. Umeda (Kanazawa University), Dr. T. Kitamura (Chiba University), Dr. E. Asaka (Showa University) and Dr. S. Makino (Gunma University). During these studies, I benefited from the collaboration of Mr. M. Kainosho and other perfumers and other staff members of the Takasago Corporation. I would like to express my deep gratitude to all these people.

In order to advance my research on olfaction, I had to study the general physiology of human and animal olfaction (Chapter I), with the cooperation of Drs. K. Ito and H. Kaise (Gunma University) and others. At the same time, I was interested the in non-scientific aesthetic aspect of olfaction and often attended formal and informal gatherings of incense groups. I was keenly impressed with the profundity of the world of incense and perfume (see Appendix). I express my sincere gratitude to two masters of the incense cult, the late Mr. K. Sanjonishi and the late Mr. S. Hachiya.

For the publication of this monograph I have to express my deep thanks for the consistent and warm encouragement of Dr. Masao Ito, emeritus professor at the University of Tokyo (The Institute of Physical and Chemical Sciences), for the devoted assistance of Dr. M. Iino in accumulating necessary documents and for the kind collaboration of the University of Tokyo Press. My sincere thanks also go to Dr. Carl Pfaffmann, Vincent & Brooke Astor Professor Emeritus at Rockefeller University, who kindly read the whole manuscript, gave me helpful advice and wrote an excellent foreword to this monograph. I also render my deep thanks to the authors who kindly allowed me to reproduce their illustrations, and to the following publishers: Academic Press, Akademiai Kiado, Alan R. Liss Inc., W. B. Saunders Co., American Medical Association, The American Physiological Society, Annual Reviews Inc., Blackwell Scientific Publications, Ltd. Elsevier Science Publishers BV (Biomedical Division), Elsevier

Scientific Publishers Ireland, Academic Press, IRL Press, Igaku-Shoin, Ltd., The Japanese Physiological Society, Invicta Press, The Society for Neuroscience; Oxford University Press, The C.V. Mosby Co., Japanese Association for the Study of Taste and Smell Kitakanto Medical Society, The Lancet, New England Journal of Medicine, Otolaryngoly Society of Japan, Japan Rhinologic Society, Nihon Keizai Shinbun-sha, Pergamon Press, The Physiological Society, Plenum Publishing Co., Presse Universitaires de France, The Japan Academy, The Rockefeller University Press and The University of Tokyo Press.

Last, I express my deep gratitude to my wife Yoshiko, whose constant encouragement, devotion and even secretarial assistance over the past five years has enabled me to concentrate on the writing of this monograph.

May, 1989

SADAYUKI F. TAKAGI

A List of Abbreviations

AC, anterior commissure
Acc, nucleus accumbens
AD, anterodorsal thalamic nucleus
AM, anteromedial thalamic nucleus
AMG, amygdala or amygdaloid complex
AON, anterior olfactory nucleus
AV, anteroventral thalamic nucleus
BA, accessory basal nucleus of amygdala
BL, basolateral nucleus of amygdala
BM, basomedial nucleus of amygdala
Ca, anterior commissure
Cd, caudate nucleus
Ce, central nucleus of amygdala
CeM, central medial thalamic nucleus
Ch, chiasma opticum
Ci, capsula interna
Cl, claustrum
CM, center median nucleus of thalamus
Co, cortical nucleus of amygdala
CP, cerebral peduncle
CPOF, centroposterior area of orbitofrontal cortex
DB, nucleus of diagonal band
ER, EnR, entorhinal cortex
FX, fornix
GLd, dorsal division of lateral geniculate nucleus
GLv, ventral division of lateral geniculate nucleus
GM, medial geniculate nucleus
Gp, globus pallidus
HPC, H, hippocampus
Hipps, hippocampal sulcus
IC, internal capsule
L, lateral nucleus of amygdala
Lats, lateral sulcus
LD, lateral dorsal thalamic nucleus
LHA, LH, lateral hypothalamic area
LOT, lateral olfactory tract
LPOF, lateroposterior area of orbitofrontal cortex

LP, lateroposterior thalamic nucleus
M, medial nucleus of amygdala
MA, medial portion of amygdala
MD, mediodorsal thalamic nucleus
MDmc, pars magnocellularis of MD
MDmf, pars multiformis of MD
MDpc, pars parvocellularis of MD
MFB, medial forebrain bundle
MM, mamillary complex of hypothalamus
MOB, main olfactory bulb = OB
MOT, medial olfactory tract
MV, medioventral thalamic nucleus
MDB, nucleus of diagonal band of Broca
NOPA, neocortical olfactory projection area
OB, olfactory bulb = MOB
OFC, orbitofrontal cortex
OT, olfactory tubercle
Pam, periamygdaloid cortex
PC, pyriform cortex
Pcn, paracentral thalamic nucleus
PeR, perirhinal cortex
Pf, parafascicular thalamic nucleus
POA, preoptic area
PPC, PPF, prepyriform cortex
PrR, prorhinal cortex
Put, putamen
R, reticular thalamic nucleus
Rhis, Rh. S, rhinal sulcus
RN, red nucleus
S, stria medullaris of thalamus
SI, substantia innominata
SN, substantia nigra
SO, supraoptic nucleus of hypothalamus
SP, septal area
Spt, septum
STems, superior temporal sulcus
St, Sth, subthalamic nucleus
T.HP, habenulointerpeduncular tract

TMT, mammilothalamic tract
TO, optic tract
VA, ventroanterior thalamic nucleus
VAmc, pars magnocellularis of VA
VB, ventrobasal thalamic nucleus
VLc, caudal part of ventrolateral thalamic nucleus
VLo, oral part of ventrolateral thalamic nucleus
VM, ventromedial thalamic nucleus
VMH, ventromedial hypothalamic area
VPI, ventroposterior inferior nucleus of thalamus
VPLo, pars oralis of ventroposterior lateral nucleus of the thalamus
VPM, ventroposterior medial nucleus of thalamus
VPMpc, pars parvocellularis of ventroposterior medial nucleus of thalamus
V III, third ventricle
ZI, zona incerta
III, oculomotor nerve

Human Olfaction

I. Olfaction in Humans

Olfaction plays a most important role in the animal world. Within its life cycle, an animal fundamentally strives to preserve itself and its species. For self-preservation, animals prey upon other animals or gather fruits and vegetables, and olfaction is most useful in the process of locating prey or food. Olfaction also helps the animal to discern whether food is detrimental or not. When an animal approaches another animal, both will use olfaction to discriminate a comrade from an enemy. If each finds the other animal to be a comrade, or a harmless animal, they cease to act cautiously toward each other and freer behavior ensues. If they find the other to be an enemy, however, they become more cautious toward each other and try to escape immediately.

To preserve their species, males and females must meet and copulate, and the females must bear offspring. Many animals live solitarily. Thus when the short rutting seasons arrive, they must seek out mates efficiently. For this purpose, olfaction is far more important than vision or audition, because sight or sound often cannot be utilized in the bush or the jungle. In the rutting season, special pheromones are emitted, mostly by females, who display their great ability to attract mates. Not only among solitary animals, but also among grouped animals, sex pheromones are essential to distinguish ovulating from non-ovulating females. When young offspring are born, they must locate their mothers among the herd in order to get milk. Conversely, mothers have to be able to find their young to protect them. All these activities depend primarily on olfaction.

Many animals maintain territorial domains around their own dwellings. An animal maintains its territory by marking its own scent along the border every day. This odor is also one of the pheromones emitted by the animal. This again demonstrates the essential role olfaction plays in self-preservation.

Olfaction has come to play various important roles in the human world as well. The most noticeable among them is that humans discovered the attractiveness of good odors and have made scents and fragrances a definite part of their daily lives. In Europe, perfumers, artists of odors, have created many masterpeices of fragrance. Europeans, especially women, spray these perfumes directly onto their bodies and transform their body odors into far more charming aromas. Utilizing pleasant odors in these ways, Europeans have enriched their lives. The Japanese, in contrast, have far less distinct body odors, so the manufacture and use of perfumes in Japan is a very recent development. Instead, the Japanese of yesteryear turned their olfactory interests to developing the pleasure of "listening to" (sniffing) incenses. (See the Appendix to this volume, "The Art of Smell.")

To understand the olfactory world of humans more clearly, let us first consider various aspects of human olfaction (Takagi, 1974, 1976).

A. Characteristics of Olfaction

1. Olfactory Acuity

The first of the characteristics of human olfaction is the remarkable acuity of odor detection. Many investigators have studied the sensitivity of human olfaction. The minimum concentrations required for the detection of different odors, the "olfactory thresholds", are shown in Table I-1. The gas chromatograph, an instrument used to measure the presence and concentration of an odor, has been developed and is constantly being improved, but it is still inferior to human olfaction in the detection of most odors. The few exceptional odors to which the gas chromatograph has higher sensitivity include acetone, 2-butanone, *n*-propanol, dimethyl sulfate, and several others. The most decisive point of difference between our olfaction and the gas chromatograph is that we have an ability not only to analyze the constituents of any one odor but also an ability to appreciate the quality of composite odors collectively. The gas chromatograph has only the former ability.

Animal olfaction is generally supposed to be superior to human olfaction. One animal that has very sensitive olfactory ability is the dog. Many wild animals are alleged to have excellent olfaction, but scientific studies

Table I-1. The lowest concentrations (threshold concentrations) for detection of various odorants in humans. (from J. Le Magnen, 1949)

Odorant	mg in 1*l* air
Mercaptan	$4 \times 10^{-8} - 4 \times 10^{-10}$
Skatole	4×10^{-10}
Ethyl ether	$1 - 1 \times 10^3$
Natural musk	$1 \times 10^{-2} - 7 \times 10^{-6}$
Synthetic musk	$5 \times 10^{-6} - 5 \times 10^{-9}$
Vanilin	$5 \times 10^{-4} - 2 \times 10^{-10}$
Carbolic acid	$4 \times 10^{-3} - 1.2 \times 10^{-4}$

Table I-2. Comparison of the threshold concentrations for detection of various odorants among humans, dogs, and fishes.

Odorant	Man, dog or fish	Number of molecules in 1 cm^3 of water or air
Acetic acid	Man	5.0×10^{13}
	Dog	5.0×10^{5}
Butyric acid	Man	7.0×10^{9}
	Dog	9.0×10^{3}
Valeric acid	Man	6.0×10^{10}
	Dog	3.5×10^{4}
β-Phenyl ethyl	Eel	1.77×10^{3}
alcohol	Minnow	2.17×10^{14}
	Rainbow trout	5.1×10^{11}
Eugenol	Eel	3.0×10^{5}
	Minnow	3.0×10^{14}

on their olfactory acuity have apparently never been performed.

In our laboratory, Kaise (1969) studied olfactory acuity in dogs using Pavlov's conditioned reflex technique. He found that the dog's olfactory threshold to clove oil is as low as 1×10^7 mg/ml. When compared to our threshold to the same odor, the dog's olfaction was found to be superior to ours by a factor of 10^{-6}. This result corresponds well with data obtained by other investigators (Table I-2). The olfactory thresholds in some fish are also shown for comparison. The eel has an olfactory threshold to β-phenyl ethyl alcohol equivalent to that of the dog, and far superior to those of other fishes.

2. Olfactory Fatigue, Adaptation, and Habituation

If we keep sniffing one odor continually, our ability to perceive that odor soon declines and disappears. This is a well-recognized phenomenon in olfaction, known as fatigue. Gas poisoning and/or gas explosion accidents are often reported in newspapers. They occur because gas leaks out gradually and the olfactory sensation of persons nearby fatigues rapidly, allowing the leak to pass unnoticed. When we enter an environment containing potent foul odors, our breathing automatically stops instantly. Soon, however, it slowly resumes and gradually returns to the original rate of respiration. This occurs because we cannot stop breathing for long and also because our olfaction fatigues easily and we soon lose our ability to smell the bad odor.

This decrement in olfactory perception is due to three factors: (1) fatigue of olfactory receptor cells (accommodation); (2) a negative feedback mechanism in the olfactory bulb (adaptation: see Bennett, 1968); and (3) decrease or loss of olfactory sensation during repeated presentations of a specific smell or smells due to diversion of our attention toward other subjects (habituation).

Even when we fail to smell one kind of odor due to the phenomenon of fatigue, we can still smell other odors. This is called "selective fatigue". We seldom live in an absolutely odor-free atmosphere, but we still perceive the air to be odor-free, despite living in an atmosphere which has a variety of odors. This is due to selective fatigue of our olfactory sense.

Taking advantage of this phenomenon, the degree of similarity between two odors can be compared. When we smell the two odors in succession, the ability to sense the second odor declines as a function of the similarity between the two odors. By repeating this procedure with many odors, we can sequentially order these odors from the most similar to the most different.

Ito (1968) examined degrees of olfactory fatigue by recording impulses from olfactory nerve twigs in normally breathing guinea pigs. When filter paper soaked in an odorous solution was put in front of the nostrils for a certain period, impulse discharges elicited during inspiration were quite numerous at first, but as inspiration was repeated, the impulses gradually declined in number. The progressive reduction of impulses soon ceased, however, and nearly half of the initial number of impulses remained (Fig. I-1). Considering that our olfactory perception entirely disappears after such a long exposure to one kind of odor, this finding suggests that our olfactory information is reduced to about half at the olfactory receptor level, while the remaining olfactory information is lost or suppressed by inhibitory mechanisms in the higher olfactory areas before they reach the final locus (or stage) where olfactory perception occurs.

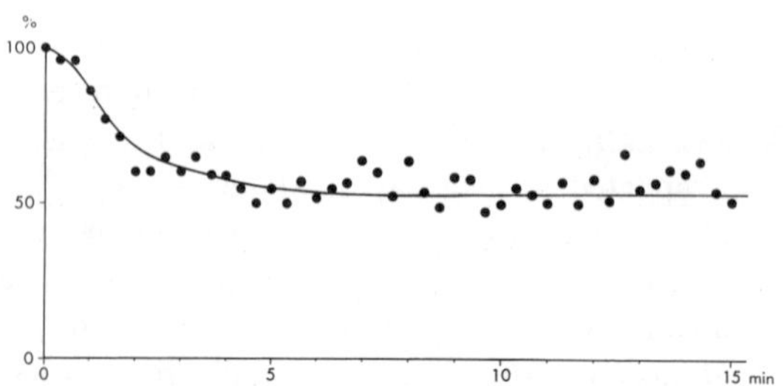

Fig. I-1. The phenomenon of olfactory fatigue at the receptor level.

Integrated records of impulse discharges elicited by each inspiration of an odor in the olfactory nerve twig of a guinea pig. The magnitude of the initial response was made 100%. The response rapidly decreased to 50% of the initial one in about five minutes, and then remained at nearly the same level. (from Ito, 1968)

3. Intensity, Quality, and Preference

Intensities of an odor have been expressed in six stages, numbered from 0 to 5, in Japan. When the concentration of an odor is gradually increased from an extremely low to an extremely high level, we smell nothing in the beginning (Intensity 0). Soon we are able to smell the odor, and eventually we become unable to tolerate the strength of the odor (Intensity 5). In this way, the concentration of an odor is expressed by one of the six numerals, as follows:

0. Odor-free, non-odorous
1. Barely detectable, but not recognizable
    (level of detection threshold)
2. Weak but easily recognizable odor
    (level of recognition threshold)
3. Clearly perceptible odor
    (odor of mediated intensity)
4. Strong odor
5. Very strong, intolerable odor

The famous Weber-Fechner's law expresses a relationship between the intensity of stimulus S and the strength of the resulting sensation R:

$$R = K \log S$$

(where K is a constant).

This equation was proven to be valid for a certain range of odor concentrations. However, Steven's law :

$$R = KS^n$$

(where K is a constant and n is a numeral determined by the kind of odor) is said to fit the relation between R and S better.

When a smell is increased in concentration, not only the intensity of the smell but also its quality becomes different. This is another important characteristic of odors. For instance, ionone smells like violets at a low concentration, but it becomes a cedar wood odor at high concentration. Indol is famous as an example of a bad odor, but when it is highly diluted, it becomes the pleasant odor of a flower. These changes in quality are said to occur because of extremely small amounts of impurities contained in the odorous substances. Skatole has been well known as an example of bad odors. When, however, the impurities of skatole were removed step by step, the skatole itself eventually became odor-free. This is a famous story concerning the chemistry of odorous molecules.

When we smell an odor, the odorous sensation is nearly always accompanied by a pleasant or unpleasant sensation. This occurs also in taste sensation, but seldom in visual or auditory sensation. In general, odors of flowers or fruits produce pleasant sensations and hence are beloved. On the other hand, burnt or putrid odors give unpleasant sensations and hence are disliked. Certainly, some neutral odors exist that are neither pleasant nor unpleasant, but those odors seem to be fairly rare.

All kinds of odors, including pleasant or neutral ones, become bad odors when their concentrations are sufficiently raised. Conversely, bad odors become less unpleasant, and some can eventually become even pleasant, when their concentrations are lowered. For instance, a secretion from the sexual gland of a musk deer emits a bad odor, but when it is diluted step by step, it eventually emits a splendid odor that can be used as an ingredient in cosmetics for women.

Preference for an odor is not equal between men and women. We found that men did not dislike the odor of ethylen brassilate, but women disliked it strongly (Chapter II, p. 27).

A hedonic scale is used to express degree of pleasantness and unpleasantness. A scale of nine steps from $+4$ through 0 to -4, is widely used in Japan, as follows:

$+4$	extremely pleasant	-1	slightly unpleasant
$+3$	very pleasant	-2	unpleasant
$+2$	pleasant	-3	very unpleasant
$+1$	slightly pleasant	-4	extremely unpleasant
0	not pleasant or unpleasant		

Sometimes, scales of seven steps (from $+3$ to -3) are also used.

4. Individual Differences and Specific Anosmia

Some people are very sensitive to odors, while others are not. Thus, various grades of olfactory acuity are found, depending on the persons examined. Even in a particular person, different grades of olfactory acuity are found, depending on the kinds of odors tested or on the persons physical condition (Berg *et al.*, 1963).

Highly sensitive olfaction is called hyperosmia; normally sensitive one, euosmia or normosmia; less sensitive one, hyposmia; and complete lack of olfaction, anosmia or general anosmia (Blakeslee and Salmon, 1931).

Le Magnen (1948) showed that adaptation to a macrocyclic musk (musk lactone) did not influence the threshold of sensitivity to a keto-musk. Guillot (1948a, b) reported consistent findings on humans with partial anosmia. He found subjects with very high thresholds for a few musks but normal thresholds for all other musks tested. These investigators concluded that the olfactory epithelium contains different types of receptors sensitive to individual musk scents, even though these musks constitute a homologous group when judged by subjective standards. Findings from partially anosmic subjects led to many later studies.

Partial anosmia was first studied by Patterson and Lauder (1948). They found six subjects anosmic only to *n*-butyl mercaptan (from the skunk) among 4,030 persons examined. Such partial anosmia only to one specific odor has been termed "olfactory blindness". Amoore (1970), however, preferred the word "specific anosmia" to olfactory blindness, because this phenomenon indicates specific loss of one component of olfactory sensation, but not the total loss of olfaction.

Specific anosmia to cyanide (CN) was studied by Mourant (1950), and Kirk and Stenhouse (1953) and later by Ikeda (1961). Kirk and Stenhouse found that 18.2% of men and only 4.46% of women were anosmic among 244 persons examined. Fukumoto *et al.* (1957) found a very similar result among 433 middle school students in Japan: 18.2% of boys and 5.5% of girls were anosmic to the CN odor. These results suggested that specific anosmia to CN odor was inherited in a sex-linked recessive pattern.

This hypothesis was questioned by Brown and Robinette (1967), who could not find a difference between the sexes in children. Amoore (personal communication) found that 6.9% of persons were anosmic irrespective of sex. Thus, the situation remains unclear and agreement has not been reached (Amoore, 1971).

According to Guillot's theory (1948 a, b), specific anosmia may be due to a defect of one or more receptive mechanisms for primary odors. If this idea is correct, then systematic studies on various specific anosmias would disclose which odors are primary. With this idea in mind, Amoore (1962a, b, 1966 a, b, 1967) commenced a series of studies to discover the primaryodors. So far eight candidates have been put forward (Table I-3) by Amoore (1968, 1975, 1977a, b, 1979), Amoore *et al.* (1968, 1976, 1977), and Pelosi (1977). Amoore had once anticipated that he might be able to find primary odors eventually. It is a regret that he had to discontinue 20–30 this line of investigation while it wat still in process.

Table I-3. Amoore's classification of eight human primary odors. (modified from Amoore's original table, 1979; Amoore *et al.*, 1975b, 1976a,b, 1977)

1. Isovaleric acid	Sweaty odor
2. *l*-Pyrroline	Spermous odor
3. Trimethylamine	Fishy odor
4. Isobutyraldehyde	Malty odor
5. 5α-Androst-16-en-3-one	Urinous odor
6. ω-Pentadecalacton	Musky odor
7. *l*-Carvone	Minty odor
8. 1, 8-Cineole	Camphor odor

5. Age and Sex

Olfactory sensitivity declines with age (Schiffman, 1979; Schiffman *et al.*, 1979; Stevens *et al.*, 1985a, b). Sensitivity to phenol was examined among 500 persons from 15 to over 55 years old (Joyner, 1963). The highest sensitivity was found between ages 24 and 34, followed by 15 to 24, then 35 to 44, 45 to 54, and finally, over 55 years. Venstrom and Amoore (1968) found that olfactory thresholds to 18 fatty acids increase between the ages of 20 and 70. Schiffman (1979) found that olfactory detection thresholds for such foods as cherry, grape, and lemon are 11 times higher on average for older persons than for younger ones. Other compounds, such as steroid hormones that are frequently described as having urine-like odors, are also less readily detected by the elderly.

In an olfactory test with pyridine, Amoore (Venstrom and Amoore, 1968) found no significant differences between men and women in general olfactory sensitivity, but a decline of sensitivity was found with age. The rate of decline between ages 20 and 70 was about one binary step for each 20 years of age.

Although studies show that olfactory thresholds to particular odors are lowest in the young, our overall olfactory ability is known to be best in our middle years. When we analyze an odor and try to identify the constituents by smelling, or when we add many odors together to synthesize a special odor, we need a lot of experience with odors. Such experience increases with age, of course, but olfactory sensitivity declines with age, so our overall olfactory ability reaches its peak in middle age.

Olfactory sensitivity to several odors has been compared between men and women. While a few investigators have reported that olfactory sensitivity is superior in men (Bailey and Nichols, 1884; Bailey and Powell, 1885) or that there is no convincing difference in olfactory thresholds between sexes (Kloek, 1961; Amoore and Venstrom, 1966; Venstrom and Amoore, 1968), most studies have shown that women are more sensitive than men to odorous stimuli (Le Magnen, 1952; Schneider and Wolf, 1955; Koelega and Köster, 1974; Amoore *et al.* 1975a; for more information see review by Doty, 1976).

Women may in fact have more sensitive olfaction than men, but Doty (1976) and Doty *et al.* (1985b) doubted this possibility on the basis of the following five points: (i) Results may differ depending on the selection of test odors. (ii) Men may exhibit higher incidences of specific anosmias than women (Griffiths and Patterson, 1970; Kirk and Stenhouse, 1953). (iii) Women are more sensitive to some odorants near the time of ovulation. Thus, hormonal influences on women's olfaction have to be considered. (iv) All these experiments have been performed by experimenters of the opposite sex. The results may become different when the tests are performed by an experimenter of the same sex. (v) Women generally have far more opportunity to sharpen their olfactory sense from childhood, by smelling cosmetics and food odors while cooking. Repeated experiences with such odors may influence and raise (a) their absolute sensitivity to the odors, (b) their ability to make accurate recognition judgements of such odors, and/or (c) their placement of the response criteria (Doty, 1976). Sex differences in sensitivity to odors is a problem which requires further study.

6. *Olfaction in Health and Illness*

Olfactory sensitivity in healthy men and women is influenced by several factors. Stone and Pryor (1967) reported that olfactory sensitivity in the evening is only 1/10 of that in the morning. Thus, fatigue reduces olfactory sensitivity. Hunger is known to raise sensitivity, especially to food odors (Glaze, 1928). Schneider and Wolf (1955) studied the sensitivity of human subjects to citral, both in hungry condition, in which six or more hours had elapsed since the last meal, and in full condition, within three hours of the last meal. The sensitivity in the hungry state was found to be twice as high as in the full state. Smokers' sensitivity to a phenol odor was found to be lower than that of non-smokers (Joyner, 1963). Moncrieff (1957) reported that smokers' sensitivity to pryidine odor was 1/20 of the non-smokers' sensitivity, but that olfactory sensitivity to other odors was not different between the two groups.

Variation in olfactory sensitivity has been reported in conditions of illness and in several diseases. When one catches a cold, one often complains that tobacco or food tastes unpleasant, and one's appetite for them declines. This phenomenon has been partly attributed to the reduced sensitivity of olfaction. The late Dr. K. Shichijo, a former professor of internal medicine at Gunma University Hospital, observed that malnutrition deprived a patient of olfaction (caused anosmia), but injection of a male hormone (Amolidine, Takeda Pharmaceutical Co.) restored it (personal communication).

Rhinitis affects olfaction in different ways. When it is difficult for odorous air to reach the olfactory epithelium, or when the nasal mucosa become dry and relatively pale or wrinkled, olfactory sensitivity declines. Conversely, sensitivity rises when the olfactory epithelium shows a relatively high degree of swelling and reddening with moisture, and if air inhalation is not obstructed.

Our health is maintained by the cooperative activities of many endocrine organs. Dysfunction of these organs often affects olfactory sensitivity. For instance, Henkin (1966) reported that hyperosmia occurred in patients with adrenal insufficiency and, conversely, hyposmia occurred in patients with hypogonadism. Schneider and Wolf (1955) reported that stress produced hyposmia to citral in men and women. They attributed this to a hormonal imbalance caused by the stress. Le Magnen (1952) found hyposmia to exaltolide in seven ovariectomized women. After administration of ethynyl estradiol to five of them, four women recovered olfactory sensitivity, but the fifth did not. Recovery or rise in olfactory sensitivity after the administration of estrogen was reported in two hypogonadal women by Schneider *et al.* (1958) and in one hypogonadal woman by Good *et al.* (1976). These findings suggest that the olfactory acuity of women is enhanced by the administration of estrogens. Schneider *et al.* (1958) demonstrated the opposite effect with the administration of androgens. Interestingly, similar contrasting results were obtained by Le Magnen (1952) when he examined the influences of both testosterone and estradiol on his own olfactory sensitivity.

Although the results of most of these studies are in good agreement, Doty (1976) felt that they could not be taken as definitive evidence, and asked for further research to establish whether or not administration of gonadal steroids influences human olfactory sensitivity.

7. Olfaction in Women

Women's bodies constantly undergo more complex hormonal changes than men's bodies. Hence, olfaction in women has generally been assumed not to be as stable as in men. In a classic experiment, Le Magnen (1952) reported an increase in olfactory sensitivity to exaltolide after menstruation (Fig. I-2). Most reports thereafter have noted an increased sensitivity in midcycle (Meixner, 1955; Vierling and Rock, 1967; Köster, 1965, 1968; Henkin, 1974). Several investigators identified another period of increased sensitivity (a secondary peak) during the midluteal phase, or near the ovulatory period (Fig. I-2; Le Magnen, 1952; Meixner, 1955; Vierling and Rock, 1967).

In contrast, Le Magnen (1952) reported a decrease in sensitivity to nearly all the odorants tested just before or at menses. A similar decrease was noted in sensitivity to different odors by Schneider and Wolf (1955), Meixner (1955), Good *et al.* (1976), and Mair *et al.* (1978).

Other researchers have reported an increase in sensitivity (Doty, 1976; Köster, 1968). Matsuzaki (1961) found hypersensitivity to Alinamine (thiamin propyl disulfide) in 52% (+5.8%), hyposensitivity in 33% (+2.9%), and no change in 20% (+2.0%) of subjects tested during menstruation. Similarly, Pangborn (1967) reported an increase in sensitivity to exaltolide in one woman, a decrease in another woman, and no change in still another woman. Kahn (1965) found no menstrual cycle phase effect for any of the odorants he used, although there was evidence of sensory

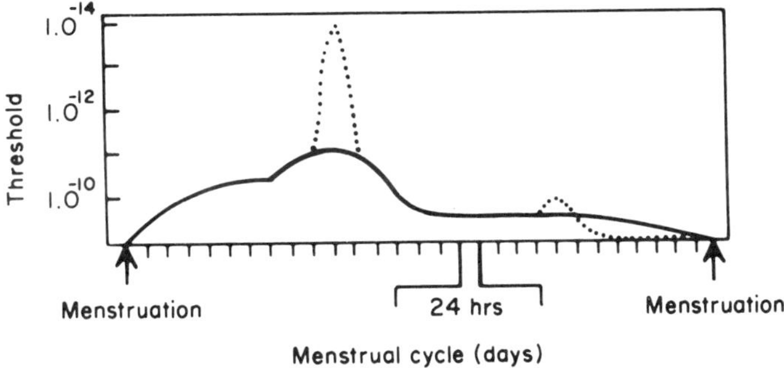

Fig. I-2. Change of women's olfactory thresholds during the menstrual cycle. (from Doty, 1976; adapted from Le Magnen, 1952)

threshold fluctuations within individual subjects. Amoore *et al.* (1975a) also found no significant differences across the menstrual cycle. Matsuda (1938) reported an occasional transient variation in olfactory sensitivity in women at menopause. While adult women are generally sensitive to exaltolide, there is a decreased sensitivity in women at menopause or in those who have undergone ovariectomy.

Recently, Doty *et al.* (1981) demonstrated that olfactory sensitivity to furfural fluctuates cyclically across the menstrual cycle, whether or not the subject is taking oral contraceptive medication. This fluctuation has no apparent relation with the plasma level of LH, FSH, estrone, estradiol, testosterone, or progesterone in women taking the pill (Fig. I-3). Doty *et al.* (1982b) extended these results by demonstrating cyclic changes in olfactory sensitivity to phenyl ethyl alcohol and auditory function over two consecutive periods in women taking oral contraceptives. They indicated that sensitivity changes related to the menstrual cycle, at least in olfaction and audition, are not directly dependent on circulating levels of primary reproductive hormones. They concluded that such changes could conceivably reflect central nervous system processes temporally related to the timing of menses and ovulation.

Quite independently, Herberhold *et al.* (1982) measured both olfactory sensitivity and serum values of FSH, LH, estradiol, and prolactin over the complete menstrual cycle in 20 women. They found no systematic change in olfactory sensitivity during the cycle, even in the ovulation or premenstrual phases. Since their results suggested that hormonal changes do not influence olfactory sensitivity significantly, they attributed the often described variations of olfactory sensitivity at midcycle or in the

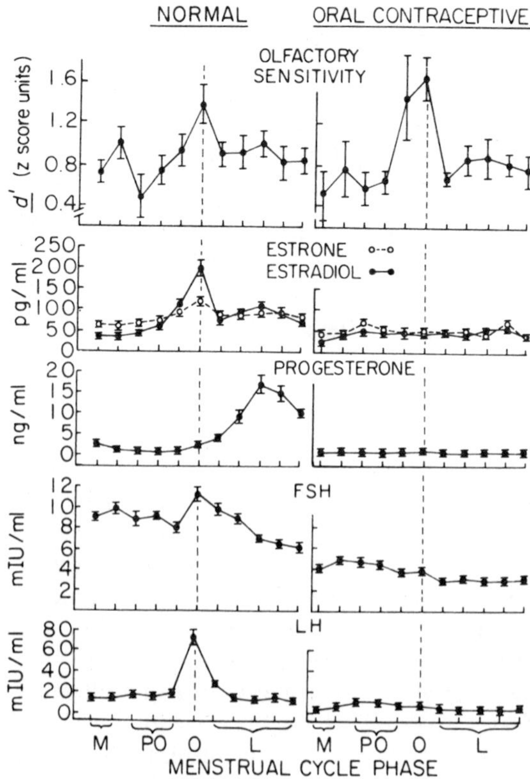

Fig. I-3. Patterns of changes in signal detection measures of olfactory sensitivity and plasma levels of five reproductive hormones across cycle phases in women taking and women not taking oral contraceptives. Data are normalized and assigned to cycle phase using the Doty procedure, where **M** = menstrual phases 1 & 2, **PO** = preovulatory phases 1–3, **O** = "ovulatory phase" (day of LH surge or day before LH surge in normal-cycle group, day 13 or 14 in oral contraceptive group, where day 1 = first day of menses), and **L** = luteal phases 1–5. Note clear fluctuation in olfactory sensitivity in both groups, and the lack of correlation between these changes and circulating levels of pituitary and gonadal hormones in the oral contraceptive group. (from Doty *et al.*, 1981, 1982b)

premenstrual phase to "psycho-vegetative" conditions of the person tested. These two recent studies have thus overturned the idea of direct hormonal influences on olfactory sensitivity.

It is commonly known that many women experience variations of olfactory sensitivity as well as changes in olfactory preference during pregnancy (Dickens and Trethowan, 1971; Doty, 1976). They feel cravings and/or aversions for particular foods when they become pregnant. Many women also experience hyperosmia during pregnancy (Zwaardemaker, 1895; Schmidt, 1925). Luvara and Murizi (1961) detected hyperosmia during the first trimester of pregnancy. Good *et al.* (1976) found it from the late second to the early third trimester. Thus, anecdotal reports of hyperosmia during pregnancy have been supported. However, some of the results of these reports may well have been due to personal aversions to certain foods. Hyposmia, on the other hand, has been reported to occur late in pregnancy (Hansen and Glass, 1936; Noferi and Guidizi, 1946; Luvara and Murizi, 1961). All of these results, however, can be attributed to nasal engorgement, which frequently occurs late in pregnancy, as shown by Mohun (1943) and by Mortimer *et al.* (1936).

B. Human Body Odors

The human body excretes or secretes various kinds of volatile substances, many of which are odorous and constitute body odors. Human body odors are considered to be combinations of sweat, urine, feces, expiration, saliva, breast secretions, sexual excretions, and skin secretions. Every person has his or her individual odor, which differs in intensity and quality, depending on ethnicity and the individual.

Dogs have especially keen olfactory abilities. Well-trained police dogs can remember the smell that a person leaves behind, and can even trace and locate the person by smell 24 hours later. We can smell odors of armpits, hands, or feet, but we consider them to be different from one another. However, police dogs can sense the odors which we feel to be different as identical. That is, dogs can recognize an individual's odor no matter what part of the body it comes from. For instance, police dogs can find a handkerchief scented with a person's armpit odor, if they smell a handkerchief with the hand odor of the same person in advance. We may say, therefore, that everyone has a basic individual body odor (Kalmus, 1958).

Body odor is not determined by food, clothing, or home environment, but by heredity. This has been demonstrated by a number of experiments on monozygotic twins (Kalmus, 1958). Dogs can discriminate without difficulty the body odors not only of persons who are mutually unrelated, but also of members of the same family who live in the same home and eat the same foods. However, police dogs have difficulty in differentiating body odors between identical twins, whose body odors were too similar for successful discrimination. In one case, a dog selected a handkerchief scented with a hand odor of one of a pair of twins after sniffing the hand odor of the other twin. In another case, a dog could not differentiate the body odors of married twin brothers who lived in different places. In fact, recent studies at the Monell Chemical Senses Center in Philadelphia have disclosed that genes regulating immunologic function in mice impart a characteristic scent to individual mice. A mouse can discriminate genetic difference among its potential mates by smell alone. Thus, it is very possible that human body odor is also determined by heredity. (Further details on the Monell group's experiments will be described in Chapter X, pp. 407–408).

As mentioned previously, body odors play a very important role in the animal world. Consequently, we may assume that odors from the human body played similar roles in primitive ages. Recent reports that menstrual cycles become synchronized among young co-dwellers have strengthened the idea that pheromones may also be at work in humans (p.17, p.25).

1. Body Odors Arising from Skin Secretions (Doty, 1981)

In humans, chemicals are secreted to the surface of the skin from three primary sources: eccrine sweat glands, apocrine sweat glands, and sebaceous glands.

The eccrine glands are commonly identified as the sweat glands in humans and are distributed over nearly the entire body. They secrete an aqueous solution that is generally considered odorless. However, dietary factors, such as garlic, can make the sweat odorous, and certain metabolic diseases clearly influence the character of these secretions.

The sebaceous glands are also located in numerous body areas and secrete most of the lipids of the skin that keep the skin supple and waterproof. Their density is greatest in the forehead, face, and scalp. Specialized sebaceous glands occur in many regions including the eyelids (Meibomian glands), the ear canal (Menous glands), the nostrils, the lips, the buccal mucosa (Fordyce's glands), the nipple (Montgomery's glands), the prepuce, and the anogenital region.

The apocrine sweat glands are the major source of "body odor" in humans. Unlike eccrine glands, the distribution of apocrine glands is limited to such skin areas as the axillae, the sternal region, the anogenital region, the mammary areola, the cheek region, the eyelid, the ear canal, and regions in the scalp. However, most of these glands are located in the axillae, where bacteria process their viscous secretions, producing a characteristic body odor. The two major axillary bacteria are the micrococci, which produce isovaleric acid, emitting an acrid sweat-like odor, and the diphtheroids, which also produce isovaleric acid in addition to other odorous chemicals that more closely resemble typical body odor (Labows, 1979). Two odorous steroids, androsterone and androstenol, have been detected in the axillae and appear to be important contributors to the axillary odor (Bird and Gower, 1980; Brooksbank, 1970; Brooksbank *et al.*, 1974; Claus and Alsing, 1976; Doty, 1981; Gower, 1972; Labows *et al.*, 1979).

2. Where in the Human Body Odors Originate
2.1. Axillary odor

Apocrine gland secretions are interesting from the perspective of chemical communication for several reasons (Doty, 1981). First, apocrine glands produce the typical body odor frequently judged as unpleasant when experienced in such situations as crowded buses, busy elevators, and sweaty locker rooms. Second, these glands become functional only after puberty but lose their function at menopause. Third, they are much larger in males than in females, although females sometimes possess a greater number of them. Fourth, they differ considerably in size and number among various races. Fifth, they excrete maximally during periods of excitement or stress. Finally, some of their components are steroids similar or identical to compounds believed, in nonhuman mammals, to influence behavioral or endocrine responses. For example, androsterone sulfate, androst-4-ene-3, 17-dione, dehydroepiandrosterone sulfate, 5α-androst-16-en-3-α-ol, and 5α-androst-16-en-3-one have been isolated from the axillary region (Brooksbank, 1970; Brooksbank *et al.*, 1974; Claus and Alsing, 1976; Gower, 1972; Labows *et al.*, 1979).

Several behavioral investigators (Russell, 1976; Doty *et al.*, 1978b) have sought to establish whether axillary odors can provide humans with information regarding gender identity and whether such odors are differentially preferred by men and women. The results of Russell's studies indicated that the stronger and/or more unpleasant the axillary odor, the more likely it was to be assigned to a male gender category, and vice versa, regardless of the true sex of the odor donor. In addition, odors believed to come from familiar sexual partners were rated on the average as more pleasant than odors believed to come from strangers.

A recent experiment on the potential role of axillary odors in human social communication claims to support the notion that the time of menstrual onset can be modified by human axillary odors (Russell *et al.*, 1980). Menstrual synchrony was first documented scientifically by McClintock (1971) and subsequently demonstrated by Graham and McGrew (1980) and by Quadagno *et al.* (1981). Russell *et al.*'s experiment supported the theory that odor is a communicative element in human menstrual synchrony, and that at least a rudimentary form of olfactory control of the hormonal system occurs in humans in a fashion similar to that found in other mammals (see p. 25).

Doty (1981), however, was very cautious about the conclusion of these authors in light of methodological limitations in their experiment. For the time being, it is probably prudent to consider the issue of the role of axillary stimuli in the production of menstrual synchrony as still unresolved. (see p. 12)

2.2. Hand odors

Löhner (1924) performed an experiment on hand odor by preparing 10 to 20 deodorized wood sticks and having his dog seek and bring back a stick that a particular person had touched. He found that if the person touched the stick with his finger for longer than two minutes, the dog could recognize it. If the person grasped the stick with his whole hand, several seconds of such grasping were sufficient for the dog to recognize the stick. Kalmus (1958) used handkerchiefs instead of wood sticks and replicated Löhner's experiment, obtaining similar results.

Based on odors from hand secretions, Wallace (1977) studied gender discrimination in humans. Women were generally found to be more accurate than men in performing the discrimination. He further studied the ability to discriminate between two related or unrelated females on the same or different diets. The results of these experiments indicated that humans can discriminate individuals on the basis of hand odor, and that women perform this task more accurately than men.

2.3. Foot odors

As secretory glands that can produce human body odors, the sweat glands on the sole of the foot contribute considerably to body odor. Socks that a person has worn for a day or so, especially in the summer, emit strong unpleasant odors.

The sole of the foot has only sweat glands, but they are present in large numbers: up to 1,000 per square centimeter. The human body sweats about 800 ml during a whole day. About 2 million sweat glands on the sole of one foot secrete 16 ml of sweat per day, about 2% of the total sweat for one day. Human sweat contains about 0.156% acid, of which about one quarter is aliphatic acid (Wright, 1964).

The odors remaining in a footprint are those foot odors that have permeated through the shoes for a long time, which consist of at least 10 fatty acids other than butyric acid. In addition, these odors appear to contain a variety of chemically different substances which may include indoxyl and other indole compounds, carbonic acid, biacetyl, and other chemicals. All of these substances are present in the sweat secreted by the sole glands.

When dogs trace a particular person, they do not track specific chemical substances particular to the person, but rather a specific odor pattern formed by a combination of various substances. If only one thousandth of the sweat secreted from the sole permeates through the bottom and seams of the leather shoes, the number of molecules of butyric acid remaining in one footprint can be calculated to be at least 2.5×10^{11}. This quantity is well beyond a million times the olfactory threshold of the dog. If conditions are favorable, dogs can even trace footprints left 24 hours earlier.

2.4. Breath odor

The oral cavity is a well-known source of both endogenous (e.g., lung) and exogenous (e.g., oral bacteria) odors. Doty *et al.* (1982a) studied whether males and females exhibit different oral odors. The results showed that the majority of men and women can correctly distinguish between male and female odors. As in the case of axillary and hand odors, women judges were more accurate than men judges in identifying the correct gender. Male odors were judged as more intense and less pleasant than female odors, and women tended to judge male odors to be more intense and less pleasant than did men (Doty, 1981).

2.5. Vaginal secretion odor

Human vaginal secretions originate from several sources in that region (Doty, 1981). It is well established that much of the odor associated with vaginal secretion arises from the microfloral bacteria in the region. The secretions have been chemically analyzed by gas chromatograph and gas chromatograph/mass spectrometry, and their changes during the menstrual cycle and during sexual arousal have been examined (Huggins and Preti, 1976; Michael *et al.*, 1974, 1975; Preti and Huggins, 1975, 1978; Preti *et al.*, 1979). Doty *et al.* (1975) attempted to establish whether vaginal odor changes markedly during the menstrual cycle, using a method that would facilitate the detection of the time of optimal fertility. On the average, secretions from preovulatory and ovulatory phases of the cycle were judged to be weaker and less unpleasant in odor than those from menstrual, early luteal, and late luteal phases. However, the results cast doubt on the likelihood that humans can reliably determine the time of ovulation on the basis of such cues.

2.6. Mother's breast odor

Many infant mammals recognize their mothers by means of odor (Rosenblatt, 1972; Alberts, 1981; Leon and Moltz, 1972; Teichner and Blass, 1976). Whether or not human newborns can recognize their mothers by odor has been the subject of several experiments (MacFarlane, 1975; Russell, 1976). These studies suggested that human infants can distinguish the odors of their own mothers' breasts from those of strange mothers' breasts and that this preference increases with age and/or testing experience. Whether or not odor imprinting is possible in humans is currently a point of conjecture open to further experimental test (Doty, 1981).

3. When Body Odors Appear

Several authors have presented data suggesting that the human neonate may be able to identify the smell of its mother. MacFarlane (1975) showed that from the age of six days onward, a baby would turn its head more often toward its mother's smell. Russell (1976) reported that babies appeared to recognize their mothers by smell at the age of six weeks. Schaal *et al.* (1980) showed that the neonate could even react to the smell of its mother's breast from the second day of life onwards. Does this ability to identify human odors remain throughout life or is it eventually lost? Recent studies have indicated that this ability is not lost even after its important early role in the mother-child relationship has diminished. In fact, it has been shown that children three to five years old can recognize the odors of their mothers.

Interestingly, Schaal *et al.* (1980) demonstrated that mothers could identify the odors of their newborns from the second day onwards. Yfantis (1980) found that mothers can identify the T-shirts of their six-month- to 17-year-old children by smell. Thus, subjects can identify and can be identified from soon after birth onwards. The fact that individual children can be recognized by odor before puberty shows that the full secretion of apocrine sweat glands is not necessary for building up an individual body odor.

Although there are no male-female odor differences in babies and small children that can be discriminated (Yfantis, 1980), sex can be clearly distinguished by odor in adults. The discrimination of sex by body odor becomes possible only after the apocrine glands begin to function at puberty. In relation to this, Schleidt and Hold (1982) put forward the interesting hypothesis that gender discrimination occurs on two levels. On the more cognitive level, when mothers were asked to label T-shirts male or female, they performed poorly. At a more emotional level, however, when asked to classify the shirts as pleasant or unpleasant, they "knew" intuitively about gender identity.

4. How Body Odors Work in Daily Human Life

Parts of this question have already been answered in the previous sections. Let us briefly summarize:

4.1. Recognition of individuality

(a) Babies have their own odors. Mothers can recognize their own babies by these smells.

(b) Mothers have their own (breast) odors. Babies can recognize their mothers by these smells.

(c) Some adults can recognize their spouses or friends by body odors.

4.2. Recognition of gender

Humans can establish gender, with above average probability, from breath, axillary, and hand odors (Doty, 1981). Thus, humans appear to have a definite mechanism for discriminating gender on the basis of body odors in adults. Women consistently perform better than men in gender identification or detection tasks based on axillary odors (Doty, 1977; Schleidt *et al.*, 1981), hand odors (Wallace, 1977), or breath odors (Doty *et al.*, 1981). Schleidt and Hold (1982) found that male odors were classified uniformly more often as unpleasant and less often as pleasant than female odor. This result is attributed to the appearance of "sex-smell" after puberty, since mothers more readily perceived boys' and girls' odors as pleasant or unpleasant without conscious discrimination of gender.

4.3. Recognition of the reproductive state of a woman

It is well known that vaginal secretions appear to be major sources of sexual arousal in a number of mammals, including primates (Doty and Dubner, 1974; Michael and Keverne, 1968). As we have seen already, Doty *et al.* (1975) wanted to establish whether or not vaginal odors play an important role in the detection of the time of optimal fertility. Their results, however, did not support the notion that vaginal odors play such a pheromone-like role in the daily lives of humans.

4.4. Influences of vaginal odors on human sexual behavior

Keverne and Michael found that vaginal secretions of female monkeys (*Macaca mulatta*) elicit sexual behavior in male monkeys (Keverne and Michael, 1971; Michael and Kevetne, 1968). They named these secreted volatile fatty acids *copulins* (Michael and Koverne, 1970). The secretion consists of acetic, propanoic, methyl-propanoic, butanoic, methylbutanoic, and methylpentanoic acids (Bonsall and Michael, 1911; Michael and Keverne, 1970; Michael *et al.*, 1971; Curtis *et al.*, 1971). These substances have been identified in human vaginal secretions (Michael *et al.*, 1975).

In light of these findings, Morris and Udry (1978) addressed the question of whether odorous volatiles from vaginal secretions can actually influence human sexual behavior. They sought to establish whether such a mixture of aliphatic acids, when applied to the chest of a woman at bedtime, influenced the self-reported frequency of intercourse, desire for sex, occurrence of sex play, or male/female orgasm among 63 married couples tested over three complete menstrual cycles. They could not find any influence of the synthetic secretion on any of these measures, so the action of pheromones to induce or promote copulatory behavior has not been demonstrated in human vaginal secretions.

C. Influences of Odors on the Human Body

1. Influences on Mental Activity

Pleasant odors refresh our moods and calm us. We feel comfortable and satisfied. Unpleasant odors, on the contrary, irritate our nerves and often produce heaviness in the head—sometimes even a full-blown headache. Bad odors disturb our concentration and disrupt our will to work. Thus, odors can exert considerable influence on the human mental state.

One such influence appears when men and women live separately for a long time. For instance, in a certain group of women's prisons, prisoners often caused trouble. When the causes were examined, the uproars were found to occur whenever men passed outside the high walls of the prisons. Therefore, when a man visits one of these women's prisons and enters a cell, he is requested to leave it as soon as possible, because the male odor that remains in the cell affects the women prisoners who live there. Thus, it is clear that when a woman lives alone for a long time, she becomes very sensitive to men's odors, and can be disturbed by them.

The same can be said in the case of men. Fishermen who work on the ocean for a long time soon become nervous and wild, and they easily quarrel with each other about trivial matters. Their mental states, however, can be soothed by smelling perfumes or pleasant odors. In these cases, women's body odors would conceivably have far better effects in calming men's mental states.

Mature women have their own body odors, but they like perfumes and cosmetics that contain musk and apply them to their bodies. Men are continuously coming in contact with women's body odors—at home, on the road, in buses, or in the office. Such odors tend to soothe men's nerves. Women are likewise exposed to men's odors both inside and outside of their homes, and can usually maintain a balance of their nerves and bodies as well. Thus, men and women affect each other with body odors which contribute, in at least some way, to peaceful human life.

This idea has been demonstrated in rats and mice by Mugford and Nowell (1970). When a male mouse sniffed the urine odor of another, aggressive male mouse, the first male mouse was observed to become very aggressive. Incidences of aggression and biting after this experience increased by 35% and 56%, respectively, as compared to control mice. However, when the aggressive mouse sniffed the urine odor of a female mouse, the incidence of aggression decreased by half. A similar phenomenon was observed by Ropartz (1968). When he applied a scent called 'Diorling' (made by Dior) to male mice, a substantial reduction of aggression occurred. This phenomenon completely disappeared when both olfactory bulbs were removed. This finding demonstrates that the aggressive behavior was the result of olfactory stimuli.

2. Influences on Respiration and Blood Circulation

Odors have various effects on animal and human respiration and can alter the respiratory, rhythm in many cases (Fig. I-6).

When we smell bad odors, our respiration immediately but unconsciously comes to a halt. Conversely, when we smell pleasant odors, such as perfumes, we automatically sniff the odor deeply, and our respiratory rhythms slow down. When we suddenly perceive an odor, we make short, rapid inspirations to investigate its quality. In this case, our repetitive short inspirations resemble a dog's sniffing behavior. When we notice unpleasant food odors, we move close to the food and sniff it in an attempt to confirm whether the food is edible or not, thereby trying to avoid food poisoning. When we encounter a novel odor, we sniff a little and then pause for consideration. If we find that it is not peculiar, we resume normal respiration. Otherwise, we keep sniffing with short, shallow breaths until its identity can be determined.

This adaptation of respiratory cautiousness has saved humans throughout history, not only from food poisoning, but also from toxic gas poisoning. When poisonous air is inhaled, not only oxygen but also odorous and nonodorous toxic gas molecules enter the arterial blood and circulate throughout the whole body. At the same time, in most cases the contaminated air stimulates olfactory receptors, eventually arousing olfactory sensations in the brain. Feelings of heaviness in the head or headaches, produced by odors, can probably be attributed to the direct action of these odorous substances on the brain as well as to olfactory sensations induced by them.

Inhaled cyanide gas enters the arteries and combines with hemoglobin in the blood, resulting in suffocation. Fortunately, most of us can avoid cyanide gas poisoning by smelling its specific odor, except for those few who are anosmic to this gas. In contrast, because it is odorless, carbon monoxide, a constituent of city and propane gases, is often the cause of gas poisoning. It also combines easily with blood hemoglobin. To protect against such an accident, a special offensive odor is added to the gas source as a warning. Despite the addition of this alarm, such gas accidents still happen frequently, because our olfactory sensation fatigues so rapidly that we cannot detect the presence of these gases when they leakg radually.

In addition to deepening and slowing down our respiration, pleasant odors also lower our blood pressure and reduce excessive stress. Thus, pleasant odors can influence our blood circulatory system. Some odors even appear to have specific effects on the circulatory system. A fit of angina pectoris results from spastic constriction of the coronary arteries that supply oxygen and nutrients to the cardiac muscles. Amyl nitrite, which has a strong, pungent odor, dramatically suppresses these spastic constrictions. Consequently, we can alleviate this crisis by sniffing amyl nitrite.

3. Influences on Digestion

When we smell foul odors, we lose our appetite, feel nauseous, and may even vomit. Bad odors from factories processing fish meal or meat-bone meal now contribute to odor pollution, giving rise to considerable social problems. On the other hand, aromas of delicious food induce feelings of hunger and stimulate digestive activity, resulting in intestinal movements and often gastrointestinal noise. Thus, good odors can arouse our appetites and promote the secretion of digestive juices.

There are two neural centers for appetite located in the hypothalamus: the food intake center is in the lateral hypothalamic area (LHA) and the satiety center is in the ventromedial hypothalamic area (VMH). They are mutually antagonistic and are sensitive to the concentration of glucose, insulin, and free fatty acid in the blood (Oomura, 1976). According to Oomura (unpublished data), unpleasant odors inhibit the activity of the food intake center in the LHA and promote that of the satiety center in the VMH, leading to a loss of appetite and cessation of eating. Pleasant odors, especially aromas of delicious food, however, have entirely opposite effects on both centers, resulting in increased appetite and greater than usual food intake.

Le Magnen (1956) reported some interesting experiments on the relation between food intake and flavor. Experimental rats were given four types of differently flavored food for 32 days. Control rats were given only one type of flavored diet throughout this period. The four types of food were presented to each of four groups of experimental rats in a sequence that was changed for each meal. Le Magnen showed that the amount eaten during a mixed meal was always greater than when only a single flavored meal was consistently given. Thus, he demonstrated that varying the flavor of diet can induce overeating, and suggested that a similar mechanism may be involved in the development of obesity in humans.

In general, unpleasant odors decrease and often kill our appetites. To the best of my knowledge, however, there is at least one exception. The durian is well known among many tropical fruits for its very delicious taste. For the uninitiated, however, it has a detestable rotten odor. Those who smell it for the first time cannot eat it, however delicious it is claimed to be. Through strong encouragement they may try to taste it while holding their noses. Once eaten, however, its truly delicious taste is discovered, and the newly initiated begin to eat it without being bothered by the bad odor. This demonstrates that the definition of a bad odor differs depending on one's past experience.

4. Influences on Reproductive Organs

For many centuries it has been thought among Japanese women that a menstruating woman can facilitate menstruation in others. In accordance with this line of thought, for example, when a daughter who has been staying in another city returns home, her menstrual cycle becomes coincident with that of her mother. Thus, Japanese women who were menstruating used to warn other women to keep away so as not to become "infected" by menstruation.

In Japan, the traditional modesty of women kept these observations strictly secret, and they had not been widely known to men until McClintock's report appeared in 1971. The influences of axillary odor on menstrual synchrony are currently being studied (p.17; Doty, 1981).

In animal experiments, influences of female and male odors on reproductive cycles have been studied by several investigators. Van der Lee and Boot (1956) reported that female mice housed in groups of four showed a higher incidence of spontaneous pseudo-pregnancy than individually housed females (the Lee-Boot effect). Removal of the olfactory bulbs greatly reduced the incidence of pseudo-pregnancy, demonstrating that this effect was dependent on olfactory stimuli.

Whitten (1956) showed that the mating frequency of female mice can be altered by the presence or absence of a male mouse or a male odor. The failure of some mice to mate was attributed to a prolongation of the estrus cycle. Whitten (1957, 1958) further observed that group housing of only female mice led to irregular estrus cycles in the majority of females but in the presence of a male kept in a small basket, the estrus cycles became shorter and more regular (the Whitten effect). On the basis of these observations, Whitten concluded that the estrus cycle of the female was modified by the presence of the male and that the effect was mediated by olfaction.

Urine has been identified as the pheromone source responsible for modifying such reproductive processes. For example, exposure of grouped females to the urine of adult males induced synchronized estrus (Marsden and Bronson, 1964; Bronson and Whitten, 1968). Furthermore, immature female mice exposed to male urine reach puberty earlier (the Vandenberg effect). However, urine obtained from grouped adult female mice retards

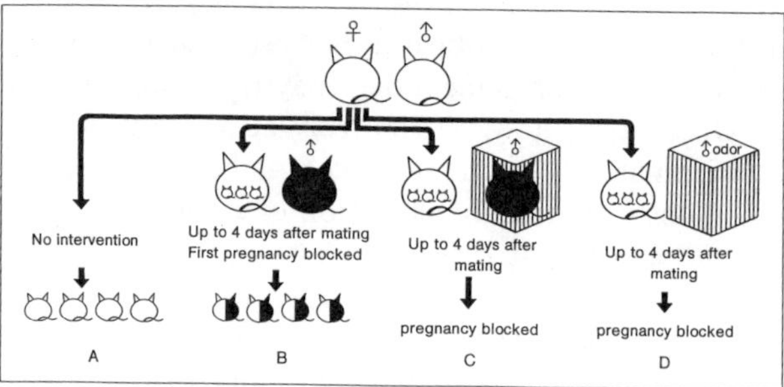

Fig. I-4. Effect of strange male rat on female pregnancy. If a female is exposed to a strange male in the first 4 days after mating, her pregnancy is blocked; she will mate with the stranger and give birth to young sired by the stranger (B). Pregnancy block also occurs if the strange male is kept in a separate box (C) or if the female is exposed to the odor of a strange male (D). (from Bruce, 1964)

the sexual maturation of growing female mice (Colby and Vandenberg, 1974).

Odors of female dogs have been known to elicit sexual behavior in male dogs. Michael and Keverne (1968) reported that rhesus monkey vaginal secretions elicit male rhesus copulatory behavior. Curtis *et al.* (1971) identified the primate sex pheromones in *copulins*.

Comfort (1971) has pointed out that during the height of the Victorian era, the onset of puberty definitely occurred at a more advanced age and the age of onset has been steadily decreasing since that time. This has occurred in both sexes and may well involve social factors, making a pheromonal effect impossible to isolate. However, Comfort's observation may be an example of the Vandenberg effect in humans.

The aborting effect of odors on pregnancy in mice have been studied by Bruce (1959, 1960a, b, 1962, 1963, 1964), Bruce and Parrott (1960), and Parkes and Bruce (1962) (Fig. I-4).

These results were extended by the finding that odors of male mice of a different strain exhibit a potent aborting effect (the Bruce effect). This blocking of pregnancy was shown to be due to the odor of urine from the male mice of a different strain (Bruce, 1965; Dominic, 1965). In the author's laboratory, we have examined the effect of odors on the uterus (Fig. I-6). When odorous stimulation was applied, spontaneous movement of the uterus was suppressed (p. 31; Takagi *et al.*, unpublished observation). In humans, pregnancy block and abortion due to odorous stimulation appear to have been neither openly reported nor studied in medical laboratories. However, the author has heard, from indirect sources, that chemicals named Musk-T (ethylen brassylate) and 2-Undecan may induce abortion in women (personal communication from A. Fiori, Givaudan Corp., Delawanna, Clifton, N.J.). We found that these odors smell especially foul to women, while men did not find them disagreeable. Whether these odors can induce abortion is a question which remains to be studied.

5. Influences on Maternal Behavior

Babies brought up on mother's breast milk are alleged to have an extremely attractive smell and an ability to tickle the maternal instinct. The baby's odor deepens the mother's affection and strengthens the bond between mother and child. Babies brought up on cow's milk, however, do not have such attractive odors.

Maternal behavior in most mammals begins with licking and cleaning of the neonates' bodies. While engaged in this behavior, mothers memorize the odors of their offspring, and the offspring remember the odors of their mothers. This well-known phenomenon is commonly referred to as "imprinting". Studies have shown that neonates separated from their mothers immediately after birth and cleaned well are usually rejected when returned to their mothers. However, if the neonates are removed from their mothers five minutes after birth, most mothers accept them even if they are returned three hours later (Zarrow et al., 1971).

When mice are made anosmic before delivery, maternal behavior disappears or becomes altered. An anosmic mouse does not display nursing behavior, often eating all of its offspring (Zarrow et al., 1971). On the other hand, when a mother mouse is removed from her offspring and a virgin mouse is introduced to the neonates, the virgin mouse displays maternal behavior in most cases. However, if olfaction is destroyed in advance, the virgin mouse does not exhibit such maternal behavior (Noirot, 1969; Gandelman et al., 1971; Zarrow et al., 1971). These experiments indicate that maternal behavior is induced by the odors of neonates.

6. Aromatherapy

The preceding sections clearly demonstrated that odors can influence our physical and mental states. Thus, it is natural that many people have considered the use of aromas for the therapeutic treatment of diseases. A French pathologist, R. M. Gattefossé, wanted to introduce essential oils of herbs and aromatic leaves into therapy of patients in the 1930s. This idea gave birth to the term "aroma therapy" (Gattefossé, 1937), which has been used commonly since then.

In Japan, for hundreds of years incenses from herbs and trees have been used as folk remedies. More recently, they have been used by such physicians as the late Dr. N. Hasegawa, a gynecologist at the Akita University of the School of Medicine (Hasegawa, 1978). His patients were 34 women all over 60 years of age who were clinically diagnosed with psychosomatic diseases. He gave each patient four pouches of incense and told them to sniff one of them as often as possible every day. For this purpose, he indicated that the pouches were to be kept in one of four locations: (1) inside a drawer of the desk which the patient was using, (2) in a pocket of her clothing which was used daily, (3) in a purse; and (4) under the pillow. The patients were asked to visit his clinic once a week, and the effect was assessed after three months. During this period of three months or longer, no other therapy was administered to these patients.

Evaluation of the therapeutic effects was made based on (a) personal interviews concerning subjective symptoms, (b) results of tests using the Cornell Medical Index (CMI), and (c) pattern analysis by means of a mirror drawing test. These tests were performed before and after the aromatherapy and a comparative analysis of the results was carried out.

The results for 34 patients showed that the treatment was highly effective in eight cases (23%), fairly effective in 17 cases (50%), and had no effect in nine cases (26%), No adverse or subsidiary ill effects were shown by any patient. Six cases of mask-depression were similarly treated. The therapy was fairly effective in two patients but had no effect in four patients.

In the aromatherapy, subjective judgements of the therapeutic effects have been well estimated, but appropriate objective measures have appeared to be deficient in many cases. The real therapeutic effects of many aromas cannot be denied by present-day medical science. However, many problems, for instance, selection of appropriate odors, the determination of their quantities, and the methods of application require further study. Furthermore, the slow appearance of the therapeutic effects has made the evaluation difficult or unclear in many cases. Consequently, some people prefer the word "aromachology" to "aromatherapy".

Also in Europe, it appears that various essential oils are being used for therapy. Readers interested in the details and practical uses of aromatherapy are directed to the books written by R. Tisserand (1977) and M. Tisserand (1985).

7. Forest Bathing

Every plant emits its characteristic gaseous substance, which protects the plant from the attack of microorganisms. The late professor Tokin of Leningrad University called this substance "Phytoncide" (Kamiyama and Tokin, 1980). Kamiyama (1983) proved that leaves from white birch, white fir, needle juniper and other trees, young roots of the wild peony, the subterranean stem of Japanese horseradish, and fruits of *Prunus padus* and other plants, when chopped into small pieces, kill protozoan-like amoebae and infusoria which were placed a few centimeters from them.

Thus, phytoncides are substances that attack other microbes and insects. In addition, it was also found that they can evoke friendly relationships with other organisms or communication with the same kinds of plants.

Recently, humans have come to pay attention to substances emitted from the forest. Crews of a ship cruising the ocean knew that the land was near when they smelled a wind from the forest. When people leave big cities and come across forests in the country, they feel the air as crisp and fragrant, and say that the air is savory and makes them feel refreshed. According to Kamiyama (1983), Father S. Kneipp, a catholic priest at Bad Worishafen in Germany advocated the so-called Kneipp therapy which included "forest therapy" one hundred years ago. As one of a group of natural therapies, he utilized the forest to remedy diseases. Wurzburg has been a center of the Kneipp therapy ever since.

Today, these subtances have been analyzed and are known to contain terpenes and their derivatives. Terpenes have been known to have pharmacological effects on the human body, for instance, anti-ulcer effects in the stomach. Kamiyama and Suzuki (1986, 1987) have shown that terpenes are effective in recovering or alleviating fatigue in human subjects, using flicker tests, pupil light reflex tests and other tests.

Thus, the importance of forest bathing has been well recognized, and many model forests have been selected in Japan. Today, many people spend holidays walking in forests for their health.

D. Objective Measurement of the Influences of Odors on the Human Body

Changes in the human body produced by the sniffing of odors can be objectively studied by various means.

1. Respiratory Movement

By fitting a belt around the chest, and by passing a weak electric current through a strain gauge on the belt, respiratory movements can be recorded as changes in electric potentials (Fig. I-5 A, A'). Alternatively, by placing a thermister in the nostril, inspiratory and expiratory air currents can be measured as changes in electric potentials, representing respiratory movements.

2. Skin Potentials

When a weak electric current is passed between small silver plates placed

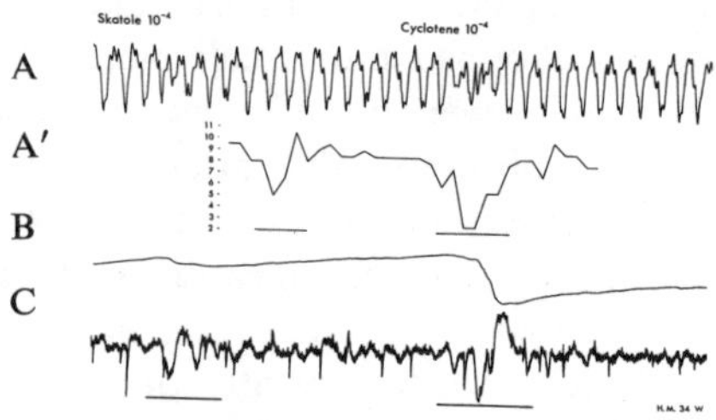

Fig. I-5. Influences of odors on respiratory rhythms (A), skin potential or GSR (B) and electro-oculogram or EOG (C).

A' shows changes of intervals between the neighboring two peaks in A. Cyclotene vapor at 10^{-4} concentration elicits major changes in A, A', B and C.

at two separate sites on the skin, changes in skin resistance elicited by sniffing odors can be registered as changes in electric potential. This measure is well known as the galvanic skin response (GSR) or skin potential (SP) (Fig. I-5 B).

3. The Electro-Oculogram*

The eyeball moves in response to various internal and external stimuli. By placing one small silver plate above the eyebrow and another below the lower eyelid, or beside the corner of the eye, eyeball movements can be recorded as changes in electric potential. When a person closes his eyes and sniffs an effective odor, eyeball movements occur (Fig. I-5 C). The electric changes recorded constitute an electro-oculogram.

4. Stomach Movement

By inserting a balloon into the stomach and inflating it, the spontaneous activity of the organ can be recorded in the form of pressure changes. The author (unpublished data) showed that an odorous stimulation modified the stomach activity in the rabbit (Fig. I-6). When we smell odors of delicious foods, we often feel a craving for food frequently accompanied by abdominal noises due to the movement of the stomach or intestine. In contrast, when we smell foul or disagreeable odors, our appetite disappears and we may even feel nauseous. Thus, it is clear that in humans, as was demonstrated in the rabbit, odors may enhance or inhibit the activities of the stomach and intestine.

---

* This word is often abbreviated as EOG in the field of ophthalmology. The reader is requested to differentiate the EOG in vision research (electro-oculogram) from EOG in olfactory research (electro-olfactogram) (Chapter I, p.33 and Chapter V, p.148).

Fig. I-6. Influences of odors upon the movement of the stomach (A) and uterus (B) in the rabbit.
Spontaneous irregular waves are suppressed by amyl acetate (AA) and ethylene dichloride (ED).

5. Movement of the Uterus

By inserting a balloon into the uterus and inflating it, the author was able to record the spontaneous activity of this organ. Application of an odorous stimulation to rabbits, resulted in the modification of uterine movement which was dependent on the kind of odors applied (Fig. I-6 B). From this finding in the rabbit, it may be inferred that sniffing odors may alter uterine movements in women, which may not be felt.

6. Evoked Potentials

The electrical activity of the brain is modified when an odor is sniffed. Such a response can be recorded as an evoked potential by placing an electrode on the vertex and a reference electrode on the auricular lobule.

Evoked potentials elicited by odor stimulation have been studied by a number of investigators (Finkenzeller, 1966; Allison and Goff, 1967; Alber, 1971; Brandl *et al.*, 1980; Giesen and Mrowinski, 1970; Giesen, 1971; Herberhold, 1971; Iizumi, 1976; Kobal and Plattig, 1978; Kobal *et al.*, 1987; Nagano *et al.*, 1976; Nakashima *et al.*, 1979; Plattig and Kobal, 1976, 1977a, b, 1979; Smith *et al.*, 1971; Tonoike *et al.*, 1979, 1982). Recently, Umeda (1981) and Ohba (1982) recorded evoked potentials elicited by each of the five odors from the T & T Olfactometer (Chapters II & III). Although evoked potentials could be recorded in normal healthy subjects, they could not be recorded in anosmic patients. Details of Umeda's study will be described in Chapter III (p.93).

7. Contingent Negative Variation

Contingent negative variation (CNV) was first devised by Walter *et al.* (1964) and later studied by Tecce and Scheff (1969) and Rohrbaugh *et al.* (1976).

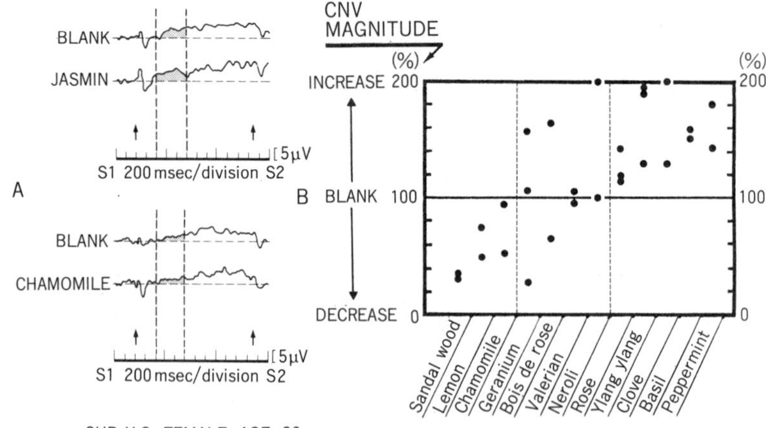

Fig. I-7. Contingent negative variation (CNV) as indicated by stippled areas in A. CNVs are observed to vary depending on stimulation with no scent, rose oil, and chamomile oil. B: Changes in amplitude of the early components of the CNVs. The negative variation decreased, did not show much change, and increased in response to the left, middle, and right groups of odors, respectively. (from Ogata *et al.*, 1986)

This method measures changes in the evoked potential occurring between two different stimuli (S1 and S2) presented in succession. S1 is an odorous stimulus presented to a subject as a warning stimulus. After an appropriate interval (for instance, 4 sec.), S2, usually a photic or acoustic stimulus, is applied. In this paradigm, subjects are requested in advance to switch off the light or sound (S2) as soon as they detect S2. Subjects are easily accustomed to this paradigm and spontaneously prepare to turn off the second stimulus after some practice. This spontaneous preparation is represented on the brain wave by a negative potential shift of about 10 microvolts between S1 and S2. This negative potential shift is called the CNV (Fig. I-7 A).

The early negative component just after S1 is maximal over the frontal cortex, indicating emotional activity, while the later negative component just before S2 is maximal over the motor cortex, indicating preparation for motor activity. Suzuki (1984a) utilized this paradigm in objective studies of olfactory sensibility. He demonstrated that olfactory disorders due to psychogenic causes or malingering can be verified by this paradigm.

Recently, Torii (1986, 1987) and his associates (Fukuda *et al.*, 1985; Ogata *et al.*, 1986) studied differences in the CNV after applying various

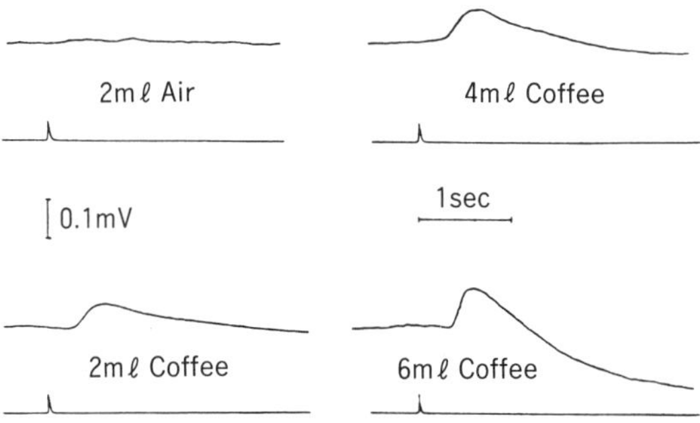

Fig. I-8. Human electro-olfactograms (EOG). The olfactory epithelium of this subject showed no response to 2 ml atomospheric air, but showed negative EOG to 2, 4, and 6 ml coffee-saturated air. (from Osterhammel *et al.*, 1969)

kinds of odors as S1. They found a large CNV when the subject sniffed jasmine (as S1), which is said to have a stimulating effect. However, they found only a small CNV when S1 was either lavender, which is said to have a sedating effect, or an odorless (blank) substance. In this way, they could objectively demonstrate differences in sensations elicited by various kinds of odors in the form of differences in brain reactions to those odors.

Comparative studies on the effects of odors of various essential oils were performed in humans. A sample result is shown in Fig. I-7 B.

8. The Electro-Olfactogram (EOG)

The slow potentials elicited in the frog olfactory epithelium by application of an odorous vapor were named the "electro-olfactogram" or "EOG" by Ottoson (Chapter V, p. 148). Osterhammel *et al.* (1969) was the first to record human EOGs (Fig. I-8), soon followed by Suga and Nakashima (1972, 1973), and recently by Osumi (1984) and Ohkado (1984). A technique to reach the olfactory cleft has recently been developed by the use of the Selfoscope (Chapter III, p.92). Thus, in the near future, otorhinolaryngologists may be able to record human EOGs in the clinic to diagnose whether the olfactory epithelium can respond to odors or not.

9. Pupillary Reflex

Nishida (1971) and Nishida *et al.* (1971, 1973) assessed the pupillary reflex for use as an objective olfactory test. They found that this reflex is activated by the minimum perceptible concentration of odors, and mydriasis was observed most frequently among them. Naturally, anosmic patients did not display any pupillary reflexes in response to odorous stimulation. Toyota (1962) precisely examined the applicability of this olfacto-pupillary reflex in the rabbit under several conditions. In measuring olfactory thresholds, he concluded that this reflex was less definite than the olfacto-respiratory reflex but was more convenient and useful than the latter.

10. Temperature of Olfactory Mucosa

Thermal changes have been observed on the surface of the nasal turbinate in both rabbits and humans when odors are applied through the nostrils (Maesaka, 1964).

They were observed in 70% of human subjects free from nasal diseases. In contrast, patients with marked hyposmia and those with prominent disturbances in the nasal turbinate rarely displayed this response.

II. Olfactory Tests

A. Standardized Olfactory Tests in Japan

The control of odor pollution became an important area for research in the 1970's, which were considered the years of environmental pollution in Japan. To cope with environmental problems, bureaus for pollution control were established in all prefectures and large cities as well as the central government. When the government embarked on the control of malodors, their first problem was the selection of olfactory panels which could properly evaluate malodors.

Quite independently, the author presumes, otorhinolaryngologists at that time began receiving increasing demands for treatment of olfactory disorders. In addition, increases in traffic accidents, and probably also the higher incidence of brain tumors and brain surgery, led to increases in the number of complaints of olfactory disorders. However, while the need was becoming greater each year, methods to treat such patients had not yet been developed.

Facing this social climate, the author keenly felt the necessity for a standardized olfactometer by which doctors in different medical fields and in different clinics could easily share their findings on patients, which would facilitate the development of methods to properly treat such patients.

1. Preliminary Studies for the Manufacture of a Standardized Olfactometer

The first step towards the solution of these two problems was to organize a research group, designated as the "Olfactory Test Committee". The next step was to obtain a four-year research grant from the Japanese government beginning in April of 1971.

1.1. The Olfactory Test Committee

The committee members and their institutes were as follows:

Drs. Bun'ichi Toyota and Ryozo Umeda, professors of otorhinolaryngology, Kanazawa University School of Medicine, Kanazawa.

Dr. Hideo Ishii, professor of otorhinolaryngology, Gunma University School of Medicine, Maebashi.

Dr. Sadayuki F. Takagi, professor of physiology, Gunma University School of Medicine, Maebashi.

Dr. Takeshi Kitamura, professor of otorhinolaryngology, Chiba University School of Medicine, Chiba.

Dr. Ryo Takahashi, professor of otorhinolaryngology, Jikeikai University School of Medicine, Tokyo.

Drs. Michiya Okamoto and Eisei Asaka, professors of otorhinolaryngology, Showa University School of Medicine, Tokyo.

Drs. Yasuro Miyoshi and Masaru Ooyama, professors of otorhinolaryngology, Mie University School of Medicine, Tsu.

Dr. Takashi Tsuiki, professor of otorhinolaryngology, Iwate University School of Medicine, Morioka.

Dr. Hatsuo Ino, professor of otorhinolaryngology, Niigata University School of Medicine, Niigata.

Dr. Kiichiro Hiroto, professor of otorhinolaryngology, Kyushu University School of Medicine, Fukuoka.

Drs. Jun Naito and Yasuo Watanabe, professors of otorhinolaryngology, Osaka University, Osaka.

Drs. Isamu Watanabe and Hidehaku Kumagami, professors of otorhinolaryngology, Nagasaki University School of Medicine, Nagasaki.

Dr. Masa-aki Yoshida, professor of human engineering, Chuo University School of Science and Technology, Tokyo.

Dr. Shuzaburo Ito, professor of physiology, Waseda University School of Education, Tokyo.

In the fall of 1971, all of these members and their staffs commenced projects to manufacture a standardized olfactometer and to develop treatment methods for olfactory disorders. Details of these activities have been published by Toyota, Kitamura, and Takagi (1978). Hereinafter this research group will be called the "Olfactory Test Committee", or simply, "the committee".

1.2. Selection of standard test odors for olfactory test

The committee asked Masayasu Kainosho, former chief perfumer of the Takasago Corporation, to select standard odors that would enable them to test human olfaction (Takagi, 1974).

The conditions that were presented to Kainosho were:

(i)   Each odor should be distinct from the others.

(ii) They should be sufficiently simple to allow easy perception of their qualities by many people.

(iii) They should not change in quality or strength over time.

(iv) They should be artificially synthesized odorants of simple struc-

Table II-1. The names of solvents that were used to dilute the ten odorants. (from Toyota *et al.*, 1978; courtesy of Igaku-Shoin, Ltd., Tokyo)

Odorant	Method of dilution	
1. *dl*-Camphor	$10^{-1} \sim 10^{-10}$	sol. in Nujol.
2. γ-Undecalactone	$10^{-1},\ 10^{-2}$	sol. in propylene glycol,
	$10^{-3} \sim 10^{-14}$	sol. in Nujol.
3. Iso-valeric acid	$10^{-1} \sim 10^{-17}$	sol. in Nujol.
4. Methyl cyclopentenolone	$10^{-1},\ 10^{-2}$	sol. in propylene glycol,
(Cyclotene)	$10^{-3} \sim 10^{-14}$	sol. in Nujol.
5. Skatole	$10^{-1},\ 10^{-2}$	sol. in propylene glycol,
	$10^{-3} \sim 10^{-17}$	sol. in Nujol.
6. β-Phenyl ethyl alcohol	$10^{-1},\ 10^{-2}$	sol. in propylene glycol,
	$10^{-3} \sim 10^{-15}$	sol. in Nujol.
7. Cyclopentadecanolide	$10^{-1} \sim 10^{-14}$	sol. in Nujol.
(Exaltolide)		
8. Phenol	$10^{-1},\ 10^{-2}$	sol. in propylene glycol,
	$10^{-3} \sim 10^{-15}$	sol. in Nujol.
9. Acetic acid	$10^{-1} \sim 10^{-11}$	sol. in Nujol.
10. Diallyl sulfide	10^{-1}	sol. in diethyl phthalate,
	$10^{-2} \sim 10^{-15}$	sol. in Nujol.

tures, because natural odorants, in many cases, are impure and have unstable components.

In response to the above requests, Kainosho selected the following ten odors (Table II-1) (Toyota *et al.*, 1978):

(a) β-phenyl ethyl alcohol (Firmenich Co.)

This has a rose-like odor and is present in the fragrance of many flowers.

(b) Methyl cyclopentenolone or Cyclotene (Dow Chemicals)

Known commercially as "Cyclotene", when it is diluted, it has a specific maple odor and is commonly used in baking. It belongs to a "burnt" odor category to which humans are very sensitive.

(c) Iso-valeric acid (Labows *et al.*, 1982)

This has the odor of sweaty feet, smelling especially like socks that have been worn by a man on warm days. It is often described as a "sweaty" or "rotten" odor. Humans are also sensitive to this odor.

(d) γ-Undecalactone

This is a fruity, peach-like odor.

(e) Skatole

This is a component of "fecal" odors. Humans are familiar with this odor because it is contained in human feces and flatus.

(f) Cyclopentadecanolide or Exaltolide (Firmenich Co.)

Commercially, this odor is well known as "Exaltolide". As a "musk" odor, it is widely used in cosmetics and is familiar to humans.

(g) Phenol

As a carbonic acid, this was once widely used as a disinfectant and its odor was prevalent in hospitals in the past. It is different in nature from the other smells, and small quantities of it exist in the human body.

(h) *dl*-Camphor

Widely used in Japan as an insect repellent and a deodorant, this substance produces a cool feeling and an anti-erotic effect.

(i) Diallyl sulfide

Alinamine®, a Vitamin B injection processed from a Korean ginseng by the Osaka-based Takeda Pharmaceutical Company, Ltd. (Fujiwara *et al.*, 1964) has been widely used in Japan to test venous olfaction. Alinamine, however, seems to vary slightly, each time it is prepared. Therefore, diallyl sulfide was selected as a test odor (see p.12).

(j) Acetic acid

In a weak solution it is familiar in Japan as vinegar. It is also widely known as a trigeminal nerve stimulant (Ito, 1968; Tucker, 1963; 1971).

These ten odors were used as standard test odors throughout our research activities. Hereinafter, they will be referred to as "the ten test odors" or, simply, "the ten odors".

1.3. Preparation of odor solutions

The original compounds, or solutions, of these ten odors were diluted successively by the tenfold method and a series of solutions was prepared: 10^{-1}, 10^{-2}, ... 10^{-x} ... 10^{-17} (10^{-x}, $x = 1$–17). The solvents used are shown in Table II-1. For the sake of convenience, this series of odorants is called the "10 odor series".

1.4. Testing rooms

The room used for olfactory testing had to be quiet, because any extraneous noise can significantly interfere with olfactory sensitivity. (see the Appendix to this volume, "The Incense Cult of Japan.")

Since the odorous vapors from the test solutions were used as stimuli, it is probable that the temperature and humidity of the test room could influence the results of the test (Stone, 1963). Comfortable settings of temperature and humidity were required for performance of the test, and pleasant room conditions were kept as constant as possible throughout the year. Problems associated with actual test performance were considered. For example, odorous vapors might spread throughout the room during a test. They may be absorbed by walls and furnitures and remain for a long time. To avoid such undesirable conditions for subsequent tests, the room was quietly and continuously ventilated. In addition, paper slips were immediately discarded after each use into an air-tight plastic envelope.

When tests were performed frequently, a specially designed deodorizer

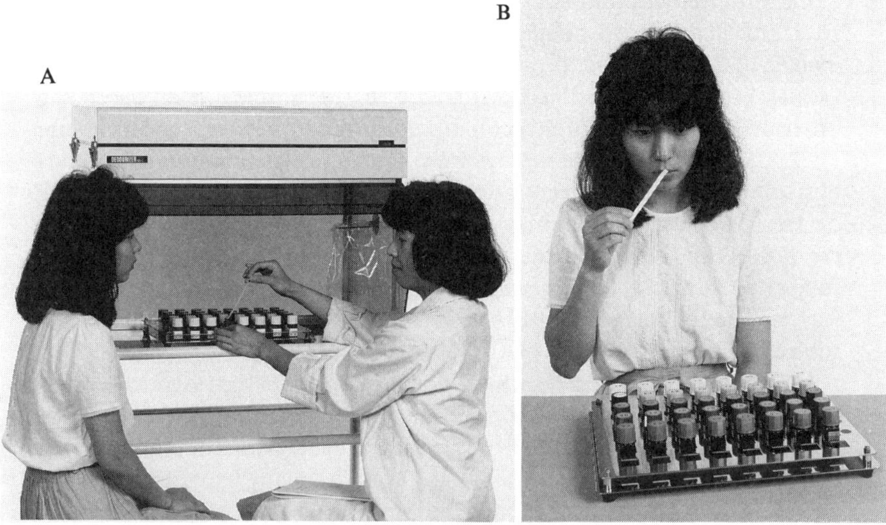

Fig. II-1. Testing with the T & T Olfactometer. A: Use of a slip of non-odorous filter paper. One end is made odorous by dipping it into an odor solution. B: Our method of sniffing the odorous end of a slip of filter paper. (courtesy of O. Fukasawa)

mounted on the enclosing walls above the T & T Olfactometer (Fig. II-1 A) was used and found to be effective.

1.5. A method for smelling test odors

After much discussion, the committee decided to use slips of non-odorous filter paper to test a subject's olfactory ability.

1.5.1. Filter paper as a medium for olfactometry: Perfumers used an odorless filter paper which is about 15cm long and 7mm wide to sniff a fragrance. Before the test, both the experimenter and the subject thoroughly washed their hands with an odorless soap especially manufactured for olfactory testing. One end of the paper slip was dipped 1cm into a test odor solution (Fig. II-1 A; cf. Doty *et al.*, 1985a). Excess solution was removed at the mouth of the bottle to avoid dripping solution on the table and polluting the room air, making subsequent tests difficult. The experimenter held the other end of the paper slip and handed it to the subject, who took it in his hand and advanced the odorous end of the paper to his nostrils (Fig. II-1 B). At this time, the subject was advised that the paper should not touch the nose directly. If the odorous end of the paper did touch the subject's nose, the nose was washed well with an odorless soap.

Each subject was told not to sniff an odor vigorously or deeply, as doing so drastically reduced olfactory acuity on subsequent trials. The experimenter told the subject to sniff lightly and gradually. Finally, the experimenter presented the subject with odorous paper slips in an ascending order from solutions of lower concentration to higher concentration.

1.5.2. Odorous solutions in small bottles: Small bottles containing the odorous solutions of the concentration series were used for olfactory testing by Henkin (personal communication in 1967). In his study, he removed the bottle cap, advanced the mouth of the bottle to the nostrils of the subject and let the subject sniff the odor directly from the bottle. This was repeated with each bottle to examine the olfactory sensitivity of the subject.

However, this method has the following shortcomings: An experimenter has to grasp each bottle in order to advance it to the patient's nostrils. This means that the experimenter's hand odor contaminates the outside of the test bottles. After many tests by many experimenters, the outsides of the bottles become odorous, making sniffing of pure test odors from inside the test bottles difficult. In short, each time a subject sniffs a test odor, he also sniffs the odor on the outside of the bottle. Consequently, the committee rejected this method of sampling.

1.6. Determination of order

In general, sniffing strong odors decreases olfactory sensitivity to subsequent odors. Thus, it is desirable to smell weaker odors before stronger ones. Sniffing bad odors in succession deprives subjects of their objectivity. Alternating pleasant and unpleasant odors is therefore preferable and effective.

Considering these factors, the order of smelling the ten odors was tentatively established as shown in Table II-2.

1.7. Determination of detection threshold

In principle, detection thresholds were sought in an ascending manner, starting from the lowest concentrations of each respective odor (between 10^{-10} and 10^{-17} in Table II-1). However, if the subject had an olfactory disorder, and his detection threshold was, for instance, 10^{-2}, reaching a concentration of 10^{-2} from that of 10^{-10} would be very time-consuming. In such cases, the experimenter had to evaluate the subject's complaint and estimate his olfactory sensitivity to determine which concentration should be tested first. If the starting concentration was too high, a pause for a few minutes was required and the test was resumed at a lower concentration. The concentration at which a subject indicated that he detected an odor, but could not identify it, was recorded as a detection threshold for the subject. In some cases, the detection threshold and the recognition threshold were equivalent.

Table II-2. Words expressing qualities of standard odors. (from Toyota *et al.*, 1978; courtesy of Igaku-Shoin, Ltd., Tokyo)

Name of odorant	Qualities of odor
β-Phenyl ethyl alcohol	Odor of rose, light sweet odor
Methyl cyclopentenolone (Cyclotene)	Burnt odor, caramel odor
Iso-valeric acid	Putrid odor, odor of long-worn socks, sweaty odor, odor of fermented soybeans
γ-Undecalactone	Canned peach odor, heavy sweet odor
Skatole	Odors of vegetable garbage, oral odor or aversive odor
Cyclopentadecanolide (Exaltolide)	Musk odor, powder odor
Phenol	Odor of disinfectants, odor of hospitals
dl-Camphor	Camphor odor, pleasant, refreshing odor
Diallyl sulfide	Garlic odor
Acetic acid	Vinegar odor, pungent odor

When subjects showed difficulty in determining the thresholds, triangle tests were performed. In this test, a slip of paper with the odor of a suspected concentration and two non-odorous slips of paper were presented as a set. Subjects were requested to select the odorous one among the three. When a subject failed in the selection, a slip of paper with a odor of one step higher concentration of odor was prepared and the triangle test was repeated. The concentration at which a subject successfully identified an odorous one for the first time was determined to be the detection threshold for the subject.

1.8. Determination of recognition threshold

After a detection threshold was determined, a recognition threshold was measured. It could usually be found by increasing the concentration by one or two steps. Recognition thresholds were recorded when the subject correctly named the odor he smelled, or when he described a name of a similar odor or the quality of the odor he perceived. In a few cases, a detection threshold equaled the recognition threshold.

1.9. Expression of the qualities and intensities of the ten odors: Selection of appropriate words for odors

For many subjects, naming an odor or describing the quality of an odor was rather difficult. For such subjects, the committee prepared appropriate expressive words for each of the ten odors (Table II-2). Then the subject could easily find one or two words to indicate his or her perception. Thus, the recognition threshold could be determined in this manner.

1.10. Intensity and hedonic scales of an odor in Japan

A six-degree scale with numerals (0 to 5) for intensity of an odor and a nine-step scale with numerals ($+4$ to 0 to -4) for pleasantness or unpleasantness were determined by the Olfactory Test Committee and are now widely used in Japan (Chapter I, pp.7–8).

2. Results of Preliminary Tests with the Ten Odors
2.1. Olfactory tests with the "10^{-x} odor series"

Using the methods described, members of the committee examined a large population and selected healthy young men and women between the ages of 18 and 25. The selected subjects presented no subjective complaints of olfactory dysfunction and were found by rhinoscopy to be free from pathology. Olfactory detection and recognition thresholds were then determined for these subjects.

Table II-3 shows the number of subjects displaying detection and recognition thresholds to each of the odor concentrations (10^{-1} to 10^{-17}). Figure II-2 shows distribution curves for the detection and recognition thresholds for each odor. The ordinate represents the number of subjects calculated as a percentage, and the abscissa indicates the odor concentration expressed as x of 10^{-x} (x, dilution number $= 1, 2, \ldots, 17$). The first five curves were obtained from tests using β-phenyl ethyl alcohol, Cyclotene, iso-valeric acid, γ-undecalactone and Skatole. The curves of these odors and the remaining five odors seem to be similar in shape to a normal distribution curve. A χ^2 test was performed to examine the null hypothesis that the distribution of threshold values was normal for each of the ten odors. The null hypothesis was rejected for most of the ten odors.

These results might be attributed to the long tailings on the right side that were found in most of the distribution curves (Fig. II-2). In fact, these long tailings were more obvious in the later five odors (Fig. II-2B). Strictly speaking, therefore, we cannot say that all these curves represent normal distribution curves. However, considering a quasi-bell curve distribution, the averages ($\bar{x}$), variances (s^2), standard deviations (s), and coefficients of variation ($100s/\bar{x}$) were calculated and are shown in Table II-4 for men and women together, and in Table II-5 for men and women separately.

2.2. Differences between men and women

Table II-5 shows that the thresholds for women were always lower than those for men. In other words, the $\bar{x}$ for women was higher than the $\bar{x}$ for men. Assuming that the distributions of the threshold values were approximately normal, t-tests were carried out to examine the null hypothesis that the population means of male thresholds were equal to those of female ones. All calculated values of t exceeded the tabular values at the 0.01 significance level. However, there may be psychological differences in responding between male and female subjects in these experiments.

Table II-3. Detection and recognition thresholds obtained by means of the 10^{-x} odor series from a large population of healthy men and women between the ages of 18 and 25. Upper and lower numerals indicate detection and recognition thresholds respectively. (from Toyota *et al.*, 1978; courtesy of Igaku-Shoin, Ltd., Tokyo)

	Odorant	Dilution number x																	Total
		1	2	3	4	5	6	7	8	9	10	11	12	13	14	15	16	17	
1	*dl*-Camphor	6	192	528	205	68	13	10	6	2									1,030
		14	248	357	32	6	2	1	1										661
2	γ-Undecalactone	1	1	30	200	475	222	62	12	2	1		1						1,007
		4	20	84	213	283	45	3	1										653
3	Iso-valeric acid		1	11	54	260	444	202	34	5	2	1		3				1	1,018
			3	13	80	236	271	55	3	1	2		2						666
4	Cyclotene			3	87	425	358	131	21	2	1	1	1						1,030
		3	9	38	146	345	111	11	1	2									666
5	Skatole		2	6	54	208	441	244	45	5	2	2	2		2	1		1	1,015
		1	9	26	106	178	255	49	2		2		1					1	630
6	β-Pheyyl ethyl alcohol	2	5	19	144	255	205	53	8										691
		10	14	94	198	115	38	2											471
7	Exaltolide	17	139	271	123	55	21	9	1				1						637
		33	159	196	20	6	3	2											419
8	Phenol	12	48	209	190	117	53	10	5	3	1	1			1				650
		9	37	201	165	13	3	4	3										435
9	Acetic acid	2	6	96	356	108	63	16	3		2	1							653
		3	23	137	240	27	5	2		1									438
10	Diallyl sulfide		6	109	283	151	51	23	7	4	2	1			1	2			640
			36	156	185	28	14	2	3		1								425

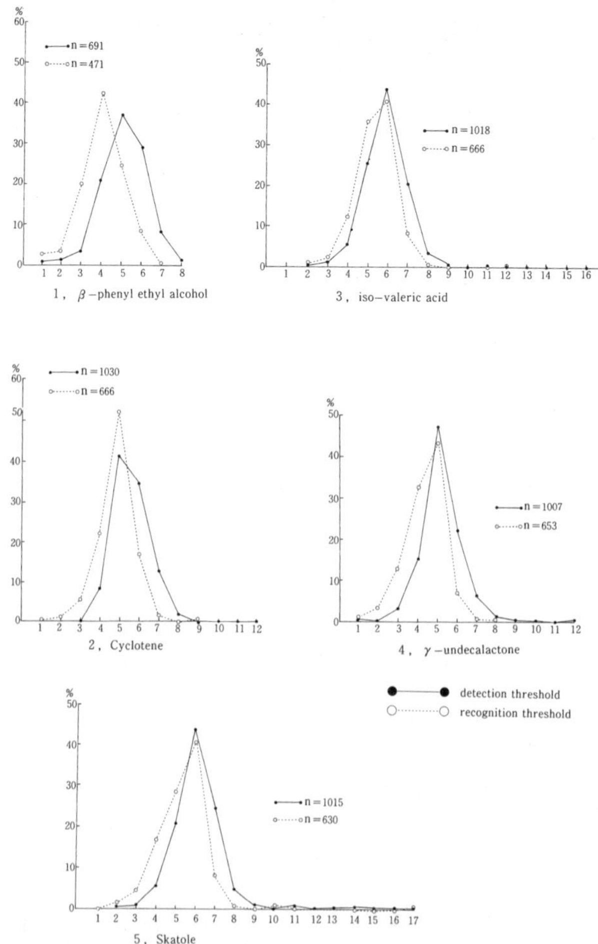

Fig. II-2. A: Distribution curves of detection and recognition thresholds of the same population of young subjects as in Table II-3. These five odors were later selected for the T & T Olfactometer. (from Toyota *et al.*, 1978; courtesy of Igaku-Shoin, Ltd., Tokyo)

Thus, it is difficult to conclude from the results of our experiments that olfaction in women is more accurate than in men.

When the F-test was performed to determine the differences of variances between men and women, half of the results were found to be insignificant (Sato, 1978). Thus, the above conclusion which questioned the apparent

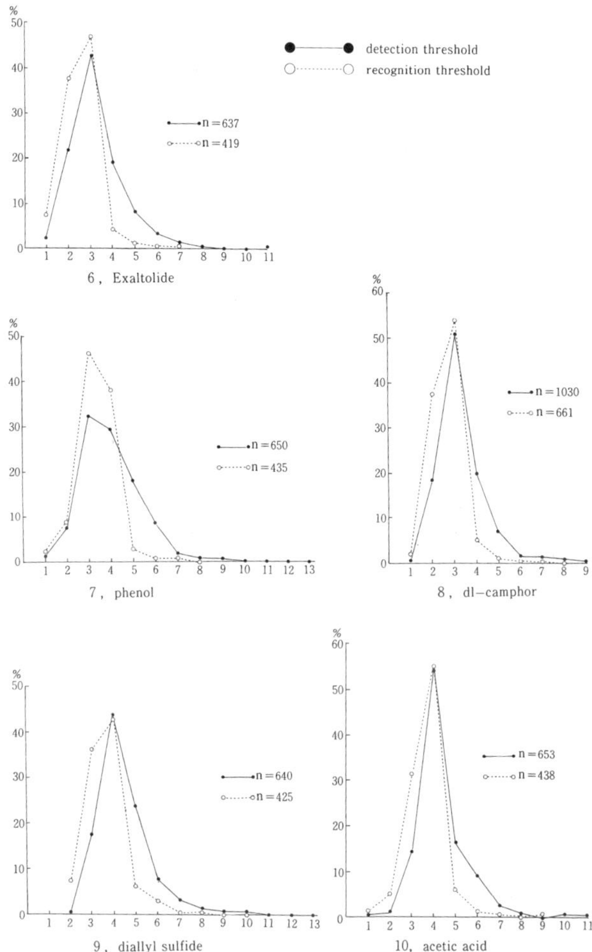

B: Distribution curves of detection and recognition thresholds of the same subjects as in Table II-3. These five odors were later judged as inappropriate as test odors and were removed from the initial ten odors. (from Toyota *et al.*, 1978; courtesy of Igaku-Shoin, Ltd., Tokyo)

superiority of women's olfaction over men's was supported by the F-test findings.

On the basis of these results the committee decided to manufacture the standardized olfactometer using the averages of detection thresholds for both men and women (Table II-4).

Table II-4. The averages ($\bar{x}$), variances (s^2), standard deviations (s), and coefficients of variation ($100s/\bar{x}$) calculated from the detection and recognition thresholds in Table II-3. Numbers 1 to 10 indicate respectively the same odorants as in Table II-3. n: number of subjects. (from Toyota *et al.*, 1978; courtesy of Igaku-Shoin, Ltd., Tokyo)

No.	Detection threshold					Recognition threshold				
	n	$\bar{x}$	s^2	s	$100s/\bar{x}$	n	$\bar{x}$	s^2	s	$100s/\bar{x}$
1	1,030	3.25	1.0956	1.05	32.31	661	2.67	0.5269	0.73	27.34
2	1,007	5.13	1.0049	1.00	19.49	653	4.38	0.9360	0.97	22.15
3	1,018	5.93	1.2719	1.13	19.06	666	5.45	1.0574	1.03	18.90
4	1,030	5.60	0.9170	0.96	17.14	666	4.82	0.9059	0.95	19.71
5	1,015	6.07	1.4376	1.20	19.77	630	5.32	1.5201	1.23	23.12
6	691	5.19	1.1037	1.05	20.23	471	4.10	1.1079	1.05	25.61
7	637	3.27	1.4553	1.21	37.00	419	2.58	0.7657	0.88	34.11
8	650	3.95	1.9516	1.40	35.44	435	3.40	0.8866	0.94	27.65
9	653	4.30	1.1509	1.07	24.88	438	3.67	0.6834	0.83	22.62
10	640	4.47	1.9239	1.39	31.10	425	3.65	1.0247	1.01	27.67

Table II-5. The averages ($\bar{x}$), variances (s^2), standard deviations (s) and coefficients of variation ($100s/\bar{x}$). These data were calculated for healthy men and women separately, from the detection and recognition thresholds obtained by means of the 10^x odor series from the same population of subjects as in Table II-3. Numbers 1 to 10 indicate respectively the same odorants as in Table II-3. M: male, F: female, n: number of subjects. (from Toyota *et al.*, 1978; courtesy of Igaku-Shoin, Ltd., Tokyo)

No.		Detection threshold					Recognition threshold				
		n	$\bar{x}$	s^2	s	$100s/\bar{x}$	n	$\bar{x}$	s^2	s	$100s/\bar{x}$
1	M	591	3.14	1.0800	1.04	33.12	314	2.55	0.4715	0.69	27.06
	F	439	3.41	1.0772	1.04	30.50	347	2.78	0.5547	0.74	26.62
2	M	577	4.94	0.8561	0.93	18.83	308	4.14	0.9455	0.97	23.43
	F	430	5.39	1.0918	1.04	19.29	345	4.60	0.8273	0.91	19.78
3	M	583	5.77	1.1979	1.09	18.89	317	5.24	1.0246	1.01	19.27
	F	435	6.14	1.2972	1.14	18.57	349	5.65	1.0096	1.00	17.70
4	M	591	5.41	0.7368	0.86	15.90	317	4.55	0.9446	0.97	21.32
	F	439	5.85	1.0472	1.02	17.44	349	5.07	0.7416	0.86	16.96
5	M	583	5.74	0.9817	0.99	17.25	284	5.14	1.1815	1.09	21.21
	F	432	6.53	1.6977	1.30	19.91	346	5.46	1.7563	1.33	24.36
6	M	395	5.08	0.9942	1.00	19.69	206	3.84	1.2538	1.12	29.17
	F	296	5.34	1.2143	1.10	20.60	265	4.29	0.9115	0.95	22.14
7	M	376	3.05	1.2292	1.11	36.39	190	2.38	0.6599	0.81	34.03
	F	261	3.58	1.6211	1.27	35.47	229	2.75	0.7952	0.89	32.36
8	M	387	3.64	1.7903	1.34	36.81	203	3.26	0.8152	0.90	27.61
	F	263	4.41	1.8390	1.36	30.84	232	3.53	0.9166	0.96	27.20
9	M	390	4.12	0.9443	0.97	23.54	206	3.46	0.6009	0.78	22.54
	F	263	4.58	1.3365	1.16	25.33	232	3.86	0.6853	0.83	21.50
10	M	384	4.31	1.7094	1.34	31.09	200	3.41	0.9174	0.96	28.15
	F	256	4.71	2.0338	1.43	30.36	225	3.86	1.0300	1.01	26.17

Table II-6. Correlation matrix among the ten test odors. Numbers 1 to 10 indicate respectively the same odorants as in Table II-3. (from Toyota *et al.*, 1978; courtesy of Igaku-Shoin, Ltd., Tokyo)

Number of odorants	1	2	3	4	5	6	7	8	9	10
1										
2	0.3998									
3	0.2986	0.5017								
4	0.3046	0.5015	0.3819							
5	0.2927	0.4930	0.4747	0.4941						
6	0.2926	0.4463	0.3657	0.4917	0.5627					
7	0.3446	0.5038	0.4847	0.5011	0.5122	0.3520				
8	0.3985	0.2779	0.6126	0.5951	0.5833	0.3700	0.5268			
9	0.4070	0.2461	0.6724	0.5684	0.6810	0.4305	0.4519	0.5248		
10	0.3821	0.5161	0.6612	0.6389	0.6984	0.3809	0.4023	0.5409	0.6449	

Table II-7. Factor loadings of the ten test odors according to principal factor method (after rotation by varimax method). Numbers 1 to 10 indicate respectively the same odorants as in Table II-3. The factor loadings with large absolute values are underlined; thus, six factors, A to F were abstracted. (from Toyota *at al.*, 1978; courtesy of Igaku-Shoin, Ltd., Tokyo).

No.	Abstracted factors					
	A	B	C	D	E	F
1	0.1898	−0.3387	0.1660	0.2618	0.1515	0.2698
2	0.1392	−0.1924	0.2332	0.7642	0.1680	0.7230
3	0.6663	−0.4037	0.1192	0.3596	0.0434	0.7524
4	0.1884	−0.3562	0.2916	0.2208	0.6481	0.7164
5	0.4253	−0.2179	0.6648	0.2206	0.2080	0.7624
6	0.1233	−0.2312	0.5713	0.2565	0.1740	0.4912
7	0.1638	−0.5688	0.2642	0.3139	0.1344	0.5368
8	0.3855	−0.6220	0.2022	0.0397	0.3017	0.6691
9	0.6886	−0.2822	0.4385	−0.0227	0.2300	0.7996
10	0.6064	−0.1348	0.2422	0.3650	0.4929	0.8209
	1.7486	1.3510	1.3115	1.1790	0.9518	total 6.5416

2.3. Classification of the ten test odors

In order to classify the ten odors, a correlation matrix between two of the ten odors (Table II-6) was computed from the detection threshold data of all subjects. Factor analysis by the principal factor method was then applied to the matrix and five factors were extracted. The factor loadings of the ten odors are shown in Table II-7. In this case, factor extraction was not complete, accounting for only about 65% of the data. Consideration of the factor loadings in Table II-7 reveals that the ten odors could be classified into six groups (A to F in Table II-8). This table later became a reference when five odors were selected from the ten odors for a standardized olfactometer.

3. Manufacture of a Standardized T & T Olfactometer

In light of the results of the preliminary test described above, the Olfactory Test Committee embarked upon a series of tests that would help establish a standard olfactometer (Takagi, 1968c, 1987, 1989).

3.1. Examination of the suitability of the ten odors for olfactory tests

Examination of the results of olfactory tests with the ten odors shown in Fig. II-2 revealed that the curves of exaltolide, phenol, and *dl*-camphor were shifted to the left by more than one unit greater than the others. This indicates that these three odors are so weak that most people can only perceive them at nearly saturated concentrations. Consequently, hyposmic subjects who are not able to detect them at a one-tenth dilution of the saturated concentration (1 on the abscissa), cannot be tested by this odor series. Therefore, these three odors were judged to be less appropriate than the others for use as test odors.

Seven odors were finally selected as appropriate test odors: β-phenyl ethyl alcohol, Cyclotene, iso-valeric acid, γ-undecalactone, Skatole, diallyl sulfide, and acetic acid.

3.2. Selection of five odors for the T & T Olfactometer

From a practical standpoint, clinicians had complained that testing all ten odors was too time-consuming. In response, the committee decided to select five odors from the above seven.

A prerequisite condition for the olfactory test was that test odors should differ from each other as much as possible. Hence, the ten odors were examined for similarity. As was stated previously, a correlation matrix (Table II-6) was calculated from the detection threshold data of all subjects with normal olfaction (Table II-4). Factor loadings were then computed (Table II-7) and the ten odors were classified into six groups (A to F in Table II-8).

In this analysis, however, the factor extraction of only 65% was not sufficient, and it was considered unwise to depend entirely upon this group-

Table II-8. Classification of the ten test odors. (from Toyota *et al.*, 1978; courtesy of Igaku-Shoin, Ltd., Tokyo)

Group	Ten test odors
A	Acetic acid
	Iso-valeric acid*
	Diallyl sulfide
B	Phenol
	Exaltolide
C	Skatole*
	β-Phenyl ethyl alcohol*
D	γ-Undecalactone*
E	Cyclotene*
F	*dl*-Camphor

* indicates the five odorants selected for the T & T Olfactometer.

ing. With this in mind, the committee examined the eligibility of the seven odors.

β-Phenyl ethyl alcohol and Skatole belong to the same group (C in Table II-8). However, this classification does not coincide with our impression that these two odors are quite distinct from each other. Consequently, both of them were included in the five test odors.

Iso-valeric acid, diallyl sulfide, and acetic acid all belong to group A in Table II-8. The first two unpleasant odors may seem inappropriate as test odors because sniffing bad odors deprives subjects of their objectivity in the olfactory test. However, iso-valeric acid is well-known as an axillary odor and may be one of the human pheromones. Thus, diallyl sulfide was excluded, but iso-valeric acid was included in the five odors.

Acetic acid is well known as a stimulant of the trigeminal nerve. Consequently, it was considered inappropriate as a test odor. In this way, by eliminating diallyl sulfide and acetic acid, the five test odors were selected for a standardized olfactometer.

The order of presenting these five odors and the words for expressing their qualities were already determined, as shown in Table II-2.

3.3. *A new concentration series for the five test odors*

Based on the averages ($\bar{x}$) of the detection thesholds in Table II-4, the committee calculated a new concentration series in eight steps, from $x \times 10^{-2}$ to $x \times 10^{+4}$ or 10^{+5}, for each of the five test odors (Table II-9). In the extreme left column of Table 9, eight numerals (from -2 to $+5$) are marked from top to bottom. They represent the concentration series of 10^{-2} to 10^{+5}. 0 represents the averages of the detection thresholds, and numerals to the right of 0 indicate the averages for the five odors (Takagi, 1989).

Table II-9. Series of concentrations of five odorants for the T & T Olfactometer. A, PEA: β-phenyl ethyl alcohol. B, CYC: methyl cyclopentenolone (Cyclotene). C, VAL: iso-valeric acid. D, UND: γ-undecalactone. E, SKA: Skatole. Numerals in the extreme left column indicate concentrations of the respective odorants in eight steps. This table wan prepared from Table II-4 in the text. (from Toyota *et al.*, 1978; courtesy of Igaku-Shoin, Ltd., Tokyo)

Test odors	A	B	C	D	E
Degrees	PEA	CYC	VAL	UND	SKA
−2	7.2	7.6	8.0	7.1	8.1
−1	6.2	6.6	7.0	6.1	7.1
0	5.2	5.6	6.0	5.1	6.1
1	4.2	4.6	5.0	4.1	5.1
2	3.2	3.6	4.0	3.1	4.1
3	2.2	2.6	3.0	2.1	3.1
4	1.2	1.6	2.0	1.1	2.1
5	0.2	0.6	1.0	0.1	1.1

3.4. Completion of the Standardized T & T Olfactometer

After the five test odors and their concentrations series had been determined, the committee commissioned the Takasago Corporation to prepare five sets of odor solutions according to the concentration series in Table II-9. The Daiichi Yakuhin Sangyo Co., Ltd. (Tokyo) and the Nagashima Medical Instruments Co., Ltd. (Tokyo) bottled the graded solutions of each of the five odors and fixed the bottles in a metal container as one set.

Although eight-step concentration series for each of the five test odors were calculated in Table II-9, a solution of methyl cyclopentenolone at 5 (namely, a concentration of 10^{+5}; B5 in Fig. II-3) could not be prepared. Consequently, each set contained eight bottles for each of the four odors (A, C, D, and E) and seven bottles for methyl cyclopentenolone, for a total of 39 bottles in one metal case (Fig. II-3). These bottles were made square in shape intentionally, and were set in square holes of the metal container. The container itself was heavy, so that bottle tops could be opened with one hand.

The standardized olfactometer was first completed in Japan in 1975. The committee named this set the "T & T Olfactometer". ("T & T" refers to Toyota and Takagi. Dr. Toyota was the first chairman of the Olfactory Test Committee.)

3.5. Authorization of the use of the T & T Olfactometer

When the T & T Olfactometer was completed in 1975, the Otorhinolaryngology Society of Japan immediately approved it as the only standardized olfactometer. Since then, it has been used not only in university hospitals

Fig. II-3. The Standardized T & T Olfactometer. (from Toyota *et al.*, 1978; courtesy of Igaku-Shoin, Ltd., Tokyo)

and laboratories for olfactometry and diagnosis of olfactory disorders, but also in many prefectures and cities for the selection of olfactory panels to check factories or sites producing odor pollution.

In 1981, the Japan Medical Association endorsed the use of the T & T Olfactometer. In 1983, the Ministry of Health and Welfare approved the olfactometer and granted 420 social welfare reimbursement points for each test conducted with it (at about 12 yen per point). Naturally, this promoted its further use.

4. Application of the T & T Olfactometer in the Clinic
4.1. The use of the olfactogram

The olfactogram was developed to record the results of clinical tests with the T & T Olfactometer. For example, olfactograms of a patient named M.M. are shown in Figs. II-4 and II-5. The detection and recognition thresholds were usually found to be different by a concentration factor (step) of 1 in normosmic subjects. In some diseases, however, the difference was found to be greater (Fig. II-5 B). One empty column with eight divisions is set at the extreme right of the olfactogram (Figs. II-4–5). This column is for an optimal test odor which the experimenter may use in the Olfactory Test. In Japan, Alinamine solution has often been used not only for venous olfactory tests, but also for nasal olfactory tests.

4.2. Classification of the degree of olfactory sensitivity

The committee classified the degree of human olfactory sensitivity into

Table II-10. Categorization of olfactory sensitivity into six classes on the basis of averages of recognition thresholds in olfactograms. (from Toyota *et al.*, 1978; courtesy of Igaku-Shoin, Ltd., Tokyo)

Olfactory loss: Average recognition threshold	Degree of olfactory disorder	Complaint of patient
~ −1.0	Hyperosmic	
−1.0 ~ 1.0	Normal or normosmic	All odors detected normally No olfactory problem in daily life
1.1 ~ 2.5	Normal or mildly hyposmic	All odors detected mildly
2.6 ~ 4.0	Moderately hyposmic	Only strong odors detected
4.1 ~ 5.5	Severely hyposmic	Odors scarcely detected
5.5 ~	Anosmic	No odors detected

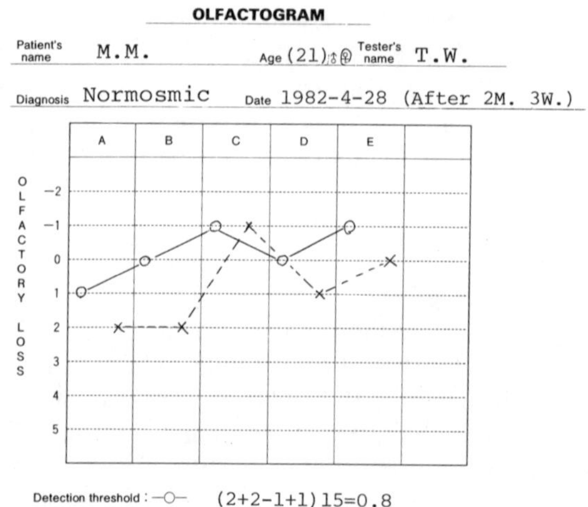

Fig. II-4. The olfactogram. This olfactogram shows the result of testing with the T & T Olfactometer in a patient after treatment for several months. The average recognition thresholds for five odors is calculated at the bottom. (Asaka, personal communication)

six categories based upon the average of the recognition thresholds for the five test odors: hyperosmic, normosmic (euosmic), mildly hyposmic, moderately hyposmic, severely hyposmic, and anosmic (Table II-10).

To obtain the average of the recognition thresholds in the olfactogram, the committee agreed on the following:

(a) To calculate the average of the recognition thresholds, all five test odors are to be used in the Olfactory Test.

(b) In the olfactogram, the olfactory sensitivity of a patient who can not identify four test odors (A, C, D, E in Table II-9) at the intensity of 5 is to be designated by 5 with a downward arrow (see Fig. II-5 A) and calculated at 6. In the case of odor B, the sensitivity of a patient who can not identify it at 4 is to be designated by 4 with a downward arrow (Fig. II-5 A) and calculated at 5.

(c) When the average is less than -1.0, the subject is diagnosed as hyperosmic. When it is between -0.1 and $+1.0$, he or she is diagnosed as normosmic. When it is between 1.1 and 2.5, he or she is moderately hyposmic. When it is between 2.6 and 4.0, he or she is severely hyposmic. When it is higher than 5.6, the subject is diagnosed as anosmic. The committee did not indicate any sub-categories within the hyperosmic category. This classification scheme is shown in Table II-10. The degree of a subject's olfactory impairment is determined by comparing the average obtained in the olfactogram with this table.

As an example, the recovery process of patient M.M. (Fig. II-4) is shown in Fig. II-5 A and B. A shows an olfactogram recorded when she visited the clinic for the first time. Even at the highest concentrations of all odors, she could neither detect nor recognize them, as shown by the downward arrows. The total of the recognition thresholds was 29, and the average was 5.8. Hence, she was diagnosed as anosmic. Her olfactogram was recorded again after 14 weeks of treatment (Fig. II-5 B). The average of the recognition thresholds was 4.2, and she was diagnosed as severely hyposmic. After a further 11 weeks of treatment her final olfactogram was recorded and is shown in Fig. II-4. Since the average of the recognition thresholds was 0.8, she was diagnosed as normosmic. It appeared that she had recovered her previous olfactory sensitivity.

4.3. Olfactory testing of individual nostrils: Monorhinal testing

The standardized olfactory test has usually been performed, using both nostrils (birhinally). Determinations of detection and recognition thresholds of each nostril independently were carried out. Since bilateral differences in sensitivity have been commonly found between eyes and ears, such a bilateral difference could be anticipated between the nostrils.

In fact, when monorhinal olfaction was measured by pressing the nose

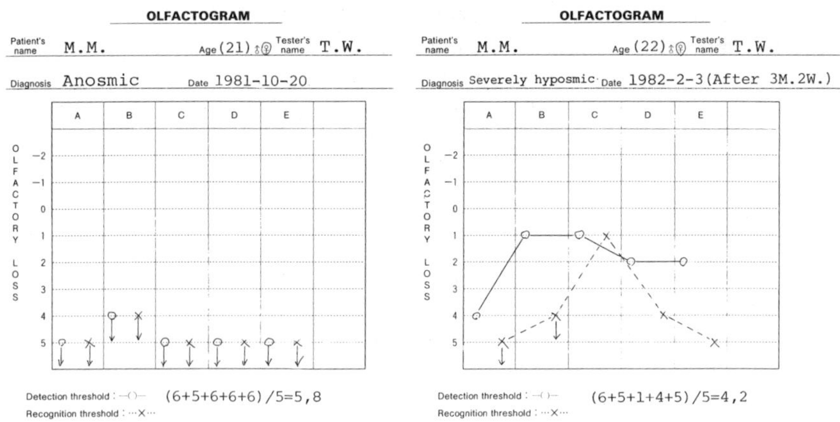

Fig. II-5. Olfactograms of patient M.M. during therapy. (Asaka, personal communication)

on one side, marked differences were found in some patients between nostrils. In these cases, birhinal olfactory tests always showed detection and recognition thresholds that were coincident with those of the more sensitive nostril. Thus, it is quite likely that unilateral hyposmia or anosmia in patients has been overlooked in the past. Monorhinal olfactory testing is, therefore, very important and should be routinely performed in clinical olfactory examinations.

5. A Standardized Venous Olfactory Test

5.1. Olfactory sensation elicited by intravenously injected odorants

Forschheimer first reported olfactory sensation after intravenous injection of neosalvarsan in 1916 and was soon followed by Kraupa-Runk (1916) and Henning (1916), who also reported this phenomenon. In 1930, Bednaer and Langfelder tested olfaction by injecting camphor, turpentine oil, and neosalvarsan intravenously. They defined intravenous or hematogenous olfaction as an olfactory sensation elicited by odor particles which reached the peripheries of olfactory nerves through the veins. Thus, intravenous or hematogenous olfaction became well known as a form of olfaction which is independent of normal usual nasal olfaction. The studies of Ishikawa (1938) supported this hypothesis.

Contrary to this hypothesis, however, Hirose (1944), Sato *et al.* (1956), Takenaka *et al.* (1958), and Ichihara *et al.* (1961) concluded that intravenous or hematogenous olfaction is actually respiratory olfaction, because they found that intravenously injected odor molecules are carried through the veins to the respiratory vesicles in the lung, enter the expiratory

Table II-11. Results of the standardized venous test. The latency period becomes longer and the duration of odor sensitivity decreases. A report from the clinic of Kanazawa University. (from Umeda, 1981a and personal communication)

	Latency period	Duration of odor sensation
Normosmia (50 sub.)	8.0 ± 0.5 sec	70.0 ± 7.0 sec
Moderate hyposmia (10 sub.)	10.3 ± 2.6 sec	65.0 ± 25.8 sec
Severe hyposmia (13 sub.)	14.2 ± 5.0 sec	42.2 ± 18.0 sec
Anosmia (45 sub.)	Alinamine odor not detectable: 18 sub. (40%) immeasurable	immeasurable
	Alinamine odor detectable: 27 sub. (60%) 17.3 ± 3.8 sec	40.3 ± 15.6 sec

air, reach the olfactory epithelium through the choanae, and stimulate the olfactory cells.

Therefore, intravenous or hematogenous olfaction is not different from usual nasal olfaction. Hence, the words "intravenous" and "hematogenous" were discarded and the word "venous" was adopted following discussion by the Olfactory Test Committee (Asaka and Umeda, 1978). Since then, "venous olfactory test" has been used as the standard name in Japan.

5.2. Alinamine as a venous test odorant

Alinamine has been widely used for over thirty years in the treatment of various diseases, and it has become well known that intravenous injection of this substance is accompanied by a sensation of garlic odor (Makino and Ishii, 1978; Zusho, 1978). Alinamine (p. 38) has been utilized in the field of otorhinolaryngology for the testing of olfaction (Asaka, 1984; Ichihara *et al.*, 1959, 1960, 1961; Kamio, 1960; Keyaki, 1964; Saito, 1966; Sato *et al.*, 1956; Umeda, 1972, 1981) since about 1957.

5.3. The Alinamine venous test

In the Alinamine test, subjects are requested to perform one light respiration every 2 sec, in time with the sound of a metronome working at a frequency of 60 beats per min. The experimenter then injects Alinamine at a slow rate into the median cubital vein of the arm, and measures both the latency period, from the beginning of intravenous injection to the onset of the garlic odor sensation, and the duration, from its initial appearance to its disappearance. Since both the latency period and the duration of odor sensation have been found to vary, depending on the olfactory disorder, they were examined in both normal healthy subjects and in patients by intravenous injection of Alinamine. The results are shown in Table II-11.

Prolongation of the latency period and shortening of the duration of the odor sensation were both found to a greater degree as the olfactory sensitivity decreased from normosmia through hyposmia to anosmia. Of significant importance was the finding that favorable prognosis correlated negatively with the latency period and positively with the duration of the odor sensation (Asaka and Umeda, 1978; Umeda, 1981).

B. Olfactory Testing in the United States

Methods of olfactory testing have recently been developed in several universities and laboratories in the United States. Amoore and Ollman (1983) devised a practical test kit to quantitatively evaluate human olfactory acuity. Doty *et al.* (1984) devised a "scratch'n sniff" test using microencapsulated odorants. Vollmecke and Doty (1985) utilized pictures for olfactory tests and produced a picture identification test (PIT). A method to measure olfactory thresholds and to identify odors was developed at the Connecticut Chemosensory Clinical Research Center (CCCRC) by Heywood and Costanzo (1985) and Cain *et al.* (1983). Developing a clinical olfactory test for children was difficult because testing for olfactory recognition thresholds in children is particularly troublesome. However, Murphy introduced a clinical olfactory test for children in 1985.

If all these test methods were standardized in the United States, as in Japan, the results of the olfactory tests in different clinics and laboratories could be compared. Heywood and Costanzo (1985) thought it essential to standardize the various olfactory tests. Many potential applications for clinical testing of olfactory threshold have been reviewed elsewhere (Henkin, 1967; Schneider, 1967a; Alter and Seltzer, 1974; Van Den Eckhaut, 1978; and Doty, 1979).

1. Amoore's Pyridine Test
1.1. Manufacture of ready-to-use olfactory test kits
Amoore's group (Sherman *et al.*, 1979) devised a practical test kit for quantitatively evaluating the sense of smell, utilizing binary-step serial dilutions of pyridine in water. However, this method was inconvenient because it required at least two hours' preparation for cleaning the glass-stoppered flasks, making dilutions for the test series, and other tasks. In the new method (Amoore and Ollman, 1983), odorless white polypropylene 8-oz (250 ml) cylindrical squeeze-bottles, like those used for shampoo, are used instead of glass-stoppered flasks, and pyridine is diluted in odorless mineral oil (White oil, U.S. Pharmacopoeia grade). Each successive binary step of the dilution series contains one-half the concentra-

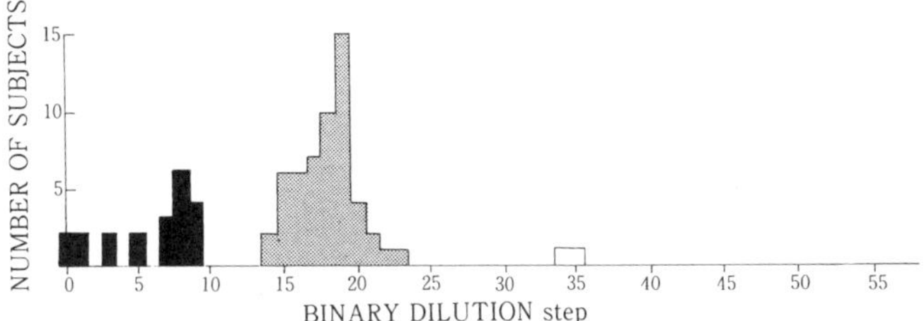

Fig. II-6. Pyridine scale of olfactory detection thresholds. Stippled histograms show the results in 54 healthy volunteers, aged 22 through 64. Solid histograms show the results in 21 patients with type I hyposmia. The open histogram indicates patients with type I hyperosmia. Their mean threshold to pyridine was at step 34.5, which was 16.6 binary steps more sensitive than the average normal subject. (from Sherman *et al.*, 1979)

tion of pyridine of the preceding step. The series is designed to correspond to the odor intensities of the standardized series of pyridine solutions in water described by Sherman *et al.* (1979). Amoore and Buttery (1978) demonstrated that the glass-stoppered flask and the squeeze-bottle methods yielded exactly the same threshold results when the vessels contained equivalent concentrations of odorants.

During the test, subjects are presented with a pair of squeeze-bottles and are instructed to bring the orifice of the bottle close to the nostrils, squeezing the bottle while inhaling through the nose. They are required to pick out the odorous, or more odorous, bottle of each pair.

The detection threshold is defined as the weakest odor step at which the patient always selects the correct bottle. Less than ten minutes are required for the questioning and testing of each subject. Since the normal range of sensitivities lies between step 14 and step 23, with the mean sensitivity at step 17.9 (SD = ± 2.0 steps), an experimenter is advised to start with binary step 14 and then use weaker, or stronger, odor intensities as required to determine the subject's olfactory detection threshold for pyridine.

1.2. Results of the pyridine test

Sherman *et al.* (1979) performed the test and analyzed the results. They obtained the following categories in the olfactory sensitivities of Americans (Fig. II-6): The normal range of sensitivities (96 % of healthy persons aged 20–60 years) was found between step 14 and step 21, with the mean sensitivity at step 18.0 (SD = ± 2.0 steps). The hyposmic range of sensitivities was found between steps 0 and 13. Sherman *et al.* found 21 hyposmia type I patients according to Henkin's (1967a) terminology (indicated by the solid histogram in the left of Fig. II-6). The observed thresholds occupied

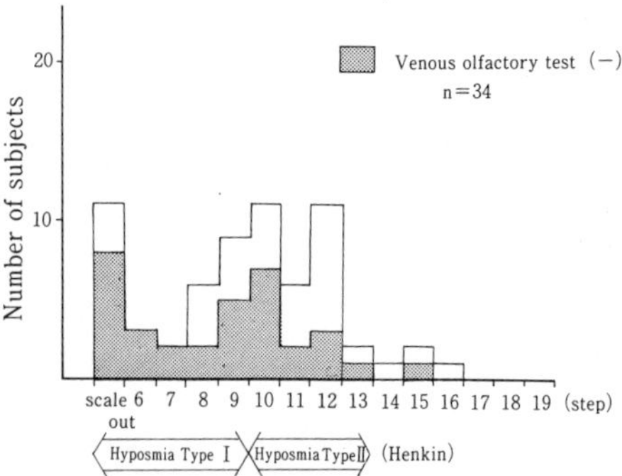

Fig. II-7. Olfactory sensitivities of 64 anosmic subjects examined by pyridine and venous olfactory tests. Ordinate: number of subjects. Abscissa: binary dilution steps. (from Kobayashi *et al.*, 1984)

a range of ten binary steps, from step 0 through step 9. These patients had an average detection threshold at binary step 5.9 (SD = ± 3.2 steps) for pyridine, which is 12.0 steps below the average for normal sensitivity. Hyposmia type I is encountered in only about 0.2 % of the population at large. Threshold sensitivities in the range from step 10 through step 13 represent hyposmia type II (Henkin, 1967a) and are often associated with clinical abnormalities (Fig. II-7). The hyperosmic range of sensitivities observed was from step 22 through step 29 (Fig. II-8). Such hyperosmia may be indicative of a clinical condition, for instance, untreated adrenal cortical insufficiency (Addison's disease) or a pituitary tumor.

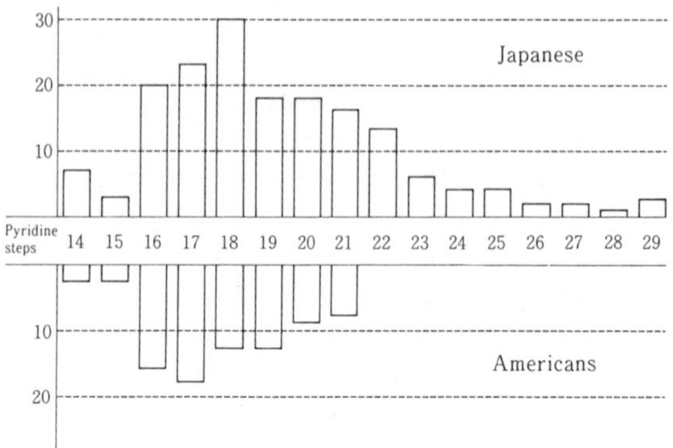

Fig II-8. Distribution of recognition thresholds of normosmic Japanese and Americans in the pyridine test. (from Zusho *et al.*, 1983a)

For research purposes, it may be necessary to employ each binary step. In clinical practice, however, test kits containing only the alternate (even-numbered) steps provide adequate precision. In fact, binary steps 6 through 29 of the pyridine scale (Sherman *et al.*, 1979) have ordinarily been found adequate to cover the maximum range in clinical practice. Amoore's binary steps offer about three times finer precision in assessing olfactory thresholds than do Henkin's and the T & T Olfactometer's decadic steps (Amoore and O'Neill, 1986).

Ready-to-use test kits in a variety of sensitivity ranges to suit particular applications were introduced commercially in 1979, and are available from Olfacto-Labs, P.O. Box 757, El Cerrito, California 94530.

1.3. Application of the pyridine test in Japan

Kobayashi *et al.* (1984) selected 64 anosmic patients after T & T Olfactometer tests in Japan. Among them, 39 subjects could detect only the five test odors, and the remaining 25 could neither detect nor recognize the five test odors. They then examined olfactory sensitivities of these 64 subjects with the pyridine and venous olfactory tests (Fig. II-7). It was noteworthy that among the 64, 30 anosmic patients showed positive results in the pyridine test. In fact, they still had olfactory sensitivities that were distributed from "scale out" to step 15. ("Scale out" means outside the scale range.)

Zusho *et al.* (1983b) performed the pyridine test on 170 Japanese subjects, composed of 146 women (86%) and 24 men (14%), with normal olfaction. Most of the women (97.3%) were teenagers (71) or in their twenties (71). Consequently, the results indicate the olfactory acuity of young Japanese women, but may not be representative of the olfactory sensitivity of all Japanese people. The distribution was spread from step 14 to step 29, in contrast to the data obtained in the United States (Fig. II-8).

In the T & T Olfactometer Test, those patients who were unable to detect the five odors at their highest concentrations were judged as anosmic. On the olfactogram, anosmia to an odor is recorded as +5 with a downward arrow, indicating "outside the range" (Fig. II-5 A). When these anosmic patients were examined by the pyridine test, however, their olfactory acuities were found to be distributed from "outside the range" to step 15. Thus, the pyridine test is superior in that it can provide much more information on anosmic subjects than the T & T Olfactometer Test. It also has the advantage that olfactory acuity can be examined for a wider range of odor concentrations than the T & T Olfactometer, as was alrready stated in B, 1.2. Moreover, the pyridine test is simple, convenient and cheap.

On the other hand, the pyridine test has the disadvantages that only one odor is used in the test and that only the detection threshold, not the recognition threshold, can be measured. Hence, it is apparently insufficient for the examination of various olfactory disorders. Zusho *et al.* (1983a) concluded, therefore, that it is desirable to use the pyridine test in conjunction with the T & T Olfactometer and venous olfactory tests in the diagnosis of olfactory dysfunction.

2. University of Pennsylvania Smell Identification Test (UPSIT)
2.1. Manufacture of the UPSIT booklets

Doty *et al.* (1984) devised an olfactory test using 50 microencapsulated odorants. Each of these odorants is attached to a brown label which measures 75mm by 13mm. Each brown label is pasted on a white sheet of paper 83mm by 182mm. Ten sheets of these papers are bound into one booklet, and five booklets compose one set. Thus, one set of booklets contains 50 odorants in total. On each page of the booklet, four words are printed above the brown label. They are names of odorant materials, and only one among them is the correct name of the microencapsulated odorant on the page. Subjects are instructed to scrape the label with an attached piece of sand paper (No. 120), break the microcapsule, and release the odor. They are requested to then sniff the label and completely fill in the circle in the right column on the last page of the booklet which corresponds to the subject's perception of the smell.

Initially, Doty *et al.* used 50 odorants, but they eventually eliminated 10 odorants which were difficult to identify and are now using 40 odorants in four booklets for the UPSIT. Since this test is simple and can be done by oneself, even people who live at a distance can be tested by mail.

This self-administered test has made it possible to rapidly and accurately assess the general olfactory function of many subjects in the laboratory, clinic, or through the mail without complex equipment or space-consuming stores of chemicals. The validity and usefulness of this "scratch'n sniff" olfactory test has already been demonstrated in experiments involving over 1,600 subjects (Doty *et al.*, 1984).

In 1986, the magazine *National Geographic* (vol. 170, No. 3, p. 329) conducted a large-scale smell survey; a small book consisting of a questionnaire attached to six scratch-and-sniff panels was enclosed in issue No. 3 of vol. 170. The odors used were androstenone, isoamyl acetate, galaxolide, eugenol, mercaptane, and rose. One point five million of the readers scratched at six smell samples of scent, sniffed them, and mailed in their reactions to the National Geographic Smell Survey. The following year, in the October issue of the magazine (vol. 172, No. 4, 1987), very significant and interesting results (though not the complete results) were

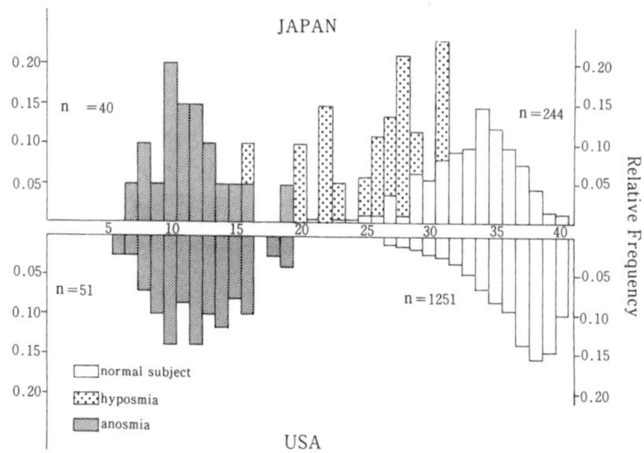

Fig. II-9. Results of UPSIT in Japan. This test was performed in Japan on 244 normal subjects and on 20 hyposmic and 20 anosmic patients. The results are shown above the upper abscissa by white, stippled, and black histograms respectively. For comparison, results obtained by the same test in the United States. (Doty *et al.*, 1984) are shown below the lower abscissa as inverted histograms. Ordinate: relative frequency. Abscissa: score of UPSIT. Further explanation in the text. (from Zusho *et al.*, 1985)

provided to the waiting readers. Readers are referred to the article "The Smell Survey Results" in that issue.

2.2. A trial of the UPSIT in Japan

Zusho *et al.* (1983b, 1985) compared this test to the T & T Olfactometer test, the intravenous test, and Amoore *et al.*'s pyridine test. The UPSIT was administered to 244 Japanese with normal olfaction (99 males and 145 females from 18 to 55 years of ages; average: 23.09, with SD = 5.00). Results showed that subjects gave between 21 to 40 correct responses to the 40 odorants, averaging 33.13 (SD = 3.39). Women (145 in number) showed more correct responses (averaging 33.54 with SD = 3.24) than men (99 in number; averaging 32.53 with SD = 3.52). In the United States, Doty *et al.* (1984) performed this test on 1,251 persons with normal olfaction (481 males and 734 females of average age of 33.69 with SD = 17.69). They obtained 27 to 40 correct responses to the 40 odorants (average: 37.55, SD = 2.42). The results from the two countries are slightly different, as shown by the upper and lower white histograms in Fig. II-9. However, since some of the odorant names are not familiar to Japanese people, this difference may not be significant.

Doty *et al.* (1984) also performed this test on a total of 51 anosmic patients (mean age: 40.76 years, SD = 20.75) and obtained 6 to 19 correct responses (average: 12.25, SD = 3.04). Zusho *et al.* (1985) examined a total of 20 anosmic patients in Japan and obtained comparable results, 7 to 19 correct responses (averaging 11.55 with SD = 2.19). They also examined 20 hyposmic patients and obtained 16 to 31 correct responses (averaging 25.50 with SD = 4.15). Some of the results of these patients overlapped with those of normosmic persons. All of these data are summarized in Fig. II-9.

Zusho *et al.* (1985) acknowledged the merits of this test as originally noted by Doty *et al.* (1984), but they also pointed out the following weaknesses. Since the odorant concentrations are fixed, this test is not suitable for determination of olfactory thresholds; the 40 or 50 odorants selected for Americans are often unfamiliar to non-Americans; and a distinction between normosmic persons and hyposmic patients cannot be made in many cases, as shown in Fig. II-9. In such cases, other kinds of olfactory tests (for instance, the T & T Olfactometer Test) are necessary.

3. An Olfactory Test from the Connecticut Chemosensory Clinical Research Center (CCCRC)

The usual clinical test of olfaction has principally sought to measure the absolute sensitivity to certain odors, with little regard to the degree of sensitivity. While threshold testing certainly provides important information on the olfactory sensitivity of a patient, it provides no useful indication of how much impairment the patient may suffer in olfactory performance in everyday life. Consequently, a properly designed test of odor identification that uses environmentally realistic odorants seems particularly appropriate. Such a test would have three necessary ingredients: highly identifiable items; enough items to measure degrees of performance; and a procedure that enables the patient to identify the items even if their names are not readily recalled. Cain and Krause (1979) have developed a test which meets these requirements.

The Connecticut Chemosensory Clinical Research Center uses a threshold test and a variation of the Cain and Krause identification test.

3.1. Threshold testing

The test uses series of aqueous solutions of butanol in which the highest concentration equals 4% (vapor phase concentration = 1000 ppm) and adjacent concentrations differ by a factor of three. The number of dilution steps ranges from nine to 13 depending on the testing situation. The solutions (60 ml) are presented in squeezable polyethylene bottles with 250-ml capacities. On each trial, the patient receives two bottles, one with and one without odorant, and must choose the bottle with the stronger smell. The patient first receives a demonstration of how to squeeze and

Table II-12. Test odorants, trigeminal stimuli, and distractors used in the odor threshold and odor identification tests. (O): test odorants. (T): trigeminal stimuli. (D): distractors. (from Cain *et al.*, 1983)

1. Ammonia (O, T)	11. Peanut butter (O)
2. Baby powder (O)	12. Pepper (black) (D)
3. Burnt paper (D)	13. Rubber (D)
4. Chocolate (O)	14. Sardines (D)
5. Cinnamon (O)	15. Soap (bar) (O)
6. Coffee (O)	16. Tobacco (D)
7. Garlic (D)	17. Turpentine (D)
8. Ketchup (D)	18. Vicks (O, T)
9. Mothballs (O)	19. Wintergreen (O, T)
10. Onion (D)	20. Woodshavings (D)

sniff. Testing then begins at or near the weakest concentration of the series and progresses upward in a manner dependent on the outcome of each trial. Four correct choices in a row constitute the criterion for determination of threshold.

3.2. Odor identification

The test uses the ten items listed in Table II-12 as odorants or trigeminal stimuli. The items are presented for monorhinic smelling from opaque plastic jars. Gauze placed over the samples precludes visual identification. The list of 20 terms in Table II-12 is available to the patient throughout testing. The list contains distractor names as well as the names of the actual test items. Some distractors represent sensations that commonly occur in dysosmia, e.g., garlic and rubber.

When presented with an item, the patient seeks its name. If the response is incorrect, the subject receives corrective feedback. If the patient identifies correctly, he or she receives credit for a correct identification. Details of olfactory tests in the CCCRC have been published together in a book (Cain and Gent, 1986; Gent *et al.*, 1986; and Goodspeed *et al.*, 1986a).

C. Olfactory Testing in Britain

1. Olfactory Tests Recommended in Neurology Manuals

Neurologists have traditionally employed odor identification tasks to test olfaction clinically. If they tested olfaction at all, neurologists wished primarily to determine whether or not the olfactory nerve and tract were functional. The test, therefore, did not have to yield a quantitative answer. Accordingly, neurology manuals have recommended testing with only a small number of odors, commonly three or four (Bickerstaff, 1968; Brain, 1969; Denny-Brown, 1974; Edwards, 1973; Mayo Clinic, 1964; Monrad-Krohn, 1958; Steegman, 1970; van Allen, 1969). Use of these odors, how-

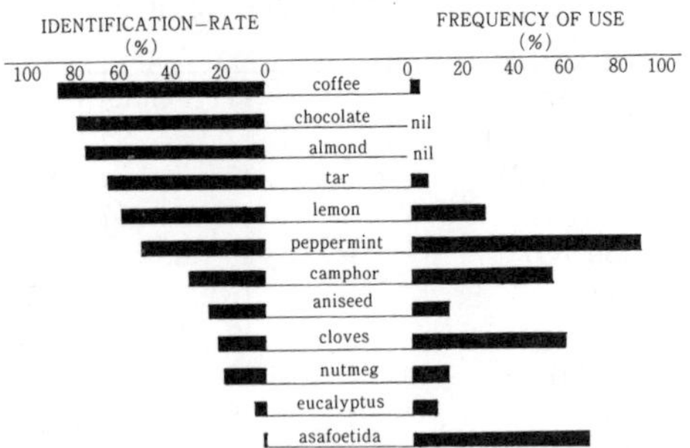

Fig. II-10. Comparison of the identification rate and frequency of use of test substances. Almond means benzaldehyde. The twelve odorous substances were arranged from coffee to asafoetida in descending order according to the identification rate. Note the generally inverse relationship between identification rate and frequency of use. (from Sumner, 1962)

ever, often resulted in poor performance even in patients with normal olfaction.

Due to the difficulties inherent in olfactory testing, Sumner (1962) carried out an investigation to determine what substances might be more readily identified. He prepared 13 numbered bottles containing test substances. Nine of the substances were those most often used or recommended in standard textbooks, three were ones which were never or only rarely used conventionally and one bottle was empty to serve as a control and a trap for the unwary (Fig. II-10). He studied the identification rate for these odors in two hundred subjects, and compared the results with the frequency with which the various test substances were used or recommended in ten clinics and three textbooks (Brain, 1955; Holms, 1960; Monrad-Krohn, 1958). Figure II-10 shows that there is an almost inverse relationship between the frequency with which a test substance is used and the ability to identify it.

At that time, the recognition and identification of test substances still remained the only practical means of clinically testing olfaction, while the quantitative methods described by other investigators were too time-consuming or too elaborate for clinicians (Elsberg and Levy, 1935; Jones, 1954; Stuiver, 1960).

Consequently, Sumner concluded that the conventional substances which had been used as test odorants were inappropriate and that coffee, benzaldehyde ("almond"), tar, and oil of lemon would be more suitable.

2. Reassessment of Odors Used in Clinical Olfactory Tests

Pinching (1977) found that 15 out of 22 (68.2%) patients with multiple sclerosis of varying duration and severity showed abnormalities when exposed to test odors. Moreover, unilateral hyposmia to three or more "pure" olfactory stimulants occurred frequently, especially in cases of short duration, while descriptive impairment was a feature in long-term cases. Furthermore, several patients had gross deficits in odor description without hyposmia or anosmia. Olfactory disorders had previously been reported to occur in 1% of cases (Herberhold, 1975). Ansari (1976) examined 40 patients with multiple sclerosis using serial dilutions of amyl acetate and nitrobenzene, and found no difference from 24 controls. Contrary to this finding, Wender and Szmeja (1971) found defective odor identification in 18 out of 52 patients with this disease; a result which correlates well with the "descriptive abnormality" group in the Pinching study. Thus, the results of olfactory tests in patients with multiple sclerosis were found to be inconsistent.

Neurophysiological studies have disclosed that all odors cause some trigeminal stimulation, which varies in degree from odor to odor (Ito, 1968; Tucker, 1971). Odors used for testing olfaction at that time (e.g., ammonia, clove, peppermint, and camphor) were those which caused considerable trigeminal nerve stimulation. Since diseases affecting the olfactory pathway are more likely to be detected during clinical examination if the test odors are not trigeminal stimulants (or are only minimally so), Pinching (1977) intended to establish a method by which the first cranial nerve might be tested separately as part of the routine clinical examination of the central nervous system. He selected musk, ketone, exaltolide, linalyl acetate, and coumarin, because musks and floral odors were considered most likely to be the "purest" olfactory (or least trigeminal) stimulants, acting virtually exclusively via the first cranial nerve. He proposed that the inclusion of musks and floral odors in "neurological trays" could improve the status of clinical olfactory testing as well as provide valuable clinical information.

D. Search for an Ideal Olfactory Test in Australia

Before the manufacture of a standard olfactometer, several factors which govern the optimum perception of odors should be defined. Since these factors had not yet been studied, the development of standard methods and instruments for measuring olfactory responses was hindered. Using three odors, pentyl acetate, l-butanol, and diethylamine, Laing (1982) studied (1) inhalation characteristics of humans during odor-threshold tests, (2) inhalation characteristics of humans during odor-intensity tests, and (3) inhalation patterns (or ranges of responses to odors) of subjects during odor-threshold tests.

Laing found that different subjects use markedly different sampling techniques, and that only three sampling parameters are consistently and significantly changed in response to odor concentration: (1) total sampling volume, (2) total sampling time, and (3) total number of sniffs. Moreover, he found that maximum inhalation rate is independent of the type of odor, olfactory task, odor quality, or odor strength. This seems contrary to the generally held belief that inhalation rate is reduced in response to strong or unpleasant odors. Values recorded for sniff volumes and inhalation rates indicate that most olfactometers in use do not account for human inhalation requirements during a sniffing episode. The data from his study therefore highlighted the need for knowledge of the conditions for optimum odor perception, as a forerunner to the standardization of methods and instrumentation in olfactory studies.

Then, Laing (1983) compared olfactory responses obtained through natural sniffing techniques with those obtained when the number of sniffs, interval between sniffs, and size of sniffs were controlled and varied. The results indicated that it was very difficult to improve on the efficacy of sniffing techniques of individuals and that a single natural sniff provides as much information about the presence and intensity of an odor as do seven or more sniffs. He found that humans achieve optimum odor perception during threshold and intensity measurements with their natural multiple-sniff technique or with a single sniff. This occurred with a sniff inhalation rate of 30/min, at volume of 200 ml and a duration of 0.4 sec. The use of several sniffs in a sniffing episode appeared to be a confirmatory action rather than a necessary one. However, when perceiving odor mixtures, several sniffs seemed necessary to aid discrimination of the components.

More refined studies (Laing, 1985), however, have now shown that near-perfect identification of odors is achieved by subjects using the shortest sniff they can physically produce (0.42 sec.), while optimum perception of intensity occurs with a sniff of between 0.39 and 0.64 sec. duration. Since his studies disclosed some characteristics of human sniffing behavior and defined the conditions under which odors are optimally perceived, Laing assumed that a basis has been provided for developing standard olfactory test procedures and instrumentation for studies of the human sense of smell. The research community awaits the completion of his standard olfactometer.

E. Olfactory Tests in the Past

Techniques of olfactometry have been in use for over one hundred years. The author does not intend to conduct a historical survey of them in this book and only wishes to introduces some important reviews.

Wenzel (1948) wrote a critical review of olfactometric techniques that appeared during the last one hundred years. More recently, Douek (1974) carried out an extensive survey of these techniques. Most of the techniques in use before 1925 were described by von Skramlik (1926) in the *Handbuch der Physiologie der niederen Sinne*. Older techniques have also been surveyed by Henning (1916) and by Gamble (1898).

III. Clinical Aspects of Olfaction

The Olfactory Test Committee was established in 1971 with the intention of developing treatment for various olfactory diseases in addition to the manufacturing of the Standardized T & T Olfactometer.

While the incidence of olfactory diseases had gradually increased in Japan prior to this time, otorhinolaryngologists had not known how to treat patients with such diseases who visited their clinics complaining of olfactory loss or lowered sensitivity to odors. In contrast to the remarkable progress in visual and auditory research, basic and applied research on the mechanisms of olfaction and olfactory disorders was not well advanced.

The circumstances were very similar in other countries. In the United States, Schiffman (1983) presented a report to the National Advisory Neurological and Communicative Disorders and Stroke Council and estimated that approximately 2 million American adults had disorders of taste and smell. She pointed out that "for 1975 and 1976 combined, a chemosensory problem was the major-presenting symptom in approximately 435,000 visits to a physician's office. In spite of the prevalence of chemosensory dysfunctions, medical textbooks provide little information on how to evaluate or treat them. Chronic disorders of taste and smell have been largely neglected because they are seldom fatal and, unlike deficiencies in sight or hearing, are sometimes not taken seriously because they are viewed as affecting the "lower" senses—those involved with sensual and emotional life—instead of the "higher" senses that serve the intellect" (Schiffman, 1983).

A. Classification of Olfactory Disorders

1. Classification according to the Literature

Schiffman (1983) collected 26 names of disorders affecting olfactory functions from a large body of literature and classified them into six groups. Herberhold (1975) classified olfactory disorders quantitatively into hyperosmia, hyposmia, and anosmia, and qualitatively into paros-

Table III-1. Classification of olfactory disorders according to etiology.

Organic section	

I. Intranasal

  A. Airway obstruction

Sinusitis	(Fein *et al.*, 1966; Ryan & Ryan, 1974)
Chronic ethmoid sinusitis	
Allergic rhinitis	(Fein *et al.*, 1966)
Vasomotor rhinitis	(Griffith, 1976; Ghorbanian *et al.*, 1978)
Polyps	(Fein *et al.*, 1966)
Adenoid hypertrophy	
Tumors	
Neurogenic tumors	(McCormack & Harris, 1955; Lindstrom & Lindstrom, 1975)
Olfactory neoblastoma	(Joachims *et al.*, 1975; Olsen & Desnato, 1983)
Deformity	
Deviated septum	(Douek, 1970; Goldwyn & Shore, 1968)

  B. Disruption of the olfactory mucosa

Chronic infectious rhinitis	(Douek *et al.*, 1975)
Influenza infection	(Douek *et al.*, 1975; Schaupp, 1967; cf. Snow, 1969)
Atrophic rhinitis	
Ozena	(Stranbygard, 1954)
Toxic fumes	(Adams & Crabtree, 1961; Douek, 1974; Doty, 1979)
(lead etc.)	

II. Intracranial

  A. Tumors

Meningioma	(Marshal, 1966)
Olfactory meningiomas	(Bakay & Cares, 1972)
Olfactory groove meningiomas	(Barraquer-Berre & Fargas, 1950)
Suprasellar meningioma	
Sphenoid side meningioma	
Cribriform plate meningioma	(Elsberg, 1935a, b, c)
Glioma	
Frontal lobe gliomas	(Elsberg, 1935c)
Pituitary adenoma	(Elsberg, 1935c)
Osteomas	
Temporal lobe tumors	(Furstenberg *et al.*, 1943)
Frontal lobe tumors	(Elsberg, 1935c)
Meningitis	
Non-neoplastic space filling lesiones	
Aneurysma	(Elsberg, 1935b, c; Schneider, 1972)
Hydrocephalus	
Neurogenic tumors in the nasal fossa	(McCormack & Harris, 1955)

  B. Trauma

Head trauma	(Caruso *et al.*, 1969; Costanzo & Becker, 1986; Goland, 1937; Hagan, 1967; Leigh, 1943; Schechter & Henkin, 1974; Schneider, 1972; Sumner, 1964; Zusho, 1979, 1982)

Fracture(Doty, 1979; Schurr, 1975)
Shearing of olfactory nerves(Schurr, 1975)
Hemorrhage
C. Infection(see also I, B)
Meningitis(Schneider, 1972)
Abscess(Schneider, 1972)
D. Vascular disorders
Vascular insufficiency or anoxia (Zilstorff-Pedersen, 1955; Cameron & Wright, 1964)
Aneurysm(Elsberg, 1935b, c; Schneider, 1972)
E. Epilepsy(Douek, 1974; Eskenazi *et al.*, 1986)
III. Congenital
Specific anosmias or hyposmias (Amoore, 1969, 1971; Blakeslee, 1918; Brown & Robinette, 1967; Doty, 1979; Fukumoto *et al.*, 1957; Kirk & Stenhouse, 1953; Patterson & Lauder, 1948)
Familial anosmia(Axelrod *et al.*, 1972; Henkin & Kopin, 1964; Lygonis, 1969; Singh *et al.*, 1970)
Hypogonadotrophic hypogonadism (Kallman's syndrome, hypoplasia of olfactory system)(Kanai, 1940; Kallman *et al.*, 1944; de Morsier, 1962; Schroffner & Furth, 1970; Sparks *et al.*, 1968; Tagatz *et al.*, 1970; Swanson *et al.*, 1971; Santen *et al.*, 1973)
Hypergonadotrophic hypogonadism   (Males & Schneider, 1972)
Congenital adrenal hyperplasia   (Henkin & Bartter, 1964)
Congenital facial hypoplasia and growth retardation
(Henkin *et al.*, 1966)
IV. Nervous diseases
Multiple sclerosis(Ansari, 1976; Catalanotto *et al.*, 1986; Peters, 1958; Pinching, 1977; Wender & Szmeja, 1971; Zimmerman & Netskey; 1950)
Parkinson's disease(Ansari and Johnson, 1975; Serby *et al.*, 1985; Ward *et al.*, 1983; Wender & Szmeja, 1971)
Aging processes(Smith, 1942; Liss &Gomez, 1958; Schiffman, 1977, 1979; Schiffman *et al.*, 1976, 1979)
Dementia(Peabody & Tinklenberg, 1985; Serby *et al.*, 1985)
Alzheimer's disease(Corwin *et al.*, 1985; Gilbert, 1985; 1986; Huff *et al.*, 1986; Serby *et al.*, 1985; Warner *et al.*, 1986)
V. Nutritional causes
Chronic renal failure(Schiffman *et al.*, 1978)
Cirrhosis of the liver(Burch *et al.*, 1978)
Vitamin B deficiency(Rundles, 1946)
VI. Endocrine diseases
Adrenal cortical insufficiency(Henkin *et al.*, 1962, 1966a, 1967; Henkin, 1975; Goodspeed *et al.*, 1986b)

Cushing's syndrome	(Henkin, 1975)
Hypothyroidism	(Schaupp & Seilz, 1969; McConell *et al.*, 1975)
Diabetes mellitus	(Jorgensen & Buch, 1961; Settle, 1986)
Gonadal dysgenesis (chromatin negative, Turner's syndrome)	(Henkin, 1967b; cf. Hamilton *et al.*, 1973)
Hypogonadotropic hypogonadism (Kallmann's syndrome)	(Kallmann *et al.*, 1944; see III)
Primary amenorrhea	(Marshall & Henkin, 1971)
Pseudohypoparathyroidism	(Henkin, 1968)
Hypogonadism in women	(Schneider *et al.*, 1958; Doty *et al.*, 1975)
Congenital adrenal hyperplasia (see III)	

VII. Viral and infectious diseases (post-inflammatory anosmia)

Acute viral hepatitis	(Henkin & Smith, 1971)
Influenza-like viral infection	(Doty, 1979; Hansen, 1970)
Viral infections	(Henkin *et al.*, 1975)

VIII. Surgery or lesions

Laryngectomy	(Heindl, 1932; Henkin *et al.*, 1968; Hoye *et al.*, 1970; Ritter, 1964; Negishi *et al.*, 1982)
Amygdalectomy	(Andy *et al.*, 1975; Chitanondh, 1966)
Rhinoplasty	(Champion, 1966; Goldwyn & Shore, 1968)
Temporal lesions	(Rausch & Serafinitides, 1975)
Amygdaloid lesions	(Andy *et al.*, 1975; Rausch & Serafinitides, 1975)
Cranio-cervical lesions	(Hinoki *et al.*, 1981; Hinoki, 1985; Nakanishi & Hinoki, 1981; Hayashi *et al.*, 1986)
Aphasic lesions	(Mair & Engen, 1976)
Eunuchoidismus	(Kallman *et al.*, 1944; Kanai, 1940)

IX. Drug reactions

Acetylcholinergic substances	(Skouby & Zilstorff-Pedersen, 1954)
Amphetamine and related drugs	(Goetzl & Stone, 1948; Guild, 1956; Turner, 1968; Turner & Patterson, 1966)
Cocaine	(Goldwyn & Shore, 1968; Zilstorff, 1965)
Methimazole (Tapazole)	(Hallman & Hurst, 1953)
Thiouracil	(Grossman, 1953; Leys, 1945; Schneeberg, 1952)
Tyrothricine	(Seydell & McKnight, 1948)
Zinc	(Asaka *et al.*, 1979; Price, 1986; Smith, 1938; Tidsall *et al.*, 1938)
Other drugs (tetracycline, streptomycin, lincomycin, *d*-penicillamine, griseofulvin)	(Alberts, 1974, Schultz, 1960)

X. Miscellaneous diseases

Sjögren's syndrome	(Henkin *et al.*, 1972)
Cystic fibrosis	(Henkin & Powell 1962; Hertz *et al.*, 1975; Weffenbach & McCarty, 1984)
Adenoid hypertrophy	(Ghorbanian *et al.*, 1978)
Bronchial asthma	(Fein *et al.*, 1966)
Leprosy	(Barton, 1974)
Multiple sclerosis	(see IV)

Parkinson's disease	(see IV)

XI. Unknown

Essential parosmia	(Douek, 1970)

XII. Volatile chemicals or pollutants from environments and industries

Hyposmia or anosmia occurs after chronic exposure to various volatile compounds or dusts. For readers who are interested in these problems, the following literature is suggested: Adams & Crabtree, 1961; Cone, 1968; Doty, 1979; Douek, 1974; Lenhardt & Rollin, 1969; Rossberg *et al.*, 1966; Stein & Ottenberg, 1958; Urbach, 1941/42.

Psychological section

Schizophrenia	(Alliez & Noshida, 1945; Bellak & Benedict, 1958; Connoly & Gittelson, 1971; Davidson, 1945; Kerekovic, 1972; Pryse-Phillips, 1971; Rubert *et al.*, 1961)
Olfactory hallucination	(Cameron & Wright, 1964; Davidson, 1945; Kerekovic, 1972; Pryse-Phillips, 1971; Rubert *et al.*, 1961)
Olfactory reference syndrome	(Pryse-Phillips, 1971)
Hysteria	(Douek, 1974)
Alcoholic Korsakoff patient	(Jones *et al.*, 1975; Mair & McEntee, 1986)
Malingering	(Douek, 1974)
Amnesic syndrome	(Mair & McEntee, 1986)

This table was prepared on the bases of data by Schiffman (1983), Douek (1970, 1974), Doty (1979), Schneider (1972), Yamada *et al.* (1978), and others.

mia and kakosmia. He further divided hyposmia and anosmia into the following three types: the respiratory, peripheral, or secondary type; the essential, central, or primary type; and the sensorineural type. The last category was further divided into prebulbar, bulbar, and postbulbar types. Douek (1974) classified abnormalities of smell according to general, peripheral, and intracranial causes. The general and peripheral causes were classified into four categories: (1) lesions of the nose; (2) lesions of the olfactory nerves; (3) physiological and hormonal changes; and (4) congenital and hereditary diseases. Intracranial causes were divided into four classes: (a) trauma, (b) tumors, (c) epilepsy, and (d) other organic causes. Furthermore, she classified psychogenic disorders into three groups: (i) illusions of smell, (ii) hallucinations of smell, and (iii) abnormal sense-memory.

Schneider (1972) classified anosmia according to organic and psychologic causes. Anosmia due to organic causes was divided into two categories: 1) The intranasal category comprises anosmia due to airway obstruction and end organ disruption. 2) The intracranial category encompasses anosmia due to tumors, head trauma, infection, vascular diseases, and congenital dysfunction. Anosmia of psychological origin includes

hysteria and malingering. All these and other classifications are summarized in Table III-1.

2. Olfactory Disorders
2.1. Disorders due to intranasal causes

(a) Obstruction of the airway: Deformity of the intranasal structure,
foreign substances, polyps, sinusitis, allergic rhinitis, and nasal tumors
(e.g., of the paranasal sinuses of nasopharynx) were all found to obstruct
the airway between the nostril and olfactory mucosa, resulting in hyposmia or anosmia (Table III-1, I, A).

(b) Disruption or dysfunction of the olfactory mucosa: In addition to
atrophic rhinitis, chronic rhinitis, and ozena, toxic fumes were also found
to disrupt the function of the olfactory mucosa or damage the mucosa
itself (Table III-1, I, B).

2.2. Disorders due to intracranial foci

When olfactory disorders due to intracranial foci are examined, it is
clear that the degree and extent of the symptoms of the disorders entirely
depend on the sites and expanses of the foci in the brain. In this respect,
discovery of two neocortical olfactory areas in the orbitofrontal cortex
of the old world monkey and identification of nerve pathways leading
to them provides important clues to localize the sites of lesions in the
brain which result in olfactory disorders. Furthermore, identification of
an olfactory area in the lateral hypothalamus and its afferent pathways

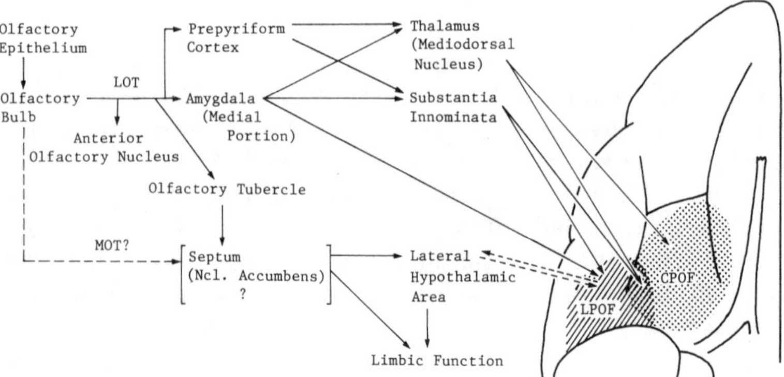

Fig. III-1. Two neocortical olfactory pathways to the LPOF
and CPOF and one subcortical olfactory pathway to the lateral
hypothalamic area in the old-world monkey. LPOF: lateroposterior area of orbitofrontal cortex, CPOF: centroposterior
area of orbitofrontal cortex. (from Takagi, 1984b)

facilitates the localization of the sites of cerebral dysfunction. These olfactory areas and nerve pathways are shown in Fig. III-1. Detailed explanations of these olfactory areas and the nerve pathways to them are provided in Chapters IX and X.

(a) Tumors: Meningiomas, especially of the olfactory groove, frontal lobe tumors, tumors and swelling around the optic chiasma, and temporal lobe tumors were indicated as causes for dysosmia by Douek (1974). Other kinds of tumors such as gliomas and osteomas, and non-neoplastic space-filling lesions such as aneurysma and hydrocephalus may also lead to olfactory disorders (Table III-1, II, A).

(b) Head trauma: Concussion, hemorrhage, and bone fractures were often found to result in lesions of the olfactory brain, leading to olfactory disorders. Shearing of olfactory nerves often follows brain concussion, resulting in anosmia or hyposmia. It has been reported that 90% of the subject's olfactory ability survives, if the olfactory nerve is spared unilaterally (Table III-1. II, B).

In patients who have become hyperosmic after craniocervical injury, vertigo and ataxia have often been found in response to odors of several substances, such as gasoline, tobacco, curry powder, garlic, onions, tar, and others. The generative mechanism of the olfactory vertigo was studied by Hinoki (1985), Hinoki *et al.* (1981), Nakanishi and Hinoki (1981), and Hayashi and Hinoki (1986).

(c) Infection in the brain: Meningitis and abscess after infection were found to cause dysosmia (Table III-1, II, C).

(d) Vascular diseases: Atherosclerosis, aneurysm, and tumors were found to cause olfactory dysfunction (Table III-1, II, D).

(e) Epilepsy: The association between epileptic seizures, olfactory aura, and lesions of the uncus was first noted by Jackson (1888). In addition, the finding that an epileptic seizure could be halted by inhaling a powerful and unpleasant odor was pointed out by Gowers (1881). The olfactory phenomena occurring in epilepsy can be considered in three parts: a) the olfactory aura (these patients complain of smelling a specific odor as an aura); b) the role of olfaction in the facilitation or inhibition of an epileptic seizure; and c) the disturbances in the sense of smell experienced by epileptics (Douek, 1974).

2.3. Disorders due to congenital causes

Specific anosmia, or olfactory blindness, has been well established as a congenital disease. Subjects with this disorder have normal sensitivities to all odors except one. Amoore (1979) has identified eight candidate odors to which these subjects are anosmic (see Table I-3). Another congenital disease, Kallman's syndrome, is also well known to be accompanied by an olfactory disorder (Table III-1, III; Kamei and Makino, 1987).

2.4. Disorders due to nervous diseases

Olfactory dysfunction was found in patients with multiple sclerosis by Wender and Szmeja (1971), but not by Ansari (1976) and Ansari and Johnson (1975). It has also been identified in Parkinson's patients. In these diseases, olfactory dysfunction may, or may not, occur depending on the location of the foci. In general, olfactory sensitivity declines with age (Table III-1, IV).

2.5. Disorders due to nutritional causes

Chronic renal failure, cirrhosis of the liver, and vitamin B deficiency were reported to cause olfactory disorders (Table III-1, V).

2.6. Disorders due to endocrine diseases

Olfactory disorders were found in a number of endocrine diseases, including adrenal cortical insufficiency, Cushing's syndrome, hypothyroidism, diabetes mellitus, gonadal dysgenesis (Turner's syndrome), hypogonadotropic hypogonadism (Kallmann's syndrome), primary amenorrhea, pseudohypoparathyroidism, and hypogonadism in women (Table III-1, VI).

2.7. Disorders due to viral and infectious diseases

Anosmia or hyposmia often occurs after influenza or influenza-like infections. Acute viral hepatitis was reported to cause olfactory disorders (Table III-1, VII).

2.8. Surgery and lesions

Alterations of olfactory function were reported after brain surgery. According to Raush and Serafetinides (1975), removal of the temporal lobe increased odor detection thresholds, whereas odor recognition thresholds remained similar to those of subjects without brain damage (Table III-1, VIII).

Brain surgeons occasionally cut the olfactory nerve unilaterally in order to lift the ipsilateral hemisphere when removing tumors, since unilateral section of the olfactory nerve has been shown to preserve 90% of the original olfactory sensitivity of the subject.

2.9. Drugs

Systemically injected, and even topically applied, drugs can influence not only nervous system neurons, but also intranasal engorgement and air flow, mucous secretion, and peripheral sensory function (Moulton and Beidler, 1967). Thus, the effects of many drugs on olfaction have been studied (Table III-1, IX).

2.10. Miscellaneous diseases

Olfactory disorders may occur in various other diseases, such as Sjögren's syndrome, cystic fibrosis, adenoid hypertrophy, bronchial asthma, leprosy and other (Table III-1, X).

Table III-2. Classification of abnormal olfaction.

Hyperosmia (total or partial)
Hyposmia (total, partial, or specific)
Anosmia (total, partial, or specific)
Parosmia
Olfactory hallucination
Kakosmia subjectiva (This is identical to an olfactory sensation which occurs spontaneously)
Aftersmell (Nachgeruch)
"Marcel Proust" Syndrome (Memories induced by odors)

(from Toyta *et al.*, 1978: courtesy of Igaku-Shoin, Ltd., Tokyo)

2.11. Olfactory disorders due to psychological causes

The aforementioned disorders are due to organic causes. However, olfactory disorders due to psychological causes have been found in various cases. Olfactory disorders including olfactory hallucinations and olfactory reference syndrome have been found in schizophrenia, hysteria, and alcoholic Korsakoff patients (Table III-1, Psychological section).

3. Classification of Olfactory Disorders according to Quantitative and Qualitative Changes

When the T & T Olfactometer was completed in 1975, the Olfactory Test Committee classified various aspects of olfactory disorders in the following ways (Table III-2).

3.1. Total anosmia or anosmia

This term refers to a loss of smell sensation due to disorder or defect of any of the olfactory receptor cells, nerves, or more central olfactory areas. Even in this situation, however, patients may be able to detect some kind of sensation due to the stimulation of the trigeminal nerve. ·

3.2. Partial anosmia

Partial anosmia only to a particular odor was once called "olfactory blindness." To this term, however, "specific anosmia" is now preferred. For further explanation see "individual difference" and "specific anosmia" in Chapter I (pp.8–9).

3.3. Hyposmia

The Olfactory Test Committee classified hyposmia into three classes: mildly, moderately, and severely hyposmic, on the basis of recognition thresholds determined by the T & T Olfactometer Test (Tables II-10 and III-5).

Henkin (1967a) found that anosmic patients who had lost the sense of smell retained an ability to detect some sensation when sniffing. He concluded that this ability derives from the functions of accessory olfactory nerves, namely, the trigeminal (Doty *et al.*, 1978a), glossopharyngeal, and vagal nerves. The olfactory function of these nerves, though very limited and poor, could be disclosed in patients with olfactory nerves severed or destroyed. Henkin thus divided hyposmia into two classes. Type I hyposmia was defined by the absence of a response at the primary olfactory area where olfactory cells function and the presence of a response at all accessory areas of olfaction. However, sensitivities of these three nerves to odors were found far inferior to the sensitivity of the olfactory nerve. They can barely detect the presence of odors, and would never be able to recognize their qualities. Shirakura *et al.* (1984, 1985a, b, 1986) supported Henkin's view and obtained further findings. Type II hyposmia was defined by the presence of responsiveness at both primary and accessory areas of olfaction, but with less responsiveness at the primary olfactory area than in normal subjects. Henkin's results are summarized as follows.

Type I hyposmia was found in patients with: (1) excision of the olfactory epithelium, trauma to the olfactory area, or infection of the olfactory area (all aquired types), (2) hypogonadotropic hypogonadism, and (3) idiopathy.

Type II hyposmia was found in patients with: (1) tumors involving the olfactory area or trauma to the olfactory area (all acquired types), (2) vitamin A deficiency (acanthocytosis, malabsorption syndromes), (3) hypogonadism, (4) facial hypoplasia, and (5) idiopathy.

The above classification of Type I and II was proposed by Henkin (1967a). Whether his hypothesis has been widely accepted is not known, however, in the field of otorhinolaryngology in the United States.

4. Classification of Anosmia and Hyposmia due to Foci

The committee classified anosmia and hyposmia into five types according to foci.

(1) Anosmia and hyposmia due to complete or incomplete obstruction of the airway between the nostril and the olfactory mucosa: Hereafter, these will be called "respiratory type" anosmia or hyposmia.

(2) Anosmia and hyposmia due to affected olfactory mucosa: Hereafter, these will be called "olfactory mucosal type" anosmia or hyposmia.

(3) Anosmia and hyposmia due to the combination of the above two types of disorders: Hereafter, these will be called "combined type" anosmia or hyposmia.

(4) Anosmia due to affected olfactory nerve function: Hereafter these will be called "olfactory nerve type" anosmia or hyposmia.

(5) Anosmia and hyposmia due to intracranial foci: Hereafter, these will be called "intracranial type" anosmia or hyposmia.

Table III-3. Sites of origin of olfactory disorders. (from Makino, 1985)

	Number	%
Intranasal origin	360	(80.36%)
Respiratory	19	(4.24%)
Mucosal	118	(26.34%)
Combined	223	(49.78%)
Olfactory nerve origin	0	(0.0 %)
Intracranial origin	62	(13.84%)
Traumatic origin	53	(11.83%)
Olfactory groove meningioma	5	(1.12%)
Others	4	(0.89%)
Congenital origin	15	(3.35%)
Kallman's syndrome	4	(0.89%)
Unknown	11	(2.46%)
Parosmia	11	(2.46%)
Total	448	

Makino (1985) analyzed their data on 448 patients according to the above and other criteria and their results are summarized in Table III-3.

5. Reports from Three Clinics in Japan

The causes of olfactory disorders treated in the clinics of Showa, Kanazawa, and Gunma Universities over a period of ten years were classified into seven groups, shown in Table III-4. This table indicates that intrana-

Table III-4. Olfactory disorders examined in three universities. (by courtesy of E. Asaka, R. Umeda, and S. Makino)

Diseases	Showa U. 1974–1983	Kanazawa U. 1974–1983	Gunma U. 1970–1983
Intranasal			
Allergic rhinitis	613 (22.1%)	73 (13.9%)	
Nasal polyposis			360 (80.4%)
Sinusitis	856 (30.8%)	199 (37.9%)	
Intracranial			
Head trauma	141 (5.07%)	41 (7.8%)	53 (11.8%)
Brain surgery		35 (6.7%)	
Tumors and others		19 (3.6%)	9 (2.0%)
Congenital			
Hypogonadotropic hypogonadism			
(Kallmann's syndrome)		10 (1.9%)	4 (0.9%)
Others		11 (2.1%)	11 (2.5%)
Infections			
Influenza (like) infections	581 (20.9%)	79 (15.0%)	
Occupational		13 (2.5%)	
Others	69 (2.48%)		11 (2.46%)
Unknown	519 (18.7%)	44 (8.4%)	
Total	2,779	524	448

Table III-5. Patients with olfactory dysfunctions in various grades. (from Makino, 1985)

Type	Number of patients
Anosmia	256 (57.14%)
Severely hyposmic	93 (20.76%)
Moderately hyposmic	64 (14.29%)
Mildly hyposmic	34 (7.59%)
Olfactory blindness	1 (0.22%)
Total	448

sal type disorders were most numerous: 52.9% in Showa University Clinic, 51.8% in Kanazawa University Clinic, and 80.4% in Gunma University Clinic.

5.1. Classification of patients according to the loss of olfactory sensitivity

In the Olfactory Clinic at Gunma University, 448 patients were examined over a period of 14 years from 1970 to 1983 (Table III-5). Patients were classified into five groups according to the degree of loss of olfactory sensitivity: mild hyposmia, moderate hyposmia, severe hyposmia, anosmia, and specific anosmia (olfactory blindness) Anosmic patients were found to be most numerous (57.1%).

5.2. Classification of hyposmia and anosmia according to the olfactograms

Asaka *et al.* (1983) classified 891 patients who visited the Olfactory Clinic at Showa University with complaints of hyposmia or anosmia into four groups according to the criteria below. Following classification, the common features of each group were extracted.

(a) A group of cases with measurable detection and recognition thresholds but high or moderate degrees of olfactory loss as shown in the olfactogram (Fig. III-2a). A total of 189 patients among the 891 (21.2%) belonged to this group. In these patients, slight morphological changes which might be attributed to slight disorders of the olfactory epithelium or respiratory pathways were often noticed. An unusual dissociation was often found between the detection and recognition thresholds. This seemed to suggest a special respiratory hyposmia due to abnormal intra-nasal air flow. Treatment for this type of hyposmia was comparatively successful.

(b) A group of patients who had detection thresholds lying between 1 and 5, but whose recognition thresholds could not be measured (Fig. III-2b). This type of hyposmia was found in 75 patients (8.5%) among the 891. No abnormal findings were found in the nasal cavities of any of these cases. In 54 cases, the olfactory epithelia were found to be normal. In 42 cases, such high degrees of hyposmia occurred after bouts of the common cold. Therefore, these hyposmias were considered to have an

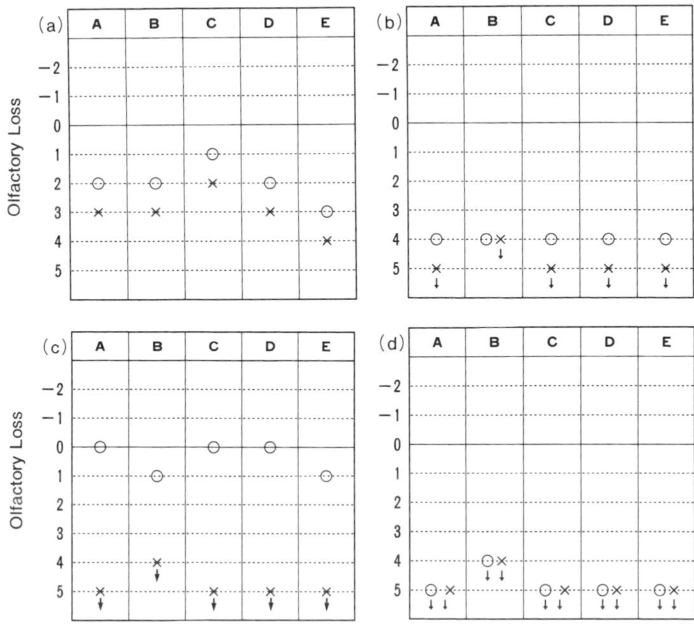

Fig. III-2. Olfactograms of patients in different dysosmic states.
(from Asaka *et al.*, 1983)

intracranial origin. Therapy for these hyposmias was entirely unsuccessful.

(c) A group of patients whose recognition thresholds were immeasurable in the olfactogram, although their detection thresholds were normal (Fig. III-2c). Thus, a striking dissociation was found between the two thresholds. A total of 33 cases (3.7%) among the 891 were included in this group. In these cases, abnormal findings were observed neither in the nasal cavities nor in the olfactory epithelia. These disorders occurred after colds in 18 cases, but the remaining 15 cases appeared to be congenital. Consequently, the dysfunction observed in this group of patients was considered to be of intracranial origin. Results of therapy for these patients were poor.

(d) A group of patients whose detection and recognition thresholds were immeasurable (Fig. III-2d). Among the 891 cases, 594 (66.7%) belonged to this group. Abnormalities were discovered in the nasal cavities of some of the patients. Various types of disorders were also found in the olfactory epithelia. A variety of causes were determined for this type of hyposmia or anosmia.

B. Diagnosis

Diagnosis of olfactory disorders has chiefly been performed using the T & T Olfactometer. However, as alternative methods, the Alinamine venous test, X-ray tomography, and endoscopy with Selfoscope (Olympus Camera Co., Ltd.) have often been employed.

1. Diagnosis with the T & T Olfactometer
1.1. Diagnosis of the degree of olfactory sensitivity

To diagnose the degree of olfactory sensitivity of a subject, the recognition thresholds are measured for each of the five test odors. Then, the average of the subject's recognition thresholds is sought. By referring this value to Table II-10, the degree of olfactory sensitivity of the subject is determined.

Makino (1985) reported that among 448 patients who visited the Olfactory Clinic of Gunma University over a period of five years from 1980 to 1984, anosmic and severely hyposmic patients made up about 78%; moderately and mildly hyposmic patients constituted only 22%, a remarkable low percentage. This difference is probably because the latter patients are more apt to discontinue their visits to olfactory clinics. Over a period of one month, the number of patients returning for therapy decreased from 448 to only 118.

1.2. Differential diagnosis of respiratory hyposmia and anosmia

In many nasal and paranasal diseases, anterior rhinoscopy shows that the walls of the olfactory cleft often touch and that the posterior region of the cleft is closed. In these cases, hyposmia often occurs due to blocking of the air passage to the olfactory epithelium. Differentiation of olfactory disorders of respiratory origin from those due to other causes is necessary to find an appropriate therapy.

The first step for the differential diagnosis is to open the closed olfactory cleft. For this purpose, an epinephrine solution (0.1%) is applied to the entrance of the olfactory cleft with a piece of cotton wool. In this case, the detection and recognition thresholds of the patient are examined before and after the cleft is opened.

If the hyposmia is due to the blocking of the air passage, the detection threshold is generally improved by a factor of 100. Such an improvement in the detection threshold serves as a criterion for the differentiation of disease origin.

1.3. Differential diagnosis of olfactory dysfunctions

While several tests to differentiate between conductive, labyrinthine, and postlabyrinthine origins of auditory dysfunction have been widely utilized, comparable tests have not yet been devised for olfactory dysfunctions. However, in the Clinic at Kanazawa University, several tests for differential diagnosis of olfactory disorders have been developed over the last ten years (Umeda, 1981a). These tests have also been used to determine the appropriate treatment of these diseases.

Patients with head injuries often complained of difficulties in making quantitative and/or qualitative discriminations of odors. They also complained of being prone to olfactory fatigue. In these cases, olfactory dysfunctions with intracranial origins were suspected. The following three tests were developed on the basis of these complaints (Umeda, 1981).

(a) An olfactory test to discriminate intensities of odors

At first, the detection threshold of the patient is determined for a test odor. Then, two concentrations of the odor which are higher than the threshold by one and two steps ($\times 10$ and $\times 100$) are selected. One end of an olfactory test paper is dipped 1 cm into the first solution and one end of another test paper is dipped 2 cm into the second solution. The patient is asked to indicate which paper has a stronger odor. This test is only applicable for those who have detection thresholds of 3 or less.

Normosmic and respiratory hyposmic subjects have been found, in most cases, to discriminate correctly two test odors which have a five-fold concentration difference. More than 80% of patients who have nasal cavity disorders (peripheral hyposmia), have been found to discriminate two odors which have a 15-fold concentration difference. However, less than 10% of patients with disorders of intracranial origin have been able to distinguish between two odors which have a 20-fold concentration difference. Consequently, when patients complain of difficulty in differentiating the concentrations of two odors, physicians may well suspect olfactory dysfunctions of intracranial origin.

(b) An olfactory test to identify odors

The patient's detection thresholds are determined for all test odors. Five odor solutions, A to E, with concentrations one stage higher than the thresholds are used. Five olfactory test papers (each 3 cm long) are dipped 2 cm into the five respective solutions and stored in five identical covered petri dishes (6 cm in diameter) for the patient. The same set of odorous papers in identical petri dishes are prepared for the experimenter. The experimenter picks up one test paper from the set for himself, and hands it to the patient. The patient smells it with two or three sniffs and is asked to remember the odor. Then the patient sniffs test papers from his own set one by one and selects the one which he believes to be identical with the one first presented to him by the experimenter. This test is repeated for each of the five test odors, with intervals of 1 minute between each answer and presentation, and the correct answers are totalled.

According to Umeda's reports (1981), subjects with normal and respiratory hyposmia have been able to correctly identify all five odors. Hyposmic patients with peripheral nervous dysfunctions have identified 3 to 5 odors correctly. In contrast, many patients with dysfunctions of intracranial origin have given only 0 to 2 correct responses.

Consequently, when a patient gives 2 or fewer correct responses, olfactory dysfunction of intracranial origin is suspected.

(c) A test for olfactory fatigue

Patients with head injuries have often complained that they were able to smell odors only temporarily and soon ceased to perceive them. This test was devised to distinguish olfactory dysfunctions of intracranial origin from those of other origins.

Test odorants D and E are used and the subject's detection thresholds are determined for both odors. Then, test solutions of concentrations two stages higher (× 100) than the thresholds are selected. The tips of two test papers are dipped 1 cm into these two separate solutions. The experimenter hands each test paper to the subject and lets him sniff it continuously for 60 sec at the rate of one nasal respiration every two seconds. The experimenter then measures how long the sensation of the smell continues (duration of smell sensation).

Immediately after odor stimulation for 60 sec, the detection threshold for the odor is determined again, and the difference between the two thresholds is calculated (change in thresholds). The recovery time of the detection threshold from the level obtained immediately after the end of odor stimulation to the prestimulus level is measured. Since test solutions of two stages higher concentrations are used in practice, this test can be applied to mildly hyposmic subjects with olfactory losses averaging less than 3.4.

With this test, 90% of the subjects with normal olfaction had durations of smell sensation longer than 60 sec, changes in the threshold of 1 to 2, and recovery times of one to two min. In 75% of slightly hyposmic subjects with disorders of the nasal cavities who showed average olfactory losses of less than 3.4, durations of smell sensation exceeded 60 sec, changes in the threshold were about 2, and recovery times were 2 to 3 min. However, among subjects with dysfunctions of intracranial origin who complained of similar mild hyposmia, only 10% showed durations of smell sensation longer than 60 sec. For these subjects, changes in the threshold were about 3, and recovery times were 4 to 6 min. These differences between the olfactory disorder of intracranial origin and those of other origins were found to be statistically significant.

Considering the variability between patients in the test results, if a patient showed a duration of smell sensation less than 50 sec and a recovery time of more than 4 min, the presence of olfactory dysfunction of intracranial origins was usually doubted.

(d) An examination of dissociation between the detection and recognition thresholds (Asaka *et al.*, 1983, 1984).

Henkin and Hoye (1966b), studied differences between the detection and recognition thresholds in 32 normal volunteers. Using pyridine solution, they found differences as large as 10^{-4}M. In a later study with 500 healthy subjects, Sherman *et al.* (1979) again found a difference of 10^{-4}M. Le Magnen (1950) tracked the olfactory sensitivity of one woman during the menstrual cycle and found that while the difference was 10^{-1} to 10^{-2}M in the luteal phase, it was as large as 10^{-6} to 10^{-7}M in the follicular phase. These data, however, were at variance with other available results. For instance, Stone and Pryor (1967) found only about a tenfold difference between the two thresholds.

The difference between detection and recognition thresholds measured with the T & T Olfactometer in subjects with normal olfaction was usually about 1. In contrast, a striking difference was found in patients with olfactory disorders. As stated above (p.83), among the 891 patients who visited the Olfactory Clinic of Showa University, the olfactograms of 33 cases (3.7%) showed that the detection thresholds were normal but the recognition thresholds could not be measured (Fig. III-2c). When these patients were examined with a rhinoscope and a fiberscope, abnormal findings were not obtained in either the nasal cavities or in the olfactory epithelia. Patients in this group were, therefore, considered to have dysfunction of intracranial origin. The dissociation phenomenon is helpful in the diagnosis of central anosmia or hyposmia, but the results of therapy for these patients are poor.

A similar dissociation was found during the recovery process in hyposmic or anosmic patients with disorders originating in the respiratory pathway to the olfactory epithelium, or in the olfactory epithelium and nerve (Asaka *et al.*, 1984). While some of these patients (Fig. III-3A) displayed marked dissociations to all five, or to two or three, test odors during the recovery process, such a dissociation phenomenon was not found during the recovery process of the other patients (Fig. III-3B). Thus, among the 226 patients, 163 (72.1%) showed one complete dissociation 36 (15.9%) exhibited incomplete dissociations to two or three odors, and 27 (12.0%) showed no dissociation. These results were obtained only in patients of respiratory or peripheral type hyposmia or anosmia.

Consequently, when these findings on anosmia or hyposmia of respiratory, peripheral, and central origin are combined with the results of other olfactory tests, for instance the olfactory test for respiratory anosmia

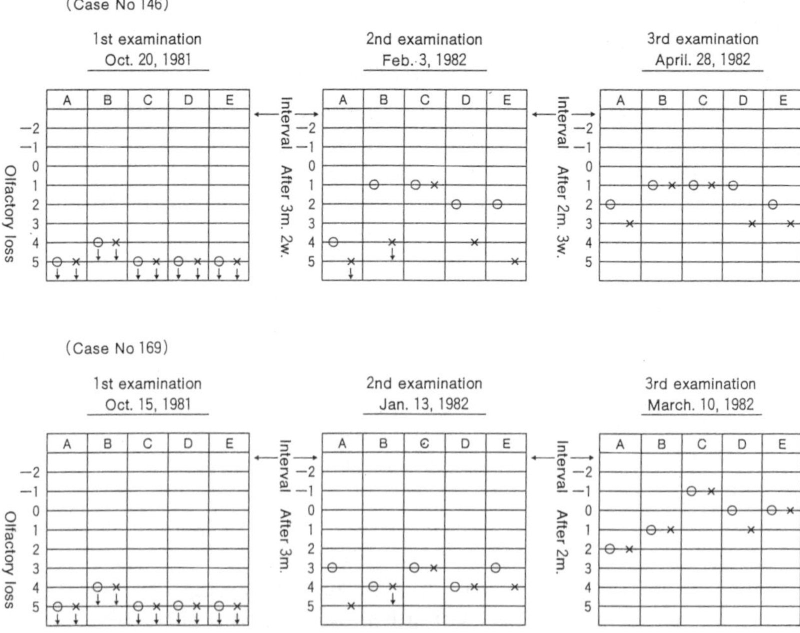

Fig. III-3. Recovery of dysosmic patients, A and B. (from Asaka *et al.*, 1984)

(Umeda, 1981a), the dissociation phenomenon observed in the olfactogram is seen to play an important role in the differential diagnosis among various types of olfactory disorders. In this regard, the discovery of the dissociation phenomenon between the detection and recognition thresholds is considered to be one of the most important results obtained with the use of the T & T Olfactometer.

1.4. Monorhinal olfactory testing (Chapter II, p.55)

In the Olfactory Clinic of Kanazawa University, monorhinal olfactory sensitivity was examined by pressing the nose on one side. Table III-6 shows seven examples of marked differences found between bilateral nostrils. It also shows the results of birhinal (bilateral) tests. These results disclosed that a subject with different olfactory sensitivities between nostrils always demonstrates the detection and recognition thresholds of the more sensitive nostril, when olfaction is tested with both nostrils. Thus, subjects with unilateral hyposmia or anosmia may have been overlooked in the past, making the monorhinal olfactory test a necessary part of a routine test in the olfactory clinic.

Table III-6. Hyposmic patients who showed different test results in left and right nostrils. (from Umeda, 1981a)

Cases	Monorhinal test on better side	Monorhinal test on worse side	Birhinal test
	A/B	A/B	A/B
1	2.0/2.6	5.2/5.8	2.2/2.6
2	1.6/2.2	4.8/5.4	1.6/2.2
3	1.4/2.2	3.4/4.2	1.4/2.2
4	1.2/2.0	5.8/5.8	1.4/2.0
5	2.6/3.4	5.8/5.8	2.6/3.2
6	2.0/2.4	5.6/5.8	1.6/2.2
7	1.4/2.0	3.8/4.4	1.4/2.0
Average	1.7/2.4	4.9/(5.2)	1.7/2.3

A: Loss of olfactory detection ability; B: Loss of olfactory recognition ability.

In this regard, Kobal *et al.*'s monorhinal evoked potential test (1987) may play an important role in olfactory clinics in the future.

2. Diagnosis with Alinamine Venous Test

2.1. Latency period and duration of odor sensation

Tokuda *et al.* (1978) performed a venous test on each of three different days in seven healthy doctors who were accustomed to this test. Variations in the latency periods were very small, with latency between eight and nine sec in these subjects (average: 8.3 sec). Thus, the day difference was only one sec. Stochastic examination showed that a difference of two sec was possible but that a difference of three sec could be attributed to a change in the threshold of the subject due to some causes.

Simultaneously measured durations of garlic odor sensation ranged from a minimum of 76 sec to a maximum of 117 sec. The average was 94.7 sec and the difference between maximum and minimum was 41 sec. Stochastic examination of the results indicated that the difference of 30 sec was possible but that the difference of 40 sec or more could be attributed to the variation of the threshold of the subject.

The latency period and the duration were measured in normal, hyposmic, and anosmic subjects, and were shown in Table II-11. When Alinamine F, a new Alinamine with reduced garlic odor, was used in the same test, the latency period was found to be 4.3 sec longer and the duration 11.7 sec shorter (Kuroishi and Zusho, 1979; Shinomiya *et al.*, 1979). These results were explained by the fact that the garlic odor of Alinamine F is 1/10 to 1/100 the strength of regular Alinamine.

2.2. Comparison of the results of venous testing with those of testing with the T & T Olfactometer (Sanada and Asaka, 1979)

Ishibashi *et al.* (1978) stochastically examined the correlation between the average olfactory losses obtained with the T & T Olfactometer test and the latency periods and durations obtained with the Alinamine venous test in 103 patients. The results were as follows:

(a) Average olfactory loss and latency period

A positive correlation (P < .01) was found between increases in the detection thresholds to each of the five test odors in the T & T Olfactometer test and prolongation of the latency period in the Alinamine venous tests. The correlation coefficient between the averaged olfactory losses to all five odors and the average of the latency period was 0.607 (P < .01).

(b) Average olfactory loss and duration of odor sensation

A negative correlation was found between the detection thresholds and the durations of odor sensation (P < .01): the durations of sensation decreased as the detection thresholds increased.

Umeda (1981a) found that 60% of the patients who visited the Olfactory Clinic of Kanazawa University Hospital with olfactory dysfunctions and were diagnosed as anosmic with the T & T Olfactometer could detect an odor sensation in the Alinamine venous test. Accordingly 40% of the patients showed negative results to both the T & T Olfactometer and venous olfactory tests (Table II-11).

2.3. Differential diagnosis of patients with intracranial foci

Patients with peripheral dysosmia displaying an average olfactory loss of about 4.5 showed durations of sensation of approximately 30 sec. In contrast, patients with intracranial lesions who had an equivalent degree (about 4.5) of olfactory loss showed, on average, far shorter durations of less than 13.1 sec. Consequently, patients presenting anosmia or hyposmia with durations of less than 15 sec were diagnosed to have intracranial foci. The venous olfactory test has been used in Japan to differentiate central dysosmia from peripheral dysosmia.

3. Other Diagnostic Techniques

3.1. X-ray tomography

The application of the T & T Olfactometer has led to the clarification of various types of olfactory disorders. Investigators have evaluated the usefulness of X-ray tomography to search for foci associated with olfactory disorders.

Initially, Iizumi *et al.* (1975) and Iizumi and Kitamura (1978) studied whether the region of the olfactory cleft containing the olfactory cells was open or not (this region will be called "true olfactory cleft" hereafter). For this purpose, they evaluated and determined the optimal direction, angle, and depth of focus required to obtain the clearest tomographic images.

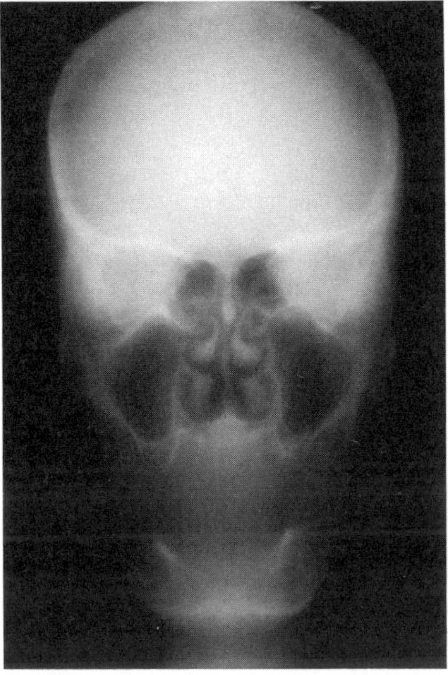

Fig. III-4. Olfactory cleft by X-ray tomography. (from Iizumi
et al., 1975)

3.1.1. *Method of examination*

(1) Occipito-frontal view: Hypocycloidal movement, 75 KVP, 20 mA,
12 sec.

(2) Occipito-mental view: Axial tomogram obtained with an angle of
12° to the plane of the canthus, 82 KVP, 20mA, 12.0 sec.

3.1.2. *Focal plane and posture*

Tomograms were obtained by adjusting the focus to the following
points:

(1) Occipito-frontal position

Frame No. 1.	1 cm frontal to the canthus
No. 2.	at the level of the canthus
No. 3.	1.5 cm occipital to the canthus
No. 4.	1/3 occipital of the distance between the canthus and the anterior wall of the external acoustic meatus

(2) Occipito-mental position

Frame No. 5.	at the level of the canthus
No. 6.	0.5 cm frontal to the canthus
No. 7.	1.0 cm frontal to the canthus

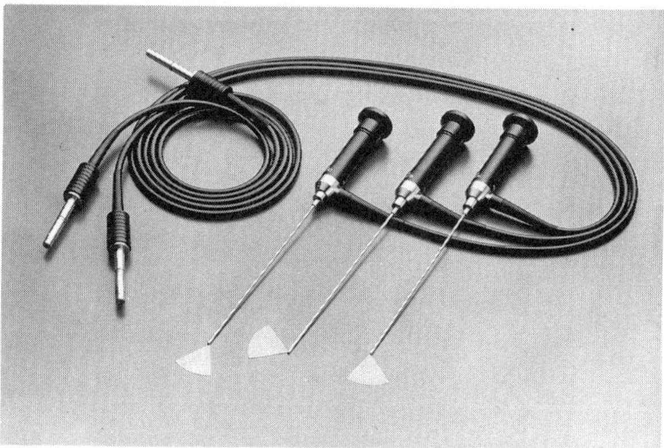

Fig. III-5. Selfoscopes. (courtsey of the Olympus Optical Co. Ltd., Tokyo)

Examination of the tomograms obtained under the above seven conditions revealed the following: (1) the clearest image of the true olfactory cleft can be obtained under condition 3 (Fig. III-4); (b) the degree of opening of the true olfactory cleft observed in the tomogram does not always correspond with the degree of opening of the olfactory cleft determined by a rhinoscope; (c) when closure of the true olfactory cleft is confirmed on the tomogram, the reduction of olfactory sensitivity observed in the olfactogram strongly suggests a diagnosis of respiratory type hyposmia or anosmia; (d) even if the true olfactory cleft is found to be open, if the airway from there to the common nasal meatus is closed, the reduction of olfactory sensitivity in the olfactogram nevertheless indicates a diagnosis of a respiratory hyposmia or anosmia; (e) when the true olfactory cleft is found to be open and the airway from there to the common nasal meatus is also found to be open, but the olfactogram nevertheless indicates a reduction of olfactory sensitivity, a diagnosis of mucosal or intracranial type hyposmia or anosmia is indicated.

These studies demonstrated the usefulness of X-ray tomography in locating the focus of dysosmia. This method is now being used as a diagnostic aid in the examination of olfactory disorders.

3.2. Endoscopy with the Selfoscope (Asaka, 1981a)

The first steps in the diagnosis of dysosmia are to inspect the nasal cavity, check the presence or non-presence of any obstacle in the airway to the olfactory cleft and, if possible, directly examine the condition of the olfactory epithelium. However, in most cases it is not possible to directly observe the epithelium with the common rhinoscope.

The Selfoscope® (Olympus Optical Co. Ltd., Tokyo) is a kind of rigid scope specially designed to inspect the olfactory epithelium with the naked eye (Fig. III-5). Its most important feature is the extremely small outside diameter (1.7 mm) which contains a rod lens having a diameter of 1 mm. The optical characteristic is

$$\alpha = 25/l$$

where α is magnification and l is the distance (unit: mm) from the tip of the lens to an object. The insertable length is 105 mm and the observable depth is 1 mm from the tip of the Selfoscope to infinity. The inserted portion includes a stainless-steel tube which protects the slender rod lens that makes the scope rigid (Asaka, 1981a).

Asaka and his collaborators (Asaka, 1981b; Nagano, 1977) have been using the Selfoscope for many years and have found that while it is not flexible, insertion of the Selfoscope into the olfactory cleft region is not difficult and does not harm the surrounding epithelium if certain techniques are employed. Following examination with the naked eye, they have succeeded in taking many color photographs of the olfactory epithelium. They classified the olfactory epithelia into six types according to moisture, surface irregularity, color, and degree of hypertrophy or atrophy (Asaka, 1981b, 1984). In consequence, the Selfoscope is now being used to aid the diagnosis of olfactory disorders in many clinics throughout Japan (Zusho *et al.*, 1981, 1984).

3.3. The Electro-olfactogram (EOG)

In humans as well as in frogs odor stimulation elicits an EOG in normal olfactory epithelia, but not in affected olfactory epithelia (Osterhammel *et al.*, 1969; Suga and Nakashima, 1972, 1973, 1978). Thus, the normal activity of the olfactory epithelium can be examined by recording the EOG. The difficulty of recording the EOG in humans had long made this method impractical, but recent use of the Selfoscope has made the recording easier. In the otorhinolaryngdogical clinic, EOGs will be used as a supplementary means for olfactory diagnosis.

3.4. Evoked potentials

Evoked potentials elicited by olfactory stimuli have been studied by many investigators (Chapter I, p.31).

Umeda (1981a,b) applied one of the five odors from the T & T Olfactometer to subjects at the rate of one for every three to five inspirations and recorded the summation of evoked potentials after 30 stimulations. In normal subjects, evoked potentials in response to each of the five odors with an intensity of 3 were observed. Furthermore, the evoked potentials increased as the odor intensity was increased from 3 to 5. However, no evoked potentials could be elicited by these odors or by acetic acid vapor, a well known trigeminal stimulant in 20 patients with dysosmia.

When the olfactory epithelium was experimentally anesthetized in normal subjects, evoked potentials ceased to appear in response to the same odors.

Consequently, Umeda (1981a) diagnosed anosmia by the loss of the evoked potential, and suggested that degrees of olfactory loss can be estimated to some extent by the amount of reduction in amplitude or loss of the evoked potential.

C. Therapy

Olfactory disorders originate from various causes, as shown in Table III-1. It is therefore important to clarify the foci or causes of olfactory disorders and find the most appropriate therapy for each of them.

1. Surgery (Kitamura and Iizumi, 1978)

Respiratory type hyposmia and anosmia is mainly caused by abnormal intranasal air flow due to polyps and swelling of nasal conchae, or from blockage of air flow due to the closure of either the entrance to the olfactory cleft or the cleft itself.

In most cases, these disorders can be cured by proper treatment of the nasal abnormality, for instance, by resection of nasal polyps or by surgical opening of the nasal cleft. In many cases, surgical treatment is accompanied by drug therapies which are described below.

2. Prednisone

Several investigators have described the use of prednisone for the therapy of olfactory disorders (Hotchkiss, 1956; Fein *et al.*, 1966; Zilstorff, 1972). Corticosteroids were used for the therapy of olfactory dysfunction by Goodspeed *et al.* (1986b).

In Japan, Shiraiwa, the late professor at Tokyo Medical College (personal communication) identified transient improvements of olfactory function in many patients when prednisone was used for the treatment of other kinds of diseases. This finding prompted the present use of a steroid hormone named Rinderon as a treatment for anosmic and hyposmic patients in Japan. Consequently, prednisone is not being used for the therapy of olfactory disorders in clinics in Japan.

3. Betamethasone Disodium Phosphate (Rinderon) and Other Steroid Hormones (Asaka et al., 1978)

While many other similar steroid hormones have been topically applied to the nasal cavity and olfactory cleft, patients have always complained of painful side-effects. Topical application of a solution of betamethasone disodium phosphate, namely a 0.1 % solution of Rinderon® (Shionogi

Pharmaceutical Co., Osaka), however, was found to have the mildest stinging effect. Since then, Rinderon has been used for the treatment of olfactory disorders in Japan.

3.1. Therapy for hyposmia or anosmia due to affected olfactory mucosa

In most cases, olfactory disorders occur due to an abnormality of the olfactory mucosa. Administering Rinderon by drops onto the olfactory mucosa with the patient's head fully inverted often produces the best effect.

A 0.1 % Rinderon solution is dropped into bilateral nasal cavities twice a day (morning and evening). After application, the patient is required to keep his or her head fully inverted for 5 minutes. With this method of application the amount of Rinderon used in one week is 5 ml. This rate of application of the steroid hormone has not produced any apparent side effects in subjects. It should be remembered that the hormone may not have a therapeutic effect if the solution is applied while the patient's head is not completely inverted.

With this therapy, olfaction often recovers with high percentages (48 %) (Fig. III-6). However, if such therapy is suddenly interrupted, the olfactory disorder often returns. Consequently, even if olfaction recovers, administration of the hormone should be continued for one or two months at a reduced frequency. For instance, in the clinic of Showa University the administration twice a day is decreased to once a day, and then to

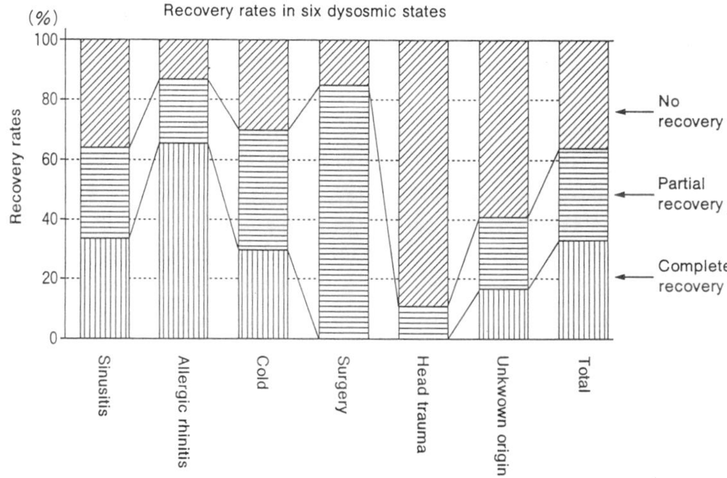

Fig. III-6. Recovery rates in six dysosmic states. (from Asaka *et al.*, 1986)

once every two days. Recurrence of the disorder may be prevented by this method.

3.2. Therapy for combined type hyposmia or anosmia

This type of disorder is a combination of affected olfactory mucosa and incomplete or complete obstruction of the airway between the nostril and the olfactory mucosa. Therefore, both surgical and pharmaceutical therapies are often necessary. If resection of nasal polyps is required, administration of Rinderon is started after the operation. If surgery is unnecessary, Tetrahydrogline hydrochloride, in the form of a sympathetic nerve stimulant, Cor-Tyzine (Taito-Pfizer Co., Tokyo) solution, is sprayed into the nasal cavity while the patient's head is fully inverted. The patient is required to maintain this position for one minute, then raise his head and blow his nose; administration of Rinderon is then initiated as described above.

3.3. Therapy for hyposmia or anosmia due to an affected olfactory nerve

This group of disorders includes the cases in which the olfactory nerve is principally affected. Using a hard fiberscope, disorders of this type can often be differentiated from diseases of the olfactory epithelium mentioned in 3.1. This type of disorder requires extended treatment with topical application of Rinderon as previously described in combination with administration of ATP (adenosine triphosphate disodium).

3.4. Therapy for hyposmia or anosmia due to intracranial foci

Hyposmia and anosmia are often due to various disorders in the brain; hence a different therapy is required for each of them. In several cases of head injury, in which detection thresholds could be measured but recognition thresholds could not, olfaction was seen to recover completely in a few, or many, weeks after injury. These phenomena may have been rare cases of spontaneous recovery. In general, however, prognoses of olfactory disorders due to intracranial foci are not promising.

3.5. Rates of recovery

The rates of recovery observed following the above therapies were classified into six types of olfactory disease and the rate were expressed in histograms (Fig. III-6). If the rate of partial recovery is added to that of complete recovery, total recovery in cases of sinusitis, allergic rhinitis, and common cold amounts to over 63 %. Such a high percentage of recovery could not have been anticipated before the introduction of the T & T Olfactometer test and therapies for olfactory disorders in 1975.

4. L-Cysteine Ethylester Hydrochloride

This drug is reported to have a normalizing effect on the mucosa. It is manufactured and sold in Japan under the commercial name of Cystanin by the Yoshitomi Pharmaceutical Co. Other companies also market it, under different commercial names.

Makino (1977) observed the effects of Cystanin on the olfactory mucosa and used it to treat olfactory disorders. Among the 25 patients treated with this drug, 23 showed improvement, of whom 11 recovered completely. For another 55 cases, in which Cystanin showed only a moderate effect, Rinderon, or other drugs, were administered in combination. Among these 55 patients, 20 showed partial recovery and 15 complete recovery. In a further 30 cases, Cystanin and Rinderon were administered from the beginning. Eight cases among them showed partial recovery and 17, complete recovery. Significantly with the exception of three cases, 20 patients who did not respond to Cystanin, also did not respond to the steroid hormone.

Makino found that recovery with Cystanin treatment is much slower than with Rinderon, but that marked side effects were not found during prolonged administration of this drug. Consequently, the recurrence of the olfactory disorders often found after therapy with the steroid hormone can be safely prevented by administering Cystanin.

5. *Vitamin A*

Standbygaerd (1954) treated ozena and rhinopharyngitis chronica sicca with vitamin A. Briggs and Duncan (1961) and Duncan and Briggs (1962) reported that intramuscular injections of vitamin A were effective for the treatment of olfactory disorders. Yasuda *et al.* (1973) treated 16 patients with neurogenic anosmia (totalling 28 nasal sides) by oral administration of 15,000 I.U. per day of vitamin A for three to six months. As a result, 71 % of affected nasal sides (20 sides out of 28) recovered olfaction after three to six months. However, Iizumi *et al.* (1979) reported recovery of only 12.5 % of patients with dysosmia (5 cases out of 40) by orally administering 1 tablet (15,000 I.U.) per day for 1 to 2 months. Iizumi suggested that administration of vitamin A for shorter periods might be the cause of such poor rates of recovery.

6. *Other Drug Treatments*

ATP or Cor-Tyzine has occasionally been administered as a supplement for the above drugs. In many cases, however, the effects of these drugs were not promising.

Recently, Juvela® (vitamin E; manufactured by the Eizai Pharmaceutical Co., Tokyo) was found to be effective for some hyposmic or anosmic patients with atrophic rhinitis (Makino, personal communication).

Henkin *et al.* (1976) reported that zinc treatment improved anosmia or hyposmia in many patients, although double-blind studies did not necessarily support this result. Hansen (1970) reported that 13 of 14 patients suffering from influenza-related anosmia improved following a series of treatments with triamcinolone acetate and a preparation of vitamin B complex, but this study lacked a control group so the level of spontaneous recovery could not be estimated. A similar problem is associated with Schaupp's (1967) report of improvement in one patient following treatment with suprarenal steroids, strychnine, and vitamin B complex.

Zilstorff (1972) reported that essential parosmia, which often occurs after influenza, was improved by repeated washings of the olfactory region with cocaine hydrochloride, but this report also lacked control subjects, and the level of spontanous recovery could not be estimated. Duncan and Briggs (1962) suggested the usefulness of vitamins A and E, as well as strychnine, in restoring olfactory function. Henkin and Kopin (1964) used methacholine for the therapy of some smell dysfunctions (Price, 1986). For parosmic (cacosmic) patients, substances increasing olfactory sensitivity were intravenously injected (Sternberg, 1931).

7. Therapy for Olfactory Vertigo

For patients who feel vertigo after sniffing some kinds of odors (p.77) cloxazolam, a minor tranquilizer which belongs to the benzodiazepine group, was found helpful. However, this drug has shown no beneficial effect for patients who have both olfactory vertigo and cerebellar dysfunction (Hinoki, 1985; Hinoki et al., 1981).

8. Ten Years of Therapies for Olfactory Disorders in Japan
8.1. A report from the Olfactory Clinic of Showa University (Asaka 1985)

The number of patients who visited the Clinic during the 10 years from 1974 to 1983 totalled 2779, ranging from 173 in 1974 to 400 in 1981. Ninety-five to 97% of these patients were found to be anosmic and only 3–5% were diagnosed as hyposmic.

After treatment with various therapies, the rate of complete recovery was found to be 29.2%, and that of partial recovery 44.6%. The remaining 26.22% of patients showed no recovery whatsoever.

8.2. A Report from the Olfactory Clinic of Gunma University

Recently, Makino (1985) examined the records of 448 patients who visited the Clinic over the past 12 years (from 1970 to April of 1983). The majority of patients had disorders of the nasal cavity (360 in total, 80.36%). The second largest group consisted of patients with post-traumatic anosmia or hyposmia (53 in number, 11.83%). There were 15 patients with congenital anosmia of intracranial origin (2.01%). In addition, parosmia was identified in 11 patients (2.46%). After treatment with various therapies, the rate of complete recovery was found to be 30%, that of partial recovery 39.6%, and that of slight recovery and no recovery 30.4%.

D. Prognosis

1. The Venous Olfactory Test

The most important clinical value of this venous test is in the judgement of disease prognosis. Eighty to 90% of the patients testing positive in the Alinamine venous test showed complete or partial recovery (Iizumi *et al.*, 1979; Makino and Endo, 1978; Fukushima and Oka, 1979).

Since the latency period and the duration of garlic odor sensation were found to vary between patients, relations between the latency periods and/ or the durations of odor sensation and the prognoses of patients were studied. Important correlations were found between a shorter latency period and a better prognosis, and between a longer duration of odor sensation and a better prognosis. The prognosis of anosmia or hyposmia in patients showing a duration of odor sensation less than 15–20 sec was generally poor (see Table II-11, p.57).

Umeda (1981a,b) often encountered patients who had never experienced any odors by sniffing through their noses, but nevertheless smelled Alinamine when injected intravenously. Their prognoses were found to be promising in many cases. Consequently, when a patient is diagnosed as anosmic or hyposmic after a test with the T & T Olfactometer, the Alinamine venous test could be used to judge the prognosis of the patient.

2. The n-Propyl Mercaptane Sniffing Test

In their study on the metabolism of Alinamine, Fujiwara *et al.* (1964) proposed that the sensation of garlic odor detected after the intravenous injection of Alinamine may be due to expiration from the lungs of an *n*-propyl mercaptane group as a product of Alinamine breakdown. Makino (1977) and Makino and Ishii (1978) used *n*-propyl mercaptane as a sixth odorant in a test with the T & T Olfactometer (Fig. II-3). They compared the results of sniffing this odor with those of the venous olfactory test. Patients who detected an odor after intravenously injection of Alinamine always detected the odor of *n*-propyl mercaptane. In contrast, among 20 patients who could not detect an odor after intravenously injected Alinamine, only seven patients detected *n*-propyl mercaptane in the sniffing test. The other 13 did not detect this odor. Among the seven patients, four subjects detected the original (10^0) solution of Alinamine, one person the solution of 10^{-1}, and 2 patients the 10^{-2} solution. In addition to *n*-propyl mercaptane, acetic acid was the only other odor these patients could smell. In conclusion, these researchers found that *n*-propyl mercaptane is more useful in determining the degree of olfactory dysfunction than the venous Alinamine test. This also shows that the *n*-propyl mercaptane sniffing test is useful in judging whether or not the patient can detect intravenously injected Alinamine by smell.

More important is the fact that the *n*-propyl mercaptane sniffing test may be most useful in predicting the prognosis of an olfactory disorder. If such a patient should recover sensitivity to the odor of *n*-propyl mercaptane during continual therapy for two or three months, the possibility of recovering sensitivity to the other odors becomes extremely high. In this case, the patient usually recovers his original olfactory ability rapidly. On the contrary, if a patient does not smell this odor at all, further continuation of the therapy would be futile.

Kamata *et al.* (1985) employed this test and found that of the patients who could detect *n*-propyl mercaptane 26% recovered, which was far higher than the rate of recovery (13.6%) in the patients who could not detect it. The *n*-propyl mercaptane (sniffing) test was found to be very useful to judge the prognosis of the disease. However, *n*-propyl mercaptane has an extremely foul odor which quickly pervades a room. Therefore, powerful ventilation of the room and use of such a deodorizer as shown at the top of Fig II-1A is necessary. It is advisable to perform this test out of doors.

3. Studies on Recovery from Olfactory Disorders

A report from the Olfactory Clinic of Gunma University by kamata *et al.* (1985), disclosed the following points on prognosis:

(a) As to relations between the foci and degree of the disorders and the rate of recovery, the rate of recovery of the combined type of disorder (23.6%) was nearly twice that of the peripheral type (13.8%). In cases with the combined type of anosmia, rates of recovery as high as about 15% were obtained.

(b) The rate of recovery of the patients who received therapy earlier was found to be higher (about 39%) than that of those who received it later (about 8.5%).

(c) The rate of recovery in the patients who experienced odor sensations during the period from the occurrence of the disease to the start of therapy was twice (21.8%) the rate of the patients who did not have such experiences.

(d) Long-term therapy was required for high rates of recovery from olfactory disorders.

IV. The Olfactory Reception Organ

The prominent central location of the nose makes it a highly distinctive facial feature. The internal structure of the nose comprises two large symmetrical spaces, called the olfactory cavities, separated from each other in the sagittal plane by the olfactory septum. Each cavity is accessible to the external environment via its respective nostril.

While the olfactory receptive organ is referred to as the "nose," the actual organ consists of the olfactory mucosa or epithelium, located in the uppermost part of the olfactory cavity in humans and in the innermost upper portion of the cavity in most other mammals. In either case, the organ proper is protected from the external environment. The olfactory epithelium contains numerous receptor cells which respond to odorous molecules and send olfactory information to higher olfactory areas. The olfactory epithelium occupies only a small region within each olfactory cavity; the remaining area is covered with respiratory epithelium.

Air inspired into the olfactory cavities is warmed and moderately humidified. Particles contained in the incoming air are removed by the hairs of the nostrils and the broad surface of the respiratory epithelium before the air reaches the olfactory epithelium. Thus, the olfactory epithelium is protected from air particles and extremes of heat, cold, and dryness, any of which can damage the receptor cells (Negus, 1958).

A. Structure of the Olfactory Organ

1. The Olfactory Organ in Humans
1.1. Olfactory mucosa
The olfactory mucosa includes the superficially located olfactory epithelium and its underlying lamina propria. The epithelium is composed of olfactory cells, which perform the actual reception, and supporting or sustentacular cells, which lie in parallel with the olfactory cells. Basal cells are found on the basement membrane which divides the olfactory epithelium from the lamina propria. The lamina is chiefly composed of Schwann cells, blood vessels, and olfactory axon bundles.

(a) Location of the olfactory mucosa

The human olfactory cavity is incompletely divided into three small passages by the lateral protrusion of three nasal turbinates: the superior, middle, and inferior meatus (Fig. IV-1). The uppermost part of the olfactory cavity between the septum and the inner surface of the superior turbinate is called the olfactory cleft. This area is covered with pigmented olfactory mucosa, and is therefore called the olfactory region. The location of the olfactory mucosa is shown in Fig. IV-1.

(b) Dimensions of the olfactory mucosa

The area of the human olfactory mucosa has been calculated and two values have been reported. Read (1908) estimated it to be about 6.4 cm^2, while Mateson (1954) determined it to be about 2.4 cm^2. This discrepancy has yet to be resolved.

Kolmer (1927) found that the olfactory mucosa is 480–500 μm thick,

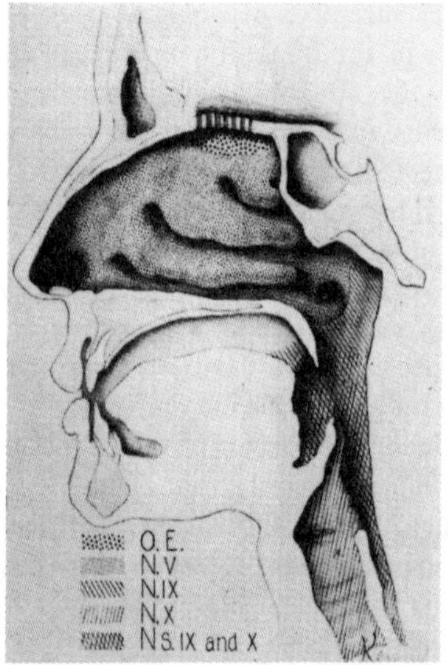

Fig. IV-1. An artist's representation of the primary and accessory areas of olfaction.
O.E.: the area of olfactory mucosa, innervated by the olfactory nerve (I). N.V.: the anterior and lateral portions of the nasal cavity which subserves olfaction, innervated by branches of the trigeminal nerve (V). N.IX.: the upper pharynx which subserves olfaction, innervated by branches of the glossopharyngeal nerve (IX). N.X.: the lower pharynx which subserves olfaction, innervated by branches of the vagus nerve (X). The cross hatched area represents sensory overlap between nerves IX and X. (from Henkin, 1967)

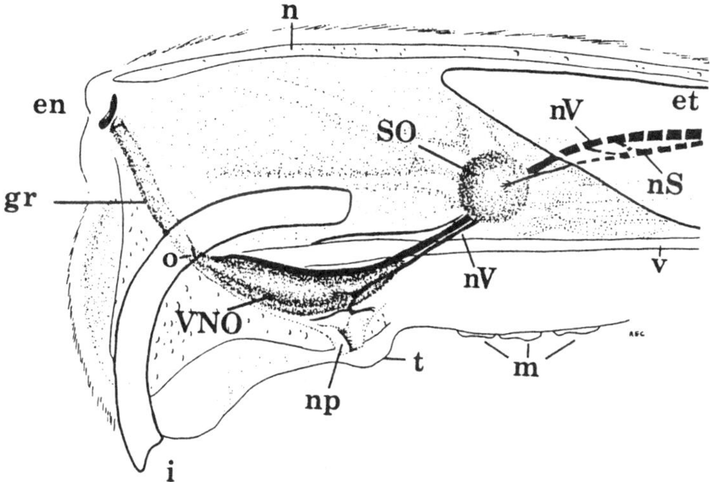

Fig. IV-2. Lateral schematic view of the nose of a guinea pig emphasizing the vomeronasal and septal organs and their relationship with the more caudal olfactory region (clear triangular region to the right). cr: cribriform plate; en: external naris; et: endoterbinates; gr: groove formed by stratified squamous epithelium (non-keratinizing); i: incisor; m: molar; n: nasal bone; np: nasalpalatine duct; nS: septal organ nerve bundle; nV: vomeronasal nerve bundle; o: opening of vomeronasal organ; SO: septal organ of Masera; t: midline tubercle (palatine papilla); v: vomer; VNO: vomeronasal organ. (from Wysocki, 1979)

and observed that thickness varies with the extent of nasal venous blood flow. He also showed that the thickness of the olfactory cell layer is about 30–60 μm. This is consistent with MacDowell's (1948) finding of about 60 μm.

1.2 Olfactory nerve

The olfactory cell becomes an unmyelinated axon at the proximal end. This nerve is called the olfactory nerve, or in the past the "fila olfactoria" (olfactory thread) according to the Jena Nomina Anatomica (JNA), and conveys olfactory information to the olfactory bulb (OB). Details of the anatomy of the olfactory nerve will be explained later (pp.136–137).

1.3. Accessory olfactory nerves

It is well known that the trigeminal nerve plays an important role in the d etection of sharp and pungent odors (Ito, 1968; Tucker, 1963b, 1971). Henkin (1967a) identified two other accessory olfactory nerves, the glossopharyngeal and vagal nerves (Fig. IV-1), and demonstrated that can detect odors in the absence of olfactory nerve function. However, the they dull sensitivities of these accessory nerves provided only minimal odordetection ability, and did not allow differentiation of odor quality (Shirakura *et al.*, 1984, 1985a).

1.4. Extrinsic innervation

Numerous investigators including von Brunn (1892) have confirmed the presence of free nerve fibers in the olfactory mucosa of a variety of vertebrates from fish to mammals (Graziadei and Gagne, 1973). These authors agreed that the free nerve endings were most likely of trigeminal origin. More recent electrophysiological studies have shown that the trigeminal nerve innervates the olfactory epithelium and has a sensory function, as demonstrated by recording electrical responses of this nerve to odorous vapors applied to the olfactory epithelium (Ito, 1968; Chapter VI, pp. 230–231).

The nervus terminalis is a ganglionated nerve that connects the forebrain and the peripheral olfactory structures in vertebrates. It has been proposed that this nerve is chemosensitive, but recent studies in elasmobranchs have not demonstrated sensitivity to chemicals applied to the olfactory epithelium (Bullock and Northcut, 1984; White and Meredith, 1985; Pearson, 1941a, b; 1942).

In humans, the nervus terminalis extends from the region of the nasal septum and projects, with the olfactory nerve, to the OB. Many neurons are distributed along its peripheral course and a ganglion is reported to be centrally located as the nerve runs medially to the olfactory bulb. The central fibers of the nervus terminalis enter the olfactory tubercle and some fibers have been traced into the septal region. They are considered a mixture of sensory and autonomic nerves (Peele, 1977). Further detailed research on the function of this nerve is required, however.

2. The Olfactory Organ in the Majority of Mammals

Many animals depend on olfaction as an essential feature of their sensory capabilities and have a highly evolved olfactory function which is far superior to that in humans (Taniguchi *et al.*, 1986).

2.1. Olfactory mucosa

The structure of the olfactory mucosa has been studied in various animals. In the dog, Okano *et al.* (1967) found olfactory vesicles and extensions of cilia in the olfactory cell that differed greatly from the findings in humans. These histological discrepancies may be related to differences in olfactory acuity. These findings are given further attention later (pp. 113–116).

2.2. Septal organ of Masera (Fig. IV-2)

This organ was discovered in mice relatively recently by Broman (1920) and was more fully described in several animals (guinea pig, rat, mouse, rabbit, and opossum) by Rodolfo-Masera (1943), after whom this organ is named. It has also been called the septal olfactory organ by Adams and McFarland (1971). It is situated bilaterally on the ventrocaudal nasal septum, generally anterior to the nasopharyngeal duct. This organ may therefore be active during periods of rest, sampling the environment at a time when stimuli are not actively brought to olfactory or vomeronasal receptors (Rodolfo-Masera, 1943; Marshall and Maruniak, 1986).

The septal organ has a small neuroepithelium containing Bowman's glands and ciliated receptor cells, both characteristics of the primary olfactory epithelium. Graziadei (1976) compared the ultrastructures of this organ with the primary olfactory epithelium and found that the septal organ epithelium has a smooth endoplasmic reticulum resembling onion rings and a prominent Golgi apparatus, while primary olfactory epithelium possesses no such features. Therefore, Graziadei did not believe that this organ was simply an ectopic area of the primary olfactory epithelium (kratzing, 1978).

The receptor cells are bipolar neurons which send their axons in small bundles to the OB. They show more cholinesterase-positive staining than olfactory nerves and are surrounded by endoneurium, much like vomeronasal fibers. The bundles, usually two per hemisphere, project through the cribriform plate with olfactory nerves and apparently terminate in the OB (Bojsen-Møller, 1975). Using anterograde transport of horseradish peroxidase (HRP), Wysocki et al. (1985) confirmed the above findings and located the terminations of these fibers in the medial and ventral glomeruli, but they did not support the finding that some nerves from the organ may also travel with the vomeronasal nerves (Bojsen-Møller, 1975; Breipohi et al., 1983; Pedersen and Benson, 1986).

2.3. Vomeronasal organ (VNO; Fig. IV-2)

This is an accessory olfactory organ found in the olfactory cavity. It was described in detail in a variety of animals by a Danish physician, Jacobson, early in the 19th century (Cuvier, 1811). This organ now bears his name.

The vomeronasal organ, or Jacobson's organ, consists of bilateral elongated tubular structures, flattened on each side. They are located on the anteromedial one-fifth to one-third of the nasal septum. This differs from the olfactory epithelium, which occupies a dorsocaudal position in the nasal cavity (Wysocki, 1979; Takagi and Shibuya, 1967).

Since the VNO plays an essential role in the lives of lower animals, the structures and functions of this organ will be described in detail in Chapter X (pp. 391–408) (Seifert, 1971a, 1972).

3. Vomeronasal Organ in Humans

This organ has often been reported to be present in the human olfactory cavity (Jordan, 1972). According to Moran *et al.* (1985), the vomero-nasal organ has a 2-mm diameter opening to a dead-end canal which extends approximately 2–3 mm posteriorly beneath the surface of the nasal cavity. Scanning electron microscopy (SEM) reveals a uniform lining of the canal which is unique among epithelia observed to date. The surface of each cell is convex and bears several short microvilli. Transmission electron microscopy (TEM) shows that the epithelium contains a single layer of slender columnar cells which do not resemble either ciliated olfactory receptors or microvillar cells present in the olfactory epithelium of humans.

The presence of a functional Jacobson's organ in humans is a controversial subject. Pearlman (1934) cited Potiquet's (1891) report that the VNO was present in approximately 25% of 200 adults, Anton (1895) reported the organ in four of seven adults, and Pearson (1941a, b; 1942) described the organ in human embryos. In his review, Pearlman (1934) suggested that the organ is present in human embryos, most infants, and many adults. In contrast to these findings, accounts claiming the absence or vestigial presence of the organ have been published (Moulton and Beidler, 1967; Crosby and Humphrey, 1939; Kreutzer and Jafek, 1980).

The presence or absence of the accessory OB has also been a subject for dispute. Pearson (1941a, b; 1942) found it in human embryos ranging in age from 5.5 to 13.0 weeks. Humphrey (1940) also identified it in human embryos, including the oldest surveyed (18.5 weeks), and concluded that the embryonic accessory OBs appeared to be undifferentiated, degenerated, or vestigial, suggesting an absent or nonfunctional VN system in the older fetus or newborn infant. Humphrey and Crosby (1938) conducted neuroanatomical investigations focusing on the VN system and demonstrated the absence of the accessory OB in adults. In fact, the presence of active vomeronasal nerves supplying the accessory OB has been demonstrated only in mammals lower than new world monkeys (Hedgewig, 1980; Wysocki, 1979; also Chapter X, p. 395). It is now generally believed that, in humans at least, adults lack a functional VN system.

4. Development of the Olfactory Organ

The olfactory organ is derived from the epithelial lining of the olfactory pit, which develops from the ectodermal olfactory placode. The development of the olfactory epithelium has been explored to some extent in the chick embryo (Robecchi, 1972; Breipohl and Fernandez, 1977) and the mouse embryo (Cuschieri and Bannister, 1975a, b; Graziadei and Okano, 1978).

Kasuga *et al.* (1978) examined the cellular differentiation of chemoreceptor cells in the olfactory organ primordium. They studied the olfactory placode of the chick embryo (50–53 incubation hours*) and the olfactory pit of the chick embryo (52–69 incubation hours*) with the aid of scanning and transmission electron microscopy. The micrographs revealed a rosette consisting of several swollen oval cellular endings (olfactory vesicles) and other intercalated cellular elements (supporting cells) on the placode surface. As chick embryo incubation continues, the olfactory placode becomes invaginated to form an olfactory pit. The surface of the epithelium lining the pit is organized by the supporting cells and by juvenile olfactory vesicles which possess several budding olfactory cilia. Groupings of several of these juvenile vesicles, which originate from groupings in the rosettes of the placode, are evenly distributed throughout the pit. Vertical sections reveal a pseudo-stratified appearance typical of olfactory epithelium. Kasuga *et al.* (1980) concluded that olfactory cells are differentiated in the epithelial lining of the olfactory pit of the 65–69-hour-old (stage 18*) chick embryo (Breipohi, 1986).

5. *Effects of Odor Deprivation on the Olfactory Epithelium*

The effects of odor deprivation on the olfactory epithelium have been studied by many investigators.

Benson *et al.* (1984) replicated Meisami's (1976) experiment in mice and electrically cauterized one naris of the rat. They found that odor deprivation produced no damage in the olfactory epithelia of the anosmic side one, three, or even 30 days after closure of the ipsilateral naris.

Similarly Farbman *et al.* (1985) cauterized the right naris in developing rats and studied the effect of such sensory deprivation on the development of the olfactory epithelium. A total of 30 counts on six animals were averaged and revealed a ratio of cell density for deprived versus nondeprived sides of 0.82 (S.D. = 0.09).

Pedersen *et al.* (1986b) employed cytochrome oxidase (CO) staining to study olfactory functional organization in rats between five and 30 days of age. On the appropriate postnatal day, rats were removed from the nest and immediately sacrificed. Rats were unilaterally deprived of odor input by occluding one nostril two to 30 days prior to sacrifice. On the control side, the olfactory epithelium was stained, and displayed heterogeneity of intensity in CO staining. In contrast, CO staining was found to be absent in the odor-deprived epithelium. Of particular significance, however, was that even though the OB on the odor-deprived side did not differ from its control in CO staining, the volume of the odor-deprived OB was reduced. Persistence of normal CO staining in the OB, but not

---

* Hamburger and Hamilton, 1951.

in the olfactory epithelium, in the absence of odor input should provide an important insight into the principles of functional organization in the olfactory system (cf. Costanzo *et al.*, 1984; Wysocki *et al.*, 1984).

Effects of odor deprivation or long exposure to an odor on the OB or higher olfactory structures will be discussed in Chapter VII (pp. 281–283).

B. Anatomical Studies on the Olfactory Mucosa

The olfactory mucosa is divided into the following three layers, from the surface inward: (i) a columnar, pseudo-stratified epithelium (olfactory epithelium) containing olfactory, supporting, and basal cells as well as free endings of the trigeminal nerve; (ii) a basal lamina (basement membrane); and (iii) a lamina propria which contains the alveoli of Bowman's glands, Schwann cells, blood vessels, and the olfactory nerve. The lamina propria adheres to the underlying bony or cartilaginous tissue.

1. Olfactory Epithelium

Many investigators have studied the structures of the olfactory epithelia in various animals by light microscope. Schultze (1862) and other histologists studied the olfactory epithelium of the human. This work revealed that the olfactory epithelium in vertebrates is composed of olfactory, supporting, and basal cells, and free nerve endings of trigeminal origin, as mentioned above (Kolmer, 1927).

Electron microscopy later disclosed the ultrastructure of the olfactory epithelium. The first observations in humans were performed by Bloom and Engstrom (1952) and those in amphibia by Bloom (1954). This work was continued by many investigators studying various animals. Okano *et al.* (1967) examined the olfactory epithelium of the dog. Moulton and Beidler (1967) provided a review of studies on the peripheral olfactory system published up to that time. Steinbrecht (1969) compared the fine structure of olfactory receptors among many species of vertebrates and insects (Fig. IV-7, see also Loo, 1977a).

The thickness of the olfactory epithelium varies from species to species. It is generally thin in warm-blooded animals——for instance, 60 μm in humans (Kopsch, 1912)——but can be as thick as 150–200 μm in cold-blooded animals, such as frogs and turtles. The nuclei of the three cellular elements are arranged sequentially in separate layers from top to bottom. Thus, the nuclei of the supporting cells are organized with a zone of columnar nuclei on top, a zone of oval nuclei (of the olfactory cells) in the middle, and a zone of basal cell nuclei at the bottom. The surface of the olfactory epithelium is normally covered with olfactory mucus, secreted by supporting cells and Bowman's glands.

Most of the author's research on the ultrastructure of the olfactory epithelium was performed on the bullfrog. Therefore, unless specified otherwise, the description that follows will refer to the olfactory epithelium of the bullfrog.

1.1. The olfactory cell

The olfactory cell is generally accepted to be a bipolar neuron. The cell body is fusiform, or columnar, with its long axis perpendicular to the

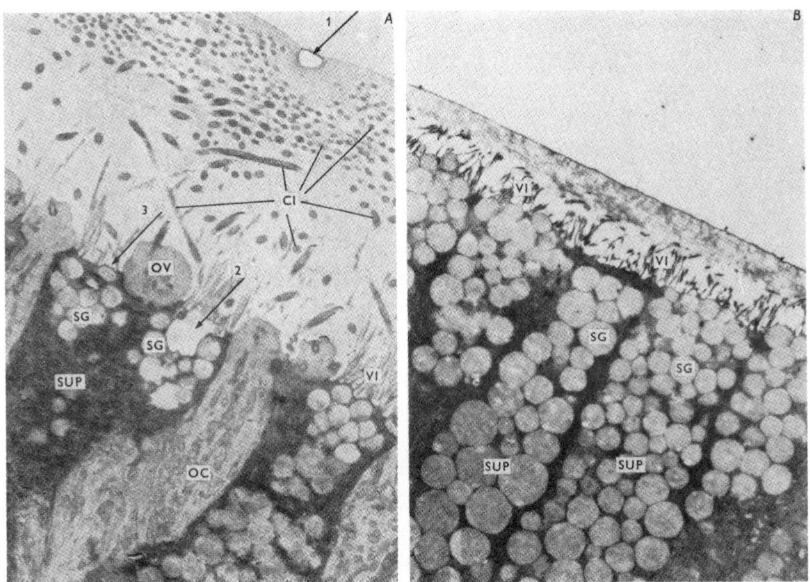

Fig. IV-3. A. An electron micrograph of the normal olfactory epithelium of the bullfrog. Olfactory cells (OC) have olfactory vesicles (OV) and long olfactory cilia (CI). They are evenly interposed among the supporting cells (SUP), which are enriched with numerous microvilli (VI) and secretory granules (SG). Arrow 1: a crypt-like structure of unknown nature on the surface of the mucus layer. Arrow 2: a pit-like structure that suggests a vestige of liberated secretory granules. Arrow 3: a secretory granule that is just about to be spontaneously liberated into the mucus layer. B. An electron micrograph of the degenerated olfactory epithelium without an olfactory cell. The supporting cells (SUP) are full of the secretory granules (SG), and are provided with typical microvilli (VI). The surface of the epithelium is covered with shallow mucus. × 6000 (from Okano and Takagi, 1974)

epithelial surface. Although it is the same height as a supporting cell, the olfactory cell is much more slender overall, except for the perikaryon which is rounded due to its spherical nucleus. A single surface-reaching dendrite arises from the apical pole of the cell body and extends to the surface. The basal pole gives rise to an axon.

Figure IV-3 is an electron micrograph of the upper half of the bullfrog olfactory epithelium. The olfactory vesicles, their cilia, and part of a so-called olfactory dendrite are shown, but oval nuclei, perikarya, or axons of olfactory cells cannot be seen. The border area between the olfactory vesicles and the olfactory dendrite is narrower in width and appears to be anchored by the invagination of the adjacent supporting cells and the terminal bars.

(a) Olfactory cell types

The olfactory cell has generally been regarded as one of the morphologically homogeneous columnar cell types (Schultze, 1856). In addition to this standard columnar type of olfactory cell, however, Dogiel (1886) found spindle-shaped and conical olfactory cells in fishes and frogs and named them "Riechstäbchen" (olfactory rods) and "Riechzapfen" (olfactory cones), respectively. Jagadowski (1901) later confirmed the coexistence of these three olfactory cell types. Recently, Muller and Marc (1984) demonstrated three olfactory cell classes in fish. Andres (1966) did not clarify whether these various olfactory cells were really different kinds of receptor cells or were simply the same kind of olfactory receptor cell at different stages of development.

As will be discussed later (pp. 124–126) different types of olfactory receptor cells have been reported in the lizard, duck and minnow (Steinbrecht, 1969). The "fourth" and "fifth" cell types were identified in the dog by Okano et al. (1967), and in the cat by Andres (1969) and Okano (1973). These cells featured microvilli with or without cilia. Receptor cells with microvilli have also been found in human olfactory mucosa (Moran et al., 1982). It is interesting that olfactory receptor cells were classified into several types, because supporting cells, which had long been considered to be homogeneous, were recently differentiated into two types (Rafols and Getchell, 1983a; also pp. 128–129).

(b) Number of olfactory cells

Allison and Turner-Warwick (1949) counted olfactory cells in the rabbit, and found 120,000 cells per mm^2 of olfactory epithelial tissue. A later study in the rabbit identified 127,000 olfactory cells per mm^2 (Clark, 1956). The area of olfactory tissue was found to be 4.5 cm^2, giving a total of approximately 50 million olfactory cells on each side. A rabbit would thus have about 100 million olfactory cells altogether. Barber and Boyde (1968), however, using a scanning electron microscope, counted only about

5,000 olfactory cells per mm² in the rabbit. This dis crepancy may arise from different counting methods.

In the pig, Gasser (1956) found 60,000 olfactory cells per mm². In the dog and cat, Andres (1969) counted 5,000–10,000 cells per mm² of olfactory epithelia. Since the number of olfactory cells equals the number of olfactory nerve fibers, Matsuzaki *et al.* (1980a, b) simply counted olfactory cell axons in sections of the olfactory nerve. From this they determined the number of cells and found about 3.9 million cells in the bullfrog, 8.5 million cells in the snapping turtle, 7.5 million cells in the four-toed tortoise (*Testudo horsfieldi*), 1.3 million cells in Reeve's turtle (*Geoclemys reevesii*), 5.8 million cells in the duck (*Anas platyrhynchos*), and some 3 million cells in the pigeon (*Columba livia*). More recent and detailed studies have been made by Hatanaka *et al.* (1982) on Reeve's turtle and by Matsuzaki *et al.* (1982) on the pigeon. Kanda *et al.* (1973) counted 80,000–100,000 cells per mm² in the guinea pig. In contrast, they found as few as 30,000 olfactory cells per mm² in humans. In his book Takahashi (1979) stated that humans have only about 10 million olfactory cells whereas dogs have 220 million.

(c) Perikaryon

Perikarya of the olfactory cells are situated at various depths from the surface of the olfactory epithelium. Some are located only 10 μm from the surface, but others reside up to 80 μm under the surface. The nuclei in the olfactory epithelium are not uniform, varying in diameter from 5 μm to 13 μm. The diameter of the olfactory cell is great in instances in which the perikaryon contains the nucleus. Kanda *et al.* (1973) thus found the diameters of the human olfactory cells to be as large as 8 to 10 μm. This is comparable in size to the olfactory cell of the bullfrog.

The nuclei and perikarya of olfactory cells resemble those of neurons in the central nervous system, although there is less development of Nissl bodies.

(d) Olfactory dendrite (Fig. IV-3)

A slender process extends from the perikaryon toward the surface of the olfactory epithelium and ends as a terminal swelling on the surface. The dendrites vary in length from slightly over 10 μm to 90 or 100 μm, depending on the depth of the perikarya. Olfactory dendrites extending toward the epithelial surface are enveloped by supporting cells in some cases, but sometimes reach the surface without touching any supporting cells, making contact with dendrites of neighboring olfactory cells instead.

Olfactory dendrites have many microtubules running along the major axis. In young dendrites, numerous centrioles migrate distally, eventually reaching the olfactory vesicles where they become basal bodies from which olfactory cilia sprout.

(e) Olfactory vesicle (Fig. IV-3)

The distal end of a dendrite protrudes 2–4 μm above the free surface and forms a small spherical swelling, called an "olfactory vesicle" (Graziadei, 1971), "olfactory rod" (De Lorenzo, 1970), "olfactory knob" (1971), "Riechkoepfe" (Seifert, 1968), or "dendritic bulb" (Yamamoto, 1976). In mammals, the olfactory vesicle has a rod-shaped appearance with very few microvilli on the surface (Frisch, 1967; Okano *et al.*, 1967).

Okano (1981) found that organelles in the vesicle were similar to organelles in the dendrite in that many mitochondria and microtubules were present, but multivesicular bodies were found only infrequently. Furthermore, root filaments and basal footlets were often poorly developed. In the frog, olfactory cilia, but not microvilli, extend from the surface of the olfactory vesicles through the mucus layer (Fig. IV-3,7). In pigeons, and ducks, on the other hand, 50 to 60 short microvilli develop and completely cover the surface of the olfactory vesicle (Graziad and Bannister, 1967; Okano, 1981).

(f) Olfactory mucus (Fig. IV-3)

The free surfaces of the olfactory and supporting cells are always covered with olfactory mucus. The mucus is composed of secretions from the supporting cell (Okano and Takagi, 1974; also Chapter V, pp. 176–178), from the Bowman's gland, and possibly from other glands, when they are present (Getchell *et al.*, 1984a, b; 1985a, b; 1986). Mucus from supporting cells is considered to be viscous whereas mucus from the other glands is thought to be serous, although this supposition as yet lacks adequate experimental verification.

The surface of the mucus is covered with a film of high electrical density. This has been referred to as a "mucus film" (Okano, 1973), "Deckhautchen" (Seifert, 1968), or "terminal film" (Andres, 1969). Immediately beneath this film, distal segments of the olfactory cilia gather and form a thick mat. Each cilium in the mat is thought to have receptive sites sensitive to odorous molecules.

From a functional standpoint, the following three roles have been considered: (1) The mucus is generally believed to protect the olfactory epithelium from dryness and extremes of temperature as well as from particulated and pathogenic contamination. (2) The mucus supplies ions necessary for the normal activities of olfactory and supporting cells. As will be shown later in Chapter V (pp. 167–170 and pp. 173–176), EOGs are elicited by Na^+ or Cl^- entry and K^+ exit (Takagi *et al.*, 1968a, b, 1969 a, b Okano and Takagi, 1974). Ca^{++} is also important in maintaining homeostasis in the cell membrane (Takagi *et al.*, 1968a). All of these ions are conti nuously available from the mucus. (3) Secretions from olfactory glands appear to affect accessibility of odorants to the receptive membrane of the olfactory cell (Getchell and Shepherd, 1978b; Hornung and Mozell, 1981; Getchell and Getchell, 1984b; Bannister, 1974; Moulton, 1976).

The thickness of the mucus layer may vary from region to region in the same animal as well as from animal to animal. Hopkins (1926) estimated that the thickness of the mucus layer ranges from 10 to 60 μm. The thickness seems to be about 5 μm in dogs and cats, more than 10 μm in birds (Okano, 1981), and 35 μm (Reese, 1965) or 30 μm (Holley and MacLeod, 1977) in frogs. According to Getchell *et al.* (1984), the mucus consists of a superficial watery layer derived from Bowman's glands estimated to be 5 μm high, and a deeper, more viscous layer derived from supporting cells estimated to be 30 μm thick in lower vertebrates such as the frog and salamander (Reese, 1965).

More precisely, the mucus layers of the bullfrog (10–15 μm in depth) were divided into the following four portions by Okano and Takagi (1974):

i) That portion which bathes the olfactory vesicles and microvilli is about 3 μm, and corresponds to the "inner layer" of Andres (1969).

ii) That part where the cilia extend straight out (5 μm in length), which corresponds to what Reese (1965) calls the "proximal segment."

iii) Just underneath the mucus surface, many cilia run horizontally in parallel with the mucus surface, piling up, and forming a seemingly rigid mat about 3 μm in depth. This part corresponds to the "distal segment" of Reese (1965) and to the "outer layer" of Andres (1969).

iv) The mucus proper, about 1 μm in depth, forms the surface of the mucus, bathes no special structures, and corresponds to the "dense terminal film" of Andres (1969).

These findings coincide closely with those obtained in the frog (*Rana esculenta* and *Rana pipiens*) by Bloom (1954), Reese (1965), and Yamamoto *et al.* (1965). In addition, Okano and Takagi (1974) discovered an entirely new crypt-like structure in the mucosal surface (arrow 1 in Fig. IV-3). This crypt can be followed into the depths of the mucus through serial sections. Although a continuity between the crypt and the secretory duct of Bowman's gland has been suspected, no positive evidence for this view, or for any relationship between the crypt and other cells, has been uncovered. At present, its function remains a mystery.

(g) Olfactory cilium

Numerous olfactory cilia extend from the olfactory vesicle. In the frog and the chicken they extend 8–10 μm obliquely, or nearly vertically, and gradually curve to run radially and horizontally as they approach the mucosal surface. In mammals, on the other hand, the olfactory cilia do not extend vertically toward the mucosal surface, but extend radially and horizontally from the start (Okano, 1965; Okano *et al.*, 1967).

In the frog, there are four to six olfactory cilia for each olfactory vesicle (Hopkins, 1926). These include long, 80–200-μm immotile cilia (stereocilia) and short, 25–50-μm motile cilia (kinocilia; Reese, 1965). For comparison, the rabbit olfactory cell has 12 to 14 cilia with lengths of 1 to 2 μm (Clark, 1956), and the olfactory cell of the guinea pig has six to eight cilia (Kanda *et al.*, 1973) with lengths of 2 to 3 μm (Arstilla and Wersäll, 1967). Markedly different from these mammals, cats have 10 to 25 olfactory cilia per cell with lengths of about 80 μm (Andres, 1969), and dogs have 100 to 150 cilia per cell with lengths exceeding 100 μm (Okano, 1965). On the basis of morphometric examination, Seifert (1970) deduced that olfactory cilia in a number of macrosomatic mammalian species are 50–60 μm long, a figure judged by Menco *et al.* (1978) to be the most accurate appraisal available. Data on human olfactory cilia will be presented later (pp. 137–139).

It is generally accepted that mature olfactory cilia are immotile. The role of ciliary movement in olfactory transduction has often been discussed, but remains a controversial issue (Adamek *et al.*, 1984; Lidow and Menco, 1984; Mair *et al.*, 1982; also Chapter VI, pp. 191–193).

Ultrastructures of the cilia have been well studied. Menco's excellent reviews (1977, 1983, 1984) have illuminated the features of the olfactory cilia. When viewed in cross-section near its base (in the proximal segment), an olfactory cilium is found to be 0.25 μm in diameter, exhibiting a pattern of nine peripheral microtubular doublets encircling two central microtubular subfibers. This arrangement disappears and the tubules decrease in number as the distal segment is entered and followed out to the tip. The transition between the proximal and distal segments of a cilium may be sharp and marked by a sudden decrease in diameter from 0.25 to 0.15 μm (Reese, 1965; Seifert and Ule, 1967; Thornhill, 1967). Near the tip, the diameter may taper to 0.06 μm and contain only two tubules (Andres, 1966; Okano, 1965; Okano *et al.*, 1967; Seifert and Ule, 1967). The length of the distal segment accounts for 80% of the ciliary length in the frog (Reese, 1965) and about 98% in the cat (Andres, 1969).

Menco *et al.* (1978) calculated the sensory membrane surface area of the olfactory epithelium in the cow to be about 20 cm^2 of sensory membrane surface per cm^2 of epithelium. Thus, as Graziadei and Bannister (1967) had previously shown, the olfactory cilia provide vast increases in receptor surface area. This device is thought to be essential for elevating the sensitivity of olfactory cells to the very high levels seen in many macrosomatic mammals.

(h) Mucociliary transport

In Dr. Beidler's laboratory in Tallahassee, the author made light microscopic observations of the olfactory mucosal surface of a land tortoise, *Gopherus polyphemus*. The mucus was seen to be flowing smoothly, rapidly transporting minute particles toward the internal nares. This observation remains fresh, because the author has not since found such a clear example of highly efficient mucus flow. Needless to say, this mucus flow is essential for the clearance of odorous molecules, dust, and other extrinsic particles from the mucus. Mucus flow, desorption into the air, and uptake into the circulatory system are the three major processes by which odorous molecules are removed from the olfactory epithelium (Hornung and Mozell, 1977).

Estimates of rates of mucociliary transport appearing in the literature range from 10 mm/min in higher animals (Moulton and Beidler, 1967), to 60 mm/min in the frog (Reese, 1965) and in the salamander (Getchell, 1984). Studies on human mucosa have found that the rate of mucociliary transport is 9.0 mm/min with a range of 5.8 to 13.5 mm/min (Karja *et al.*, 1982) and 44.8 mm/min with a range of 0 to 23.6 mm/min (Anderson *et al.*, 1972).

Interested readers are directed to the excellent review by Getchell *et al.* (1984b) for further information (Rossman *et al.*, 1984).

(i) Olfactory receptive site

Information about odors is thought to be received on the ciliated surface of the olfactory vesicle. It is therefore assumed that many receptive sites are distributed on the surface. Odorous molecules that enter the mucus layer from the air may reach deeply situated olfactory vesicles, but they are far more likely to contact the olfactory cilia that form a thick mat just below the interface between air and mucus (Fig. IV-3). For this reason most odorous molecules are adsorbed by the very broad surface membrane of olfactory cilia. Adsorption is believed to be the first step in chemo-electrical transduction at the receptive sites of the olfactory cell.

Research on the receptive sites of other sensory receptors has advanced well ahead of that on olfactory receptors. The rhodopsin receptors of the retina (Jan and Revel, 1974) and the acetylcholine receptors of skeletal muscle (Hartzell and Fambrough, 1972) and the electroplax organ (Orci *et al.*, 1974) have been visualized in freeze-fractured preparations by electron microscopy. These receptors were found to be related to particles 8 to 12 nm in length, embedded in the surface membranes of the sensory cells (Menco *et al.*, 1976).

Using a similar freeze-fracture technique, Menco *et al.* (1976) found such intramembrane particles (IMPs) in the surface membranes of bovine olfactory vesicles and cilia (Fig. IV-4). Similar membrane particles were found in mouse olfactory cilia by Kerjaschki and Hörandner (1976) and in newt olfactory cilia and cells by Usukura and Yamada (1975).

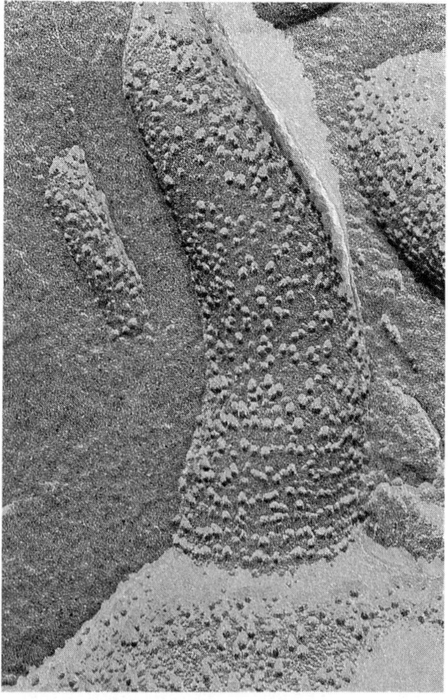

Fig. IV-4. Membrane features of an olfactory cilium. Numerous intramembranous particles and part of a ciliary necklace (circular strands at the base of the cilium) are visible on the exposed P-face, which is the fracture plane directed toward the cytoplasm. × 100,000. (courtesy of B.P.M. Menco, 1988)

Freeze-fractured olfactory epithelia in the reports of Menco and his colleagues (Menco *et al.*, 1976; Menco, 1980b) displayed intramembranous particles 8 to 11 nm across, in abundance on the olfactory cilia and over the whole of the exposed sensory surface of the dendritic ending. Particle densities were the same on the proximal parts of the cilia, on the long narrow distal parts of the cilia, and on the dendritic endings themselves, i.e., about 1,000 to 1,500 particles/μm^2 for the membranous P-faces (fracture plane containing membrane monolayer directed toward cytoplasm). In contrast, the cilia of the neighboring nonsensory respiratory epithelium bore few membrane particles, 200–300 particles/μm^2. Particle densities at the E-faces of the membranes (fracture plane containing membrane monolayer directed toward exterior of cell) of either the olfactory or sensory cilia were about 20 to 30 % of those of the P-faces.

From measurements made on freeze-fracture transmission and scanning electron micrographs, Menco (1980a, b, and personal communication) calculated that each sensory ending had a surface area of 300–1,000 μm^2, bearing approximately 0.5×10^6 to 10^6 particles pooled over both fracture faces. Five to eight million endings were situated on each cm^2 of epithelial surface. Allowing for the volume occupied by receptor and supporting cell processes, the concentration of particles within the confines of the 6-μm (mammals) to 30-μm (frog) layer of mucus at the epithelial surface was estimated to be 0.2 (frog) to 3×10^{19} (mammals) particles per liter, representing a molarity of 3 (frog) to 25 μM (mammals), if each particle is assumed to be a single molecule.

The rather high packing density in the olfactory vesicles and cilia is a feature shared with other chemoreceptor systems such as the acetylcholine receptor membranes and also membranes of the discs of outer segments of rods and cones in the retina, both of which have even higher particle densities than olfactory vesicles and cilia. Unlike the situation in those two systems, where particles are known to represent acetylcholine receptor molecules and rhodopsin, respectively, in olfaction it is not yet known whether the particles represent receptor molecules and/or also transducer molecules, such as channel proteins. Assuming that the former is the case, it can be speculated that the presence of the intramembrane particles in such high density could be associated with the sensitivity of sensory cells to low concentrations of odorants, and also with the hypothesis that many classes of receptive sites, responding to different odorants, may be present on individual sensory cells (Gesteland *et al.*, 1965; Menco *et al.*, 1976).

In a recent review, Menco (1983) cited many more findings supporting the view that intramembrane particles of olfactory sensory endings are involved in odor-receptor interactions, which are summarized in the following:

i) Particle densities in these endings and the electrophysiological response to odors decrease simultaneously as a result of prolonged fasting (Holley and MacLeod, 1977; Masson *et al.*, 1977).

ii) Deciliation and subsequent regeneration of cilia are closely followed by disappearance and reappearance of the summated receptor potential (Bronshtein, 1977; Cancalon, 1980; Gesteland *et al.*, 1980; Shibuya and Takagi, 1963).

iii) Particle densities in sensory endings show dramatic increases during dendrite maturation (Kerjaschki, 1977; Menco, 1988).

iv) Insect olfactory cilia display higher particle densities in regions underneath antennal exoskeletal hair pores than in lower regions of those cilia. Densities underneath these pores are in the same range as the densities of vertebrate olfactory cilia (Menco and Van der Wolk, 1982; Van der Wolk and Menco, 1980; Menco and Van der Wolk, 1980; Steinbrecht, 1980a, b).

v) Nonsensory respiratory cilia (Menco, 1977, 1980b; Kerjaschki and Hörandner, 1976; Kerjaschki, 1977; Menco *et al.*, 1976; Menco, 1980a; Usukura and Yamada, 1978; Carson *et al.*, 1981) and mechanosensory cilia (Jahnke, 1976) display considerably lower intramembrane particle densities in their P-faces than olfactory sensory endings and cilia. This makes it clear that the higher particle density of olfactory cilia is apparently not required for motility. Furthermore, other types of cilia, in particular kinocilia, usually have lower particle densities (Satir, 1977; Boisvieux-Ulrich *et al.*, 1977, 1980; Inoue and Hogg, 1977).

vi) Various binding studies suggest that olfactory receptor processes involve proteinaceous moieties (Den Otter, 1981; Getchell and Gesteland, 1972; Menevse *et al.*, 1977a; Delaleu and Holley, 1980; Ma, 1982; Persaud *et al.*, 1981; Shirley *et al.*, 1981). The intramembrane particles, abundant in olfactory receptor structures of vertebrates and insects, could in part represent such proteinaceous moieties (also Lancet, 1986). Nevertheless, some of the particles may represent lipidic rather than proteinaceous entities (Verkleij and Ververgaert, 1978; Verkleij *et al.*, 1979), since lipidic entities are also important in the olfactory reception mechanism (Punter *et al.*, 1981).

Usukura and Yamada (1979) provided further evidence on the characteristic features of the olfactory receptive membrane. They fixed the olfactory and respiratory epithelium in a solution containing ruthenium red and found that the membranes of the olfactory vesicle and olfactory cilia therefrom were as thick as 20 nm, and that the outer membrane was thicker than the inner one. In contrast, the dendrites of olfactory cells surrounded by supporting cells had no such characteristics. The membranes of the supporting cells and the cells of the respiratory epithelium were as thin as 8 nm and showed little difference in thickness between outer and inner membranes.

After their initial studies on intramembrane particles using the freeze-fracture technique, Menco *et al.* (1976, 1980) modified their techniques in several ways and reexamined their data in light of older findings (Menco, 1980a, b, 1983, 1984; Lidow and Menco, 1984).

Goldberg *et al.* (1979) tried a different approach to clarify the distribution of receptive sites. They reported that the olfactory responses of mice can be blocked by topical application of an antibody directed against what appear to be olfactory receptor sites. Since the molecular weight of the antibody was 160,000, the effect was probably not on a site inside the cell or beyond the tight junction between cell apices in the epithelium. Their study provided evidence that olfactory receptive sites lie no deeper than the apical surfaces of the cells, which were 30 to 50 μm below the mucus surface. Thus, these data taken together with other lines of evidence (Tucker, 1967; Bronshtein and Minor, 1977) indicate that the olfactory receptor sites are located on the cell membranes of the receptor cells in the region of the apical knob of the dendrite and the proximal portion of the cilia (Getchell *et al.*, 1980).

Finally, studies on the molecular weight of presumed receptor molecules for odorants must be referred to Fesenko *et al.* (1979), who reported a molecular weight of approximately 120×10^3 for the camphor binding molecule, and Gennings *et al.* (1977), who reported a value of 23×10^3 for a porcine pheromone binding molecule. There was an approximate correspondence between the diameter of presumed receptor molecules (80 to 100 Å, Fesenko *et al.*, 1979) calculated from experimentally determined molecular weights and the diameter of IMPs (7 to 11 nm; Menco, 1977; Miragall, 1983). Menco (1977, 1983) and Chen and Lancet (1984) have proposed that the IMPs described in freeze-fracture studies correspond to receptor molecules with high dissociation constants (K) determined from binding studies.

(j) Degeneration of olfactory cells

According to Clark (1957), Nagahara (1940) may have been the first histologist to demonstrate degeneration and regeneration of the olfactory cell in the olfactory epithelium of the mouse. Finding that some of the receptors remained after complete removal of the OB, he claimed that these "resting" elements could by mitotic proliferation lead to a reconstitution of olfactory epithelium of normal appearance. Several experiments were then conducted to address this possibility; but the outcome of these studies was largely equivocal (see a comprehensive review by Takagi, 1971).

Takagi and Yajima (1964, 1965) sought to clarify the relation between histological and electrophysiological change in the olfactory epithelium after sectioning the olfactory nerve. After axotomy in the bullfrog, olfactory cells decreased in number and seemingly disappeared. As a result, an interstice was formed between the slender nuclei of supporting cells above and the round nuclei of basal cells below (Fig. IV-5). Correspondingly, the electronegative EOGs to odors (Chapter V, p. 182) decreased in amplitude and eventually disappeared. These experiments were later duplicated and the same process of degeneration was confirmed in the salamander (Simmons and Getchell, 1981a, b). Further details of these experiments will be covered later (p. 182).

Several investigators have sectioned the olfactory nerve in the frog (Exner, 1877; Hoffmann, 1867; Nakamura, 1916); and a similar study has been carried out in fish (Quastler and Weingarten, 1930). Numerous

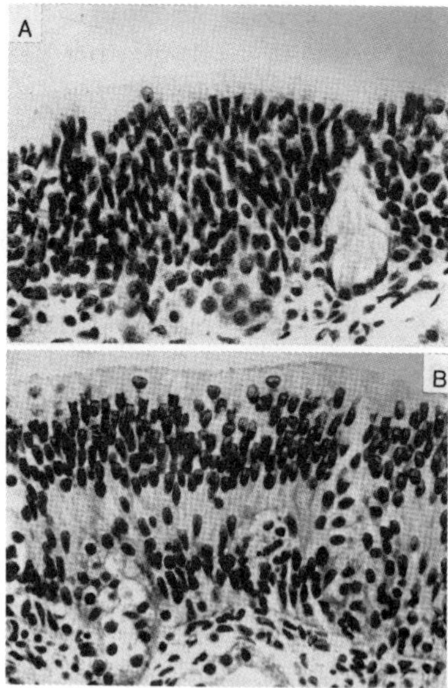

Fig. IV-5. Degeneration of olfactory cells from A to B after olfactory nerve section. (from Takagi and Yajima, 1965)

studies have been conducted in rabbits in which the olfactory bulb was totally or partially ablated (Baginsky, 1894; Clark, 1951; Clark and Warwick, 1946; Exner, 1877; Früwald, 1935; Haffman, 1867; Lustig, 1924). Takata (1929) and Sen Gupta (1967) did the same experiment in the rat. It is important to note that none of these experiments showed regeneration of the olfactory cells. In fact, most of these studies found degeneration of all the receptor cells, but a few reported varying degrees of receptor cell survival. In this regard, it might be recalled that retrograde degeneration of the neuron is generally considered to occur more severely the closer the axon is sectioned to the cell body, and that the neuron is destroyed completely when the sectioning is performed in the immediate vicinity of the cell body. Consequently, the survival of neurons observed in some experiments could be attributed to this phenomenon.

It is significant to note that the degenerated epithelium was found to be replaced by nonsensory epithelium. Exner (1877) in the frog, and Smith (1938) in the rat, found that nonciliary columnar epithelia replaced the olfactory epithelia. Takata (1929) observed that ciliated respiratory epithelia supplanted the olfactory epithelia in the rat. Furthermore, it is well known that mammals are frequently affected by acute or chronic rhinitis and that the affected areas are replaced or repaired by nonsensory epithelium in many cases. It is therefore essential to determine whether or

not such changes occurred solely due to the experimental procedures.
(k) Regeneration of olfactory cells
The concept of regeneration of the olfactory epithelium and cell after degeneration was already confirmed by some indirect evidence many years before experimental studies were performed on animals. Epidemic encephalitis viruses were, at one time, presumed to enter the nasal cavity and invade the brain along the olfactory nerve, causing epidemic encephalitis. The author heard from his father, a pediatrician that for many years before and after World War I as a prophylactic against this disease the olfactory cavity was irrigated with zinc sulfate solution to effect degeneration of the olfactory epithelium and nerve, thereby preventing viruses from entering the brain. Although the zinc sulfate solution would destroy the olfactory epithelia, these patients never complained of chronic anosmia after these treatments. This was considered to be entirely due to the recovery of the olfactory epithelium to its original state, completely restoring olfactory sensitivity (Takagi, 1971).

Later, experimental bases for the regeneration of olfactory epithelium after treatment with zinc sulfate were provided by Smith (1951) in the frog and by Schultz (1941) in the monkey. Later, Schultz (1960) found in the monkey that, as in the frog (Smith, 1951), the degenerated olfactory epithelium separated from the lamina propria and a single layer of new flat cells covered the lamina. Soon after the tenth day, normal-looking olfactory cells with dendritic processes began to appear. They became numerous in most areas by the 20th day, and by six months to a year after the treatment, the general structure of the epithelium was found to be essentially normal. These histological findings agreed with the observation that the resistance provided by an intranasal irrigation with zinc sulfate solution a day or two prior to intranasal inoculation with a given neurotropic strain of poliomyelitis (MV strain) was usually followed in three or four months by a loss of resistance to intranasal inoculation. The fact that the initial resistance was transient allowed us to presume that restoration of the destroyed olfactory epithelium had taken place, even before the recent definitive experimental studies had been conducted (Takagi, 1971).

Since 1970, many histological experiments have been reported and provided evidence for post-axotomy regeneration of the olfactory cell (Oley et al., 1975; Graziadei, 1973b, 1975, 1977; Graziadei and DeHan, 1973; Graziadei and Metcalf, 1971; Graziadei and Monti Graziadei, 1978b, 1979, 1980, 1986a; Graziadei and Okano, 1979). For instance, Graziadei and Monti Graziadei (1978a) observed the process of degeneration and regeneration in the pigeon after sectioning the olfactory nerve. The axotomy resulted in an "olfactory epithelium" devoid of olfactory cells, but at the

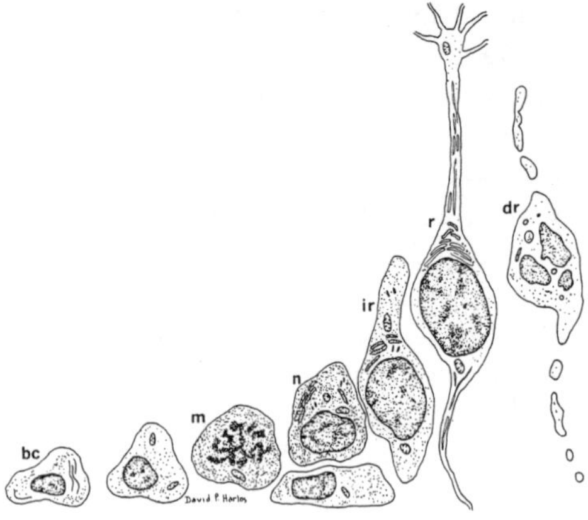

Fig. IV-6. A scheme to illustrate the process of neuron differentiation, maturation, and senescence. The staminal or immature cells (bc) are close to the basal lamina (b), and correspond to the so-called basal cells. The first sign of differentiation is an increase in cell volume and appearance of a clear cytoplasmic matrix and numerous free ribosomes. These cells are often seen undergoing a mitotic process (n and m). Immature receptors (ir) maintain the characteristics of (n) cells but a stumpy dendritic process appears where many basal bodies are localized. The mature neurons (r) have a characteristic bipolar appearance with a ciliated dendrite and a fine unmyelinated axon. Senescence figures are also seen in the epithelium and appear as morphologically deteriorating receptors (dr) which are eventually phagocytosed by macrophages penetrating into the epithelium. (from Graziadei, 1977)

same time, mitotic division of the basal cells became active. Olfactory cells started to regenerate six to 10 days after the section. Newly generated olfactory cells, gradually extending their dendrites and axons, soon took the form of mature cells. The perikarya of cells migrated from the basal to the middle region, and many round nuclei were seen in the middle layer of the epithelium about 10 to 18 days after the operation (Fig. IV-6). By 30 to 50 days after the operation, regeneration of the degenerated olfactory epithelium became so complete that it became difficult to distinguish between the olfactory epithelia of the treated and control sides.

Takagi and Yajima's study (1964, 1965) initiated a series of several axotomy experiments using electrophysiological and histological techniques in parallel. Thus, Kiyohara and Tucker (1978) in pigeons, Burns *et al.* (1981) in the frog, Simmons and Getchell (1981a, b) and Simmons *et al.* (1981) in the salamander, and Samenen and Forbes (1984) in hamsters confirmed the recovery of electrical responses to odors corresponding with the regenerating olfactory epithelium (Chapter V, pp. 182–183, Booth *et al.*, 1981).

Harding *et al.* (1977) studied biochemical changes in the olfactory epithelium and bulb after olfactory nerve section. They measured three biochemical markers of the olfactory neurons: carnosine, carnosine synthetase activity, and the olfactory marker protein (OMP; pp. 139–140, p. 189 and p. 214). These markers decreased rapidly in both olfactory tissues, reached a minimum by the end of the first week after surgery, and then slowly returned to 80% of the control values by one month. Nevertheless, carnosinase activity in the epithelium was essentially unaffected. This may be a very important finding. These biochemical observations coincided temporally with the onset of degenerative changes seen morphologically in both the olfactory epithelium and bulb. Because of the temporal coincidence of these observations, Harding *et al.* suggested that they are manifestations of the same process.

(l) Turnover of olfactory cells (Fig. IV-6)

Olfactory cells degenerate and regenerate not only in the artificially or pathologically damaged olfactory epithelium, but also, in the normal olfactory epithelium under ordinary conditions. Andres (1966, 1969), studying the olfactory epithelia of the rat, cat, and dog, found that rims of olfactory vesicles were disintegrating, as if eroded, while young neurons were simultaneously developing among supporting cells. He further noted that young neurons were light and showed poor development of organelles, and that the dendrites, which had not yet formed olfactory vesicles, contained many centrioles. The co-existence of undeveloped blastema cells together with mitotic cells in the basal cell layer led him to propose that the basal cells were destined to be differentiated into olfactory cells.

The most decisive experiments to prove that the basal cell is the precursor of the olfactory cell were performed using electron-microscopic autoradiography. Studies by Graziadei and Metcalf (1971) and Graziadei (1973b) followed the process whereby a basal cell develops into an immature neuron and then finally to a completely differentiated olfactory cell (Fig. IV-6). Graziadei (1973a) called the basal cell a "staminal cell." Yamamoto (1976) examined the olfactory epithelium of the bat and obtained micrographs demonstrating the turnover of the olfactory cell to be a continuously active process. Turnover times in the mouse were estimated at 30 days for receptor cells and 8 to 16 days for basal cells, based on experiments with colchicine and tritiated thymidine (Moulton, 1974, 1975).

Turnover of olfactory cells in normal epithelium, i.e., the adult pattern of olfactory cell kinetics in which mitosis takes place in the basal layers and the cells then migrate toward the epithelial surface (Thornhill, 1970; Graziadei and Metcalf, 1971; Moulton and Fink, 1972; Graziadei, 1973b), differs from the pattern seen in prenatal animals. Smart (1971) demonstrated that during fetal life olfactory epithelial cells undergoing division migrate toward the surface, aggregate, and divide. Daughter cells then move back into deeper layers of the epithelium. This phenomenon has also been identified in the vomeronasal organ, where mitotic forms have been observed in all regions of the organ (pp. 396–397; Cuschieri and Bannister, 1975a, b).

(m) Olfactory neurons and the blood-brain barrier

Foreign substances such as proteins, viruses, particles, dyes, and most drugs, when injected into the vascular system, have rather limited access to the central nervous system. The term "blood-brain barrier" has been applied to this exclusion mechanism. The olfactory system has been considered to have a unique relationship with the blood-brain barrier, because dyes placed on the olfactory mucosa are found to stain the olfactory lobes and brain in relatively short periods of time. Bodian and Howe (1941a, b) applied polio viruses intranasally to monkeys and found that polio virus traveled within the axoplasm of olfactory axons rather than along sheaths of nerve bundles. The fact that virus was recovered from OBs demonstrated that the olfactory mucosa and nerves are a port of entry to the central nervous system (p. 121).

De Lorenzo (1970) placed herpes simplex and vaccinia viruses as well as particulate tracers such as ferritin and colloidal gold into the noses of squirrel monkeys. Electron micrographs indicated that the introduced material appeared to be incorporated into the cytoplasm of the receptor cells by pinocytotic vacuoles in 15 minutes; within 30 minutes the material was in the axons comprising the fila olfactoria, and by one hour it had entered the olfactory glomerulus. These particles thus moved quite rapidly. For example, the colloidal gold particles moved about 2.5 mm/hour in the olfactory nerve. Transsynaptic movement of these foreign substances to the dendrites of mitral cells was also demonstrated.

Thus, a new means to apply drugs or nutrients directly to brain cells unhampered by the blood-brain barrier could conceivably be developed by utilizing this olfactory neuronal pathway. This possibility should provide impetus for an entirely new research field in the future.

(n) Olfactory cells in other vertebrates

The receptor cells are bipolar sensory cells directly connected to the OB by their axons, which are extremely thin and packed in bundles of naked fibers by glia cell processes. Opposite the axon hillock, a single dendrite, always thicker than the axon but rarely exceeding a diameter

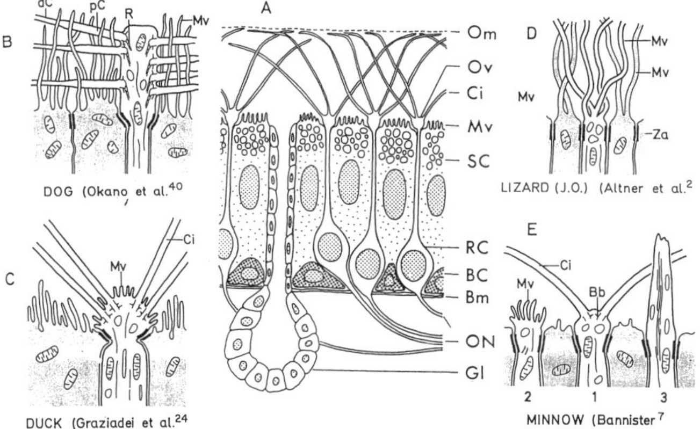

Fig. IV-7. Schematic representation of vertebrate olfactory epithelia. A: The general composition of the olfactory mucosa. (Takagi, unpublished drawing). B-E: Various modifications of mucosal surface and receptor endings. (from Steinbrecht, 1969. Reproduced from *Olfaction and Taste*, III, 1969, p.5 by copyright permission of the Rockefeller University Press, New York) Bb: basal body, Bc: basal cell, Bm: basal membrane, Ci: cilium, dC, pC: distal and proximal part of cilium, respectively, ON olfactory nerve, Om: mucus, Mv: microvillus, R: ciliary root let, Rc: receptor cell. Sc: supporting cell, Za: zonula adhaerens, 1, 2, 3 in E; the three types of receptor ending in the minnow.

of 1 μm, extends to the surface of the epithelium. The termination of the dendrite is contiguous with the overlying mucus and usually rises slightly above the surrounding supporting cells to form a "terminal swelling," or "olfactory vesicle." The fine structure, dimensions, and arrangement of olfactory receptor cells have been found to be remarkably similar among many series of vertebrates (Fig. IV-7A; Andres, 1966, 1969; Bloom, 1954; De Lorenzo, 1957, 1960, 1963, 1968; Frisch, 1964; 1967; Graziadei, 1971, 1973a; Reese, 1965; Steinbrecht, 1969; Yamamoto, 1976).

Steinbrecht (1969), however, reviewing these studies, pointed out that the olfactory vesicle is that part of the sensory cell with the greatest structural variability (Fig. IV-7 B, E). In the dog, the olfactory vesicle is up to 4 μm long and 1.3 to 1.8 μm thick (Fig. IV-7B; Andres, 1966; Okano *et al.*, 1967). In the duck, the vesicle takes the shape of a hemisphere up to 3.5 μm in diameter (Graziadei and Bannister, 1967). In frogs and toads, it is a sphere of about 2 μm diameter (Bloom, 1954). In contrast, the olfactory vesicle in some species of fish is only a slight, convex protrusion of the membrane (Bannister, 1965; Thornhill, 1967; Tryjilloo-Cenóz,

1961). In both the olfactory receptor of the carp, *Carassius carassius* (Wilson and Westerman, 1967), and Jacobson's organ in *Lacerta* (Altner and Müller, 1968), the dendrite does not rise above the epithelial surface.

In most cases, the olfactory vesicle bears cilia as the only form of surface differentiation. Steinbrecht (1969), however, has indicated some exceptions in the following cases: i) The few species of birds examined so far (Andres, 1968; Brown and Beidler, 1966; Graziadei and Bannister, 1967), and the fish *Carassius carassius* (Wilson and Westerman, 1967), in which there are numerous microvilli in addition to cilia (Fig. IV-7C). ii) Jacobson's organ in reptiles, studied in *Anguis fragilis* (Bannister, 1968), *Lacerta agilis* (Altner and Müller, 1968), and *Natrix natrix* (H. Altner, unpublished), in which the dendrites bear no cilia but do have numerous microvilli (Fig. IV-7 D). iii) The minnow, *Phoxinus,* whose dendrite endings may be ciliated (Fig. IV-7 E, type 1) or may bear microvilli (type 2), and sometimes have neither cilia nor microvilli, but instead rise as simple rods about 4 μm into the mucus (type 3; Bannister, 1965).

Glees and Spoerri (1978) found a rather long type of microvillus on the Rhesus monkey sensory ending. They proposed that the actual receptor sites are on such microvilli, while the cilia merely propel odors towards microvilli. This proposition, however, has not been supported by various subsequent freeze-fracture studies (Menco, 1977; Kerjaschki, 1977; Usukura and Yamada, 1978). Intramembranous particle distribution and various other subcellular morphological features clearly demonstrate the sensory nature of these cilia, which distinguishes them from nonsensory cilia (Menco, 1977).

1.2. The supporting or sustentacular cell

The supporting cell has an irregular, columnar, slender body. The cell extends from the free surface of the olfactory epithelium to the basement membrane, and is a type of ependymal cell (Okano and Takagi, 1974; Takagi and Okano, 1971; Getchell, 1977a, b; Masukawa *et al.*, 1983; Rafols and Getchell, 1983). The oval nucleus is situated one third of the way between the free surface of the epithelium and the basal membrane. These nuclei are consistently found at this depth, forming a band of oval nuclei within a few μm of one another. The part of the cell immediately below the nucleus is irregular and uneven in shape and embraces axons and somata of olfactory cells. The basal part of the cell is divided and forms foot processes. The basal surface of the foot process is connected to the basement membrane by hemidesmosomes (Yamamoto, 1976).

The supporting cell is characterized by bushy microvilli, many cytoplasmic organelles, and dense secretory granules. The cytoplasmic organelles includ mitochondria, both smooth and rough endoplasmic reticula,

Golgi apparatus, microtubules, and microfilaments. In general, the supporting cell resembles the shape of the goblet cell in the epithelium of the small intestine (Neutra and Leblond, 1969).

In contrast to olfactory cells, it seems that the number of supporting cells has not been counted yet. Tangential sectioning of the olfactory epithelium of the rabbit showed that olfactory cells group around each supporting cell (Fig. 2 in Allison, 1953b). Hence, it seemed that the olfactory cells outnumber the supporting cells by a factor of more than 5. In fact, Takagi and Yajima (1965) counted the ratio of the number of the olfactory cells to that of the supporting cells in the bullfrog and found it to be 5.2 to 5.7.

Numerous cytological studies have shown that the supporting cell has a secretory function in lower vertebrates (Bloom, 1954, in the toad and frog; Porter and Bonneville, 1964, in the toad; Bronshtein and Ivanow, 1965, in the lamprey; Reese, 1965, in the frog; Wesolowski, 1967, in the turkey; Graziadei, 1972, in the box turtle; Okano and Takagi, 1974, and Takagi and Okano, 1974, in the bullfrog). The secretory granules are found in the supranuclear part of the cytoplasm, especially just under the free surface of the cell.

With the exception of mice (Frisch, 1967), cows (Gladysheva and Martynova, 1982), and monkeys (Saini and Breipohl, 1976), secretory function of supporting cells in mammalian species including humans (Moran *et al.*, 1982b; Jafek, 1983) has not been confirmed. Yamamoto (1976), however, did reveal a unique and delicate structure on the plasma membrane of the microvilli of supporting cells in both the bat and the rabbit. This hitherto unknown structure was characterized by innumerable, fine bubble-like, or drumstick-shaped, processes protruding perpendicular to the outer leaflet of the plasma membrane. Although signs suggesting the formation of secretory granules in Golgi complexes have yet to be confirmed, the abundance of smooth endoplasmic reticulum in the cytoplasm strongly suggests that special unknown substances might be formed in mammalian supporting cells. The smooth endoplasmic reticulum has been credited with an essential role in the elaboration of steroid hormones in endocrine cells, such as adrenocortical cells (Long and Jones, 1967), interstitial cells of the testis (Yamada, 1965; Nagano, 1965), and ovarian lutein cells (Sato and Kurosumi, 1965). The smooth endoplasmic reticulum also participates in carbohydrate metabolism (Yamamoto, 1976). The supporting cell in most mammals may thus secrete unknown substances more or less different from ordinary mucus by a secretory mechanism of the microapocrine type. Yamamoto's finding of bubble-like microprotrusions in mammalian supporting cells strongly suggests that supporting cells secrete various kinds of substances.

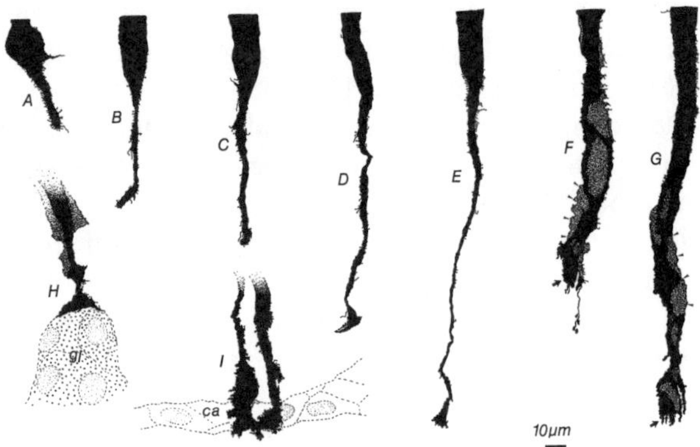

Fig. IV-8. Two types of sustentacular cells in the olfactory epithelium. A-E are type I sustentacular cells. F and G are type II sustentacular cells. H shows the basilar expansion of a type II sustentacular cell in apposition to an acinus of a Bowman's gland (gl). I shows the central stalks of two type I sustentacular cells with their basilar expansions in apposition to each other and to the wall of a capillary (ca). (from Rafols and Getchell, 1983)

In a study of bullfrog olfactory epithelium, Okano and Takagi (1974) and Takagi and Okano (1974) often encountered a pit-like structure in the free surface of the olfactory mucus (arrow 2 in Fig. IV-3). This was an entirely new finding that was interpreted as vestiges of liberated granules. The secretory process of these granules will be described in Chapter V (p. 176).

Turnover can also be observed in the supporting cell, although the frequency of turnover is lower than in the olfactory cell. Since mitotis is found in the zone of the oval nuclei, the supporting cell presumably divides and repeats turnover *in situ* (Graziadei and Monti Graziadei, 1978 a, b).

On the basis of the early Golgi studies by Grassi and Castronuovo (1889), Van Gehuchten (1890), Retzius (1892), and Ramón y Cajal (1911), the sustentacular cells had been regarded as a morphologically homogeneous cell population until recently. A study by Rafols and Getchell (1983) identified two morphologically distinct types of sustentacular cells in the olfactory epithelium of the salamander (Fig. IV-8). The similarities between type I and type II cells included a central stalk spanning the entire depth of the epithelium, filiform appendages, and basilar expansions at the base of the epithelium. The expansions were juxtaposed to the walls of capillaries and acinar cells of Bowman's glands located in the most superficial region of the lamina propria. The main difference between these two types of sustentacular cell was that type I had an un-branched central stalk traversing the depth of the epithelium, whereas the central stalk of type II sustentacular cells gave rise to numerous rib-like processes and protoplasmic veils enveloping the cell bodies of receptor neurons.

The functions of the supporting cell may be summarized as follows (Getchell, *et al.* 1984b): (a) isolation of the receptor neurons from one another (e.g., Graziadei, 1971); (b) glial type function; (c) mucus secretion (Okano and Takagi, 1974; Takagi and Okano, 1974; Getchell and Getchell, 1977b); (d) transepithelial transport of molecules (Rafols and Getchell, 1983); and (e) guides for migration of postmitotically active cell bodies and dendrites of receptor neurons through the depth of the olfactory epithelium (Rafols and Getchell, 1983).

1.3. The basal cell (Fig. IV-7)

These cells are small with round nuclei, and lie at the bottom of the epithelium on the basement membrane. Mitotic cells are often found in this deep region. Andres (1966) called these cells "Blastema Zellen" and stated that they are precursors of receptor cells. Both Graziadei and Metcalf (1971) and Graziadei (1973b), using electron-microscopic au-toradiography, demonstrated that the basal cell is the stem cell for the olfactory cell, and discounted the possibility that the basal cell differen-tiates into a supporting cell. Details on the functions of this cell were already covered (pp. 121–124).

1.4. The fourth cell type

Okano *et al.* (1967) found, through electron microscopy, a cell in the dog that did not belong to any of the above three cell types. This "fourth cell" is located in the olfactory epithelium of the dog. A homologous cell was also found in the cat, the turtle, and the rat (Andres, 1969, 1970a, b).

These cells are, on the whole, similar to the supporting cell (Fig. IV-9A; Okano *et al.*, 1967), but differ in some respects. Their cytoplasms appear to be clear and are deficient in organelles. Their microvilli differ in shape, are shorter, and are less numerous. However, since turnover of the supporting cell occurs *in situ*, these cells are probably newly generated, immature supporting cells (Okano *et al.*, 1967).

1.5. The fifth cell type

Following the discovery of the fourth type cell, a "fifth type" of cells, very different from the other cells, were found in the olfactory epithelia of the cat (Andres, 1969) and of the dog (Okano, 1973). Such peculiar cells (Fig. IV-9 B, Okano, 1981) have short, strong, finger-like microvilli. Core structures of these microvilli extend into the cytoplasm, grad-ually becoming fine fibers which scatter radially before terminating.

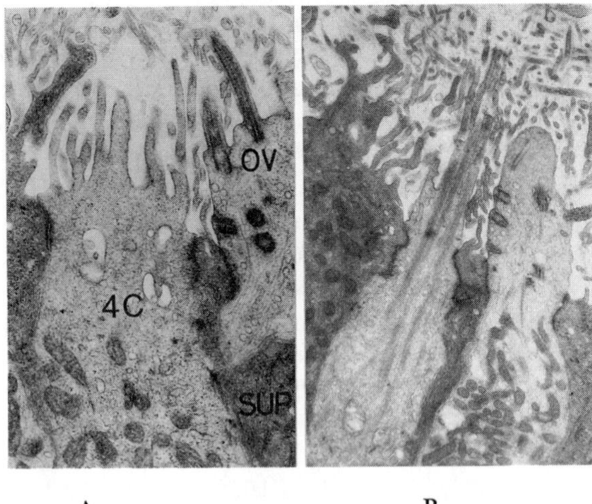

Fig. IV-9. The fourth (A) and fifth (B) cell types. (courtesy of M. Okano)

These cells probably function as mechanoreceptors that synapse with extrinsic nerve fibers (Andres, 1970a, b).

1.6. Dissociation of olfactory epithelium

To study the properties of cells in a tissue, many workers have devised ways to dissociate the tissue and produce cell suspensions. Ash *et al.* (1966), for example, used a homogenization technique, Hirsh and Margolis (1979) used enzymes that hydrolyze connective tissue, Cancalon (1978) employed trypsin, Kleene and Gesteland (1981) worked with *N*-ethylmaleimide, and Kurahashi and Shibuya (1986a) used collagenase (Sigma Type IA) and mechanical manipulation.

Kleene and Gesteland (1983) reported a new method for producing olfactory cell suspensions that used no degradative enzymes, except for a brief DNAase incubation. Dissociation was produced by treatment of the epithelia in a buffered solution with a pH of 10.3 and a free calcium ion concentration of 10^{-6} M. The resulting cell suspension contained single olfactory cells, sustentacular cells, glandular cells, and respiratory epithelial cells. Examination of vital staining, ciliary motility, and RNA synthesis led these researchers to conclude that the cells were viable, but some cell types lost their normal columnar shapes and became rounded in suspension.

Suzuki (1986) isolated olfactory receptor cells of bullfrogs by combined use of an enzyme (Dispase I) and ethylendiaminetetraacetic acid (EDTA).

Isolated cells had high morphological integrity and were easily identifiable in the heterogeneous mixture of different cell types by the presence of motile cilia on the olfactory vesicles, thin dendrites (12 to 20 μm long; 2 to 3 μm in diameter), and oblate, sphere-shaped somata (10 to 20 μm in the short axis). He used these isolated cells in voltage, current, and patchclamp experiments (Chapter VI, pp. 204–205).

Dionne (1986) isolated single cells from the adult mudpuppy by incubation at room temperature in zero-calcium amphibian saline with trypsin (0.5 mg/ml, 30 min; in later studies papain), followed by a Ca-containing saline with collagenase (2 mg/ml) and bovine serum albumin (1 mg/ml). Cells isolated by this method retained their morphology and six cell types were distinguished, only one of which was neuronal. Columnar supporting cells showed a poor rate of survival after the dissociation treatment (Chapter VI, p. 206).

A recent development in the patch-clamp technique (Neher *et al.*, 1978; Hamill *et al.*, 1981) has allowed measurement of transmembrane currents with high spatial resolution. This method, however, requires a suspension of single cells where the surfaces of the cells are free of connective tissue. The above procedures have appeared to meet these requirements (intracellular and patch-clamp experiments in dissociated cells in Chapter VI, pp. 204–205).

2. The Basement Membrane or Basal Lamina

This membrane forms a border between the olfactory epithelium and the lamina propria. Basal cells lie on this membrane, and the long, slender proximal processes of the supporting cells are attached to it, forming half-desmosomes.

3. The Lamina Propria

The lamina propria beneath the basement membrane contains many small bundles of olfactory axons of various sizes, Schwann cells, blood vessels, and olfactory glands.

3.1. The Schwann cell

Nonmyelinated axons from each olfactory cell form small clusters in the deep part of the epithelium, and then converge into small nerve fascicles. These in turn join nerve bundles in the deep layers of the lamina propria. Schwann cells in the lamina propria are found to embrace small bundles of the olfactory nerve fibers. The typical relationship among many fascicles of unmyelinated nerve fibers, a Schwann cell, and mesaxons were already diagrammatically illustrated in a transverse section of the olfactory nerve (De Lorenzo, 1970). The Schwann cell is also said to embrace Bowman's glands.

Following the discovery of the glial marker protein, S-100, and the neuronal marker protein, 14–3–2 (nerve-specific enolase), in the olfactory mucosa (Hirsch and Margolis, 1981), the presence of glial fibrillary acidic protein (GFAP) was demonstrated in Schwann cells of the olfactory nerves (Barber, 1981; Barber and Lindsay, 1982). This substance had also been identified as an immunocytochemical marker for astrocytes (Bignami *et al.*, 1972). Hence, the properties of Schwann cells in the olfactory nerve were found to resemble more closely those of astrocytes in the central nervous system than of Schwann cells in the peripheral nervous system. The presence of GFAP in cells surrounding the unmyelinated olfactory axons suggests that they may be similar to astrocytes and should be evaluated for their possible contribution to the guidance of olfactory axons to their synaptic targets in the OB. The extent to which these prop-

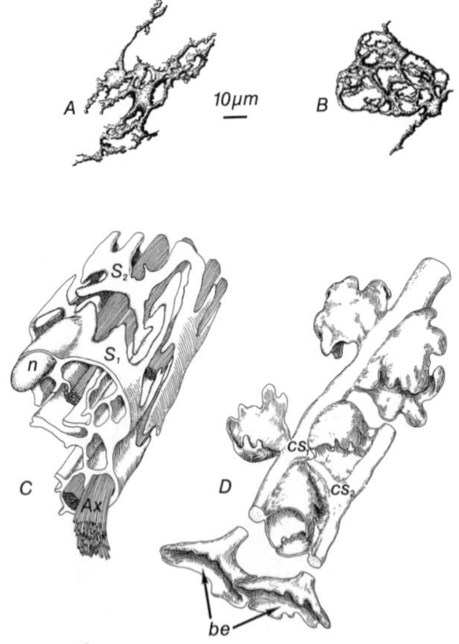

Fig. IV-10. Golgi impregnated sheath Schwann cells in the olfactory nerve bundles of the lamina propria. A is from a longitudinally sectioned olfactory bundle, and B is from a frontally sectioned bundle. C: Morphological relationships between Schwann cells and the axons (Ax) of receptor neurons. D: Morphological relationships between adjacent type II sustentacular cells. The cytoplasmic veils arise directly from the central stalks (CS_1 and CS_2) or from rib-like processes. Thee dges of adjacent veils interdigitate and form compartments for the cell bodies of receptor neurons. The sides of the basilar expansions (be) of adjacent sustentacular cells also interdigitate with each other. (from Rafols and Getchell, 1983)

erties affect the neurophysiological activity of the receptor axons as well as the system's unique regenerative capacity is now being investigated (Rafols and Getchell, 1983). Incidentally, this specialized neuroglial arrangement can probably explain the capacity of olfactory and vomeronasal axons to reinnervate central neurons as an exception to the rule that peripheral axons cannot grow into the central nervous system of adult mammals (Raisman, 1985).

Rafols and Getchell (1983) studied the morphological characteristics of the sheath Schwann cell in the olfactory mucosa of the salamander (*Ambystoma maculatum et tigrinum*). They found that the cell body of the sheath cell was very irregular in shape due to many processes (Fig. IV-10 A, B) which formed elaborate partitions surrounding groups of axons within the nerve bundles. In addition, the partitions were found to be formed by the interdigitating processes of adjacent sheath cells aligned in tandem along the nerve bundles (S_1 and S_2 in Fig. IV-10 C). The spatial compartmentalization of the sheath cell processes thus produced a continuous network of cavities through which the axons of receptor neurons coursed. The composite stereogram shown in Fig. IV-10 C, which Rafols and Getchell skillfully constructed from their analysis of longitudinal and frontal sections, allows readers to conceptualize easily the spatial organization of these cavities and their relations to the olfactory axons.

A thick bundle, or bundles, of olfactory nerve fibers leave the lamina propria and run toward the OB.

3.2. The olfactory gland, or Bowman's gland (Seifert, 1971b)

The olfactory gland, once called Bowman's gland, is located in the lamina propria and is a branched, tubulo-alveolar type of gland. The gland extends a narrow duct perpendicularly through the layer of olfactory and supporting cells to the surface of the epithelium, where the duct widens and releases secretions. Secretions from Bowman's gland had generally been assumed to constitute the major source of olfactory mucus (Hopkins, 1926; Gompper, 1950; Andres, 1966; Frisch, 1967; Moulton and Beidler, 1967) until Okano and Takagi (1974) and Takagi and Okano (1974) clarified the important contribution of secretions of the supporting cell. Pyridine (Wesolowski, 1967) and chloroform and other irritative odorants were shown to induce marked secretion from the supporting cell (Okano and Takagi, 1974). At present, it is generally accepted that the serous or watery component of the mucus comes from the olfactory gland, while the viscous component of the mucus derives from the supporting cell (Getchell *et al.*, 1984a; Getchell *et al.*, 1987).

The secretory mechanism of the olfactory glands, however, has received very little attention, and the factors which induce secretions from these glands have yet to be determined. Stimulation of the epithelial surface with pyridine failed to induce secretion from Bowman's gland (Wesolowski, 1967).

The salamander, *Ambystoma tigrinum* provides a model system for studying the olfactory system (e.g., Graziadei and Monti-Graziadei, 1976; Rafols and Getchell, 1983). Getchell *et al.* (1984a) found that, in the land-dwelling adult animal, Bowman's glands with ducts projecting to the epithelial surface lay subjacent to olfactory epithelium containing supporting cells in both dorsal and ventral mucosae. They identified two additional layers of olfactory glands, the middle olfactory and deep olfactory glands, in the ventral mucosae below the level of Bowman's glands. The acinar cells of these two glands contained secretory granules resembling those seen in typical serous glands, suggesting a glycoprotein secretion.

Using histochemical techniques, Getchell *et al.* (1984a) demonstrated that supporting cells contained both neutral and acidic mucopolysaccharides, and small amounts of sulfated mucopolysaccharides, whereas Bowman's glands and the middle and deep olfactory glands contained only neutral mucopolysaccharides.

Experimental findings obtained in other laboratories have shown that secretory cells in the olfactory mucosa are rich in the enzyme carnosinase (Margolis *et al.*, 1983) and fluorescein-conjugated peanut lectin specifically binds to the cells or cellular secretions of olfactory glands (Hempstead and Morgan, 1983a).

Although odorants, or other substances which could induce secretion from these glands, were unknown, Getchell and Getchell (1984) demonstrated that the β-adrenergic agonist isoproterenol reduced the number of secretory granules in the acinar cells of these glands while attenuating the amplitude of the negative electro-olfactogram (EOG). These effects were blocked by the β-adrenergic antagonist propranolol.

These results suggested for the first time that secretion from olfactory glands can be stimulated by β-adrenergic drugs and that the effect of isoproterenol might be mediated through β-adrenergic receptors on the acinar cells. Direct autonomic innervation to the glands in the nasal mucosa had already been reported by Änggaerd and Denset (1974) in the cat and by Cauna (1982) in human nasal glands.

Getchell *et al.* (1985a) then demonstrated that ephedrine and guaiacol elicit secretory granule release from the superficially located Bowman's glands, and elicit water and electrolyte secretion from the more deeply situated glands. They (Getchell *et al.*, 1985b) also showed that pyrazine has multiple effects both on the olfactory cell and on the whole secretory mechanism of the olfactory mucosa (supporting cells, Bowman's glands, and deep olfactory glands). Furthermore, Getchell *et al.* (1986) obtained evidence to suggest that 2-isobutyl-3-methoxypyrazine may activate cholinergic mechanisms that regulate secretion from olfactory glands, by either direct action on the acinar cells or indirectly through a secretomotor reflex. Research on the secretion of Bowman's and other glands in the lamina propria is continuing and much more about the functions of these glands may be revealed in the near future.

Although Getchell and Getchell (1984) discovered two kinds of olfactory gland in the lamina propria of the tiger salamander, either, or both, of these glands remain to be identified and examined in humans.

4. The Olfactory Pigment

A distinctive feature of the olfactory mucosa is its pigmentation: light yellow in humans and monkeys, and darker yellow or brown in macrosomatic animals. The pigment is said to occur as discrete granules in the supporting cells of the olfactory epithelium and the acinar cells of Bowman's glands (Allison, 1953b). This pigmentation is the only distinguishing feature of the olfactory epithelium which allows easy discrimination by the naked eye from the surrounding respiratory epithelium.

The function of the olfactory pigment in olfaction has long been a subject for discussion. Some indirect evidence suggests that the pigment is important or essential in olfaction (Gerebtzoff and Phillippot, 1957; Moncrieff, 1951; Wright, 1961). All-black pigs are reared in parts of Virginia (U.S.A.) and all-black sheep in Trentino (Italy) because white pigs and sheep inevitably die from toxic grasses that the black animals leave alone. Thus, the albino animals, in which the olfactory pigment is lacking, are particularly vulnerable to poisoning (Allison, 1953b). A connection between albinism and anosmia has also been described in man.

For many years, the olfactory mucosa was thought to contain a yellow pigment that would induce the yellow color of the olfactory mucosa, but the location of such a pigment was not precisely known. Several investigators (Le Magnen and Rapaport, 1951; Vinnickow and Titow, 1949) have suggested that vitamin A or carotenoids are related to olfaction because stores of carotenoids are found in the olfactory epithelium of cattle (Milas et al., 1939; Briggs and Duncan, 1961). Some substantial reports can in fact be cited that assert that the sense of smell is impaired in persons or animals suffering from vitamin A deficiency (Chapter III, p. 97). In this connection, Briggs and Duncan (1961) proposed in an analogy with vision that carotenoids are the receptors of energy from odorous molecules. Rosenberg et al. (1968), on the other hand, suggested a transduction mechanism based on the fact that carotenoids are semiconductors whose conductive properties can be modified if odorous molecules become weakly bound to them. Gerebtzoff and phillippot (1957), however, found no carotenoids in rabbit or sheep olfactory epithelia. Furthermore, Moulton (1958) examined the olfactory epithelia from one cow, albino rats, black rats, and pigs, and found β-carotene and other carotenoids only in the cow (Moulton, 1962, 1971).

Pigments other than carotenoids have also been reported, including lipofuscins, phospholipids, and melanin. These noncarotenoid pigments are apparently responsible for the yellowish-brown color of the rat and pig olfactory epithelia. Jackson (1959, 1960) isolated and fractionated the fat-soluble olfactory pigment from rabbits, dogs, and opossums. He found that the phosphatidic acid and lecithin fractions all contained the pigment. Comparing the properties of the pigment with other phospholipids, he concluded that the yellow-brown color was due to auto-oxidation products of phospholipid origin. Thus, the pigment may be a waste product of lipid metabolism peculiar to unmyelinated nerve fibers (Goodwin, 1952). Jackson thereby disproved the supposed relationship between albinism and anosmia.

Melanin is a rather inert polymer. The function of melanin in the skin is to protect the underlying tissue from excessive exposure to ultraviolet radiation. This led Wright (1978) to consider whether the pigmentation functioned to shield the receptor from photo effects. According to Yamamoto (1976), numerous large and small dense bodies, probably lipofuscin granules, can be found in the infranuclear cytoplasm of the supporting cell. These infranuclear lipofuscin granules might play an essential role in the yellow coloring of the olfactory mucosa.

Some investigators have hypothesized that the darker the olfactory pigmented area, the more acute is the animal's olfaction (Hosoya and Yoshida, 1937; Onagawa, 1957). In fact, the degree of pigmentation varies both intra- and interspecifically. Jackson (1960), however, could not find any difference in pigmentation between albino and pigmented rabbits and rats, and assumed, as previously mentioned, that the pigment may be a waste product of lipid metabolism peculiar to unmyelinated nerve fibers. Some reports even went so far as to say that the pigmented and olfactory areas do not coincide (Brunn, 1892), and that pigmentation is lacking from the olfactory area of some species with known olfactory powers (Gerebtzoff and Schkapenko, 1951).

Much more information concerning the location, chemical composition, and distribution of pigment complex in representative vertebrate species is needed before we can come to a definite conclusion about the role of olfactory pigments.

5. Olfactory Axons, or Olfactory Nerve Fibers

An unmyelinated olfactory axon arises from the proximal end of the olfactory cell. These axons have a modal size of 0.2 μm with a range of less than 0.1 μm up to 0.4 μm or 0.5 μm (De Lorenzo, 1957; Gasser, 1956). Each fiber has a neurilemma about 10 nm thick, enclosing a light axoplasm. These axons feature microtubules and mitochondria within the axoplasm, and low electron density. Axonal regions containing mitochondria are thicker than other parts, often as thick as 0.8 μm, making the overall axon appear rugged.

Many axons aggregate into small fiber bundles (about 1,000 fibers per bundle according to Moulton and Beidler, 1967) in the spaces between basal cells. These bundles are enveloped by Schwann cells when they penetrate the basement membrane and enter the lamina propria. In some lower vertebrates, such as frogs, all the olfactory axons of one side gather into a single thick bundle and leave the lamina propria for the OB. In mammals, the olfactory axons gather into many small bundles which pass through numerous pores (cribriform foramina) in the ethmoid bone and enter the cranial cavity to terminate in the OB. Evans and Christensen (1979) found as many as 300 such foramina in the dog, the largest foramen having a diameter as large as 1.5 mm. Thus, the dog has as many as 300 olfactory nerve bundles.

The existence of a one-to-one relation between nerve fibers and olfactory cells deduced by earlier histologists was confirmed by both Gasser (1956) and Clark (1956). The olfactory axons neither branch nor synapse anywhere along the olfactory nerve from the receptor cell to the OB.

The relation between the Schwann cell membrane and the nerve fibers is fundamentally like that in other unmedullated fibers. The plasma membrane of the Schwann cell forms a common mesaxon for a number of fascicles (Gasser, 1958; De Lorenzo, 1970; pp. 131–133), so that in some cases considerable areas of cytoplasm may separate neighboring fascicles, each of which is about 0.48 mm in diameter in the rabbit (Clark, 1956). Within each fascicle, however, the fibers are separated from each other by about 10 to 15 nm. This proximity is thought to allow some sort of interaction among neighboring fibers.

The sheath surrounding the bundle of olfactory fibers is retained until the fibers arrive at the OB. Here, the fascicles disperse into a complex network so that individual fibers from one fascicle reach widely scattered glomeruli, thus forming the olfactory nerve layer, the outermost layer of the OB.

C. The Human Olfactory Mucosa (Nakashima *et al.*, 1984)

Bloom and Engström (1952) and Engström and Bloom (1953) may have been the first investigators to study human olfactory epithelium with the electron microscope. They reported that human olfactory cilia were 1 to 2 μm in length, about 0.1 μm in diameter, and that one olfactory cell contained about 1,000 cilia. Since the olfactory epithelium occupies only a very narrow region, these investigators could conceivably have mistaken the respiratory epithelium for the olfactory epithelium. A finding of 1,000 olfactory cilia per single olfactory vesicle is extraordinary if we compare this to the findings obtained in other mammals.

Naessen (1970, 1971b) identified and topographically localized the olfactory epithelium of man and examined it extensively with light and electron microscopy. He also studied changes of cells in the olfactory epithelium, especially of receptor cells with age (Naessen, 1971a). He identified a process of olfactory cell degeneration with age, leading to receptor cell depletion and subsequent changes to supporting cells.

In a later study, Kanda et al. (1973) differentiated two types of olfactory cilia in the human olfactory epithelium, one slender and short (2 to 3 μm in length), and the other thick and long (about 10 μm in length). They determined the number of cilia to be about 10 per olfactory cell. These data were comparable to those in guinea pigs (Kanda et al., 1973). In addition, they showed that the number of olfactory cells was about 30,000/mm^2 of olfactory epithelium. Since the area of human olfactory mucosa has been calculated as 6.4 or 2.4 cm^2 (pp. 102–103), extrapolation reveals that humans have about 19.2 or 7.2 million olfactory cells depending on which figure is used. This roughly agrees with the estimation of about 10 million cells by Takahashi (1979) and Takasaka at al. (1980).

Polyzonis et al. (1979) performed electron microscopic examinations of the fine structure of human olfactory mucosa. For this purpose, they took their own mucosa using an intranasal forceps. Their important and interesting findings on two satisfactory biopsy specimens showed that the end of the olfactory dendrite was flat or dome-shaped, not a bulb-like projection (olfactory vesicle), and that the cilia were positioned on the long axis of the olfactory cell, perpendicular to the epithelial surface, not parallel to it as described in some species.

Douek et al. (1975) used a long toothed forceps to perform biopsies on four patients, but this study lacked the necessary control specimens obtained from normal subjects. Lovell et al. (1982) developed an instrument to procure small biopsy specimens of human olfactory mucosa suitable for electron microscopy. Electron micrographs showed olfactory cells with olfactory vesicles protruding over the free surface of the olfactory epithelium. They concluded that Polyzonis et al. (1979) missed the olfactory region entirely and obtained electron micrographs of the respiratory epithelium instead.

Moran et al. (1982a, b), using a special biopsy instrument and technique, took fresh biopsies of olfactory epithelium. Transmission electron microscopy revealed, in addition to the usual three types of cell (olfactory, supporting, basal), that the fourth cell type, the microvillar cell, is conspicuous, abundant, and a consistent constituent of the human olfactory epithelium. The cell body is neuronal in fine structure, and closely resembles bipolar neurons that serve as primary sense cells. Furthermore, an axon-like cytoplasmic extension often emerges from the basal pole of the cell. Since the existence of microvillar olfactory receptors is common in vertebrates (Graziadei, 1973a, b), we can reasonably suppose that the microvillar cells in the human olfactory epithelium may be chemoreceptors. If so, the current concept of human olfactory epithelium will have to be revised to include two morphologically distinct classes of olfactory receptors.

Also important were the findings that the olfactory cell has 10 to 30 cilia that lack dynein arms, and that the supporting cell, markedly different from the goblet cell of the respiratory epithelium, is not specialized for mucus secretion, but is equipped to contribute materials to, and remove materials from, the surface mucus.

Lastly, findings of clinical and laboratory studies will be mentioned. These studies suggest that the volume and rheological properties of secreted mucus vary under a variety of experimental, pathological, and therapeutic conditions (e.g., studies on humans by Brofeldt *et al.*, 1983, and Malm *et al.*, 1981; on birds by Wesolowski, 1967; and dogs by King and Vires, 1979). For example, stimulation of the parasympathetic innervation of the nasal mucosa produces a watery rhinorrhea through the activation of nasal glands (Eccles, 1982) and presumably, olfactory glands. Similarly, independent stimulation of supporting cell secretion, perhaps by a local irritant (Chapter V, p. 177), could result in a mucoid rhinorrhea. As a result, the height and rheological properties of the mucus layer would be altered. These changes would play important roles in olfaction as major determinants of odorant access. For further information on this matter, readers are advised to read excellent review by Getchell *et al.* (1984).

D. Immunohistochemical Studies on the Main Olfactory System

1. Studies on Olfactory Marker Protein (OMP)

Since the pioneering effort of Moore and McGregor (1965) to isolate any protein specific to the nervous system, many proteins specific to mammalian brains have been discovered. Margolis (1972)searched for such protein markers in the olfactory nervous system and examined extracts of soluble mouse brain proteins by polyacrylamide gel electrophoresis (PAGE).

He found a major protein band in extracts of mouse olfactory bulb which was apparently absent from all other brain regions studied. This protein has been named the olfactory marker protein OMP, (pp. 125 and 214), and was also found in the vomeronasal organ.

The OMP isolated from OBs of mice and rats is a soluble, acidic protein, with a low molecular weight (Margolis, 1972, 1985; Keller and Margolis, 1975, 1976). The protein is unique to the primary olfactory pathway, synthesized by primary receptor neurons in the olfactory epithelium, and transported along the axon to its termination (Margolis and Tarnoff, 1973). An antibody to OMP has been prepared in goats (Margolis, 1972), and immunological studies have shown that it is widely distributed phylogenetically in garfish, frogs, rodents, herbivores, carnivores, and man (Keller and Margolis, 1975). The level of OMP was found to decrease rapidly in the olfactory epithelium and then to return to near control levels following olfactory nerve section (Harding *et al.*, 1978; Margolis, 1977). For a comprehensive survey of OMP, see the review by Margolis, 1982.

Immunohistochemical studies by indirect techniques using fluorescein-labeled antibodies (Hartman and Margolis, 1975) and peroxidase-anti-peroxidase-labeled antibodies (Monti-Graziadei *et al.*, 1977), have shown the protein to be localized in olfactory receptor cell perikarya, dendrites, and axons, all the way to their terminations in the glomerular layer of the OB. OMP is localized in the cytosol of receptor cells and appears mostly in the older cells of the epithelium (Monti-Graziadei *et al.*, 1977).

Farbman and Margolis (1980) tracked the appearance of OMP during ontogeny in rats and mice. They found that at embryonic day 18, OMP was first identified solely in rat olfactory receptor neurons, but at 21 days it was first identified in the OB, thereby demonstrating that only mature cells stain for OMP in the main olfactory organs. This was also the case in vomeronasal olfactory organs (Chapter X, pp. 391–408). Hinds and Hinds (1976) showed in mice that olfactory axons first enter the OB 11 days after conception, but synaptogenesis is not detectable in their electron micrographs until day 13. Farbman and Margolis (1980), studying the relation between maturation of olfactory receptor cell terminals and synapse formation in the OB, noted the appearance of OMP in mouse receptor cells one day after synapses were first observed, or 14 days after conception. This suggests that synaptogenesis induces receptor cells to synthesize OMP, possibly by a trophic retrograde phenomenon (cf. Cuschieri and Bannister, 1975b). In fact, Chuah and Farbman (1983) have indicated that the precursor to the OB specifically increased the OMP level by a mechanism requiring direct contact between the receptor cells and the bulb cells.

2. Subclassification of Olfactory Cells

In each mammalian sensory organ, receptor cells typically differentiate in form and function into several subclasses, and each subclass of receptor cells makes distinct connections with separate subsets of neurons at successive central levels of the sensory pathways. In most mammals, olfac-

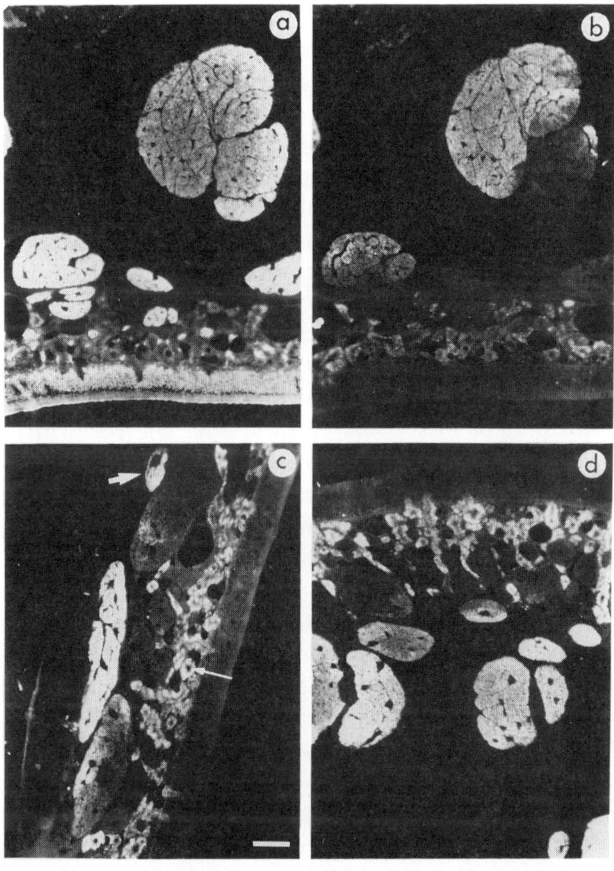

Fig. IV-11. Immunofluorescent staining of coronal sections of the olfactory epithelium with MAbs R2D5 (a) and R4B12 (b–d). a and b: Neighboring sections of the medial part of the naso-turbinate, c: septum, d: dorsal portion of the first endoturbinal. Note the predominance of stained nerves in the deeper part. The Bowman's gland cells (thin arrow in c) have orange autofluorescence easily distinguishable from the specific green immunofluorescence over olfactory nerve bundles (thick arrow). Scale bar = 100 μm (from Fujita *et al.*, 1985)

tory systems employ two classes of receptor cells (olfactory and vomero-nasal) with distinct central connections, namely the main and accessory OBs, respectively. Olfactory receptor cells may be further classified into subclasses with distinct functional properties and distinct connections to the OB (Clark, 1956, 1957).

Fujita *et al.* (1985) and Mori *et al.* (1985) used an immunohistochemical method involving monoclonal antibodies (MAbs) to examine the organization of the olfactory nerve projection to the OB in the rabbit. Out of 42 MAbs raised against the homogenate of the OB, they selected two types of MAbs that strongly stained the olfactory nerve fibers and analyzed their staining patterns in detail. MAbs of one type (MAb R2D5) specifically labeled all olfactory receptor cells in the nasal epithelium and all olfactory nerve fibers (Fig. IV-11a) and their terminal portions (glomeruli) in the OB. The MAb R2D5-positive axons formed bundles in the submucosa and ran rostrocaudally under the epithelium. The distribution of R2D5 immunoreactivity closely resembles that of OMP.

In contrast, MAb R4B12 labeled the olfactory axons but not the cell bodies of the receptor cells. Comparison of neighboring sections stained with MAbs R2D5 and R4B12 showed that MAb R4B12 labeled only one subgroup of olfactory fibers (Fig. IV-11a, b). The R4B12-positive fibers were distributed over the ventrolateral areas, but not the dorsomedial areas of the epithelium. In the OB, the R4B12-positive fibers terminated in the glomeruli in the ventrolateral and caudal regions, but not in the dorsomedial region. MAb R4B12-positive and -negative fibers tended to be segregated both within and between the bundles (Fig. IV-11a, b). These results demonstrated, for the first time at the molecular level, cellular heterogeneity among olfactory receptor neurons. The segregated distribution of the subtypes of olfactory receptor cell axons in both the epithelium and the bulb indicated a defined topographical organization of the olfactory nerve projection. These results also implied a functional division between dorsomedial and ventrolateral areas in both the olfactory epithelium and the OB (section E below).

The hybridoma technique has proved to be very powerful in defining subsets of neurons, not only in the lower olfactory nervous system, but also in various parts of other neural systems (Barnstable, 1980; Dodd *et al.*, 1984; Fujita and Obata, 1984; Hawkes *et al.*, 1984; Hendry *et al.*, 1984; Lawson *et al.*, 1984). Considering the above results, the future may witness the production of MAbs that recognize subsets of OB neurons and olfactory cortex neurons as well as further subsets of olfactory receptor cells. This approach has the prospect of becoming indispensable in elucidating the functional organization of the olfactory nervous system.

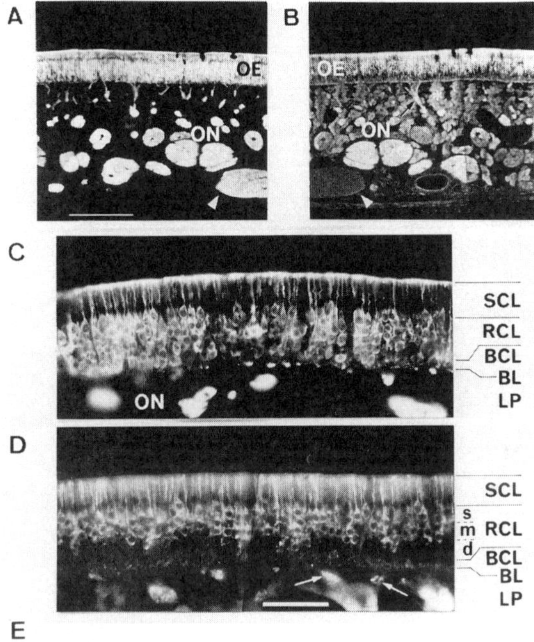

Fig. IV-12. Immunostaining of the adult rabbit olfactory mucosa with MAbs 2D5 (A, C) and 4G12 (B, D). In adjacent sections, (A) and (B), olfactory epithelium (OE) and olfactory nerve bundles (ON) are stained by MAbs; arrowheads point to the vomeronasal nerve positive to MAb 2D5 (A) and negative to MAb 4G12 (B). The autofluorescent Bowman's gland is shown by an asterisk in B. (C) and (D) are magnified micrographs of the OE. Scl: Supporting cell nuclear layer, Rcl: receptor cell layer, BCL: basal cell layer, BL: basal lamina, LP: lamina propiria. The receptor cell layer is subdivided into the superficial (s), middle (m), and deep (d) compartments. Arrows in (D) point to the olfactory nerves (ON). Scale in A, B: 200 μm, scale in C, D: 50 μm (from Onoda & Fujita, 1987)

3. Further Studies Using Monoclonal Antibodies

Onoda and Fujita (1988) found that one monoclonal antibody MAb 2D5 (same as MAb R2D5 above), stained all of the receptor cells, whereas a new MAb 4G12, stained the upper two thirds to three fourths of the receptor cell layer, leaving the deep compartment unstained. MAb stained neither the supporting cells, basal cells, nor Bowman's glands (Fig. IV-12). In the OB, MAb 2D5 stained the olfactory nerve layer and glomeruli, whereas MAb 4G12 stained the whole of the OB, particularly the glomeru-

Table IV-1. Immunocytochemical characteristics of biochemical markers in different regions.

Regions (species)	Markers		
	OMP (mouse, rat)	2D5 antigen (rabbit)	4G12 antigen (rabbit)
Olfactory mucosa			
Receptor cell	mAll	All	2/3–3/4
Supporting cell	–	–	–
Basal cell	–	–	–
Olfactory nerve	mAll	All	2/3–3/4
Vomeronasal organ	+	All	–
nerve	+	All	–
Olfactory bulb ONL	All	All	+
GLL	All	All	+ +
EPL	–	–	+
MCL	–	–	MC
GCL	–	–	+
Pyriform cortex	–	–	Ia, IIIS
Cerebellum	–	–	Pu
Hippocampus	–	–	Py
Ependymal cell	–	+ +	–

All: All of receptor neurons were positive, mAll and 2/3–3/4: mature receptors and 2/3 to 3/4 of receptors were positive, respectively. +: Weakly positive, + +: strongly positive. –: absent. MC: mitral cells. Ia: layer Ia. IIIS: soma in layer III. Pu: Purkinje cells, Py: pyramidal cells, ONL: olfactory nerve layer, GLL: glomerular layer, EPL: external prexiform layer, MCL: mitral cell layer, GCL: granule cell layer. (from Onoda and Fujita, 1987)

li and the mitral cells. Thus, these two MAbs, especially MAb 2D5, showed distribution patterns that were very similar to OMP in the staining of the olfactory epithelium and the OB. Incidentally, the pyriform cortex was not stained by MAb 2D5, but layer Ia of the cortex showed the highest immunoreactivity to MAb 4G12. Comparisons of the respective staining patterns are summarized in Table IV-1.

Onoda (1988a) studied the development of olfactory neurons using these two MAbs. MAb 2D5 stained all of the receptor neurons in the olfactory epithelium and vomeronasal organ (VNO) during development. Staining of MAb 4G12, on the other hand, showed characteristic developmental patterns in the olfactory epithelium that were similar to the expression of OMP. In the fetus at embryonic day 17, 4G12-positive cells appeared dispersed throughout the epithelium. At embryonic day 25 or 26, 4G12-positive cells were found in the superficial receptor cell layer. The receptor cell arrangement appeared "superficialpositive" and "deepnegative." A gradual increase in "superficial-positive" cells was accompanied by a decrease in the "deep-negative" cells, which continued until postnatal day 30. The MAb 4G12-positive deep compartment remained in the adult rabbit. Thus, MAb 4G12 was demonstrated to be a useful probe for studies on neurogenesis in adult animals. The supporting cells, the basal cells, and the VNO showed no immunoreactivity to 4G12 throughout development.

Onoda (1988b), using these two MAbs, studied degeneration and regeneration of the olfactory receptor neurons over various time periods (12 hours to 6 months) after unilateral OB removal in adult rabbits. MAb 2D5 stained all receptor neurons, including degenerating olfactory neurons, but MAb 4G12 showed a rapid decrease in immunostaining so that 4G12-positive cells disappeared within seven days after the lesion. MAb 4G12-positive cells then reappeared four weeks after the lesion and progresively increased in number, recovering to normal levels by 6 months. Onoda concluded that MAb 4G12 may be the first marker that can be used to characterize certain embryonic traits during degeneration and regeneration of the olfactory epithelium.

4. Identification of Cell Surface Glycoproteins in the Receptor Neurons

Allen and Akeson (1985a) independently developed a monoclonal antibody, designated 2B8, which recognizes a family of cell surface glycoproteins found abundantly in rat olfactory receptor neurons. This MAb 2B8 was produced by the fusion of X63-Ag8.653 myeloma cells and spleen cells of a mouse immunized with PC12 rat phenochromocytoma cells. Immunofluorescence analyses of cryostat sections of neonatal olfactory epithelium showed prominent 2B8 binding to receptor neurons. Within the olfactory bulb only the glomerular and olfactory nerve layers showed 2B8 binding. Analyses of 2B8 binding to particulate protein preparations from several central and peripheral nervous system components demonstrated the highest 2B8 antigen-specific activity in the olfactory bulb and epithelium. Using two distinct assay methods, Allen and Akeson (1985a) suggested that the 2B8 antigens recognized in the olfactory system are glycoproteins having sialic acid and D-galactosyl components. Thus, they showed that the 2B8 MAb recognizes a group of glycoproteins which are the first identified cell surface components of olfactory receptor neurons, and that 2B8 immunofluorescence in the nervous system could be found only on cells of the olfactory system.

Allen and Akeson (1985b) then studied development of olfactory cell populations and found that the MAb 2B8 recognized cells early during development and appeared to recognize a subclass of olfactory receptor cells and their axon terminals.

E. Relationship between the Olfactory Epithelium and Olfactory Bulb

The axons of the olfactory receptor cells come together to form the olfactory nerve fasciculi which pursue a parallel course in the deeper layers of the mucosa directly toward the ethmoid bone; once they have passed into the anterior cranial fossa, however, the bundles are interwoven in a most complex fashion in the outermost layer of the olfactory bulb. In histological sections of the olfactory bulb impregnated with silver, bundles of fibers are seen entering each glomerulus from several angles. Topographical localization of the olfactory epithelial projections onto the olfactory bulb has thus been one of the central subjects in olfactory research.

Before we address this issue, quantitative observations on the olfactory system of the rabbit must be introduced. Allison and Warwick (1949) determined the following numerical data:

(1) The total surface area of olfactory mucous membrane in the nasal cavity of each side in four-month old rabbits is about 4.5 cm^2.

(2) There are on average 1.2×10^5 olfactory receptors per mm^2 of epithelium. The total number of receptors is, therefore, on the order of 50×10^6 in the nasal cavity of each side.

(3) In each OB there are about 1900 glomeruli, 45,000 mitral cells, and 130,000 tufted cells (cellules a panache). Thus each glomerulus receives impulses from something like 26,000 receptors and relays them through 24 mitral cells and 68 tufted cells.

Thus, olfactory information from receptor cells first converges on the glomeruli and then diverges in olfactory tract fibers.

V. Receptive Mechanism for Odors—Studies on the Electro-olfactogram (EOG)

A. Electrical Responses of the Olfactory Epithelium to Odors

1. Pioneer Studies

Working with the dog, Hosoya and Yoshida (1937) were the first investigators to study the electrical activity of the olfactory epithelium *in vitro*. They found that the outer surface of excised olfactory epithelium is electronegative to the inner surface by about 1 to 6 mV. A positive correlation was found between the magnitude of this potential difference and the amount of epithelial pigmentation. Furthermore, the potential difference gradually decreased in amplitude and disappeared over time, and mirror galvanometer (Shimazu Co., Kyoto) records showed various slow changes in response to the odors applied. These studies were performed in the physiological laboratory of the University of Taipei, Taiwan, and an abstract of Hosoya and Yoshida's work was published in German in 1937. Their experiments were discontinued as a result of World War II, and most of the data were destroyed by fire during the hostilities.

The surviving data were published by Yoshida in 1950. He noted that pure odorants like lemon oil, naphthalene, guaiacol, xylol, and menthol elicited monophasic action potentials (Fig. V-1 A); weak irritating odorants like camphor and carbon bisulphide generated biphasic potentials (Fig. V-1 B); and strong irritating odorants like dog urine and ammonia elicited triphasic or oscillatory potentials (Fig. V-1 C). Latencies of these action potentials were 0.18 to 0.5 sec for pure odorants and 0.002 to 0.39 sec for irritating odorants.

He also studied the relationship between these action potentials and the pigmentation of the olfactory epithelium. The amplitudes of the action potentials were found to increase in relation to the increase in pigmentation. Among 122 dogs, he found parallels between the gradual darkening of the olfactory epithelium color and the range of external body colors; from brown to light brown to yellow to blackish brown to

Fig. V-1. EOGs in the excised olfactory epithelia of the dog. A: lemon, B: camphor, C: acetic acid. Tracings of the original photographs obtained by a mirror galvanometer (Shimazu Co. Kyoto). (courtesy of Dr. H. Yoshida. Reproduced from *Olfaction and Taste* III, 1969, p.72. by copyright permission of the Rockefeller University Press, New York)

black. The color of the olfactory epithelium was darker when a dog had white areas in its body color than when it had none.

About 17 years later, Ottoson (1954) took up this subject and studied the electrical activity of the olfactory epithelium through the ethmoid bone in the rabbit. In 1956, he carried out extensive investigations on the frog. In both species he recorded a negative monophasic potential with a quick rise followed by a slow decline in response to a short puff of odorized air. He designated this potential the electro-olfactogram (EOG) (see p.33).

Ottoson (1956) identified the following properties of EOGs: The EOG appears only in the yellow pigmented area of the olfactory eminentia, but not in the other nonpigmented (pink) region of the olfactory cavity. It is temporarily abolished by a puff of ethyl ether. The amplitude of the EOG is, within certain limits, proportional to the logarithm of the stimulus intensity and to the volume of the stimulating air at a given stimulus strength. Equal amounts of odorous materials distributed in different volumes of air evoke EOGs of equal amplitude. The EOG varies in shape and time course according to the stimulus strength, differences in the wave form of the stimulating air current, and differences in the physicochemical properties of the odorants. With an increase in odor intensity, the potential rises at a faster rate, the crest of the response broadens, and the decay time lengthens. When an olfactory stimulus of long duration is applied, the EOG shows a shape with an initial peak followed by a plateau that lasts throughout the stimulation at a reduced level, and ending with an exponential decline to the baseline. When an odor is applied repetitively at short time intervals, adaptation occurs to that odor, but not to other odors. Thus, a phenomenon of selective adaptation was also found (pp. 164–165). Finally, by covering the olfactory epithelium with a thin plastic membrane which transmits only infrared radiation, Ottoson demonstrated that the EOG cannot be elicited unless the odorous material is brought into contact with the epithelium. Ottoson (1959a, b, c) continued these studies and published more data. Comparing these properties of the EOG with those of the other receptors, he proposed that the EOG acts as a generator potential (Ottoson, 1956, 1963a).

In 1958 Ottoson studied the relationship between the olfactory stimulating effectiveness and physicochemical properties of odorous compounds. He made comparative studies of the slow potentials in the olfactory mucosa and in the olfactory bulb, using rabbits (1959a) and frogs (1959b). Ottoson (1963a, b, 1970, 1971) has reviewed the results of his total research. A particularly insightful review on olfactory physiology was written by himself and Shepherd (1967).

In 1959, when Ottoson published his last three original papers (Ottoson, 1959a, b, c) and virtually halted his experiments on the EOG, the author and his collaborators initiated a series of experiments on the EOG (Takagi, 1969a, b) and published their first report (Takagi and Shibuya, 1959). The results of these studies will be introduced in the following sections (Takagi, 1978).

2. Kinds of Odorants Used as Olfactory Stimuli

Beginning with the study by Lord Adrian (1942), various kinds of odorants have been used by many investigators; amyl acetate being a most commonly used odor. Nearly all investigators have used this odor in their olfactory research chiefly because Adrian (1950a) first used it in his original studies on olfactory research. Indeed, amyl acetate has always been found to elicit robust responses in the olfactory system of the frog and other vertebrates including mammals.

Hosoya and Yoshida (1937) and Yoshida (1950) used lemon oil, geraniol, naphthalene, guaiacol, xylol, and menthol as pure odorants, camphor and carbon disulfide as weak irritating odorants, and ammonia and dog urine as strong irritating odorants. In addition, they used acetic acid, musk, and borneol. In frog experiments, Ottoson (1956) used amyl acetate, butyl alcohol, oil of clove, a series of aliphatic alcohols, and several other odors, but the number of odorants used did not constitute a large number. Later, Gesteland (1964) used 62 odors as stimuli. Takagi *et al.* (1969b) selected 122 odorants as stimuli from Amoore's (1962a) classification. Reviewing these experiments, it is clear that various kinds of odors have been used without any firm theoretical basis.

Recently, using photoactivable odorant, Delaleu and Holley (1983) were able to characterize two distinct categories of odorants in the olfactory epithelium of the frog: an "aromatic group" that included acetophenone, anisole, benzene, and nitrobenzene; and a "camphoraceous group" that included camphor, cineole, and isoborneol (Chapter VI, p. 228).

They were the first to provide a theoretical basis for the selection and use of odorants as stimulants in olfactory research. Their results supported the hypothesis that the compounds identified with each category interact selectively with two different types of receptor sites.

3. Methods of Odor Application

In neurophysiological research, an electrical stimulus with an accurate square form can easily be applied to the nerve tissue allowing precise control to vary freely the intensity and duration of the stimulus. An entirely similar situation exists in studies on vision and audition: the onset and offset of visual and acoustic stimuli can be precisely determined and their intensities and durations can be easily and accurately altered. In the case of odorous stimulation, however, the situation was entirely different and investigators in olfactory research have tried different approaches to apply odorant stimuli with an accuracy equivalent to that of visual or auditory stimuli. Ottoson (1956) used syringes to apply odors, soaking a small filter paper in an odorous solution of a desired intensity and placing it in a syringe. After the air inside the syringe was saturated with the odor, the plunger was pushed and the odorous air was projected onto the olfactory epithelium. The use of a filter paper soaked in different odorous solutions of different concentrations allowed variations in the kind and intensity of stimulating air. With this method, however, it is not possible to change the concentration of an odor continuously. Even to increase or decrease the concentration of an odor stepwise is time consuming and is often inaccurate.

Later, many investigators adopted this syringe method, including the author. In order to push the plunger, the author used a DC motor and drove it at a constant velocity using a square-waved direct current. The duration of a stimulus could thus be altered at will. Even so, the odorous jet vapor produced by this device was far from perfect. When the vapor was applied through a small diameter, odorless teflon tube, the stimulus had, in a very strict sense, a slowly increasing shape up to a steady level at the onset and at the end a slowly decreasing shape down to the zero level. The author, however, used this syringe method throughout all his experiments chiefly due to its convenience. Employing 32 syringes, eight kinds of odors were applied at four different concentrations in one of the monkey experiments (Chapter IX; Fig. IX-13B and p. 363).

Along this line of research Kauer and Shepherd (1975) devised their stimulating and monitoring methods. Later, along the same line, Getchell and Shepherd (1978a, b) designed an apparatus to produce step pulses of odorous vapors at different concentrations and of varying durations.

Some investigators constructed a multi-channel olfactometer using numerous glass tubes, joints, flowmeters, teflon tubes, three-way cocks and valves, and an air compressor (Tucker, 1963a). Keeping original solutions of odorants in a large bath containing water of a fixed temperature, vapor pressures of odorants could always be made constant regardless of the room temperature. By mixing saturated vapors of odorants with an odorless air passed through activated charcoal and silica gel, concentrations of odors could be changed at will although within a certain limit. In addition, odorous vapors could be supplied at optional velocities.

This apparatus, however, has several drawbacks. It requires much time and money for construction and occupies a large space inside a laboratory. Space limits usually restrict the number of odors for stimulation to two, three, or four kinds at most. In addition, once chosen, it is not easy to change the kinds of stimulating odors. If one wants to change the stimulating odor in the system, all glass and teflon tubes, three-way cocks, etc. in the line of the old odor must be cleaned .completely. The time and labor required for this process cannot be underestimated.

In conclusion, olfactory research still awaits the development of an ideal apparatus for the application of odors, and for convenience the glass syringe method remains the method employed in many laboratories.

4. Concentrations of Odors in the Mucus

It is well known that the olfactory cell has very low thresholds to many odors; studies have found that threshold concentrations of many odors calculated in the gas phase are very low in various animals. However, the concentrations of odorants in the mucus overlying the olfactory receptors or in the lipid phase of the cell membranes are presumed to be far higher because the odorant compounds will preferentially partition into the liquid phase. Senf *et al.* (1980) calculated the mucus concentrations of the odorants in their study by assuming that the air/mucus partition coefficients are the same as the air/water partition coefficients. In each case, the concentrations were at least 100 times higher in the mucus than in the air. This is a notable finding since the olfactory sensitivities of animalsand humans have been reported mostly on the basis of odorous concentrations in the air.

5. EOGs as Biological Phenomena

Hartman (1954) reported that sustained potentials may be recorded from nonbiological systems in response to certain odorants. This suggested the possibility that the EOGs recorded by many workers may be nonbiological phenomena. Mozell (1962) showed that nonbiological potentials responded to increases in stimulus intensity, decreased with the introduction of cocaine, and even showed adaptation. Thus, it was found to be impossible to differentiate between nonbiological and biological potentials. However, Kimura (1961) reported that if Ringer-agar type electrodes were used, such nonbiological potentials were not recorded. Kimura (1961) and Mozell (1962) warned that platinum, tungsten, and nichrome electrodes, and 3M KCl micropipettes were all able to record nonbiological potentials from Ringer-soaked cotton when an odorant was puffed toward the electrode-cotton junction, and recommended the use of Ringer-agar (silver-silver chloride) electrodes. Since Ottoson (1956) and most other investigators including the author used the recommended types of electrodes, it is not necessary to consider the intervention of such nonbiological potentials in their experiments (Getchell, 1974a).

6. EOGs Elicited by Various Kinds of Odors

As briefly stated before (p. 149), Gesteland (1964) applied 62 odors onto the olfactory epithelium of the "frog immobilized with *d*-tubocurarine," and found that 35 odors elicit negative EOGs, while the remaining 27 odors elicit strong initial positive potentials followed by negative potentials. Takagi *et al.* (1969b) applied vapors of 122 odorants in three concentrations onto the olfactory epithelium of the "decapitated bullfrog." Among them, 106 odorants (87%) elicited only negative EOGs (group I odors). The remaining 16 odorants, however, elicited genuine positive or negative-positive type EOGs (Table V-1). Among the 16, six odorants (group II odorants in Table V-1) elicited only positive EOGs. Increasing the concentrations only resulted in increased amplitudes (Fig. V-2). All the odorants in group III (Table V-1, III), except eugenol, elicited negative EOGs at low concentrations and, in three cases, these were accompanied by positive afterpotentials. Increasing the concentrations resulted in the appearance of positive afterpotentials in all cases and increased the amplitude of those seen previously. The negative EOGs increased, decreased, or did not change following concentration increases. This type of EOG has also been found in the rabbit (MacLeod, 1959). Three odorants in Group IV elicited only negative EOGs at low concentrations (1/36), and only positive EOGs at high concentrations (1/6 and 1/1; Fig. V-3). It is noteworthy that 1, 2-dichloroethane (Matheson, Coleman & Dell Co.), which is equivalent to ethylene chloride (Kokusan Chemicals Co., Tokyo), elicited only positive EOGs and was included in group II, while the latter was included in group IV.

In degenerated olfactory epithelia preparations, the majority of odors in these three groups did not elicit any negative EOGs, and only small-amplitude negative EOGs were elicited in three cases. However, they did elicit clear positive EOGs as in normal epithelia which increased in amplitude with increasing concentrations.

Table V-1. The magnitudes of negative and positive EOGs elicited by groups II, III and IV odors.

Group	Odor	Normal						Degenerated					
		Neg EOG			Pos EOG			Neg EOG			Pos EOG		
		1/36	1/6	1/1	1/36	1/6	1/1	1/36	1/6	1/1	1/36	1/6	1/1
II	Acetonitrile	0	0	0	1	2	3	0	0	0	1	2	3
	Carbon disulphide	0	0	0	1	1	1	0	0	0	1	1	1
	Chloroform	0	0	0	2	3	4	0	0	0	2	3	5
	Dichloromethane	0	0	0	1	2	4	0	0	0	1	2	4
	1, 2-Dichloroethane	0	0	0	1	2	3	0	0	0	1	2	3
	Trichloroethylene	0	0	0	1	1	1	0	0	0	1	1	1
III	*n*-Butyl alcohol	2	2	3	0	0	2	0	0	1	0	0	2
	iso-Butyl alcohol	1	1	1	1	2	3	0	0	0	1	2	3
	sec-Butyl alcohol	2	2	3	0	0	2	0	0	1	0	0	2
	tert-Butyl alcohol	1	1	1	1	2	5	1	1	1	1	2	4
	Ethyl acetate	2	4	4	0	0	1	0	0	0	0	0	1
	Eugenol	0	0	1	0	0	1	0	0	0	0	0	0
	Tetrahydrofuran	1	3	4	1	2	4	1	1	1	1	2	4
IV	Formic acid	2	0	0	0	2	3	0	0	0	0	2	3
	Methylamine	4	1	0	0	1	3	0	0	0	0	1	3
	Ethylene chloride	3	0	0	0	1	2	0	0	0	0	1	4
	Ethyl ether												

Fractions 1/36, 1/6, and 1/1 indicate dilution rates of saturated vapors. (from Takagi *et al.*, 1969b. Reproduced from *Olfaction and Taste*, III, 1969, p.97, by copyright permission of the Rockefeller University Press, New York)

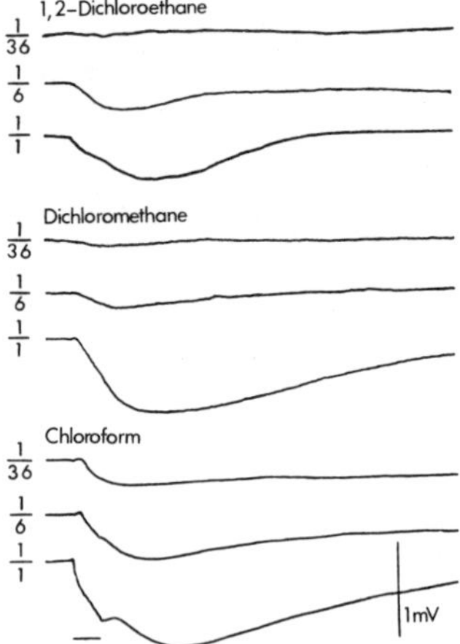

Fig. V-2. Positive EOGs elicited by three odorants in group II at three different concentrations. (from Takagi *et al.*, 1969b. Reproduced from *Olfaction and Saste,* III, 1969, p.98. by copyright permission of the Rockefeler University Press, New York)

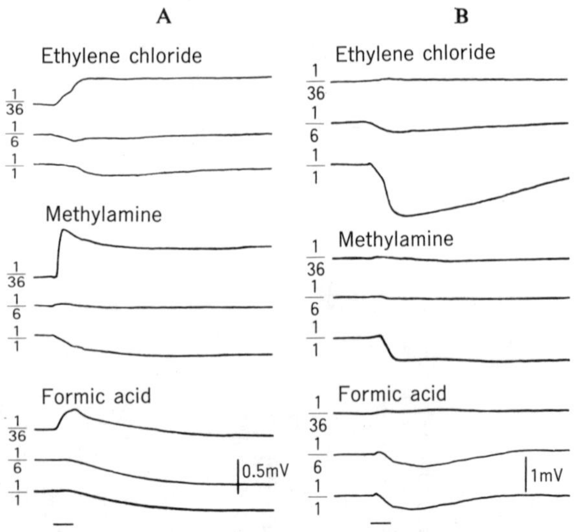

Fig. V-3. Negative and positive EOGs elicited by group IV odorants. A: EOGs in normal olfactory epithelium, B: in degenerated olfactory epithelia after olfactory nerve sectioning. Positive EOGs remained while negative EOGs were not elicited. (from Takagi *et al.*, 1969b. Reproduced from *Olfaction and Taste* III, 1969, p.99. by copyright permission of the Rockefeller University Press, New York)

7. *EOGs Recorded under Different Experimental Conditions*

When EOGs were recorded in the olfactory eminentia of "decapitated bullfrogs," vapors of chloroform and other odors that belong to group II to IV often elicited positive on- and negative off-potentials accompanied by large positive afterpotentials. When EOGs were recorded in the same area of the olfactory eminentia in bullfrogs "immobilized with *d*-tubocurarine," the same odorants elicited negative on- and off-potentials, and the positive afterpotentials were relatively small in amplitude or nonexistent. Thus, it is clear that the EOGs change greatly in shape and polarity depending on the experimental conditions. Differences between the two frog preparations were that oxygen was supplied through blood circulation and *d*-tubocurarine was applied to immobilize the animal in the latter, whereas EOGs were recorded without supply of oxygen through circulating blood and without the drug in the former. To remove the possible effect of *d*-tubocurarine, bullfrogs were immobilized by high spinal transection. EOGs recorded in this noncurarized and oxygen-supplied condition were found to be entirely similar to those recorded in curarized bullfrogs.

In other trials, EOGs were at first recorded after high spinal transection. Then, with the recording electrode *in situ*, blood circulation was stopped by sectioning the carotid arteries or by decapitation. The negative EOGs that had appeared when circulation was maintained disappeared and positive EOGs appeared or increased after blood supply ceased.

From these experiments, the author concluded that negative EOGs are vulnerable to a shortage of oxygen and hence they decrease in amplitude when O_2 is not sufficiently supplied. As a result, positive EOGs which had been masked by negative EOGs and/or newly elicited by the lack of O_2 appeared.

Another important contribution of oxygen to the generation of the EOG was demonstrated by other investigators. Dmitriew *et al.* (1984) studied the effect of hyperoxia on the total electrical activity of the olfactory epithelium of the frog. Applying odorous vapors of amyl acetate, butanol, naphthalene, and acetic acid onto the frog olfactory mucosa, they recorded EOGs in air and oxygen media under conditions of normal and high oxygen pressure (0.2 MPa and 0.7 MPa). They found that the EOG amplitude was higher in oxygen, probably due to oxygen saturation of the olfactory mucosa as well as to the activation of processes related to EOG generation.

These experiments clearly indicate that EOGs should be recorded in animals with constant oxygen supply. Furthermore, all these findings support the author's hypothesis that EOGs are composed of negative and positive potentials and that the final shape of the EOG is determined by competition between these two opposing potentials.

EOGs have been studied in the excised olfactory epithelium of the dog (Hosoya and Yoshida, 1937; Yoshida, 1950), in decapitated frogs (Ottoson, 1956, 1958; Takagi *et al.*, Takagi *et al.*, 1969b) and in frogs or turtles immobilized with *d*-tubocurarine or other anesthetics (Ottoson, 1959a, b; Kimura, 1961; Mozell, 1962; Gesteland *et al.*, 1964, 1965; Takagi *et al.*, 1969b). The results of these studies were often contradictory and no one has yet attributed such large differences in the EOGs to the differences in the experimental conditions of the animals.

Incidentally, an interesting finding which puzzled the author was obtained while EOGs were recorded when applying 122 odors. Methanol and ethanol elicited only positive EOGs in bullfrogs immobilized with *d*-tubocurarine. When these bullfrogs were decapitated, the same odors elicited only negative EOGs (Iino and Takagi, unpublished data). Gesteland (1964) also obtained positive EOGs in response to these two alcohols in pithed or curarized frogs, but he did not record EOGs to them in decapitated frogs.

Reversal of positive into negative EOGs after arterial oxygen supply was interrupted cannot be simply explained by the above hypothesis. In the present stage of experiments, the author still finds it difficult to explain this reversal phenomenon. The reversal may be attributed to some stimulating properties characteristic of methanol and ethanol. This finding again indicated that EOGs appear differently depending upon the experimental conditions. Above all, the author believes that this finding is important because it indicates the complexity of the EOG generating mechanisms which remain to be clarified.

8. EOGs Recorded at Different Loci

When EOGs were recorded at seven different loci in the olfactory epithelium, EOGs of different shapes and magnitudes were recorded (Shibuya and Takagi, 1962). These findings suggested that processes generating the negative and positive potentials are unevenly distributed in the olfactory epithelium. This subject was later studied extensively by Mustaparta (1971), Thommesen and Döving (1977), and Mackay-Sim and Shaman (1984).

When the EOGs recorded in the olfactory eminentia were compared with those recorded in the ceiling epithelium of the same decapitated bullfrog, the same odors elicited potentials of different shapes and magnitudes (Fig. V-4). In general, positive EOGs were elicited markedly in the excised ceiling epithelium, while the same odors elicited negative EOGs far more strikingly in the olfactory eminentia. In the olfactory eminentia of bull-

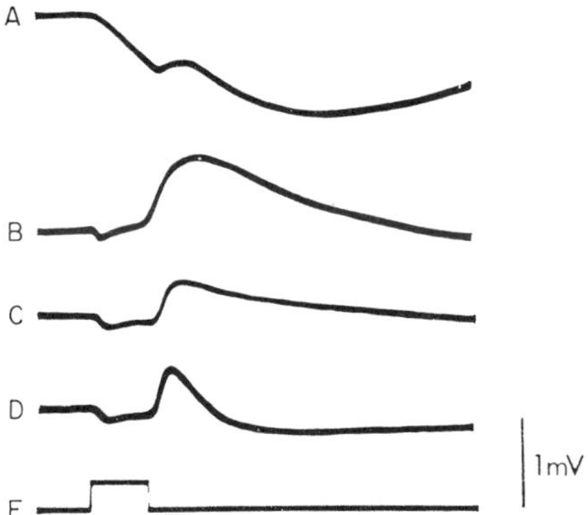

Fig. V-4. Positive and negative EOGs of various types elicited by chloroform vapors under different conditions. E: Time of stimulation (4 sec) is indicated at bottom left. (from Takagi *et al.*, 1969b. Reproduced from *Olfaction and Taste,* III, 1969, p.100. by copyright permission from Rockefeller University Press, New York)

frogs immobilized with *d*-tubocurarine, chloroform vapor that elicited large positive EOGs in the excised ceiling epithelium (Fig. V-4 A) and also in olfactory eminentia of decapitated bullfrogs elicited only small positive on- and afterpotentials and large negative off-potentials (Fig. V-4 B, C, and D). It is thus apparent that positive potentials are wholly or partly masked by predominant negative potentials in the curarized bullfrog.

9. Latency of EOGs

EOGs recorded from the vertebrate olfactory epithelium are characterized by long latencies. When Ottoson (1956) attempted to measure the latency of the frog EOG, he encountered two difficulties. The first one was to determine the moment when the stimulating air reached the receptive membrane of the olfactory cell. He estimated the time of this event by noting the contact between the olfactory epithelium and a gold leaf bent by a puff of an odorous vapor. The second difficulty was to determine the time when the EOG started. In an effort to reduce error, he took three readings from each record, and obtained the mean values from 12 experiments. Thus, the latencies of EOGs varied from approximately 400 msec at a low stimulus intensity to about 200 msec at the highest stimulus intensity. However, these must be regarded as only approximate values for latencies.

It is generally agreed that in order to produce excitation odorous particles must be brought in direct contact with the receptors. Before such contact is established, the particles must pass through the layer of watery mucus covering the olfactory cells. The time required for this passage depends on a number of factors, among which the solubility of the odorous substance and the thickness of the mucus layer may be the most important. The first stage in the excitatory process may be assumed to involve the adsorption of the odorous molecules onto the chemosensitive membrane. This stage may be followed by several intermediate processes, which finally lead to a depolarization or hyperpolarization of the olfactory cells. Among such intermediate processes, long latencies may be primarily due to the diffusion of odorous molecules from the surface of the olfactory mucus deep into the receptive sites in the region of the apical knob of the olfactory dendrite and the proximal portion of the cilia (Getchell *et al.*, 1980).

10. Effects of Polarizing Currents

We have seen that research demonstrated the presence of negative and positive EOGs. Their generative mechanisms became the next subject for inquiry. The first step was to examine the effects of polarizing currents upon EOGs. The electronegative on-EOG elicited by a general odor like amyl acetate increased its magnitude in accordance with the increase in anodal current, while it decreased its magnitude with the increase in cathodal current (Fig. V-5 A). Similar relations were found in the case of the vapors of an organic solvent like ethyl ether at "low" concentrations. Conversely, the on-EOG elicited by the vapors of ethyl ether of "high" concentrations decreased its magnitude in accordance with the increase in anodal current, while it increased its magnitude with the increase in cathodal current (Fig. V-5 B). Thus, depending on the kinds and concentrations of odors, an entirely opposite relation was found in the resulting EOGs. These results led to the postulation that two receptive processes are present in the olfactory epithelium: One is an ordinary excitatory process which produces a negative EOG; the other is a different kind of process activated only by the vapors of highly concentrated organic solvents. This process evokes an entirely opposite reaction to that generally found in excitable tissues when an electrotonic current is applied (Gesteland *et al.*, 1965).

In this connection, an interesting reversal phenomenon of the potential amplitude-electrotonic current relation must be added to these data. When another organic solvent, chloroform, was applied at a low concentration, the relationship shown in Fig. V-5 A was observed. However, when the concentration was increased to higher levels, the relationship began to

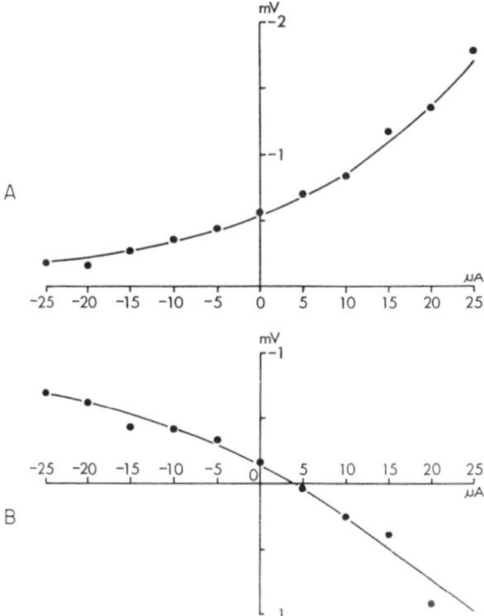

Fig. V-5. Analysis of EOG by polarizing currents. (from Higashino and Takagi, 1964. Reproduced from *The Journal Physiology,* 1964, Vol. 48, pp.328 and 329. by copyright permission of the Rockefeller University Press, New York)

change and finally the reversed relationship shown in Fig. V-5 B was observed. This reversal was taken to indicate that chloroform vapor had the characteristics of a general odor at low concentrations, but that the characteristics of an organic solvent increased and surpassed those of a general odor as the concentration increased. Later, Takagi *et al.* (1966, 1968a, 1969b) demonstrated that the former derives from the activity of the olfactory cell, while the latter derives from the secretory activity of the supporting cell (pp. 162 and 170).

11. Impedance Changes of the Olfactory Epithelium during EOGs

The same conclusions obtained by polarizing current experiments were reached by an entirely different approach. Lettvin and Gesteland (1965), Gesteland *et al.* (1965), and Gesteland (1967) measured impedance changes of the olfactory epithelium during EOGs. They found that the positive-drive process could be identified at a low frequency with a change in phase and the negative-drive process with a change in magnitude. They also found linear and nonlinear interactions between the EOGs elicited by the simultaneous or successive applications of two different odors. From these findings, they were able to distinguish two opposing processes: the positive-driving, or resting-level-seeking process was identified with an inhibitory action of active sites in the receptors, and the other was identified with excitatory action of active sites (Gesteland *et al.*, 1965).

12. Finding of an "Off-potential" in EOGs

The appearance of an off-potential in the olfactory epithelium of the decapitated frog was first reported by Takagi and Shibuya (1959). The off-potential often appeared when saturated vapors of amyl acetate, amyl alcohol, isopropyl alcohol, and butyl alcohol were applied (Takagi and Shibuya, 1960a).

In the course of EOG studies on the frog and toad, Takagi *et al.* (1960) found that anesthetics, such as ethyl ether, chloroform, and chlorethyl, and organic solvents, such as acetone, petroleum ether, carbon tetrachloride, toluol, and benzene almost always elicited the off-potential when they were applied at saturated concentrations.

13. Anesthetic and Stimulating Effects of Ether, Chloroform, and Others

The effects of ethyl ether and chloroform were studied extensively and some interesting findings were obtained. These vapors naturally have anesthetic effects on the olfactory epithelium, and when applied for a certain duration they reversibly depress the subsequent odor-elicited EOGs (Ottoson, 1956). Such a reversible anesthetic effect was also demonstrated by the following experiment. EOGs and induced waves elicited at a certain interval by repetitive application of amyl acetate vapor were depressed during simultaneous application of ethyl ether, but they gradually began to reappear soon after the cessation of ethyl ether application (Takagi *et al.*, 1960). Thus, the anesthetic effect was transient in nature.

Since the anesthetics have their own stimulating odors, it is also clear that they elicit EOGs in the olfactory epithelium and induce waves in the olfactory bulb. Takagi *et al.* (1960) often applied saturated vapors of these anesthetics for 1 to 10 sec or longer, but they found that EOGs and induced waves elicited by the same odors were as large as before when a rest period of a few minutes was interposed. Thus, they confirmed that short application of these anesthetics and organic solvents can be repeated without any harm to the olfactory epithelium.

When ethyl ether was applied at the concentration of 1/16 of the saturated vapor, an on-EOG appeared. When it was applied at the concentration of 1/4, on- and off-EOGs appeared. When it was applied at the saturated concentration (1/1), the on-EOG disappeared, or markedly decreased in amplitude, while the off-EOG appeared. Although the on-EOG entirely or almost completely disappeared, the on-induced wave continued to appear as powerfully as before, while corresponding to the appearance of off-EOGs, off-induced waves appeared (Fig. V-6). Important findings in this study included (1) the appearance of the inducedwave in the OB regardless of the on-EOG and (2) appearance of the off induced wave in the OB before the onset of the off-EOG at the end of stimulation (Fig. V-7). These two results presented challenges to the theory proclaiming the EOG to be a generator potential (Ottoson, 1956; p. 178).

When ethyl ether was applied successively with longer durations of 1, 3, 5, and 7 sec, the off-EOG that was masked at the beginning (1 sec) manifested increasingly. Correspondingly, off-induced waves in the OB also appeared.

14. Potential Oscillations Appearing on the EOGs

Ottoson (1956, 1959a) found that rhythmic waves, 20 per sec, infrequently superimpose on the crest of the EOG when strong odor stimulation is applied to the olfactory epithelium.

In contrast, Takagi and Shibuya (1960b, d, e) found that such potential oscillations appeared markedly in the toad (*Bufo vulgaris Japonicus*) very frequently and only in winter. They extensively studied the properties of these interesting oscillations and the conditions which elicited them (Takagi and shibuya, 1961; Adrian, 1957).

15. EOGs in Various Animals

EOGs found in the dog and frog by pioneer workers were further studied in some fish (carp, Channa, and catfish) by Shibuya (1960), in the newt (Shibuya and Takagi, 1962, 1963), in the land tortoise and turtle (Shibuya, 1964; Tucker and Shibuya, 1965; Shibuya, 1969), in the land snail (Suzuki, 1967), in vultures (Shibuya and Tucker, 1967), and in the rabbit (MacLeod, 1959). The presence of the electro-antennogram (EAG) found in insects shows that the slow potential, such as the EOG and the EAG, is a common response of the olfactory receptor organ to odors in the animal world (Boeckh, 1967; Kaissling, 1987).

16. EOGs in Humans

Osterhammel *et al.* (1969) were the first to record EOGs in humans (Fig. I-8). Suga and Nakashima (1972a, b, 1973) recorded EOGs in many patients in the otorhinolaryngdogical clinic of the University of Kyushu, and found little difficulty in obtaining recordings. In particular, the introduction of the Selfoscope (Chapter III, p. 93) into the Japanese clinic appeared to facilitate greatly the insertion of a recording electrode onto the olfactory epithelium (Chapter I, p. 34 and Chapter III, B, p. 93).

B. Negative EOGs

In their initial experiments, the author and his collaborators (Takagi and Shibuya, 1959, 1960a, b, c, d; Takagi *et al.*, 1960) used saturated vapors of amyl acetate, butanol, ethyl ether, chloroform, and others as stimuli. The results obtained were criticized by many sources (e.g., Ottoson, 1970, 1971). In response the author duplicated all of his old experiments applying the same odorants at lower concentrations in decapitated frogs, as Ottoson (1956) did, and later in immobilized frogs and bullfrogs, as Gesteland (1964) and Ottoson (1959a, b) did. In some cases, the concentrations used were 1/2, 1/4, 1/8, and 1/16 or 1/10, 1/100, or 1/1000 that of saturated vapors. In other cases, saturated vapors were diluted to 1/6 and 1/36 with deodorized air. The veracity of the results of the earlier experiments were reconfirmed.

Since EOGs often appeared differently depending on the kinds of odorants, Takagi *et al.* (1969b) studied shapes, magnitudes, and polarities of EOGs by applying 122 odors onto the olfactory epithelia of decapitated bullfrogs. Negative EOGs were elicited by 106 odors (87%) and increasing the concentrations (from 1/36 to 1/1) of these odors simply increased the amplitudes of the negative EOGs. They were classified as group I odors. These 106 odors did not elicit negative EOGs in the degenerated olfactory epithelia many days after sectioning the olfactory nerves. The results were analyzed in comparison with the EOGs elicited in normal epithelia (Table 1 in Takagi *et al.*, 1969b).

1. Magnitudes
(a) Strength of odorous stimulation

Using butanol, Ottoson (1956) studied the relationship between the amplitude of the EOG and the strength of a stimulating odor. The EOG increased in amplitude by almost 10 fold, showing an elongated S-shaped curve when the stimulus strength was increased by a factor of 100 from 0.001 M to 0.1 M. Takagi *et al.* (1969b) replicated Ottoson's experiment, applying amyl acetate and butanol as stimulants and obtained the same results.

(b) Application of an odor at different velocities

Takagi (unpublished data) applied a fixed volume of an odorous vapor of 0.01 M amyl acetate solution at three different velocities. Although the duration of odorous stimulation became shorter as the velocity was increased, the negative EOG increased in amplitude with the increase in velocity. Naturally, negative EOGs successively increased in amplitude when the vapor was applied at higher velocities for the same duration.

Application of ethyl ether vapor at increasing velocities had interesting effects on the EOG (Fig. V-6). For instance, application at a velocity of 0.07 cc/sec produced an on-EOG. A velocity of 0.14 cc/sec elicited on- and off-EOGs. Finally, when it was applied at a velocity of 1.4 cc/sec, the

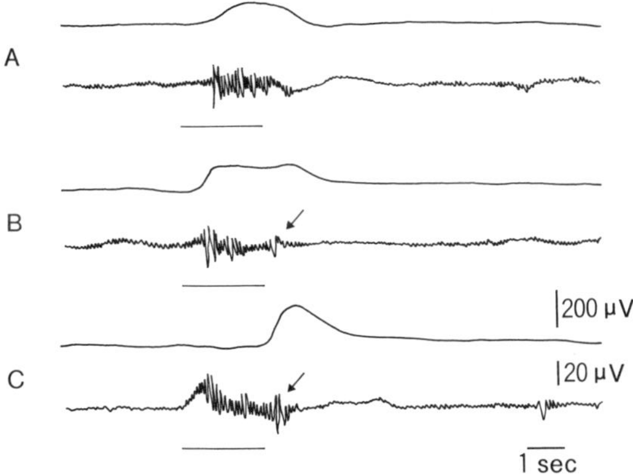

Fig. V-6. Transition of potential types when ethyl ether vapor was applied at increasing velocities from A to C. In each group, the top and bottom record show the potentials of the olfactory epithelium and bulb, respectively. Arrows indicate the corresponding off-induced waves which appeared. (from Takagi *et al.*, 1960)

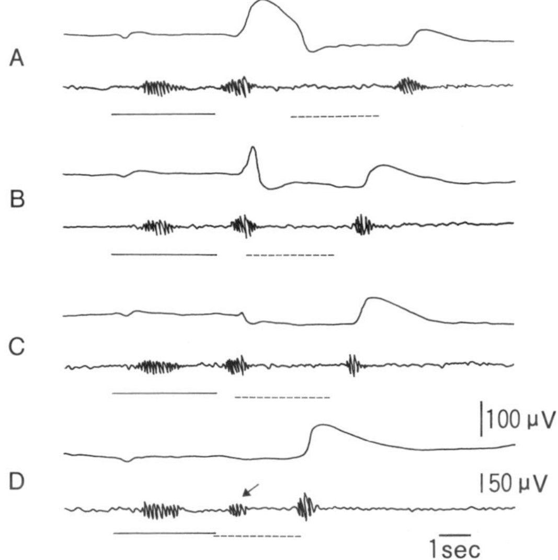

Fig. V-7. Gnawing effect of chloroform vapor on the off-EOG elicited by a preceding ethyl ether vapor. In each group, the top record indicates EOG, middle induced-wave in the OB, and bottom durations of odorous stimulation. (from Takagi *et al.*, 1960)

on-EOG disappeared and the off-EOG increased. Although the on-EOG disappeared, the on-induced wave appeared independently of the on-EOG. Corresponding to the off-EOG, off-induced waves also appeared and again they began to appear before the onset of the off-EOGs (Fig. V-7).

(c) Gnawing effect of anesthetic vapors on the negative EOG

Ethyl ether and chloroform vapors show "gnawing" effects on the negative EOGs elicited by an odor when they are applied while the preceding negative EOGs are present. When the interval between the application of an odor and the subsequent application of ether or chloroform vapor is gradually shortened, the gnawing effect becomes increasingly stronger, resulting in the complete elimination of the negative EOG. Figure V-7 shows such a case: the negative off-EOG elicited by ethyl ether vapor is gradually gnawed by the subsequent application of chloroform vapor at decreasing intervals. In the last stage, the off-EOG is completely abolished, but the off-induced wave elicited by the preceding ethyl ether persists. This finding again contradicts the hypothesis that the negative EOG is a generator potential (p. 178; Takagi *et al.*, 1960).

2. Shapes

(a) Changes in stimulus strengths

A short puff of an odorous vapor elicits a negative potential with a steep rising phase followed by an exponential fall toward the baseline. Ottoson (1956) studied changes in the shape of EOGs by applying the same odor at different strengths. The EOGs thus elicited differed not only in magnitude but also in shape. The rising phase was found to be somewhat steeper with increasing strengths, and the greatest change was found in the shape of the falling phase of the EOGs, which was prolonged by the application of a stronger odor.

(b) Changes in stimulus duration

When the duration of a stimulus was gradually increased from a short puff, EOGs initially increased in amplitude, and the falling phase became elongated. A plateau then began to appear in the middle of the falling phase and was maintained as long as the stimulation continued. The plateau phase was different in magnitude according to the strength or kind of the applied odor (Ottoson, 1956).

3. Adaptation, Selective Adaptation, and Recovery

When the olfactory epithelium is repeatedly stimulated with the same odorous vapor at short interstimulus intervals, EOGs successively decrease in amplitude. The rate of amplitude decrement increases with increases in the strength of the odor or decreases in the interstimulus interval (Ottoson, 1956). The period required for recovery of the EOG after adaptation is prolonged as the intensity of the stimulus increases. Van Boxtel and Köster (1978) also found a negative relationship between stimulus intensity and recovery rate of the EOG after adaptation, and between the recovery rate and the duration of the rising and decay phases of the EOG. They also found that the olfactory receptors were very resistant to adaptation and that the recovery was almost complete after 25 sec despite the strong stimulus concentrations. Of related interest, Mozell (1962) found that adaptation of the frog EOG was accompanied by a decrease in integrated olfactory nerve activity.

Even when adaptation occurred to one odor, an entirely different odor could elicit an EOG with its original amplitude which the new odor could produce when applied singly. Greater similarities between this odor and the first odor led to decreases in the ability of the second odor to produce the EOG. Thus, a phenomenon of selective adaptation was observed (Ottoson, 1956).

4. Relationships between EOGs and Physicochemical Properties of an Odorous Compound

Ottoson (1958) studied the relationship between an odor's effectiveness in producing olfactory stimulation and its physicochemical properties for a number of odorous compounds. The magnitude of the EOG in the frog was used as a measure of the stimulatory efficacy and its relationship to the chemical properties of a given compound was examined. The results revealed that chemical properties such as unsaturation, residual valencies, or functional (osmophoric) groups were not necessary for a substance to elicit the EOG. In light of findings from other investigations linking the olfactory stimulative effectiveness of a compound to its molecular architecture (Dethier and Chadwick, 1950; Moncrieff, 1951; Mullins, 1955), these results lend further support to Wright *et al.*'s (1956) claim that the EOG generating process is physical in nature.

Ottoson (1958) also studied the relationship between the physical properties of an odorous molecule and the magnitude of the elicited EOG. In this connection, he took advantage of the fact that as homologous series are ascended there is a gradual change in physical properties, while the chemical properties remain unaltered. When the number of carbon atoms (c) is increased in homologous alcohols, aldehydes, or ketones of equimolecular concentrations, he found that the EOG increased in magnitude up to a certain maximum. This finding was generally supported by Higashino *et al.* (1961). Ottoson, however, could not find such a relationship in the series of fatty acids. When water solubility, vapor pressure, and the magnitude of the EOG were compared, an inverse relationship was found between the former two and the magnitude of the EOG. In addition, he found that the partial vapor pressure (Pt) for alcohols at concentrations which elicit EOGs of equal magnitude increase approximately linearly with the saturated vapor pressures (Ps). Furthermore, in terms of thermodynamic activities (Pt/Ps) for a substance in the vapor phase, alcohols of intermediate chain length have equal stimulating power, that is, an ability to elicit EOGs of equal magnitude, while short-chain alcohols (C_1, C_2, C_3) are less effective.

Further investigations of these phenomena in aqueous and vapor phases by Higashino *et al.* (1961) agreed with the earlier results for alcohols of intermediate chain length, but contradicted the results obtained for the short- and long-chain alcohols (C_1-C_3 and C_7-C_8). These findings suggest that the Brink-Posternak rule (1948), which states that substances of equal thermodynamic activity have biologically equal stimulating effectiveness, is only applicable for a small range of values (A$\,=\,$0.01 and 0.1) of thermodynamic activity.

Higashino *et al.* (1961) compared the olfactory stimulating effectiveness of homologous alcohols on an equimolar basis. Equimolar alcohol vapors elicited the largest EOG at C_6, when the saturated vapor pressure of C_8 was used as a standard. When the values of the thermodynamic activities were compared with the magnitudes of the EOGs, a striking correspondence was found from C_1 to C_6. A peak in stimulatory efficacy at C_6 found in all of their experiments may be attributed to the finding that alcohols are water soluble at C_1 and C_2, oil soluble at C_7 and C_8, and both water and oil soluble at C_3 to C_6. This property of dual solubility was thought to be a causative factor resulting in larger magnitude EOGs at C_3 and C_6 than at other chain lengths.

This line of investigation may play an important role in clarifying the receptive mechanism of the olfactory cell in the future.

5. Comparison of the Stimulatory Efficacy of Various Odorants Using Negative EOGs

Negative EOGs are elicited by various odorants. The amplitudes of these EOGs are thought to reflect the stimulatory powers or efficacy of given odorants. Working from this hypothesis, Higashino *et al.* (1961) compared the stimulatory efficacy of two odorants by applying them successively in one order and then in reverse order. The interstimulus-interval was set at 5 sec so that the second stimulation could occur in the relative refractory period of the first stimulation. Thus, if the stimulatory efficacy of the two odorants were the same, the second EOG would decrease relative to the first. A second EOG larger or equal to the first would indicate that the second one had stronger stimulatory efficacy than the first. For example, C_3-alcohol was applied after C_2-alcohol, and after a sufficient pause the order was reversed and the stimulatory efficacy of C_3 was shown to be larger than that of C_2-alcohol.

By repetition of this method, the stimulatory efficacy was determined for homologous series of alcohols, acetates, and ethers. The results were as follows:

In alcohol series,

$$C_2 \leqq C_1 \leqq C_8 < C_7 < C_4 < iC_3 < C_3 < iC_5 < iC_4 < C_5 < C_6.$$

In acetate series,

$$C_1 < C_2 < iC_3 < C_3 < C_4 < C_5 \leqq iC_4 \leqq iC_5.$$

In ether series,

$$C_6 < iC_5 \leqq C_5 < C_2 < C_4 < iC_3 \leqq C_3.$$

A similar result was obtained in the cases of the induced wave in the OB.

6. Origin of the Negative EOG

In a series of studies, Takagi and Yajima (1964, 1965) attempted to identify the origin of the EOG. By sectioning the olfactory nerve, they induced olfactory cell degeneration in the olfactory epithelium of the bullfrog, and found that electrical responses to odors disappeared in the degenerated epithelium. Thus, they concluded that the electrical responses to odors, namely the negative on- and off-EOGs, originate from the activity of the olfactory cell (further details of this experiment will be explained on p. 182).

An entirely different approach to this problem was taken by Suzuki (1967). He removed the receptor cell layer of the tentacular epithelium in the olfactory organ of the snail under a dissecting microscope with a pair of sharpened forceps.

In the olfactory epithelium in which the olfactory cell layer was removed but the supporting cell layer was left intact, the negative EOGs elicited by amyl acetate vapor were reduced to $18.5 \pm 6.3\%$ of the original amplitude; the positive EOGs elicited by ethanol were only reduced to $64.5 \pm 7.7\%$ of the original. Considering the difficulties of the mechanical manipulation, the results strongly support the author's conclusion stated above. Suzuki's results also suggested that the positive EOGs are produced by the supporting cells (p.173).

7. Ionic Mechanisms of Negative EOGs

Takagi (1968) and Takagi and Wyse (1965) attempted to clarify the ionic mechanisms of negative EOGs by changing the ionic environment

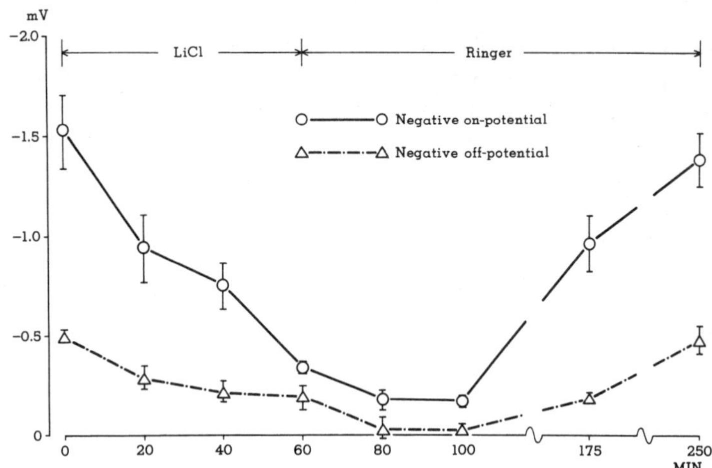

Fig. V-8. Substitution of Li⁺ for Na⁺ in Ringer solution. Decrease and disappearance of EOG in Li⁺ Ringer solution (from
0 to 60 min) recovered in normal Ringer solution (from 60 to
250 min). (from Takagi *et al.*, 1968b. Reproduced from *The
Journal of General Physiology,* 1968, vol 51, p.559. by copyright
permission of Rockefeller University Press, New York)

of the olfactory epithelium. A piece of the olfactory mucosa was excised
from the ceiling of the olfactory cavity and was spread flat, receptor side
upward, on a filter paper mounted on a perforated lucite platform which
was suspended across a groove made in paraffin wax inside a Petri dish.
The olfactory epithelium was thus immersed into various solutions to
change the ionic environment of the excised mucosa (ion substitution
experiment).

(a) The role of Na⁺

First, the role of Na⁺ in the generation of the negative EOG was examined by replacing Na⁺ with Li⁺. When the olfactory epithelium was
immersed in this Li⁺ Ringer's solution, the negative EOG decreased in
amplitude and showed a tendency to disappear. When this solution was
replaced with a normal Ringer's solution, the negative EOG recovered
to the previous magnitudes (Fig. V-8). All mono-, di-, and trivalent cations were thus examined for their ability to substitute in the role of Na⁺.
Sucrose, choline⁺, tetraethyl ammonium (TEA), and hydrazine were also
examined, but none of them could substitute for Na⁺ in the generation of
the negative EOG (Takagi *et al.* 1968a, b, 1969a).

An interesting reversal phenomenon was found when Na⁺ was replaced
by K⁺, and vice versa, in the Ringer's solution. In the K⁺ solution, negative EOGs reversed their polarity. When the K⁺ in the solution was replaced by Na⁺, the reversed EOG returned to its original negative state.

In other studies, K^+ was not replaced by Na^+ but by one mono-, di-, and trivalent cations and also by either sucrose, choline$^+$, TEA$^+$, or hydrazine. In all of these solutions, the reversed potentials showed no recovery whatsoever. These two types of experiment thus demonstrated the essential role of Na^+ in the generation of negative EOG.

(b) The role of K^+

When K^+ was removed from the Ringer's solution, the negative EOGs first increased and then decreased in amplitude. When the concentration of K^+ in the Ringer's solution was increased within a certain limit, the negative EOGs increased in amplitude. In the former case, the increase was thought to indicate a contribution of the increased membrane potential to the generation of negative EOGs. The increase in the latter case was explained by a decrease in K^+ exit due to a decrease in the internal to external K^+ concentration ratio. These two effects indicate the important role of K^+ in the generation of negative EOGs.

(c) The role of Cl^-

When the olfactory epithelium was immersed in a Ringer's solution containing 1mM Ba^{2+}, the positive EOGs elicited by chloroform vapor were selectively depressed, while the negative EOGs were slightly depressed and were never observed to increase in amplitude. As will be shown later on p. 173, it has been demonstrated that positive EOGs are elicited mainly by the entry of Cl^- and the exit of K^+. If the exit of K^+ were affected by Ba^{2+}, the negative EOGs should have increased in amplitude. Since that was not the case, it follows that Cl^- entry was selectively inhibited by Ba^{2+}. Thus, if Cl^- entry contributes to the generation of negative EOGs, and if it were inhibited by Ba^{2+}, then addition of Ba^{2+} should result in an increase in negative EOG amplitude. However, the negative EOGs never increased in the Ba^{2+} solution, which leads to the conclusion that Cl^- does not contribute to the generation of negative EOGs (Takagi *et al.*, 1968a).

(d) Contribution of Na^+ and K^+

Since the participation of Cl^- in the generation of negative EOGs was discounted, it may well be concluded that the negative EOGs are elicited primarily by an increase in permeability of the olfactory receptive membrane to Na^+ and K^+. It is clear, then, that negative EOGs resemble other receptor potentials, particularly the muscle endplate potential, in many respects (Takeuchi and Takeuchi, 1960; Takeuchi, 1963a, b). Thus, it is highly probable that all, or at least most, negative EOGs are composed of the generator potentials of many receptor cells.

(e) The role of Ca^{2+}

Takagi *et al.* (1968a) studied the possible role of Ca^{2+} in the generative mechanism of the EOG.

When the olfactory epithelium was immersed in Ringer's solution without Ca^{2+}, negative and positive EOGs gradually decreased in amplitude with no sign of an initial increase. Furthermore, amplitudes never recovered even after immersion in normal Ringer's solution for 3 hr. The lack of recovery suggests that Ca^{2+}-free Ringer's solution irreversibly damages the olfactory receptive membrane, and indicates that Ca^{2+} is essential to the normal activity of the olfactory receptive membrane.

The importance of Ca^{2+} in the function of excitable tissues is well known (Brink, 1954; Koketsu, 1965). Nonmyelinated and myelinated axons lose excitability in Ca^{2+}-free solutions (Frankenhaeuser and Hodgkin, 1957; Frankenhaeuser, 1957), and the same phenomenon has been found in the Purkinje fiber (Weidmann, 1956). The role of Ca^{2+} as a charge carrier has been shown in lower animals (Fatt and Ginsborg, 1958; Hagiwara and Naka, 1964) and in the heart muscle of the frog (Hagiwara and Nakajima, 1966). An investigation of the ionic changes of the end-plate potential during acetylcholine activation in a low-sodium medium indicated a slight increase in membrane permeability to Ca^{2+} (Takeuchi, 1963a, b). The similarities in the ionic mechanisms of the EOG and the endplate potential suggest that Ca^{2+} may play a similar role in the generation of both processes. However, the precise role played by Ca^{2+} in the generation of the EOG has yet to be studied. Nonetheless, whatever the outcome of further investigations, the essential roles of Na^+ and K^+ cannot be disputed.

C. Positive EOGs

1. Historical Notes—Positive EOGs as Biological Phenomena

Electropositive potentials have been observed in the olfactory epithelia of several animals in response to odorous stimulation. Studying the olfactory epithelium of the frog, Ottoson (1956) found that often a brief, small electropositive potential preceded a large electronegative potential in response to odorous stimulation. He attributed this positive potential to the water vapor content in the stimulating air and thus concluded that the electropositive potential recorded from the olfactory epithelium was a nonbiological phenomenon. He did not admit the existence of a biological positive potential in the olfactory epithelium.

Since then, however, genuine positive potentials have been found by several investigators: a positive on-EOG accompanied by a negative off-potential was found in the frog (Takagi et al., 1960; Fig. V-4); and later studies have demonstrated similar positive potentials in the frog (Gesteland, 1964; Gesteland et al., 1965; Higashino et al., 1961; Fig. V-9, 3), in the newt (Shibuya and Takagi, 1962, 1963), and in the tortoise

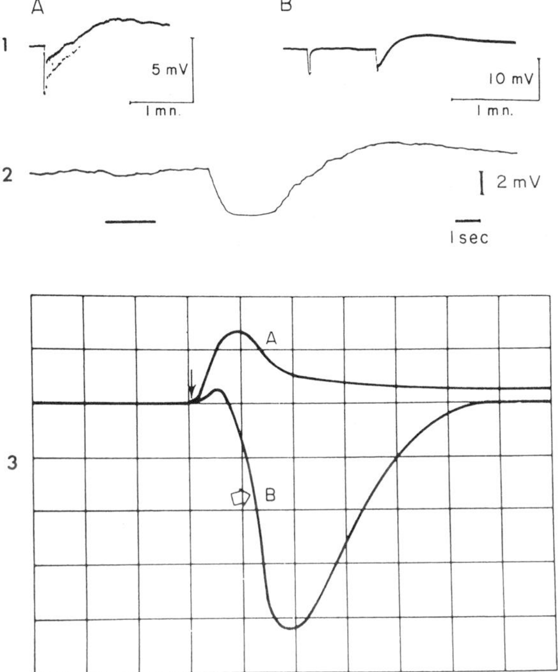

Fig. V-9. 1. EOGs in the rabbit. A: Negative EOGs (downward deflection) accompanied by positive afterpotentials elicited by repeated application of ethyl acetoacetate. B: Negative EOG by ethyl acetoacetate (left), and negative EOG followed by a positive afterpotential by benzene (right) (MacLeod, 1959).
2. EOGs of the positive-off type elicited at the cessation of odorous stimulation in *Channa argus* (Shibuya, 1960).
3. A: Positive EOG to methanol in the frog. B: Negative EOG to *n*-butanol. The solid arrow indicates the times of odor application. The open arrow indicates an inflection point in the negative rise (Gesteland *et al.*, 1965).

(Shibuya, personal communication). In addition, a positive off-EOG was identified in the fish (Shibuya, 1960; Fig. V-9, 2), and later in the frog (Gesteland, 1964).

A negative EOG followed by a positive afterpotential was found in the olfactory epithelium of the rabbit (MacLeod, 1959; Fig. V-9, 1). Takagi and Okano (1971, 1974) and Okano and Takagi (1974) showed that in the degenerated olfactory epithelium after olfactory nerve section a long-lasting positive EOG, or afterpotential type EOG, survived even though the negative EOG had disappeared. Throughout this period of investigation the existence of a biological potential with positive polarity was a highly debated issue (Ottoson and Shepherd, 1967). Higashino and Takagi (1964), applying polarizing currents on EOGs, discovered two different processes which produce negative and positive potentials (Fig. V-5). Takagi *et al.* (1966) later identified the origin and the ionic mechanism underlying positive EOGs (pp.173–176). Thus, data have accumulated that indicate the presence of true biological positive potentials elicited by odorous stimulation in the olfactory epithelium.

2. Three Types of Odors That Elicit Positive Potentials

As already shown on p. 152, 16 odors among the 122 produced genuine positive or negative-positive type EOGs and were classified into groups II, III, and IV (Table V-1). These 16 odors were applied onto normal as well as degenerated olfactory epithelia of decapitated bullfrogs, and the results of this study are summarized in Table V-1.

In the course of many experiments on decapitated bullfrogs, the author found that often ethyl ether vapor elicited a simple negative on-EOG at low concentrations, a negative off-EOG without a preceding negative on-EOG at intermediate concentrations, and positive on-, negative off-EOGs at high concentrations (Takagi *et al.*, 1960). On the basis of these data, ethyl ether was included in group IV. In other cases, however, ethyl ether produced only negative on- (1/36, 1/6), or negative on-off- (1/1) EOGs.

3. Positive EOGs Recorded under Various Conditions
(a) Positive EOGs recorded in the eminential and ceiling epithelia

As shown in Fig. V-4, positive EOGs were far more prevalent in the excised ceiling epithelia than in the olfactory eminentia of a decapitated frog (pp. 155–157). In addition, considerable differences in the shape of positive EOGs were found between the EOGs of the two epithelia. During repetitive stimulation, however, changes in the shape of the potentials were occasionally observed among the EOGs recorded in the eminentia. For example, small positive EOGs recorded in the eminentia at the beginning of experiments increased in amplitude, and the typical chloroform EOG, usually found in the ceiling epithelium (Fig. V-4 A), was often found in the eminentia. It appeared, therefore, that the EOGs recorded in these two types of epithelia are not fundamentally different.

(b) Positive EOGs recorded *in* vivo *and* in vitro

Gesteland (1964) showed in frogs immobilized with *d*-tubocurarine (*in vivo*) that 27 odors among the 62 applied elicited strong initial positive potentials followed by negative potentials. These results differed from the responses elicited by 122 odorants in the decapitated frog olfactory epithelia *in vitro* (Takagi *et al.*, 1969b).

As was already stated on pp. 155–156, in the olfactory eminentia of curarized bullfrogs (*in vivo*), chloroform and other organic vapors (groups II, III, and IV in Table V-1) elicited predominantly negative on- and off-potentials, and only rarely elicited relatively small positive potentials. After decapitation however, with the recording electrode in the same site on the epithelium, the same odor elicited predominantly positive potentials *in vitro*. These differences were explained by the presence or absence of oxygen supply (pp. 155–156).

4. Origin of Positive EOGs

Takagi and Yajima (1964, 1965) found that several days after olfactory nerve section, negative EOGs decreased in amplitude and disappeared, but positive EOGs survived in the olfactory eminentia. After the axotomy, the same phenomena were observed in the ceiling epithelium. The degeneration of olfactory cells in the epithelium, and the fact that basal cells situated at the bottom of the epithelium were not exposed to odorous stimuli, led to the conclusion that one or both of the remaining elements, the supporting cells and Bowman's gland, were involved in the generation of positive EOGs.

5. Ionic Mechanisms of Long-lasting Positive EOGs

Electropositive or hyperpolarizing potentials have been found at various synapses and their ionic mechanisms have been extensively studied. For example, the inhibitory postsynaptic potential is generated by the entry of Cl^- and the exit of K^+ in the cat spinal motoneuron (Coombs *et al.*, 1955), in the crayfish stretch receptor (Edwards and Hagiwara, 1959; Hagiwara *et al.*, 1960), and probably in some crab muscles (Fatt and Katz, 1953). Cl^- plays the principal role in the production of hyperpolarizing potentials in the crayfish endplate (Boistel and Fatt, 1958), in snail ganglion cells (Kerkut and Thomas, 1963, 1964), and in the Mauthner neuron (Asada, 1963), while K^+ predominates in the endplate potential of frog and dog heart muscles (Harris and Hutter, 1956; Trautwein and Dudel, 1958).

By ion substitution experiment (p. 167, Takagi *et al.*, 1969) studied the ionic mechanisms of long-lasting positive potentials elicited in the normal and degenerated olfactory epithelium by application of chloroform or other odorant vapors listed in Table V-1.

(a) Role of Cl^-

When Cl^- in Ringer's solution was replaced, for instance, by SO_4^{2-}, the

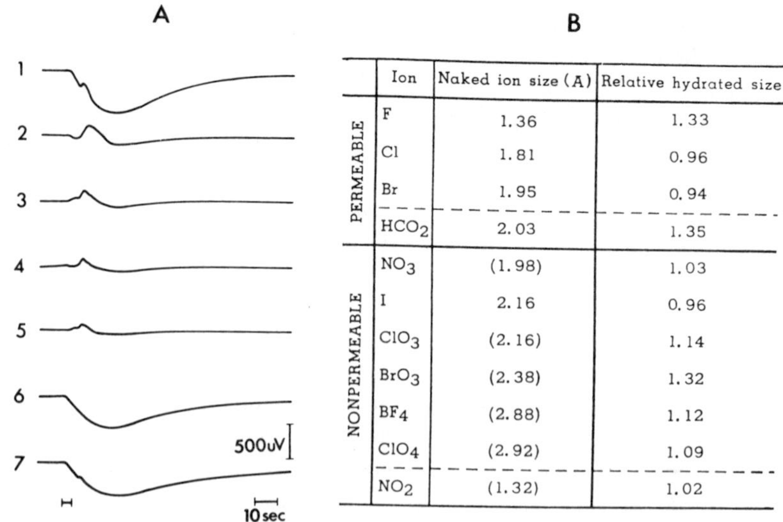

	Ion	Naked ion size (A)	Relative hydrated size
PERMEABLE	F	1.36	1.33
	Cl	1.81	0.96
	Br	1.95	0.94
	HCO_2	2.03	1.35
NONPERMEABLE	NO_3	(1.98)	1.03
	I	2.16	0.96
	ClO_3	(2.16)	1.14
	BrO_3	(2.38)	1.32
	BF_4	(2.88)	1.12
	ClO_4	(2.92)	1.09
	NO_2	(1.32)	1.02

Fig. V-10. A. Positive EOGs (1) decrease in amplitude in Cl⁻-free Ringer's solution (2 to 5), but recover in normal Ringer's solution (6 and 7). B: Among all anions listed, only F⁻, Br⁻, and HCO⁻₂ were found to be substitutes for Cl⁻. Other anions whose naked radii are larger than 1.98 A were nonpermeable. (from Takagi *et al.*, 1966. Reproduced from *The Journal of Gerneral Physiology*, 1966, Vol. 50, pp.478 and 487. by copyright permission of Rockefeller University Press, New York)

positive EOGs showed marked decreases in amplitude (Fig. V-10 A). When Cl⁻ was replaced by propionate ion, the positive EOGs decreased, reversed their polarity, and became negative EOGs.

Appropriate substitutes for Cl⁻ were sought in trials with many anions. The positive EOGs recovered only when SO_4^{2-} or propionate ion was replaced by Cl⁻, Br⁻, F⁻, or HCOO⁻. Hence, these anions were thought to be permeable to the receptive membrane of the supporting cell. Other anions showed no such substitute action for Cl⁻. When all the anions used in the experiment were arranged according to size of the naked ion, it was found that permeable anions were those with naked ion sizes smaller than 1.95 A, while nonpermeable anions had naked ion sizes larger than 1.98 A as shown in Fig. V-10 B. It has been explained that HCOO⁻ is permeable because of its ellipsoid shape (Araki *et al.*, 1961).

(b) Cl⁻ permeability

Araki *et al.* (1961) and Ito *et al.* (1962) studied the permeability of the receptive membrane to anions in the inhibitory postsynaptic potential (IPSP) of the spinal motoneuron. They arranged anions according to the relative sizes of the hydrated ion and found that anions smaller than MnO_4^- were permeable, but anions larger than BrO_3^- were nonpermeable. They concluded that Bernstein's classic sieve hypothesis (1912) is applicable to the inhibitory postsynaptic membrane on the basis of the relative sizes of hydrated ions, but not on the basis of the sizes of naked ions. In contrast, from the findings on the Cl^- permeability of the supporting cell, Takagi *et al.* (1966) concluded that the sieve hypothesis is applicable to the receptive membrane of the supporting cell on the basis of the sizes of naked ions, but not on the basis of the relative sizes of hydrated ions (Fig. V-10 B). It is interesting that the sieve hypothesis was found to be valid for both types of membranes, but on entirely opposite bases.

$HCOO^-$ was found to be exceptionally permeable to both membranes. Araki *et al.* (1961) explained that $HCOO^-$ has an ellipsoid shape which allows it to penetrate the receptive membrane of the motoneuron using a shorter radius. It is quite probable that the same explanation is valid in the case of the positive EOG (Fig. V-10 B).

(c) Role of K^+

As stated before, K^+ plays a principal role in the hyperpolarizing potentials of some synapses. The small residual positive potential usually found in Cl^--free Ringer's solution (Fig. V-10) may be due to the exit of K^+ from the receptor cell.

The effect of removing K^+ from Ringer's solution was studied in the normal and degenerated olfactory epithelia. In K^+-free Ringer's solution the positive potentials initially increased in amplitude and then began to decrease. The initial increase suggested the participation of K^+ in the generation of the positive potentials.

In K^+-free Ringer's solution, external Na^+ enters through the excitable membrane of muscle in exchange for the exit of internal K^+ (Desmedt, 1953). Since negative potentials depend upon the entry of Na^+ (p. 168, Takagi *et al.*, 1968a), such an increase in the internal Na^+ concentration may explain the gradual decrease in negative potentials in later stages. These changes in the distribution of Na^+ and K^+ across the receptive membrane probably result in a change in the Cl^- concentration ratio across the receptive membrane, which may explain the decrease in positive potentials in the later stages.

6. Secretory Mechanism of the Supporting Cell

Using electron microscopy, Okano and Takagi (1974) and Takagi and Okano (1974) demonstrated that application of chloroform and some other odorous vapors elicited vigorous protrusion of the distal cytoplasmic

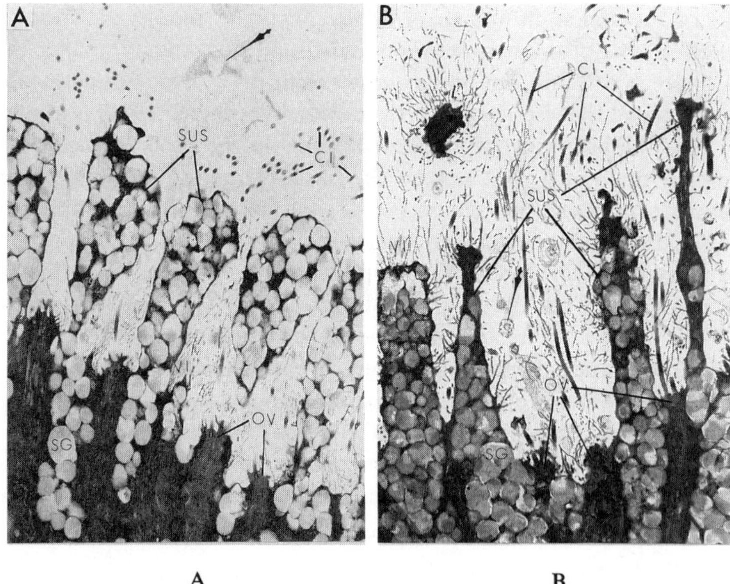

A   B

Fig. V-11. Secretion of supporting cells in normal (A) and degenerated (B) olfactory epithelia. (from Okano and Takagi,
1974)

portion of the supporting cell containing secretory granules. It is
important to recall that pyridine induces secretion from the supporting
cells, but not from Bowman's glands (Wesolowski, 1967; pp. 133–134).
Electrophysiological examinations should that application of those odors
simultaneously and in parallell generated long-lasting positive potentials
in the olfactory epithelium.

(a) Secretory process

The supporting cell showed the following sequence of secretory processes: The first detectable indication is the protrusion of the apical portion of the supporting cell (Fig. V-11); the protruded part then detaches
from its maternal supporting cell and floats as a droplet in the mucus;
and finally, the secretory granules inside the droplet disintegrate into
the mucus.

(b) Spontaneous and sporadic secretory activity of the supporting cell

Electron micrographs from Okano and Takagi's (1974) study demonstrate that nonstimulated olfactory epithelia liberate secretory granules
spontaneously and/or sporadically (Fig. IV-3).

(c) Odors which elicit heavy secretion

Okano and Takagi (1974) initially used chloroform vapor to elicit vigorous secretion from the supporting cell. They then selected *n*-butyl alcohol and *t*-butyl alcohol from the 16 odorants in Table V-1 and examined their effects on secretory activity. These odors also elicited heavy secretion from the supporting cell. Since secretory activity and a long-lasting positive potential occurred in parallel, most, if not all, of the odorants in Table V-1 were presumed to elicit heavy secretion from the supporting cell.

(d) Odors which elicit less secretion

The group I odorants (p. 152) were shown to elicit only negative EOGs (Takagi *et al.*, 1969b). As representative of these odorants, amyl acetate vapor was applied onto normal and degenerated olfactory epithelia. Electron micrographs showed that amyl acetate did not elicit active secretion. Since a parallel relation was found between the occurrence of negative EOGs and lack of active secretion, most, if not all of the group I odorants were presumed to elicit no, or only slight, secretion from the supporting cell. In the case of these odorants, however, seemingly negative results do not necessarily indicate that no secretion whatsoever occurred (Okano and Takagi, 1974).

(e) Effects of Cl^- and Ba^{2+} on secretory activity

Positive potentials showed striking decreases in amplitude or often reversed polarity after the immersion of normal or degenerated olfactory epithelia in Cl^--free Ringer's solutions (p. 174). Correspondingly, electron micrscopic study disclosed that secretory activity was completely arrested when Cl^- was removed from the environment of a supporting cell (Okano and Takagi, 1974; Takagi and Okano, 1974).

Those investigators examined the effect of Ba^{2+} on positive EOGs by immersing the olfactory epithelium in Ringer's solution containing 1 mM Ba^{2+}. This procedure resulted in selective decreases in the amplitude of positive EOGs with very slight decreases and no increases in the amplitude of negative EOGs as mentioned previously (p. 169).

In one study, Ringer's solution containing 1–2 mM Ba^{2+} was dripped onto the olfactory epithelium. Electron microscopy disclosed that under these conditions the supporting cells liberated their secretory granules en masse from their apical regions, and electrophysiological techniques revealed that positive potentials occurred simultaneously. The protrusions had a club-like shape and closely resembled those produced by chloroform vapor. Thus, Ba^{2+} was found to have a stimulating effect on the secretory activity of the supporting cell.

Subsequent application of chloroform vapor resulted in further disintegration of the secretory granules, but it elicited neither liberation of new granules nor a new positive potential.

(f) Secretion in the degenerated olfactory epithelium

Chloroform vapor applied to the olfactory epithelium after olfactory nerve section, in which the olfactory cells had degenerated leaving the supporting cells intact, elicited both the liberation of the secretory granules (Fig. V-11B) and the positive EOG (Fig. V-3), as it did in the normal olfactory epithelium (Okano and Takagi, 1974; Takagi and Okano, 1974).

D. What Are EOGs?

1. Are Negative EOGs Generator Potentials?

Ottoson (1956) studied the EOG in the olfactory epithelium of the frog and assumed that the electronegative EOG is a generator potential which elicits afferent discharges in the olfactory nerve, although data were insufficient then.

In sections A and B of this chapter the author described various properties of negative EOGs, and clarified their origin and ionic mechanisms. In particular, his findings that negative EOGs originate in olfactory receptor cells and that the ionic mechanism of the negative EOG resembles that of the endplate potential, and Aoki and Takagi's (1968) recording of spike discharges superimposed on slow negative potential strongly supported Ottoson's hypothesis. However, when the EOGs were recorded together with the induced wave in the OB, some questions were raised about this hypothesis (Takagi *et al.*, 1960; Takagi, 1967, 1969).

(a) Changes in amplitude of negative on-EOGs did not parallel those of induced waves in the OB (Fig. V-6). Even when a negative EOG was encroached on by a second stimulus, an induced wave consistently appeared in the OB (Fig. V-7).

(b) Even when an electropositive EOG appeared in the olfactory epithelium, an on-induced wave appeared in the olfactory bulb (Fig. V-7 C).

(c) The off-induced wave in the olfactory bulb preceded the onset of the negative off-potential in the olfactory epithelium (Figs. V-6, -7).

(d) Shibuya (1964) succeeded in showing that olfactory nerve discharges can be elicited independently of the EOG in the olfactory epithelium. He removed olfactory mucus from the surface of a restricted small region of the epithelium. The same odor could elicit olfactory nerve discharges in a nerve twig derived from that small region after the removal, as before, but the EOG that had been recorded in the region disappeared after the removal. This dissociation of the EOG from the olfactory nerve discharges seemed to be a most definite finding against the generator potential hypothesis. Facts which are supportive and nonsupportive of the hypothesis have to be taken into consideration to answer the question of whether negative EOGs are generator potentials or not. Olfactory and supporting cells are presently being studied by intracellular electrodes and patch-clamp recording methods, and many new data have been obtained. A complete answer to the question of negative EOGs, however, has yet to be found.

2. What Are Positive EOGs?

The author together with his collaborators mainly studied the properties, origin, ionic mechanism, and function of the long-lasting positive EOGs or positive EOGs of the afterpotential type (p. 170). They concluded that this EOG originates in the supporting cell and is a secretory potential of the supporting cell. The secretory mechanism of the cell can be explained as follows: First, application of chloroform and other vapors induces entry of Cl^- through the surface membrane of the supporting cell. Second, Cl^- entry elicits long-lasting positive EOGs on the one hand and, on the other hand, triggers liberation of secretory granules from the supporting cell. Third, the secretory granules are discharged en masse together with K^+ into the mucus layer.

The author supposed at first that the entry of Cl^- into the supporting cell occurs passively due to the concentration gradient and on the basis of the sieve hypothesis (Takagi *et al.*, 1966). In the light of the active transport mechanism of Cl^- found in the sublingual gland (Lundberg, 1957a, b), it is also conceivable that the entry of Cl^- is partly or wholly due to the active transport mechanism initiated by the adsorption of chloroform (or other odor) molecules on the receptive membrane of the microvilli and the supporting cell. Whether the secretion occurs passively or actively remains to be studied in the future.

3. Four Kinds of Positive EOGs

In the studies on EOGs from 1956 to now, three kinds of positive EOGs have been found and another one has been assumed: A positive on-EOG was elicited accompanied by a negative off-potential in response to ethyl ether vapor (Fig. V-6, Fig. V-12; also pp.172–173). A long positive afterpotential was observed after a negative or positive on- and after a negative off-potential when vapors of chloroform and other organic solvents were applied (5 in Fig. V-12). A positive off-potential was found by Shibuya (1960) in some fish (4 in Fig. V-12).

In this connection, the author has to assume that the olfactory cell generates not only an excitatory potential, but also an inhibitory potential, because he has observed that extracellular recording of olfactory cell discharges often shows inhibition of spontaneous or induced discharges in response to odors that have not been identified. This view of an

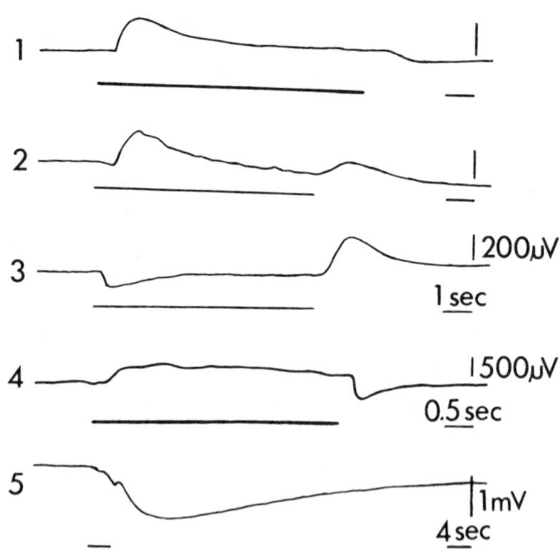

Fig. V-12. Five fundamental types of EOGs recorded in olfactory epithelia of decapitated frogs. The horizontal line under each EOG indicate duration of odorous stimulation. (from Takagi, 1984a; Takagi, 1969)

inhibitory potential in the olfactory cell conforms well with Gesteland *et al.*'s view (1965) (Boeckh, 1967). Whether this inhibitory potential is equal to the positive on-EOG (Fig. V-12, 3) or not is still a subject to be studied in the future. It is concluded, therefore, that the positive EOG is composed of the four kinds of potentials mentioned above.

4. Relationship between Negative and Positive EOGs

Takagi *et al.* (1968a) frequently observed that the amplitudes of the negative and positive EOGs are inversely related: When the negative potential is relatively large, the positive potential is relatively small, and vice versa. For instance, when the ionic mechanisms of negative and positive EOGs were studied (p. 167), the excised ceiling epithelium was immersed in Ringer's solution in which 10% of the Na concentration was replaced with sucrose. After 20 min, the negative and positive EOGs showed interesting changes in their amplitudes. When the negative EOG decreased in amplitude, the positive EOG strikingly increased in amplitude. When the negative EOG increased considerably in amplitude, the positive EOG increased only slightly. Often an intermediate change was observed between the above two cases: Both the negative and positive

EOGs moderately increased in amplitude before they eventually decreased in amplitude (Fig. 10 in Takagi *et al.*, 1968a). These inverse relations again support the author's hypothesis that the EOG is composed of negative and positive potentials and that the final shape of the EOG is determined by competition between the two processes that generate the positive and negative potentials.

5. Five Fundamental Types of EOGs

In a series of experiments, the author and his colleagues have observed various types of EOGs. Although EOGs take many forms, they can be classified into five fundamental types.

The first EOG type is the negative on-EOG (Fig. V-12, 1). In response to an odor, a simple electronegative potential most frequently appeared in the olfactory epithelia of the dog (Hosoya and Yoshida, 1937; Yoshida, 1950), the frog (Ottoson, 1956), the fish (Shibuya, 1960), and the newt (Shibuya and Takagi, 1962). The second, electronegative on-off-EOG (Fig. V-12, 2) was first found in the frog by Takagi and Shibuya (1959) and was confirmed later by many workers Shibuya, 1960; Higashino *et al.*, 1961; Ai and Takagi, 1963; Shibuya and Takagi, 1962, 1963; Higashino and Takagi, 1964; Gesteland, 1964; Byzov and Flerova, 1964; Takagi and Yajima, 1964, 1965; Gesteland *et al.*, 1965; Suzuki, 1967, Takagi *et al.*, 1968a, b, 1969a, b). The third, the electropositive on-EOG (Fig. V-12, 3) was identified by Takagi *et al.* (1960) and later studied by Higashino and Takagi (1964), Takagi *et al.* (1966, 1969b), Takagi (1969), and Suzuki (1967). The fourth, the positive off-EOG (Fig. V-12, 4) was found in the fish by Shibuya (1960) and in the frog by Gesteland *et al.* (1965). The fifth, the positive EOG of the afterpotential, or long duration, type (Fig. V-12, 5) was found in the rabbit by MacLeod (1959; Fig. V-9, 2) and in the bullfrog by Takagi *et al.* (1966, 1968b, 1969b), These five types of EOGs were first described by Takagi (1969).

Takagi (1969) believed that the various shapes of EOGs observed in many experiments (for instance, Figs. V-2, 3, 4) are formed through a combination of some or all of these five types of potentials.

E. EOG and Electrical Activity of the Olfactory Epithelium during Degeneration and Regeneration of Olfactory Cells

1. Electrical Activity during Degeneration

Takagi and Yajima (1964, 1965) studied electrical activity (EOG) and histological changes in degenerating olfactory epithelium after olfactory nerve section. The electrical responses to odors, or EOGs, began to decrease in amplitude and disappeared within 8 days in the summer, 11 days in the early autumn, and 16 days in the early winter. The average temperature was 22°C during the summer experiments and 10.9°C during the early winter. If the disappearance of the electrical response is assumed to be due to the retrograde degeneration of the olfactory cells, the Q_{10} of the velocity of nerve degeneration can be estimated to be approximately 2.

Examination of the degenerating olfactory epithelium revealed a striking decrease in the number of olfactory cells, but not in the number of supporting cells (see p.119). The ratio of olfactory cells to supporting cells was found to decrease from 5 or 6 to below 2 after the nerve section. At a ratio below 2, electronegative EOGs to odors disappeared, but electropositive EOGs to odors survived. As mentioned above, we thus concluded that electronegative EOGs originate from the activity of the olfactory cell, but electropositive EOGs are generated by the activity of other cells, most likely supporting cells.

Several investigators have since studied this degeneration process (e.g., Simmons and Getchel, 1981a, b), but few studies have examined the detailed relation between the decreases in amplitude of negative EOGs and decreases in the number of olfactory cells which was disclosed in the paper by Takagi and Yajima (1965).

One exception is provided by Burns *et al.*'s (1981) scanning electron microscopy study showing changes of the olfactory epithelial surface after olfactory nerve section. By day 10 after the axotomy, the epithelial surface showed degeneration and loss of the dense ciliary matrix and olfactory knobs, exposing the microvillar surface of the sustentacular cells. During those 10 days the EOG was found to systematically decrease in amplitude.

2. Electrical Activity during Regeneration

Kiyohara and Tucker (1978) studied the recovery of EOGs in regenerating olfactory epithelia after bilateral transection of the olfactory nerves in 26 pigeons. No sign of neural function was found 6–15 days after neural resection. Spontaneous neural activity but no response to amyl acetate was recorded 8–15 days after surgery. Detectable responsiveness to amyl acetate ($10^{-1.0}$ vapor saturation content) was seen 17–28 days postoperatively. Normal sensitivity and responsiveness to amyl acetate were found approximately 30 days after surgery. These results indicate a temporal course of functional recovery and suggest that a normal population of functioning receptors can be reestablished within about 30 days.

Simmons and Getchell (1981a) studied recovery of EOGs in the salamander (*Ambystoma tigrinum*) at 24, 45, 80, and 100 days after nerve section. At each stage of recovery, the wave and time course of the EOGs were characteristic for each stimulus. The simultaneous loss and recovery of negative and (short) positive components of the EOGs indicated that both are dependent on the presence of functionally mature olfactory receptor neurons.

Adamek *et al.* (1984) analyzed changes of EOGs during ciliary growth following detergent-induced ciliary removal in the frog. The EOG initially increased in amplitude proportionally with ciliary length, and reached a maximum after 2 days of growth, when cilia were 40 μm long. Consequently, it was thought that the olfactory transduction sites are located primarily on cilia rather than on terminal dendrites, and that most of the receptor current enters through the proximal portion of the cilium.

Adamek *et al.* (1984) then induced selective necrosis of the olfactory epithelium by $ZnSO_4$ lavage of the nasal cavity. During epithelial regeneration, EOGs increased linearly over time from 13 days after zinc lavage, the time of first cilium emergence, through 30 days. Rates of increase differed for different odorants. At 30 days, EOG amplitudes increased sharply within a period of a few days, then reached asymptote. Thus, the develop ment of receptors for different substances occurred at different rates and in two steps. The transition between the two developmental states was coincident with the arrival of receptor axon terminals in the central nervous system and the immobilization of the ciliary contractile apparatus. Since there is a continual regeneration of new receptor neurons through out life, EOGs recorded in a normal olfactory epithelium reflect a complex combination of the differing receptor processes of cells at differing developmental stages.

F. Aspects of EOGs Studied by Other Investigators

1. Quantitative Aspects of EOGs

van As *et al.* (1985) applied 13 different odorants and recorded EOGs in the tiger salamander. EOGs elicited by all these odors had similar characteristics with regard to concentration dependency, time course of adaptation, and time course of the rising phase, despite the fact that different odors were examined in different preparations. The data were analyzed by evaluating how well EOG characteristics could be described by two theoretical descriptions (occupation theory by Beidler, 1954 and rate theory by Heck and Erickson, 1973) of stimulus-receptor interaction in the absence of significant partition and diffusion barriers to stimulus access. Although their data were inconsistent with the original occupation theory, they found that a modified occupation theory (Kamo *et al.*, 1980) provided a useful mathematical framework in which to describe the primary olfactory process as measured by the EOG when stimulus access limitations were unknown. Their data, analyzed in this theoretical framework, suggested that there might be a conversion of the stimulus receptor complex from an active to an inactive form.

$$S + A \rightleftharpoons (SA)_{active} \rightleftharpoons (SA)_{inactive}$$

where S is a stimulus molecule, A is a receptor site, SA is the stimulus molecule-receptor site complex, $(SA)_{active}$ is the active conformation of the stimulus molecule-receptor site complex, and $(SA)_{inactive}$ is the inactive conformation of the stimulus molecule-receptor site complex.

2. Determination of Odor Affinities Based on the Dose-response Relationships of the Frog EOG

Senf *et al.* (1980) recorded EOGs from the frog olfactory epithelium for 11 odorous compounds, seven of which were aliphatic *n*-alcohols. Using these EOGs and Beidler's taste equation (1954), they calculated dissociation constants which served as an index for the affinity between odorant and receptor site. These constants were calculated with and without a correction for the odor partition between water and air. These values were found to decrease as a homologous series of seven *n*-alcohols was ascended and reached a minimum with 1-heptanol, indicating that it is the most effective stimulus in the series.

3. The Role of 3′, 5′-AMP in the EOG

Minor and Sakina (1973) studied the effects of cyclic adenosine-3′, 5′-monophosphate (3′,5′-AMP) and of substances modifying the rate of its breakdown (inhibitors and activators of phosphodiesterase) on the EOG. Cyclic 3′,5′-AMP and its dibutyryl derivative were found to excite the olfactory receptors effectively, but both 5′-AMP and cyclic 2′, 3′-AMP were ineffective. Phosphodiesterase inhibitors, methylxanthines or paraverine were found to increase the duration of the EOG. Conversely, the negative component of the EOG was shortened under the influence of the phosphodiesterase activator, imidazole.

Thus, they considered that cyclic 3′,5′-AMP plays the role of mediator in the mechanism of excitation of the olfactory receptor (Chapter VI, p. 216).

G. Electrogenic Activity of the Olfactory Epithelium

1. A Different Approach to the Study of Electrogenic Sources of EOGs

Getchell (1974a) attempted to identify the electrogenic sources of EOGs using a technique employed by Pearson *et al.* (1970). By means of a source resistance matching technique and by loading the epithelium with very weak resistors, he was able to determine the order in which negative or positive EOGs were shunted out. He estimated the source resistance of the electrogenic mechanism which generated the negative EOG to be approximately 1 Megohm. The source resistance for the positive EOG were estimated to be higher, approximately 1.3 megohms. Positive and negative EOGs were not further reduced into subcomponents by shunting the epithelium. This suggested that each process either had a single electrogenic source, or multiple sources with similar source resistances. Although his technique did yield results which distinguished between these two possibilities, he proposed an equivalent electric circuit for the olfactory epithelium (a voltage divider model named after him), which, he believed, would serve as a useful framework in conceptualizing the underlying generative mechanisms. In the circuit, E_n and E_p represent two electrogenic sources for the negative and positive EOGs, and R_n and R_p represent their source resistances. An electrochemical source, tentatively identified as E_c with its source resistance R_c was added in parallel with the other sources. E_c, accounts for physicochemical processes which may contribute to the EOG under certain specific recording conditions (cf. Takagi, 1969). Extracellular current, i_m, derived from any or all of the three electrogenic sources allows voltage to be measured across the extracellular resistance, R_m, of the olfactory epithelium.

Recent studies indicate that olfactory receptor neurons undergo continual cell renewal during the normal life span of the animal. This suggests that neurons at various stages of development, maturity, and senescence are present within the epithelium at all times (cf. Simmons and Getchell, 1981b). In addition, the supporting cells are also mitotically active. Consequently, it is apparent that ionic currents measured by Getchell's method are composed of currents from very diverse sources. In light of these histological findings, this approach appears to have limitations in its explanatory capacity. In fact, neither the E_n in the olfactory cell nor the E_p in the supporting cell could be specified by this method.

Although this approach has been tried more recently, it did not yield the conclusions on electrogenic sources of EOGs which had already been attained by previous degeneration and ion substitution experiments (Takagi and Wyse, 1965; Takagi *et al.*, 1966, 1968a, b, 1969a, b).

2. *Electrogenic Active Ion Transport across the Frog Olfactory Mucosa in vitro*

Heck *et al.* (1984) excised a piece of the bullfrog olfactory epithelium from the roof of the olfactory cavity (Takagi *et al.*, 1968) and studied electrogenic ion transport across the epithelium in an Ussing-type chamber (Ussing and Zerahn, 1951). They found that the ciliated side of the olfactory epithelium was electronegative ($V = -5.2$ mV $\pm$ 0.7 mV) and the average value of the short circuit current was 35.9 μA/cm^2. The magnitude of specific resistance of the olfactory mucosa was found to be comparable to the values reported for various actively transporting respiratory and oral cavity epithelia. Substitution of NO_3^- for Cl^- in the bathing Ringer solution suggested that cation absorption plays a role in the polarity of the short-circuit current.

3. The Voltage-clamped Frog Olfactory Mucosa: Odor-stimulated Current Transients

DeSimone *et al.* (1985, 1986) recorded standing transmucosal current and current transients associated with odorant stimulation of the frog olfactory mucosa *in vitro*. The tissue was mounted in Ussing-type chambers and bathed symmetrically in amphibian Ringer's solution. In the steady state, the system developed a transmucosal potential (blood side electropositive) indicating continuous active ion transport. The direction of positive current was from the ciliated toward the nonciliated side. Seven preparations gave the following mean ($\pm$ SEM) values for the potential (Voc), short-circuit current (Isc), and resistance (R).

$$Voc = 3.8 \pm 0.6 \text{ mV}$$

$$Isc = 56.0 \pm 16.8 \ \mu\text{A/cm}^2$$

$$R = 73.0 \pm 15 \text{ Ohm cm}$$

The standing short-circuit current served as a baseline for observing the odorant-stimulated current transients. A dose-dependent increase in inward positive current was observed when the ciliated surface was stimulated with 2-methoxy-3-isobutyl-pyrazine added to the Ringer's solution. The magnitude of the current change from baseline ranged from a mean of 0.2 μA/cm^2 at the lowest concentration to 3.7 μA/cm^2 at the highest. Odorant-induced current transients of 5 μA/cm^2 were consistent with the magnitude of the voltage and resistance changes. The diuretic furosemide, a potent blocker of sodium chloride co-transport systems, eliminated both the standing and the stimulated short-circuit current. This suggested that the odor-stimulated voltage transient and current were dependent on a specific transcellular chloride flux. Although it was not possible to assign a precise locus to the ion pathways involved, it was likely that both receptor and sustentacular cells comprise part of the local circuit. The investigators concluded that whatever the topographical arrangement of the ion pathways, the voltage-clamp method should permit the identification of the transmucosal current carriers mobilized by various classes of odorants, and, in addition, that this method was well suited to the investigation of secretagogues promoting ion translocation.

The voltage-clamp method was originally used to measure ionic currents across the membrane of a single neuron. Hodgkin and Huxley (1952) obtained their illuminating and clear-cut results only because they conducted intracellular recording from a single neuron in conjunction with this method. As already stated in section G (p. 185) the olfactory epithelium is a nervous tissue with diverse electrogenic sources in different developmental stages of cells, and hence, currents recorded from this tissue with only an extracellular electrode would be too complicated to analyze.

VI. Neurophysiology of Olfactory Receptive Mechanisms and Responses

A. Neuroanatomical Studies on the Olfactory Receptive Mechanisms

1. Development of Olfactory Cells
1.1. Appearance of electrical responses and olfactory marker protein (OMP, a 19 kD cytoplasmic protein) during development

Gesteland *et al.* (1980) measured stimulus-related potential changes in single cells of fetal rat olfactory epithelium by means of extracellular electrodes. In addition, Farbman and Margolis (1980) identified OMP, a protein unique to the olfactory receptor neuron. The time course of its appearance provided information about events in the cell's life cycle. The findings of these studies helped to elucidate the development of olfactory function.

In the rat fetus, olfactory neurons were first seen 12 days after conception (E-12). The OMP first appeared in some of these neurons 6 days later (E-18) and birth occurs 3 days after that (E-21). From E-15 to E-19, about 85% of the stimuli presented were effective in evoking responses from virtually any receptor cell. On day E-20, however, there was a sudden change in selectivity, as only about half of the stimuli presented were effective in evoking responses from most cells (Gesteland *et al.*, 1982). Mair (1986) found similar results and Cuschieri and Bannister (1975a, b) reported that the olfactory epithelium contains Bowman's glands and exhibits signs of histochemical maturity by E-17.

Thus, electrical activity of the receptor cells can be demonstrated several days before they achieve the apparent capacity to discriminate among odors. In the rat, the most mature cells probably establish synapses in the olfactory glomeruli by day E-18. OMP appeared simultaneously with the first formation of connections capable of transmitting messages along recent neural projections to the OB. This implies that the protein marked those neurons which were mature and functional. The marker protein was identified in patches, suggesting that several thousand nerve cells initiated development together.

Farbman and Margolis (1980) found that OMP is first expressed on day E-18 in the rat, and on day E-14 in the mouse. They proposed four criteria to identify olfactory receptor maturation: (i) Ciliogenesis, (ii) OMP accumulation, (iii) synapse formation, and (iv) receptor response selectivity.

1. 2. *Differences in morphology and response between immature and mature olfactory cells*

Lidow *et al.* (1985, 1986) succeeded in producing olfactory epithelia with developmental synchronization of receptor cell populations. In this preparation, the epithelia are first ablated by the action of $ZnSO_4$, and are then allowed to regenerate freely for about 10 days. Generation of new cells in the epithelia is suppressed by treatment with hydroxyurea. They first recorded EOG and single-unit activity from epithelia which were allowed to regenerate for 8 days before hydroxyurea treatment. A 10-day period of regeneration followed by hydroxyurea produced synchronized development of epithelia which could be used for electrophysiological recordings (Lidow *et al.*, 1987).

Here, it is important to recall Hemstead and Morgan's (1983a) demonstration that fluorescin-conjugated concanavalin A preferentially binds to mature receptor neurons in the olfactory mucosa.

"Immature" olfactory cells whose axons had not yet reached the olfactory bulb were characterized by perikarya with poorly developed rough endoplasmic reticulum, dendrites containing an abundance of ribosomes and mitochondria in their shafts and olfactory knobs, and by short motile cilia. In contrast, "mature" cells had perikarya with well-developed rough endoplasmic reticulum; their dendrites contained no ribosomes, and their olfactory knobs had no mitochondria. Cilia in these cells were long and immotile.

EOGs recorded from olfactory epithelia with immature receptor cells were not different in polarity or shape from those from epithelia with mature receptor cells. However, low levels of spontaneous activity were observed in the majority of immature receptor cells, while a greater variety of frequencies was found in the spontaneous activity of the mature olfactory cells.

1. 3. *Differentiation of olfactory cells in organ culture and* in vivo

Farbman (1977) studied differentiation of olfactory receptor cells 'in organ culture. He excised presumptive olfactory mucosa from the heads of rat fetuses in the 11th and 12th days of gestation and explanted the tissue into organ culture. He observed the progressive differentiation of olfactory cells and concluded that olfactory cells, as defined by morphological criteria, were able to differentiate in suitable organ culture conditions in the absence of any central nervous system tissue.

From a physiological perspective, however, the situation becomes very complicated at this stage of development. The most fascinating finding to olfactory investigators may be the ability to maintain consistent responsiveness to a given odorant even though receptor cells are constantly being replaced and newly formed cells are able to synapse with principal cells in the OB.

Response can be achieved in at least two ways, anterograde or retrograde specification. In the first case, receptor cells are inherently tuned to respond to certain stimuli and perceptual constancy is achieved through the ability of receptor cells to synapse with the appropriate mitral cell. In the case of retrograde specification, receptor cells are tuned to respond to specific stimuli from the mitral cell with which they happen to synapse. At the present stage of olfactory research, the data seem most consistent with the retrograde hypothesis. For instance, if mitral cells control led the response properties of the receptor cells with which they synapse, then developing receptor cells would be expected to exhibit an abrupt change in response specificity after synaptogenesis, which was demonstrated in a study by Gesteland *et al.* (1980, 1982). Similarly, the abrupt transition in other properties of the receptor cells observed at about the same time is consistent with the idea that synapse formation with mitral cells has a fundamental impact on the maturation of these neurons (Mair, 1986).

1. 4. Regeneration and embryonic development of olfactory cells

Generally speaking, if the olfactory epithelium of an adult animal is removed, it regenerates over a period of weeks. The sequence of events seen during regeneration is similar to that which occurs during embryonic development. Indeed, the embryo can be viewed as a model of the events continually occurring throughout adult life as the olfactory receptors are replaced (Gesteland, 1982).

2. Development and Regrowth of Olfactory Cilia

2. 1. Development of olfactory cilia

In mice, the olfactory placode is distinguishable by the 10th day of gestation (E-10) and primodal axons sprout and grow into the OB by day E-11 (Cuschieri and Bannister, 1975a, b; Hinds, 1972a, b). Cell axons have penetrated deep into the OB and formed synapses with second-order neurons. These cells begin to express the olfactory marker protein near day E-14, and almost simultaneously begin to show an increase in the number and the elongation of cilia on the dendritic knobs of the olfactory cells. A similar developmental pattern was found in rats with a lag of several days (Farbman and Margolis, 1980).

2. 2. Regrowth of olfactory cilia

Gesteland *et al.* (1980) and Mair *et al.* (1982) have studied changes in the morphology and physiology of olfactory cilia during development.

These investigators irrigated the nasal passages of adult frogs with a solution of 0.05% Triton-X detergent and removed the cilia on the receptor neurons with little or no damage to the cells. The cilia regenerated over the next few days, increasing in length at a rate of about 25 μm per day.

The cilia regenerated in a two-stage process. The first stage begins with the initiation of regrowth 2 hours after removal, and the new cilia were short and immotile. The first recordings of the EOG were obtained at 16 hours. At 24 hours, three morphologically distinct types of receptor cilia are distinguished: short, rapidly moving wigglers; longer, gracefully moving strokers; and the longest, immotile streamers. After 4 days, the EOG amplitude had increased to its limit. At this time the longest cilia measured 100 μm. The second stage of maturation consisted of cilia growth with no change in EOG amplitudes. This stage was complete at 8 days, when streamers reached 200 μm in length. Since the EOG appeared prior to complete cilial differentiation and reached maximal levels before full cilial maturation, the clear implication is that the majority of the stimulus transduction sites are located on the proximal portions of the cilia.

In a separate experiment, regeneration of the epithelium was observed after ablation with $ZnSO_4$. The first receptor cilia were seen 12 days after the $ZnSO_4$ treatment, and were the short and highly active wigglers. The first EOG recordings were obtained at this stage. By day 16, ciliated cells had spread throughout the tissue and longer cilia had motion parameters like strokers rather than the short wigglers. All three cilia types were present by day 27. Thus, the morphology of the cilia growing from a receptor depended upon the developmental age of the cell. The oldest cells had streamers, younger cells had strokers, and the most recently differentiated cells had wigglers (Gesteland, 1987). The membranes of the different cilia types differ in their permeability and excitability. Wigglers responded dramatically to stimulus substances and to chemicals which affected calcium concentrations. Strokers were much less sensitive to their environment, and streamers were insensitive to both odorant stimulation and application of calcium reagents.

Mair *et al.* (1982) argued that the life span of olfactory receptor cells in the frog can be divided into two stages. These authors expressed a slightly different view on the functions of cilia: During the initial stages of development, olfactory cells bear motile cilia which are involved in the removal of odorous molecules from the olfactory epithelium. When cells mature and synapse with principal neurons in the OB, they have long, immotile cilia which have a larger surface area for olfactory transduction and hence function as discriminating receptors. In any case, the membrane properties of the regenerated cilia depend upon the developmental age of the cell from which they sprout, not upon the developmental stage of the cilium.

Gesteland *et al.* (1980) characterized unit responses in these regenerating epithelia. Soon after deciliation, the cells fired bursts of action potentials with long intervening periods of quiescence. After a few hours, the slow, irregular activity of normal neurons developed and the activity changed with odorous stimulation. In conclusion, they made the very important point that studies on coding of olfactory information in receptors will have to consider the developmental states of the responding neurons. The impact that this view may have had upon past studies of information processing of receptor cells cannot be estimated.

2. 3. Olfactory cilia of the newt in water and on land

Shibuya and Takagi (1963) found a correlation between the length of the olfactory cilia and the EOG of the olfactory epithelium in newts living in water and on land. The olfactory cilia grew when newts were transferred from water onto land. The cilia in the olfactory bud grew to their maximum length 108 hours after the transfer and then became shorter, while those in the interstitium only gradually elongated. EOGs were evoked in the epithelium by the application of odorous fluids but not by odorous vapors for 20 hours after the transfer. The EOG to vapors began to appear and reached maximal amplitude between 60 and 70 hours after the transfer, and was not evoked by odorous fluids in this period. In the later stage, the EOG to vapor decreased in magnitude and disappeared 120 hours after the transfer, while it began to reappear in response to odorous fluids. When these changes in the EOG were compared with the development of the cilium, a discrepancy was found between the period of maximal potential magnitude and that of maximal cilium length (Fig. VI-1). These findings contradict Matthes' hypothesis (1927) that the development of sensitivity to odorous vapors in the newt depends on the growth of the olfactory cilia.

Shibuya and Takagi (1963) found no evidence to suggest that the growth of olfactory cilia is related to the reception of odorous molecules and proposed that the growth of the olfactory cilia in the newt is a secondary phenomenon. For instance, enhanced metabolism may occur in the olfactory cells after the transfer from water onto land which may induce cilial growth. The results of recent experiments on olfactory cilia, however, suggest that this experiment may require reinterpretation.

3. Location of the Receptive Site of Olfactory Cells
3. 1. Are cilia necessary for reception of odors?

For many years, olfactory cilia were thought to play a central role in odor reception. Challenging this view, Shibuya (1964) found that removal of both the olfactory mucus and the olfactory cilia from a small restricted area of the olfactory epithelium resulted in elimination of the EOG, but survival of olfactory nerve discharges when recording in the area.

In a later study, Tucker (1967) stripped the olfactory cilia in the box turtle with aqueous solutions of detergents, Triton, and other agents. Exposure to the solution for only a few seconds removed more than 99% of the cilia. Nevertheless, there was quick recovery of the neural response to amyl acetate, although the EOG was impaired to a far greater degree. These results appear to support the idea that responses of olfactory cells occur independently of olfactory cilia. However, electron microscopy revealed that the olfactory cells retained the ciliary basal bodies (Tucker, 1967), which may also have been the case in Shibuya's (1964) experiment.

Bronshtein and Minor (1977) replicated Tucker's experiment, destroying the olfactory cilia of frogs. Although this abolished the electrophysiological response of the olfactory epithelium to butyl acetate, the responses recovered to 100% of the pretreatment values when the cilia regenerated to only one-fourth of their original length.

These results imply that only the proximal portion of the cilia, or one-fourth of the original length, is required for the reception of odorous molecules and generation of olfactory impulses. This theory corresponds well with the morphological findings of Shibuya (1964) and Shibuya and Tucker (1967).

The role of olfactory cilia was also studied during the process of olfactory cell regeneration after olfactory nerve section. Using scanning electron microscopy, Burns *et al.* (1981) examined the regeneration of the olfactory epithelial surface of the frog after axotomy. This treatment resulted in loss of the dense ciliary matrix and olfactory knobs from the epithelial surface. Partial renewal of the ciliary matrix began to occur by day 40 after the axotomy, at which time the olfactory epithelium became responsive to odors. Substantial replacement of the ciliary matrix was observed by day 100, but recovery of the EOG was delayed beyond this time, implying that reappearance of olfactory knobs and cilia is causally related to the recovery of the EOG.

More recently, Adamek *et al.* (1984) studied the relationship between the regeneration of olfactory cilia and reappearance of the EOG. During ciliary regrowth following detergent-induced ciliary removal, EOG amplitudes initially increased proportionately with ciliary length and reached a maximum when cilia were 40 μm in length. They concluded, therefore, that olfactory transduction sites are located primarily on cilia rather than on the dendrite terminal.

In the electrophysiological study on olfactory receptive sites performed by Ottoson (1956), EOGs in the olfactory mucosa were recorded in re-

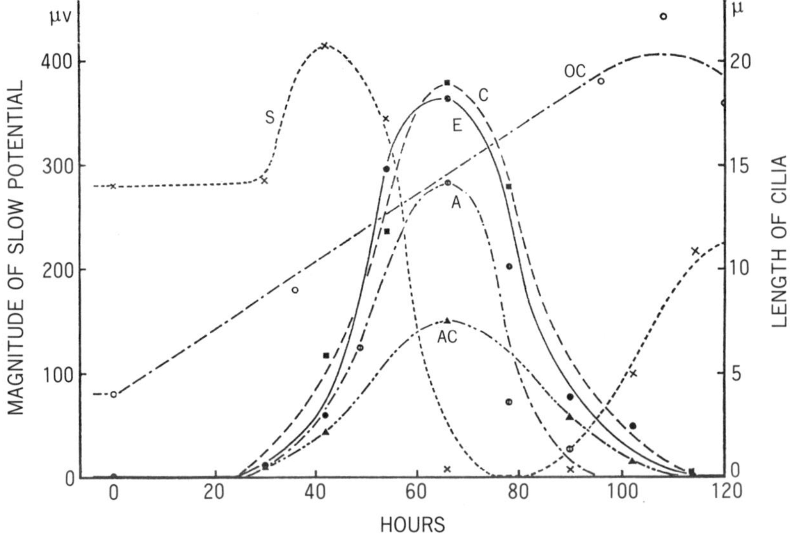

Fig. VI-1. Development and disappearance of the slow potentials in relation to the growth of the olfactory cilia in the olfactory bud of newts (in October). The curves indicate, respectively, the magnitudes of the slow on-potentials evoked by amyl acetate (AC), amyl alcohol (A) vapors, and odorous fluid (S) and those of the slow off-potentials evoked by ethyl ether (E) and chloroform (C) vapors. OC: olfactory cilia; refer to scale at the right. Curve S shows that an odorous fluid evokes a maximum slow potential when odorous vapor first becomes effective, and that an odorous fluid is nearly ineffective when odorous vapor evokes maximum potentials, although the olfactory cilia are growing at a uniform rate during this time. (from Shibuya and Takagi, 1963. Reproduced from *The Journal of General Physiology*, 19 63, vol. 47 p.79 by copyright permission of Rockefeller University Press, New York)

sponse to odorous stimuli and showed systematic decreases in amplitude as the recording electrode was gradually lowered from the surface to the bottom of the mucosa. This result was later confirmed in the salamander olfactory epithelium by Getchell (1977a, b), and in a dissociated single olfactory cell of the newt by Kurahashi and Shibuya (1986a; pp. 205–206).

When all of these findings are considered together with other neuro-anatomical (Kerjaschki and Hörandner, 1976; Menco, 1977; Reese, 1965; Takagi and Yajima, 1964, 1965), neurophysiological (Byzov and Flerova, 1964; Getchell, 1974a; Ottoson, 1970), and biophysical (Getchell *et al.*, 1980) results, it becomes clear that the proximal part of olfactory cilia and knobs are the sites of odorant interaction and transduction which lead to the generation of EOGs (Burns *et al.*, 1981).

3. 2. Receptor site protection studies using group-specific protein reagents

Electron microscopic studies have indicated the presence of olfactory receptive sites by identifying intramembranous particles in the receptive membranes of olfactory vesicles and cilia (pp. 115–119). In addition, neurophysiological investigations have demonstrated the existence of specific receptive sites in the olfactory receptive membrane (Getchell and Gesteland, 1972) and evidence for the existence of specific receptor sites has mainly been generated by neurophysiological studies (Price, 1984). Odor-induced EOGs in frog olfactory epithelium were found to disappear after treating the epithelium with *N*-ethylmaleimide (NEM), a modifier of protein sulfhydryl groups (Getchell and Gesteland, 1972). If an odorous substance, e.g., ethyl-*n*-butyrate, was present in the nasal cavity before and during a brief exposure to NEM, the epithelium, after a wash with Ringer's solution and a recovery period, responded in nearly normal fashion to vapors of ethyl butyrate, but not to vapors of other odorants, excepting those closely related to ethyl-*n*-butyrate. They concluded that special proteins were required for olfactory neurons to be stimulated and that ethyl-*n*-butyrate was able to protect the specific protein through which it interacts with the receptor cells.

Menevse *et al.* (1977a) extended these findings and demonstrated that other protein modifying reagents were also olfactory inhibitors. They concluded that the receptor proteins are on, or in, the cell membrane, because mersalyl, a sulfhydryl modifier that does not generally enter cells, had effects essentially identical to those of NEM. Delaleu and Holley (1980) made similar observations, reporting that an alkylating agent applied to the frog olfactory epithelium inhibited EOGs to esters, but not to amines. This inhibition can be prevented by amyl acetate (Criswell *et al.*, 1980; Meneves *et al.*, 1978).

Taken together, these studies strongly suggest that specific receptor sites exist on, or in, the olfactory cell membrane.

B. Neurophysiological Studies on the Olfactory Receptive Mechanisms

1. Experiments on Olfactory Cell Models

For many years, the minute size of the olfactory cell was an insurmountable obstacle neurophysiological research. Consequently, in stead of studying the olfactory cell directly, several investigators performed indirect experiments on model cells.

Arvanitaki *et al.* (1967) examined the naked apexes of a few giant nerve cells of the helix ganglia and applied several odors at various low concentrations. These cells had the advantage of approximately 150-fold increase in dimensions, as compared with the frog olfactory cell. Intracellular recording revealed a specific and reproducible biphasic local osmoreceptor potential (ORP) in a given cell when activated by a given odor molecule. An accompanying membrane current (i_{rec}) was correlated with the ORP. The final amplitude of the ORP, and the corresponding i_{rec}, varied as a function of the odor molecule concentration, and concentrations above a certain threshold value elicited a spike discharge which was propagated down the axon.

Responses to odorants have also been observed in non-olfactory systems, such as the vomeronasal organ, trigeminal nerve, the Nitella internodal cell, the fly taste nerve, the frog taste cell (Kashiwayanagi and Kurihara, 1984a).

Kashiwayanagi and Kurihara (1984, 1985) also found that mouse neuroblastoma cells (N-18 cells) are reversibly depolarized by various odorants in a way highly resembling that seen in olfactory cells. The minimum (threshold) concentrations of odorants which induced depolarization varied greatly between odorant classes. They found a good correlation between the order of the threshold concentrations for various odorants in the N-18 cell and that in the frog olfactory cell, which led them to conclude that this cell would serve as a good model.

2. New Hypotheses on the Mechanism of Odor Reception

In the past, the olfactory receptive membrane was viewed as a part of the ordinary excitable membrane which has ionic channels for Na^+, K^+, and Ca^{2+}. This line of thought held that the odor-induced opening of Na^+ channels makes the depolarized region a sink which primes an electric current, particularly from the source in the axon hillock of the cell where afferent spike discharges are generated.

If true, experimental alterations of the ionic environment over the olfactory epithelial surface should influence receptor responses to odorous stimuli. In the lamprey olfactory receptors, however, Suzuki (1978) found that up to 10 mM alterations of both Na^+ and K^+ concentrations over the epithelial surface did not influence the receptor responses to L-arginine. This result indicates that the apical receptor membrane is functionally different from other parts of the olfactory cell membrane which respond to stimuli with permeability changes, and has electrogenic properties (Minor, 1980). This result could be attributed to the specific receptive membrane of fish which is constantly exposed to water. Furthermore, the above result was contradictory to the data on ionic mechanisms of EOGs (Chapter V, pp. 167–170).

Based on the above findings, however, Suzuki (1982) proposed a gen-

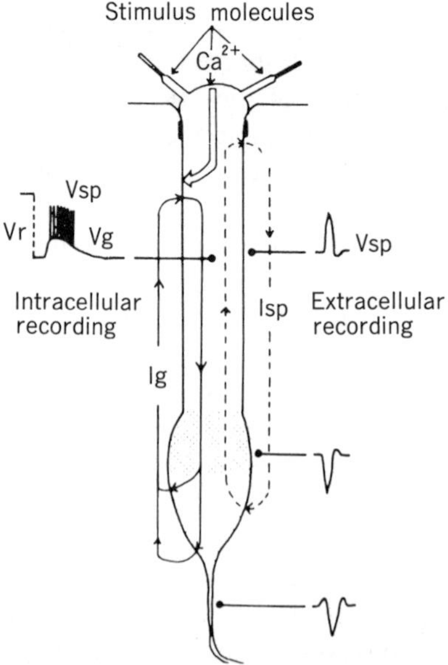

Fig. VI-2. Hypothetical model of the olfactory transduction process. The hollow arrow indicates the sequence of enzymatic reactions mediating the permeability changes in the dendrite membrane. Vr, Vsp, and Vg denote resting potential, spike potential, and receptor potential, respectively. Ig, receptor current for generation of spike potential in the spike initiation zone (dotted region); Isp, the underlying current evoked by electrotonic spread of the spike potential. The extracellularly recorded spike shapes at different loci are based on the analysis by Getchell (1973). (from Suzuki, 1982)

eral model to explain olfactory receptor mechanisms, not only in fish, but also in other vertebrates (Fig. VI-2). His model consists of a cell membrane geometrically and functionally separated into two divisions: the membrane zone for reception of odorous molecules and the electrogenic membrane zones in which ionic permeability changes occur. The former is composed of the membranes of the olfactory vesicle, and olfactory cilia (and perhaps microvilli), while the latter includes the membranes of the olfactory dendrite (underneath the olfactory vesicle), soma, and axon hillock. These two divisions are separated by tight junctions which connect the distal regions of olfactory cells with neighboring supporting cells. Included in the diagram of Suzuki's model is a spike initiation zone located in the proximal segment of the axon including the hillock region (Fig. VI-2, Kuffler and Eyzaguirre, 1955; Getchell, 1973).

According to Suzuki's hypothesis, however, the receptive membranes are activated or inhibited by odorous stimuli, while the electrogenic membranes are induced to elicit electronegative or electropositive EOGs (Takagi, 1969). To account for these effects there must be some underlying reactions, presumably a sequence of enzymatic reactions between the receptive and the electrogenic membranes. Suzuki (1986, 1987, 1988) presented new evidence to support the above intracellular messenger hypothesis for the olfactory transduction process, where the cyclic nucleotides play a role in the coupling of the receptive membrane activated by olfactory stimulus molecules and the effector membrane, to generate receptor potentials. He (1987b) showed that cGMP and cAMP effectively induced a phasic increase in membrane conductance, but that structurally different cyclic nucleotides were ineffective. The conductance increases induced by cGMP and cAMP were not influenced by the presence of di- or triphosphate nucleotides, indicating a direct action of these nucleotides on the membrane. The cGMP-induced conductance increase was identified as being due to an increase in cationic permeability of the receptor membrane. Incidentally, evidence for cAMP and cGMP involvement in membrane permeability changes to Na^+ has been obtained in studies on photoreceptors (Bitensky et al., 1971; Miller et al., 1971; Miller and Nicol, 1979). Since Nakamura and Gold (1987) found a similar cyclic nucleotide-gated conductance change in the olfactory cilia, dendrites, and soma, Suzuki's most recent finding (1987, 1988) should be applicable to all parts of the olfactory receptor cell.

Applying Suzuki's hypothesis to the EOG would indicate that the electronegative EOGs were generated by the electrogenic membranes of the olfactory dendrites and soma, but not by the receptive membranes of olfactory cilia, and vesicles (and probably microvilli).

Ionic experiments carried out in the bullfrog (Takagi, 1968; Takagi et al., 1965, 1968a, b, 1969a) indicated that the amplitude and often the polarity of the EOG changes, depending on the concentrations of the external ions, and that Na^+ is essential for the generation of negative EOGs. These facts are entirely contradictory to Suzuki's findings in lamprey olfactory epithelium. It would be difficult to suppose that the EOG is generated in the dendritic region below the olfactory vesicles and soma. In fact, Ottoson (1956), Getchell (1977a), and Byzov and Flerova (1964) have already shown that the EOG is generated in the distal portion (including olfactory vesicle and possibly cilia) of the olfactory cell. Again, Suzuki's hypothesis cannot explain these results. Recently, Nakamura and Gold (1987) demonstrated the existence of cyclic nucleotide-gated conductance, not only in olfactory cilia, but also in the dendrites and cell soma, making it difficult to separate the olfactory cell membrane into receptive and electrogenic divisions as Suzuki did.

On the other hand, Getchell (1977a, b) and Juge *et al.* (1979a, b) presented conventional local circuit models with homogeneous membranes and explained their data obtained by stimulation of the olfactory mucosa with odorants, and anodal current and intracellular current injection, respectively.

A similar but different hypothesis was proposed by Kurihara and his collaborators. Yoshii and Kurihara (1983a) studied the effects of modified ionic environments on olfactory responses in the carp, the rainbow trout, and the bullfrog. They found that olfactory responses were greatly reduced by removal of ions from the surface of the olfactory epithelia but recovered on addition of ions. Salts of cations of various species including inorganic cations (Na^+, K^+ Ca^{2+}, etc.) and high molecular weight organic cations ($Tris^+$, $choline^+$, etc.) were effective in supporting the olfactory responses. These results suggested that cations do not act as current carriers across the apical membranes for the generation of the receptor potential. Again, these results were entirely contradictory to the results on ionic mechanisms of EOGs (pp. 167–170).

Kashiwayanagi and Kurihara (1984, 1985) observed similar phenomena in the N-18 cell. Replacing Na^+ and Cl^- in the bathing solution with impermeable ions, or reducing the calcium concentration from 1.8 mM to 0.1 mM, had virtually no effect on the magnitude of the depolarization response to odorants in this cell. This and other results suggested that depolarization induced by odorants occurs without changes in ionic permeabilities.

In a series of papers on the taste receptor (Kamo *et al.*, 1974; Kurihara *et al.*, 1981, 1988; Yoshii and Kurihara, 1983a, b, c), Kurihara and his associates proposed that taste receptor potentials mainly stem from changes in the phase-boundary potential at the outer surface of taste cells and then discussed the conditions under which changes in the phase-boundary potential contribute directly to the transmembrane potential (Miyake and Kurihara, 1983a, b). They thought that a similar mechanism could be applied to the olfactory receptor potentials on the basis of findings obtained in the N-18 cell. From analogy, Kurihara and his associates speculated that action potentials in olfactory cells are not generated by odorants at the apical cell membrane where the odorants are adsorbed, but are generated at the axon hillock of the olfactory cell. The potential change elicited at the apical membrane was assumed to spread electrotonically to the hillock of the olfactory cell, which would activate voltage-dependent ionic channels (Kurihara, 1988).

As described above, Suzuki and Kurihara *et al.* proposed new hypotheses that the olfactory cell surface is composed of two kinds of heterogeneous membranes. However, Nakamura and Gold's (1987) recent demonstration of the presence of cyclic nucleotide-gated conductance in the olfactory cilia, dendrites, and cell body indicates that the olfactory cell surface is covered with a homogeneous membrane.

Criticizing these cAMP experiments, however, Kurihara (personal communication) questions whether such conductance changes can be physiologically elicited by cAMP *in vivo*. Consequently, the resolution of this issue rests on the outcome of further experiments in the future.

3. Intracellular Recording of Olfactory Cell Activity

For the analysis of membrane properties of olfactory cells, intracellular recording had been thought urgent for many years. Aoki and Takagi (1968) attempted intracellular recordings of odor-induced responses in bullfrog olfactory cells. Although they recorded large spike potentials superimposed on a slow depolarizing potentials, their attempts to record intracellular potentials appeared to be unsuccessful.

3.1. Intracellular recording of lamprey olfactory cells

Suzuki (1977) succeeded in the intracellular recording of resting potentials and action potentials in response to odorous solutions in the olfactory cell of a fish, the lamprey. The sizes of lamprey olfactory cells had been shown to be larger than those of other higher vertebrates (Thornhill, 1967). Suzuki injected 4% Procion yellow MX4R into these assumed olfactory cells and indicated that about 10% of the recorded cells were olfactory by histological examination. In a subsequent study, Suzuki (1984b) irrigated lamprey and frog olfactory epithelia with Procion yellow MX4R and Lucifer yellow VS. One of Suzuki's most important findings may be that these dyes stained only olfactory cells, but not other cell types. Thus, he demonstrated the usefulness of these fluorescent dyes as cell markers for the identification of olfactory receptor neurons in dissociated cell mixtures.

His intracellular study revealed several important electric properties of the olfactory cell membrane: A marked membrane rectification was observed in the depolarizing current range, whereas an ohmic relation was found in the hyperpolarizing current range (Fig. VI-3 A, B, C). Analysis of voltage transients evoked by hyperpolarizing current pulses by Rall's (1969) method revealed that the transients consist of at least two exponential functions, and supported the idea of the olfactory receptor-

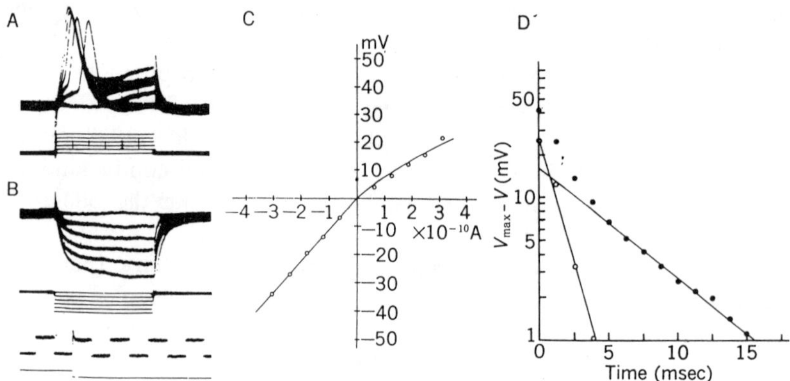

Fig. VI-3. Intracellulary recorded receptor responses to short current steps. A: response to the depolarizing currents. B: responses to the hyperpolarizing currents. The traces under the records indicate applied current intensities. Calibration marks at bottom left indicate 10 mV at 100Hz and 1×10^{-10} A. C: current-voltage relation of the data shown in A and B. D: semilogarithmic plot of voltage transient versus time. (from Suzuki, 1977)

membrane heterogenity, as proposed in Suzuki's hypothetical olfactory cell model (pp. 197–199). In a semilogarithmic plot of voltage transient versus time (Fig. VI-3 D), the voltage plots could be fitted by two straight lines, from which two time constants for the receptor membrane were determined. The first one was found to vary between 2.23 and 5.59 msec and the second one from 0.36 to 1.54 msec.

Another important finding was that the olfactory cell showed a depolarizing response to 10^{-2} M L-arginine and a hyperpolarizing response to 10^{-2} M L-glutamic acid. While spike discharges were elicited by the depolarizing potential and were superimposed on it, spontaneous discharges were inhibited by the hyperpolarizing potential.

This phenomenon where the same receptor cell can produce a depolarizing or hyperpolarizing potential, depending on the type of stimulus, had previously been found in the electroantennogram, (EAG; Kaissling, 1971). Demonstrations of the electropositive EOG by Takagi *et al.* (1960), Shibuya and Takagi (1962, 1963), and Gesteland (1964) were once strongly questioned (Takagi, 1969). This challenge, however, is discounted by Suzuki's above-mentioned discovery of both potentials in the olfactory cell.

3.2. Intracellular recording of salamander olfactory cells

Quite independently of Suzuki's research, Getchell (1977a, b) also succeeded in intracellular recording of the activities of the olfactory and sustentacular cells in the land-dwelling salamander. He chose this animal after study the dimensions of olfactory cells in many vertebrates.

Two categories of intracellularly recorded responses were found in the salamander. When ethyl *n*-butyrate vapor was delivered to the epithelial surface at $f_t = 1.2\%$ for 1 sec, intracellular recordings showed 9 spikes about 50 mV in amplitude superimposed upon a membrane depolarization approximately 8 mV in amplitude. A logarithmic dilution of the stimulus evoked a membrane depolarization about 6 mV in amplitude upon which were superimposed 4 spikes. Anisole, delivered at $f_t = 1.2\%$ evoked a negative EOG of about -3 mV, but did not elicit a response from the cell.

A more important aspect of Getchell's study was the recording of the second category of intracellular responses (Fig. VI-4), which were concluded to originate in the supporting cell. The results of intracellular recording from this cell will be discussed in detail later in this Chapter (pp. 208–209).

Trotier and MacLeod (1983) took recordings from a different species of salamanders. They considered the occurrence of an intracellular spike in response to the antidromic stimulation of the olfactory fibers as a physiological criterion of a neuronal impalement. The input resistance of the olfactory cell was shown to change from more than 180 MΩ at rest to less than 90 MΩ when the maximal depolarization was reached. Appropriate olfactory stimulation evoked a spike potential which presented an inflexion in the rising phase indicating a two-stage depolarization of the somato-dendritic membrane.

3.3. Intracellular recording of olfactory cells in vitro

An *in vitro* preparation provides a significant improvement in the ability to record from and analyze the properties of epithelial cells. For this reason, Masukawa *et al.* (1983, 1985a) excised segments of the dorsal wall of the olfactory sac of the salamander and pinned them, mucus side up, in a recording chamber (Takagi *et al.*, 1966). Their intracellular recording disclosed two electrophysiologically distinct cell types. Type I cells did not discharge action potentials, but had high resting membrane potentials of -50 to -140 mV (-86 mV on average) and relatively low input resistance (0–30 MΩ). Type II cells had resting membrane potentials of -24 to -52 mV (-36 mV on average), high input resistances, and discharged action potentials upon penetration and in response to depolarizing current steps.

Masukawa *et al.* (1985a) identified another type of cell which they encountered least frequently. These cells had an intermediate resting potential, high input resistance, and never generated action potentials.

They tentatively classified this nonspiking type as immature receptor neurons in the process of differentiating from basal stem cells in the epithelium.

Masukawa *et al.* (1985b) also studied changes in the electrical properties of olfactory cells in the tiger salamander after olfactory nerve transection. The percentage of intracellular penetrations of olfactory cells was greatly reduced 1 week after nerve section, but gradually returned to normal within 4 weeks. These cells had a resting membrane potential between -30 and -50 mV and high input resistance (100 to 600 MΩ) characteristically seen in normal epithelium. However, at 1 week, the receptor neurons were able to generate only a single spike in response to injected current, and did not reacquire their ability to respond repetitively until 4 weeks. The number of cells with these immature receptor neuron properties (resting membrane potential between -30 and -50 mV and high input resistance, 100 to 600 MΩ, but unable to generate spikes) increased significantly in the post-transection period. This correlates with the burst of mitotic activity giving rise to new receptor neurons after nerve transection.

Masukawa *et al.* (1986) later studied changes in excitable properties of olfactory cells in normal and degenerating olfactory epithelia. The threshold for impulse generation in response to injected current increased after transection from 74 ± 46 pA in the control epithelium to 176 ± 41 pA two weeks, and 254 ± 80 pA 4 weeks after olfactory nerve transection. The changes observed following transection appeared to be associated with an increased potassium conductance, as suggested by prominent membrane rectification and greatly reduced amplitudes of membrane action potentials in the spike trains. These investigators proposed that the changes might reflect the role of potassium conductances in neural mitogenesis following nerve transection. Further advances in this promising line of research are eagerly awaited.

4. Intracellular and Patch-clamp Studies on Dissociated Olfactory Cells
4.1. Recording of ionic currents in bullfrog olfactory cells

Suzuki (1986), adopting the dissociation technique, overcame the minute dimension of bullfrog olfactory cell which had restricted the progress of intracellular studies (Chapter IV, p. 130). With enzyme (Dispase I) and EDTA treatments, he isolated olfactory cells from the olfactory mucosa and then using gigaseal technique (Hamill *et al.*, 1981), studied the electrical properties of those dissociated cells. These cells had a high morphological integrity and were easily identifiable in the heterogenous mixture of different types by their motile cilia on the olfactory vesicles, thin dendrites

(12–20 μm in length; 2–3 μm in diameter), and oblate sphere-shaped somata (10–12 μm in long axis; 7–8 μm in short axis).

In the cell-attached configuration, a patch pipette directed to the center of the soma frequently recovered current wave forms due to intracellular spontaneous action potentials. The amplitude and frequency of these action potentials were altered by making the pipette potential more positive or more negative than the bath potential. After rupture of the membrane patch (in the whole-cell recording configuration) the pipette-cell seal remained stable for longer than 10 min. The resting potential measured at a zero-current level in normal Ringer's solution ranged from -33 to -39 mV, the input resistance was about 6 gigaohms, and the input capacitance was between 2.5 and 5.0 pF. A current clamp experiment showed that the receptor cell elicited a single action potential in response to a short depolarizing current and a single anode break action potential at the cessation of a short hyperpolarizing current. A voltage clamp experiment in the wholec-ell recording configuration showed that the membrane currents of receptor cells associated with positive voltage pulses in normal Ringer's solution were characterized by an initial inward current followed by a slowly developing outward current. The currents were tentatively identified as an inward Na^+ current and an outward delayed-rectifier K^+ current, thus supporting the conclusions reached by Takagi *et al.* (1968a, b, 1969a, b) on the ionic mechanism of the negative EOG.

It is interesting that the magnitude of the Na^+ current was found to be quite variable between receptor cells, with many showing no Na^+ current. Suzuki attributed this variation in Na^+ current to the differences in maturation of the receptor cells in the olfactory mucosa.

4.2. *The odor response and odor-induced current in solitary olfactory cells*
    isolated from the newt

Kurahashi and Shibuya (1985, 1986a) isolated olfactory cells from newts by dipping them in 0.5–1.0% collagenase (Sigma) solution for 5 to 10 min and analyzed responses to odors by using the giga-seal suction pipette in the whole-cell recording confuguration. Odorous stimuli (*n*-amyl acetate or limonene) were applied with a glass micropipette or double-barreled glass pipette.

In the current-clamp mode, the resting potentials were -41 ± 7.0 mV (mean $\pm$ S. D.; $n = 70$). Depolarizing the membrane by an injected current elicited spike discharges. Increasing the magnitude of injected current led to increased frequency of spike discharges. Application of 1 mM *n*-amyl acetate depolarized the cell membrane in a dose-dependent manner with amplitudes ranging from 5–40 mV (Kurahashi and Shibuya, 1986b).

Maximum amplitude and minimum latency of the odor response was achieved when the stimulus pipette tip was placed near the apical portion of the dendrite. It is very likely that the sensitivity of the olfactory cells to odors was restricted to regions such as the terminal swelling or cilia (pp. 194–195).

When the membranes were clamped at the resting potential, odorous stimuli induced an inward current. The time courses of the odor-induced current resembled those of the odor responses observed in the same cell. The odor-induced current showed a linear current to voltage relationship when the membrane potential was clamped from 36 mV to -45 mV. The reversal potential was $+7\pm3$ mV ($n=4$). These results suggest that the responses to odors were mediated by an increase in the membrane conductance.

Under the voltage-clamp condition, Kurahashi and Shibuya (1986c) identified four types of ionic current, I_{Na}, I_{Ca}, $I_{k(v)}$, and $I_{k(Ca)}$. These currents were similar to those found in other excitable membranes (see section 4. 5).

On the basis of their results, Kurahashi and Shibuya (1986b) described the process of olfactory reception when odorous substances contact the ciliary membrane or terminal swelling as follows: (1) ion channels are activated and the membrane is depolarized; (2) Ca^{2+} and Na^+ channels are activated by membrane depolarization inducing a sharp increase in potential; and (3) the large depolarization and Ca^{2+} influx open two types of K channels resulting in membrane repolarization. They proposed that repetitive spike discharge is probably due to the repetition of steps (2) and (3).

4.3. Membrane conductance mechanisms in dissociated cells of Necturus
      olfactory epithelium

Dionne (1986) dissociated olfactory and nonneuronal cells from the olfactory epithelium of the adult mudpuppy, *Necturus maculosus* (his method is described in Chapter IV, p. 131). Isolated cells retained their morphological integrity, allowing him to distinguish six types of cells, only one of which was neuronal. Whole-cell membrane currents were recorded from olfactory neurons and from a non-neuronal cell type that was spherical and partially covered with asynchronous, motile cilia ("Medusa" cells). Columnar supporting cells did not survive the dissociation treatment and were not studied.

Sustained outward membrane currents (presumably K^+) were elicited from Medusa cells by depolarizing voltage steps. The time course of the outward current in olfactory neurons resembled that of the delayed potassium current seen in Medusa cells. The inward current in olfactory neurons, which had a rapid onset and was transient, was predominantly carried by Na^+; it was blocked by TTX and was fully inactivated at -20

mV. This experiment again demonstrated the role of Na^+ in the generation of the action potential of the olfactory cell, and, hence, of the negative EOG.

Anderson and Hamilton (1987) also performed intracellular recordings from isolated receptor neurons of the tiger salamander. They found similar electrical properties in the receptor membranes and confirmed previous studies indicating that transduction of olfactory stimuli leads to a depolarization of vertebrate olfactory receptor neurons.

4.4 Preparation of isolated mammalian olfactory neurons

Maue and Dionne (1987a, b) described a method for producing isolated olfactory neurons from adult mouse nasal epithelium. The dissociated neurons and other cell types were judged to be intact and viable by trypan blue dye exclusion, the presence of OMP, and a variety of electrophysiological measurements indicating the presence of substantial membrane potentials, low levels of intracellular Ca^{2+}, and the ability to fire action potentials. These cells were found to be amenable to the patch-clamp technique and to immunohistological studies.

4.5 Patch-clamp studies on olfactory cell responses

Firestein and Werblin (1985, 1986), using the whole-cell patch-clamp technique, analyzed the response properties of enzymatically isolated olfactory receptor cells of the tiger salamander. Odorants puffed into the vicinity of the cells elicited a slow depolarization from the resting level of -65 mV to the spike threshold level near -55 mV when a single, large action potential (60–80 mV) was generated. This initial response was followed by one to six attenuated action potentials riding on a sustained depolarization.

Firestein and Werblin (1987) advanced their research in isolated olfactory cells of the larval tiger salamander. They were able to separate voltage-dependent currents into three ionic components: a transient inward sodium current, a sustained inward calcium current, and an outward potassium current. Furthermore, they could identify three components of the outward current by their gating and kinetics: a calcium-dependent potassium current [IK(Ca)], a voltage-dependent potassium current [IK(V)], and a transient potassium current (Ia).

Taken together, these findings led them to conclude that the high sensitivity and slow adaptation commonly ascribed to olfactory receptors can be mediated by the gated conductances in the ciliary and soma membranes. In addition, more precise analyses of membrane properties by injection of current led them to suggest that olfactory receptor neurons may be acting primarily as detectors and do not encode stimulus intensity. This view has important implications and will surely be examined in the future.

Trotier (1986) also performed patch-clamp analysis of membrane currents in isolated and cultured olfactory cells of the adult salamander. Using different channel blockers and ion substitutions, several distinct components were separated in the overall whole cell current: In most cells inward current reflected the activation of a TTX-resistant conductance which was blocked by cobalt ions; in a minority of cells, a TTX-sensitive sodium current was also observed. The outward K^+ current was blocked by cessium and tetraethylammonium, and could act as a depolarization limiter in cases of intense odor stimulation.

5. Intracellular and Voltage- and Current-clamp Recordings of Olfactory Cells in the Spiny Lobster

Anderson and Ache (1985) attempted intracellular recordings from spiny lobster olfactory cells, using the whole-cell configuration of the patch recording technique, and showed that isolated somas of olfactory receptor cells are electrogenic, producing fast overshooting action potentials when depolarized. *In situ*, these cells produced graded receptor potentials and action potentials when stimulated chemically. Anderson and Ache voltage clamped the receptor potential at -40 mV and -70 mV and found that the extrapolated reversal potential for the inward current was $+40$ mV, and that the receptor potential in these cells is produced by a net inward current, consistent with the observation for the vertebrate EOG (Takagi *et al.*, 1968a, b, 1969a). Thus, the electrical properties of these crustacean receptors were found to be consistent with the data already obtained in other animals.

6. Intracellular Recording of Supporting Cell Activity
6.1. Getchell's experiment

Getchell (1977a, b) carried out intracellular recording in the olfactory epithelium of salamanders. As mentioned above (pp. 202–203), he identified two categories of potentials; one type was ascribed to the olfactory cell, while the other was found in cells which never showed a spiking discharge or spontaneous activity upon penetration. Odor-induced responses of the second category were characterized by gradual changes in potential which did not result in the generation of action potentials. Cells exhibiting these properties had resting potentials of -20 to -90 mV (average -49 mV), which were higher than those of olfactory cells, and stable recordings were easier to obtain for longer periods of time. Considering these characteristics in light of the finding that sustentacular cell bodies were far larger than olfactory cell bodies ($25.6\ \mu m \pm 1.8\ \mu m \times 7.8\ \mu m \pm 0.9\ \mu m$ in 31 cells), it is highly probable that recordings of the second category were made from sustentacular cells.

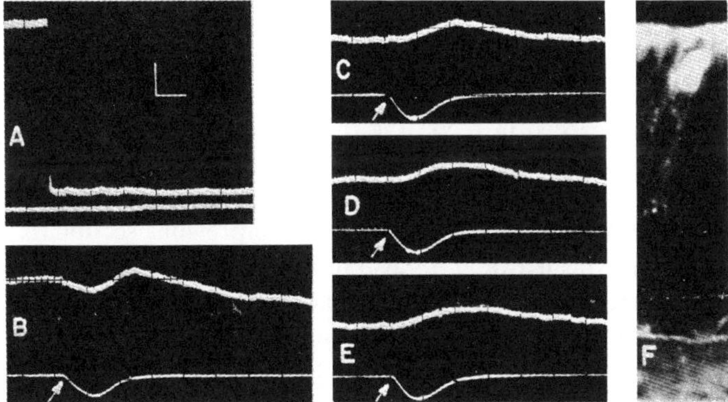

Fig. VI-4. Intracellular recordings from a sustentacular cell and negative EOG recorded from the epithelial surface. Penetration shown in (A) and responses evoked in sequence by ethyl *n*-butyrate (B, E), amyl acetate (C), and anisole (D) delivered to the epithelial surface. Procion marked cell shown in F. Calibrations: top traces, 10 mV; bottom traces, 5 mV; time, 2 sec. Arrows indicate onset of negative EOG. (from Getchell, 1977a)

Although action potentials were not found in these cells, Getchell identified three basic odor-induced response types (Fig. VI-4): (i) membrane hyperpolarization; (ii) membrane depolarization elicited by amyl acetate and anisole; and (iii) an initial hyperpolarization followed by a depolarization and then a second hyperpolarization, elicited by ethyl *n*-butyrate. These membrane responses were typically 12 sec in duration, required about 4 sec to reach peak amplitude, and were initiated about 0.5 sec after the onset of a negative EOG. Intracellular recordings of potentials from glial cells showed similar characteristics (Kuffler and Nicholls, 1966; Somjen, 1975). As Okano and Takagi (1974) noted in their paper, the sustentacular cell belongs to the ependymal cell group and can thus be classified as a kind of glial cell.

Getchell (1977a, b) did not find a large hyperpolarizing potential accompanied by a release of secretory granules when odorants were applied to sustentacular cells. It is unfortunate that Getchell used odors different than those listed in Table V-1 (p. 153), and thus may not have effectively stimulated the sustentacular cells.

6.2. Masukawa et al.'s *experiments*

Masukawa *et al.* (1983) obtained intracellular recordings from two types of cells in an *in vitro* preparation of the salamander olfactory epithelium. The superficial location in the epithelium of the type I cells (depth, 0 − 350 μm) suggested to Masukawa *et al.* that these were sustentacular cells with glial-like electrophysiological properties.

Following Getchell (1977a), this study represents the second intracellular examination of sustentacular cell membrane properties. While Getchell (1977a) succeeded in recording three types of responses elicited by some odors, Masukawa *et al.* failed to record action potentials induced by depolarizing the membrance with an injected electric current. As with Getchell's (1977a) experiment discussed in the previous section (6.1), it is regrettable that Masukawa *et al.* also did not apply odorants that would elicit both electropositive EOGs and release of secretory granules from the sustentacular cells (Table V-1; Takagi *et al.*, 1969b).

Masukawa *et al.* (1985a) studied the electrophysiological properties of supporting cells in an *in vitro* preparation of the tiger salamander olfactory epithelium. Cells that were identified as supporting cells by Lucifer yellow injections had high resting membrane potentials, relatively low input resistances, and never discharged impulses, either spontaneously or in response to injected current. These results support the previous suggestions regarding the glial-like properties of supporting cells.

In an associated study, Masukawa *et al.* (1985b) found changes in the electrical properties of supporting cells in the tiger salamander after olfactory nerve transection. The results revealed different changes in two types of cells: One type (A) showed a decrease in resting membrane potential and a small increase in input resistance. A second type (B) showed a very large increase in input resistance. According to these investigators, the findings imply that the degenerating receptor neurons transmit a signal which induces changes in the functional properties of the glial-like supporting cells. These changes may involve adjustments in membrane properties or electrical coupling between cells. The changes may also be mediated by alterations in trophic connections which are thought to exist between supporting and receptor cells.

Similar to Getchell's (1977a, b) studies, Masukawa *et al.* did not find the large hyperpolarizing potential accompanied by release of secretory granules in the supporting cell which was demonstrated by Okano and Takagi (1974) and Takagi and Okano (1974). The hyperpolarizing potentials recorded by Getchell (1977a, b) were smaller and shorter than Okano and Takagi had anticipated. Nevertheless, the author expects that striking hyperpolarizing potentials accompanied by the release of secretory granules in the sustentacular cell will soon be demonstrated using intracellular recording techniques in response to chloroform or other vapors listed in Table V-1.

C. Neurochemical Studies on Olfactory Receptive Mechanisms

1. Studies of Olfactory Receptive Membrane in Lipid Bilayer Membranes

Nomura and Kurihara (1987a, b) studied responses to odors of the lipid bilayer membrane as a new model of the olfactory receptor membrane.

1. 1. Responses of liposomes to odorants

Those workers found that liposomes having certain lipid compositions are depolarized by various odorants. There was a good correlation between the threshold concentrations of odorants in the liposomes and porcine olfactory cells. This suggests that the reception of odorants in the olfactory cell is primarily due to adsorption of odorants on the hydrophobic region of the cell membrane (Kashiwayanagi *et al.*, 1987).

1. 2. A mechanism of odor discrimination——Neurological studies

Altering the lipid composition of liposomes led to great changes in specificity of the responses to odorants (Kashiwayanagi *et al.*, 1987). For example, four out of six odorants tested included depolarization in phosphatidylethanolamine (PE)-enriched phosphatidylcholine (PC; $PE/PC = 0.25$). A further increase in PE content ($PE/PC = 0.54$) led to depolarization in response to only two odorants. Liposomes containing a mixture of sphingomyelin (SM), PE, and PC exhibited depolarizing responses to four odorants tested and addition of increasing amounts of cholesterol (cholesterol/(PC + PE + SM) = 0.05 and 0.11) led to corresponding decreases in the number of odorants capable of eliciting depolarization (two and one odorants, respectively).

On the basis of the above results, the following mechanism of odor discrimination was proposed: The lipid composition of the receptor membrane is postulated to be different for each olfactory cell. Hence, each olfactory cell is differentially sensitive to various odorants. Different response profiles at the cellular level are transformed into specific firing patterns among various olfactory axons, thus encoding the quality of an odor which can be recognized by the brain. Variation of lipid and protein combinations in the membrane provides many different adsorption sites for odorants.

It is known that a single olfactory cell can respond to various odorants, and that each cell has a specific response spectrum. Such response variation can be explained by assuming that the composition of receptor membranes is specific for each olfactory cell. However, Kurihara and his collaborators did not explain how adsorption at different membrane sites of the same cell can be transduced into different patterns of afferent nerve impulses, or projected to different olfactory areas. More precisely, the mechanisms underlying the transduction of different response profiles into different patterns of nerve impulses in the same or different olfactory axons were not addressed in their papers.

2. Research on Specific Receptor Proteins

It is generally agreed that the initial event in olfactory cell responses to odorous stimuli occurs by interaction of the stimulant with a macromolecule, the receptor protein, which is thought to exist on the receptive membrane. The receptor protein presumably has the ability to change its conformation upon binding the stimulant which somehow initiates the cell response (Price, 1984).

2. 1. Anisole binding protein

Using affinity chromatography, Goldberg *et al.* (1979) isolated an anisole binding protein from the olfactory epithelium of dogs. Extracts of the tissue were passed through a column, with *o*-methylphenol added to the eluting stream to displace protein from the column. Obviously, the displaced protein had a high affinity for *o*-methylphenols. Because the odorant, anisole, is the simplest *o*-methylphenol, the displaced protein was named "anisole binding protein." It was present in the olfactory epithelium, but not in the adjacent respiratory epithelium, indicating that it was not a ubiquitous component of cells, mucus, or cilia. A detergent was needed to extract it, demonstrating that it was a membrane constituent.

The most difficult problem for investigators attempting to isolate olfactory receptor proteins has been to prove that the isolated protein is really from the olfactory receptor. Most workers simply use binding of odorants by the protein as a criterion, along with tissue specificity, and evidence that the protein originated in membranes and/or cilia (Lancet, 1986). Price (1986) has introduced a functional criterion: Antibodies directed against the protein are prepared and applied to functioning olfactory epithelium. Effects of the antibodies, if any, on the olfactory responses are then determined, and inhibition of the response is expected if the protein is a receptor molecule (Price, 1986).

In a subsequent study Goldberg *et al.* (1979) immunized rabbits with the anisole binding protein and studied the effects of the sera on the EOG of mice. While a control serum had no effect, the immune serum was as effective in inhibiting the EOG to amyl acetate as it was against anisole. He hypothesized that anisole binding protein is a receptor protein for compounds similar to anisole and that other receptor proteins are structurally related to it; that is, that they have a constant and a variable region, similar to proteins of the immune system. Thus, the immune serum raised against anisole binding protein would inhibit responses to other odorants reacting with the constant region of other receptor proteins.

In this connection, Price and Turpin (1980) predicted that there is much more protein in the olfactory epithelium reactive to the antibodies than can be accounted for by anisole binding protein alone. In fact, columns prepared from the antibodies bound about five times as much protein from dog olfactory epithelial extracts as could be accounted for by anisole binding protein.

2. 2. Benzaldehyde binding protein

Price and Wiley (1987) predicted that other odorant binding proteins will have similar structural properties. A procedure essentially identical to that used to isolate anisole binding protein was used to isolate a benzaldehyde binding protein from the dog olfactory epithelium. Anisole binding protein was removed from the extracts prior to affinity chromatography on the column to which the benzaldehyde derivative was coupled, thereby ensuring that the two proteins were different. Antisera against benzaldehyde binding protein inhibited the EOG responses to benzaldehyde, anisole, and amyl acetate (Price and Willey, 1987).

3. Studies with Monoclonal Antibodies

Antisera contain many antibodies directed against different regions of the antigen. Consequently, the question of whether or not there is a structurally related class of receptor proteins in olfactory neurons could not be answered with antisera. To overcome this problem, Price (1986) used monoclonal antibodies, obtained from clones of single cells that each produced a specific antibody, expecting that the monoclonal antibodies directed against the binding proteins would show three kinds of specificity when tested for effects on the EOG:

i) Those directed against the variable region would be specific inhibitors of responses to the odorants binding to that protein.

ii) Those directed against the intramembrane protein would be unable to interact with the protein and have no effect on the EOG.

iii) Those directed against the exposed part of the constant region would be general inhibitors of olfaction.

Among 26 cell lines tested (Price, 1986) three (12%) produced antibodies that specifically inhibited responses to the odorant which bound to the antigen, and had no effect on responses to the other two test odorants (benzaldehyde, anisole, and amyl acetate were used as test odorants). Another nine (35%) produced antibodies equally effective against responses to all three odorants. The remaining 14 (54%) produced antibodies that had no effect on the EOG. These results were expected if there was a class of structurally similar receptor proteins in olfactory neuronal membranes. Of the 12 monoclonal antibodies that were inhibitory, most reduce the EOG by 25% or less and none by as much as 40%, which led them to conclude that most of a response to an odorant arises from processes independent of receptor proteins. In other words, it is clear that although anisole and benzaldehyde binding proteins from dog olfactory epithelium are olfactory proteins, a significant fraction of the total olfactory response is mediated by mechanisms other than receptor proteins of this class (Price, 1986).

Monoclonal antibodies and lectins have been used in fluorescence labeling of sensory neurons (Allen and Akeson, 1985a; Hempstead and Morgan, 1985), nerve fascicles (Fujita *et al.*, 1985), surface epithelial structures (Hempstead and Morgan, 1983b), and specific ciliary proteins (Chen *et al.*, 1986a). Further use of these reagents could facilitate investigations of olfactory function (Lancet, 1984).

4. Olfactory Marker Protein (OMP)

As already described in Chapter IV (p. 123 and pp. 139–140) OMP was found to be expressed in developing epithelium only after olfactory cell axons form synapses with the apical dendrites of mitral cells in the glomeruli of the OB. Therefore, it constitutes a highly useful probe for neuronal differentiation (Farbman and Margolis, 1980). The recent isolation of the mRNA species coding for this protein (Rogers *et al.*, 1985) may help to resolve its yet unknown function, and possibly relate it to receptor mechanisms.

5. Criteria for Olfactory Receptor Identification

As previously stated, Price (1981) pointed out that the mere binding of one or a few odorous compounds is not sufficient proof of receptor identification. In answer to this criticism, Lancet (1986) proposed the following set of criteria that stems from the notions that olfactory receptor proteins of different specificities share "common denominator" molecular properties, and unless otherwise proven, these general properties can be assumed to be similar to those of other well-studied membrane receptors:

(1) Binding of odorous compound
(2) Tissue specificity
(3) Enrichment in the cilia (vs. epithelium)
(4) Glycosylation
(5) Transmembrane disposition (integral membrane protein)
(6) Correct bilayer concentration (major compound)
(7) Diversity (sequence heterogeneity)

(8) Specific recognition by function-modulating reagents (antibodies, lectins)

(9) Interaction with transductory proteins

(10) Reconstitution of odorant modulation of enzymatic activities.

Lancet (1986) suggested that fulfillment of most or all of these criteria should be taken as evidence for olfactory receptor identification, even in the absence of ligand binding data or clear-cut pharmacological correlations. These criteria should be applied to distinct polypeptide species that constitute candidate receptor proteins. In fact, some of them have been proposed and utilized previously. For further information see Lancet (1986).

6. Nonspecific Receptor Proteins in the Olfactory Receptive Membrane

Kashiwayanagi and Kurihara (1985) used the N-18 cell model (Kashiwayanagi and Kurihara, 1984) to examine a mechanism of odor discrimination. Using a fluorescent dye, rhodamine 6G (Rh6G; Aiuchi *et al.*, 1980; Tanabe *et al.*, 1980), they observed dose-dependent depolarization in response to 19 odorants. Since the N-18 cell is a neuroblastoma independent of any olfactory receptor cell, it is unlikely that it carries specific receptor molecules for these odorants. Rather, it was thought that the reception of odorants very likely occurred via nonspecific molecules. This idea is supported by the fact that various nonolfactory systems have been found to respond to various odorants (p. 196). Price (1984) has attempted to isolate the receptor proteins for odorants, and has argued that odor reception can occur via nonspecific proteins.

7. Receptor Candidates in Isolated Olfactory Cilia

Isolated olfactory cilia constitute a cell-free membrane preparation rich in receptive compounds. The first preparation of isolated olfactory cilia was obtained from fish olfactory organs and was shown to contain binding activity toward amino acid odorants (Rhein and Cagan, 1980, 1981). A more recent study by Chen and Lancet (1984) identified several specific proteins, four of which were glycosylated, in a preparation from olfactory cilia of the frog, *Rana ridibunda*. A similar result was reported in a study on bullfrogs (Anholt and Snyder, 1986). Evaluating effects of lectins on olfactory responses, Lancet (1986) proposed that one or more of them was a functional surface component. Chen *et al.* (1984, 1986b) identified and characterized a high molecular weight (95 kD) glycoprotein, gp 95, isolated from olfactory cilia. Lancet (1984, 1986) suggested that major olfactory membrane proteins, such as gp 95, corresponded to the intramembranous particles (IMPs) located on the P-face of the ciliary membrane (Menco *et al.*, 1976). Since gp 95 fulfilled

3, 5, 6, and 8 of the above criteria, they concluded that gp 95 constitutes a plausible candidate for an olfactory receptor protein.

It may be important to recall here Chen and Lancet's report (1984) that the olfactory cilia and the glycoprotein isolated from them preferentially binds to wheat germ agglutinin. On the other hand, Hemstead and Morgan (1983a) had demonstrated that fluorescin-conjugated concanavalin A preferentially binds to mature receptor neurons in the olfactory mucosa.

8. Role of Cyclic AMP in Olfactory Reception and Transduction

During the past fifteen years, many investigators have studied the role of cAMP in olfactory transduction: Kurihara and Koyama (1972) showed that the olfactory epithelium has a relatively high level of adenylate cyclase activity; Margolis (1975b) discovered that its phosphodiesterase isozyme pattern changes when the neurons degenerate; Minor and Sakina (1973) and Menevse et al. (1977b) found that cAMP and its dibutyl derivative (but not cGMP or its analogues) can induce the appearance of EOG-like potentials, and that phosphodiesterase inhibitors can modulate the electrophysiological response to odorants (Chapter V, p. 184).

Pace et al. (1985) demonstrated that a cell-free olfactory cilia preparation has an extremely high activity of adenylate cyclase (10 times that of brain membranes), which was specifically enhanced by odorants. This finding fulfilled the last of Sutherland's criteria to identify cyclic nucleotide mediation (Robinson et al., 1971). Thus, they characterized an odorant-sensitive adenylate cyclase localized in olfactory cilia and confirmed the above mentioned reports indicating a role for ATP, cyclic AMP, and other nucleotides as second messengers in olfactory transduction. It is noteworthy that odorant-induced transductory events may also result in the modulation of other membrane enzymes, such as Na/K ATPase (Lancet, 1986; Dreesen and Koch, 1982).

Pace et al. (1985) showed that odorant stimulation of ciliary adenylate cyclase occurs only in the presence of GTP, suggesting the involvement of a signal-coupling GTP-binding protein, or G-protein (Gilman, 1984; Schramm and Selinger, 1984; Stryer, 1986). In fact, Lancet and Pace (1984) and Pace et al. (1985) have directly identified the olfactory G-protein in olfactory cilia (but not in respiratory cilia), via labeling with cholera toxin (Snyder et al., 1988).

9. Probable Mechanism of Olfactory Transduction

Extrapolating from a variety of data, Lancet (1986) proposed a very plausible molecular model for olfactory reception and transduction (Fig. VI-5). In Lancet's model, the receptor molecule (R) is an integral mem-

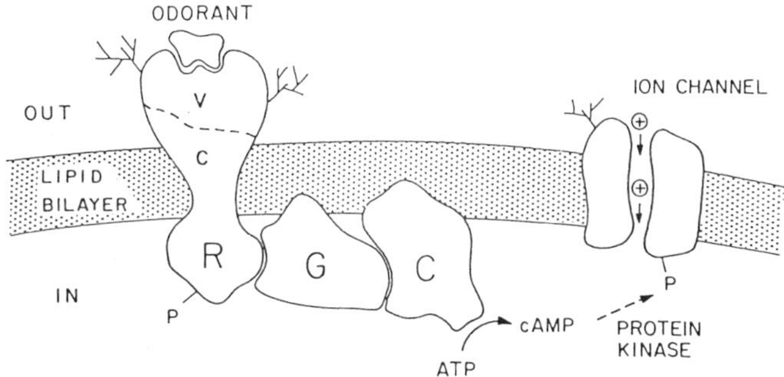

Fig. VI-5. The most probable molecular model for olfactory reception and transduction, based on data and speculation as described in the text. (from Lancet, 1986)

brane glycoprotein with a constant (c) and a variable (v) region. Coupling of an odor molecule to the receptor activates a GTP binding protein, or G-protein (G), which modulates the activity of adenylate cyclase (C), and the essential enzyme for synthesis of the second messenger, cAMP. The cAMP molecule activates cAMP-dependent protein kinase resulting in the phosphorylation of other proteins. Phosphorylation (p) of ion channel polypeptides leads to changes in membrane potential, while receptor polypeptide phosphorylation constitutes a negative feedback mechanism underlying adaptation.

10. Discovery of a Cyclic Nucleotide-gated Conductance in Olfactory Cell and Cilia

As mentioned above, olfactory transduction is thought to be initiated by the binding of odorants to specific receptor proteins in the cilia of olfactory receptor cells. The mechanism by which odorant binding initiates membrane depolarization was a mystery until the recent discovery of an odorant-stimulated adenylate cyclase in purified olfactory cilia (Pace *et al.*, 1985; Sklar *et al.*, 1986) suggested that cAMP may serve as an intracellular messenger for olfactory transduction. If this is the case, conductance across the ciliary plasma membrane may be controlled by cAMP.

10. 1. Suzuki's experiment on olfactory cells

In the olfactory transduction process, cyclic nucleotides have been proposed as intracellular messengers which translate the chemical activation of the olfactory receptive membrane (i.e., olfactory cilia or vesicle membrane) into ion channel activity resulting in the generation of depolarizing potentials (Pace *et al.*, 1985). Examining this hypothesis Suzuki (1986) studied the effects of candidate nucleotides on enzymatically isolated receptor cells from bullfrog olfactory epithelium by means of a whole-cell patch-clamp perfusion technique. With voltage clamped at -75mV, introduction of cGMP into the receptor cell induced a slow monophasic inward current, showing a fast rising response which gradually declined. The latency and rising time constant for 500 μM c-GMP were 400 msec and 530 msec, respectively. Similar inward currents were also induced by cAMP, although cAMP was less effective. The response thresholds were 20 μM for cGMP and 50 μM for cAMP. The inward currents in the receptor membrane were not affected by these cyclic nucleotides, indicating that the responses to the nucleotides were due to increased cationic conductances in voltage-sensitive currents. The other nucleotides (5′-GMP, 5′-AMP, 5′-GDP, 5′-ADP, 5′-GTP, and 5′-ATP) had no effects on receptor membrane conductance. This suggested that both cGMP and cAMP act directly on cyclic nucleotide-sensitive channels in the receptor membrane and that the conductance changes are not mediated by phosphorylation in the receptor cell. Thus, Suzuki provided strong evidence that cyclic nucleotide-sensitive channels are distributed in the olfactory receptor cell membrane, as was found in the vertebrate photoreceptor cell (Ohmori, 1985).

10. 2. Nakamura and Gold's experiment on olfactory cilia

Nakamura and Gold (1987) identified a direct cAMP-gated conductance ciliary plasma membrane, obtained from dissociated receptor cells of toads (*Bufo marinus*). This finding was consistent with evidence that the cilia are the site of olfactory transduction (Rhein and Cagan, 1980; Adamek *et al.*, 1984). They also observed a cAMP- and cGMP-gated conductance in patches excised from the dendritic and cell body regions, implying that the same or a similar conductance is present in the cilia, dendrites, and soma.

This conductance resembled the cGMP-gated conductance found to mediate phototransduction in rod and cone outer segments (Fesenko *et al.*, 1985; Haynes and Yau, 1985), but differed in that it is activated by both cAMP and cGMP. Nakamura and Gold's findings indicate a mechanism by which an odorant-stimulated increase in cyclic nucleotide concentration could lead to an increase in membrane conductance and, therefore, membrane depolarization.

Suzuki's and Nakamura and Gold's data are significant because they not only support the molecular model of olfactory transduction outlined in section 9, but also suggest a remarkable similarity between the mechanisms of olfactory and visual transduction, indicating considerable conservation of sensory transduction mechanisms.

D. Sensory Activities of the Olfactory Epithelium

Odorous stimulation elicits responses in the olfactory and trigeminal nervous systems. They have been recorded by means of a pair of silver wire electrodes from these nerves. In the OB, these wire electrode as well as glass pipette or tungsten microelectrodes with tip diameters of 1 μm or so have been used to record induced waves and impulse discharges.

1. Odorous Stimulation of Olfactory Cells
1.1. Spontaneous discharges—Influence of mucus on olfactory cell sensitivity

It is well known that olfactory cells elicit spontaneous discharges. These discharges are most likely elicited by the action of the olfactory mucus, which is secreted by Bowman's glands and the supporting cells. Although the exact composition of mucus from either source is unknown, there is histochemical evidence that it differs (Getchell and Getchell, 1977b). Okano and Takagi (1974) found differences among odorants in stimulation of release of granules from supporting cells into the mucus of the bullfrog olfactory epithelium. Taken together, these findings suggest that mucus composition will change if Bowman's glands and supporting cells are differentially stimulated by chemicals.

The pH, ionic, and colloidal components of the mucus are likely to fluctuate with time, which will influence the resting membrane potentials of olfactory cells. Lindvall (1977) altered mucus pH by presenting aerosols of buffered solutions along with odorants, and found that sensitivity to hydrogen sulfide increased as the buffer pH rose from 5.3 to 9.3. Many organic acids and bases are odorous, and application of them as stimuli will alter the pH of the mucus.

Bourne (1948) and Baradi and Bourne (1951) identified alkaline phosphatases and esterases in the nasal mucosa. Therefore, esters in contact with the mucosa will be hydrolyzed to form acids and alcohols (Cushieri, 1974; Gladysheva and Martynova, 1982; Gladysheva *et al.*, 1982). Getchell *et al.* (1980) observed that some olfactory neurons are stimulated by carbon dioxide, presumably via pH effects. Thus, olfactory cell sensitivity is continuously fluctuating under the influence of pH changes.

In addition, it is likely that odorants affect neuronal metabolism and change the properties of the receptive membrane. Furthermore, extracellular fluid levels of oxygen, carbon dioxide, nutrients, and waste products may be constantly affected by mucus flow on the epithelial surface as well as by local blood flow. All of these environmental changes influence the resting membrane potentials of the olfactory cell, and will thus elicit sporadic spike potentials.

1.2. Responses to odorous stimuli—Extracellular recording

Gesteland *et al.* (1963) were the first to perform extracellular recording of spike discharges of single frog olfactory cells. They devised a special metal microelectrode (Gesteland *et al.*, 1959) and studied the odor-specific properties of olfactory cells by applying 26 odors. Quite independently, Takagi and Omura (1963) conducted extracellular recording of spike discharges of single bullfrog olfactory cells by means of a glass pipette microelectrode and identified seven response types. Since then, many investigators have recorded single olfactory cell activities and studied cell responses to odors in frogs, salamanders, and other animals (Gesteland, 1971, 1976; Gesteland *et al.*, 1965; Getchell, 1973, 1974b, c; Kauer and Shepherd, 1975, 1977; Shibuya and Shibuya, 1963; Shibuya and Tonosaki, 1972).

Getchell and Shepherd (1978a, b) analyzed the impulse activities and adaptive properties of salamander olfactory receptors by applying step pulses of odors at different concentrations. Shibuya and Tucker (1967) studied single-unit responses of olfactory cells in vultures. They recorded short trains of spikes during inspiration and long trains of spikes during expiration. Spikes increased in frequency but decreased in duration with increasing odor concentrations and were only elicited by some odors, indicating a degree of odor selectivity. Shibuya (1969) also studied activities of single olfactory cells in the gopher tortoise, *Gopherus polyphemus*, in Florida.

1.3 Recording of impulses from an olfactory nerve twig

Tucker (1963a, b) separated small strands of the olfactory nerve under Ringer's solution and detached them centrally to obtain peripherally directed nerve twigs. Twigs were taken from the intracranial portion of the olfactory nerve and were placed on a pair of Pt-Ir wire electrodes spaced about 1 mm apart. Mineral oil was substituted for Ringer's solution to permit recording of impulse discharges. The diameter of nerve twigs in this preparation ranged upward from about 10 μm, which was about the lower size limit to permit recording. Some nerve twigs with diameters larger than 100 μm yielded excellent records. Tucker estimated that a twig as small as 10 μm in diameter could contain an upper limit of 2,500 fibers, and usually as many as 2,000 fibers, on the basis that unmyelinated fibers are uniformly 0.2 μm in diameter (Gasser, 1956). This constituted Tucker's most common method of recording responses to odors in the olfactory nerve as well as the trigeminal and vomeronasal nerves. In a successful research career, Tucker and his collaborator(s)

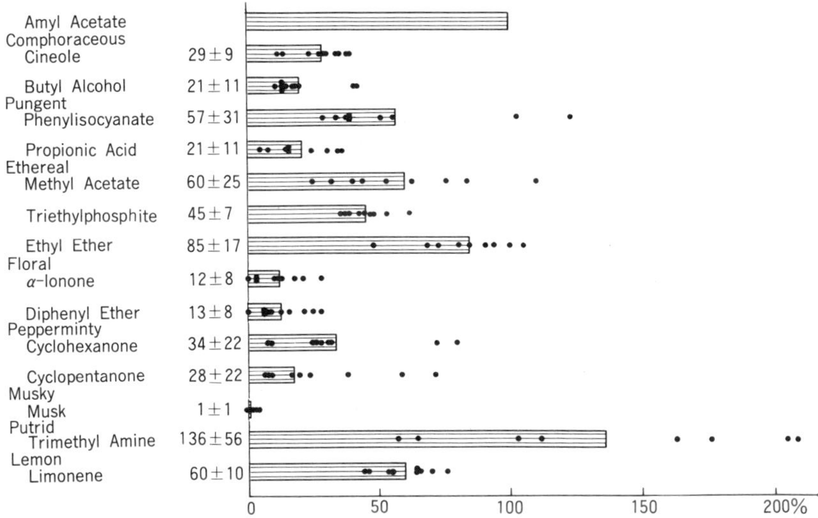

Fig. VI-6. Spike discharges elicited by odors in olfactory nerve twings of guinea pigs. (from Ito, 1968)

have published many important papers on the peripheral olfactory system (Tucker and Shibuya, 1965; Tucker and Kiyohara, 1978; Tucker, 1978).

(a) Magnitudes of responses to different odors

Using an olfactory nerve twig preparation, Ito (1968) recorded responses to 19 odors in the guinea pigs. Spike discharges elicited by the odors were passed through an electronic integrator to compare the magnitudes of the responses among the 19 odors. As shown in Fig. VI-6, the magnitude of the response to amyl acetate was designated as 100% and, on this basis, relative magnitudes of responses to the other odors were calculated as percentages. Numerals on the left side of the vertical line indicate the average and standard deviation of the results of 10 samples for each odor. Each horizontal histogram represents the averaged response magnitude for each odor. The response magnitude to a given odor recorded in a twig was found to be the same if the odor was reapplied after a sufficient interval. Comparison of response magnitudes recorded from two olfactory twigs of the same animal revealed various degrees of difference according to the kind of odor applied.

(b) Different response patterns to successive stimuli

Successive application of the same odor elicited different patterns of impulse discharges. Impulse discharges disappeared rapidly when stimulation with either ethyl alcohol or methyl acetate was discontinued after three successive applications. However, impulse discharges survived the cessation of stimulation in different degrees when applying limonene, cyclohexanone, or cyclopentanone. Thus, Ito (1968) clearly indicated that the after effects of odorous stimulation depend on the kind of odor applied.

(c) The phenomenon of olfactory adaptation

Ito (1968) studied olfactory adaptation in the guinea pig. When an odorous paper was placed in front of the nostrils, impulse discharges appeared corresponding to each inspiration. While the discharges were striking in the beginning, they soon began to decrease down to a certain level (Fig. I-1). The curve indicates that about half of the olfactory adaptation can be attributed to the olfactory receptor cells, but the remaining adaptation, if it exists, occurs in higher olfactory areas.

1.4 Olfactory cell activity under electrical polarization

Juge *et al.* (1979a) studied electrical activity of single olfactory receptor cells under electrical polarization of the olfactory epithelium in the frog. They found that the spontaneous discharge frequency varied as a linear function of the polarizing current. In addition, surface-positive polarizations caused the spike activity to increase, while surface negative polarizations suppressed spike activity. They also observed transient after effects such as rebound suppression after positive polarization, and rebound excitation after negative polarization.

They (Juge *et al.*, 1979b) then studied the interrelations between electrical polarizations of the olfactory epithelium and responses to odorous stimulation by examining the extracellular spike activity of receptor cells in the frog. Surface positive polarizations were found to enhance the excitatory responses, while negative polarizations suppressed these responses. The response of receptor cells to electrical polarization was markedly reduced or suppressed for several seconds following olfactory stimulation. This effect and the time course of the recovery period depended on the nature and the concentration of the olfactory stimulus. They suggested that the olfactory stimulation caused changes in the distribution of the total constant current in the different cell pathways. Changes in conductance induced by olfactory stimuli imply that the supporting cells were involved. They developed a model of receptor cell function that assumes a deep, axo-somatic localization of the action potential trigger zone.

The polarization technique facilitates the detection of cell units, making unit sampling more representative of the receptor cell population. In addition, this technique confirmed that the olfactory cell displays the main features of the classical concept of the neuron. The results observed by Juge *et al.* (1979a, b) are similar to the effects of polarizing currents on the EOG described in Chapter V (pp. 158–159).

1.5. Ionic and other non-odorous stimulations

(a) Stimulation with ions (Ooshima and Takagi, 1973)

The olfactory epithelium in terrestrial animals and fish has generally been viewed as a receptor of odorous substances in vapor and aqueous phases, respectively. However, this assumption has often been challenged. In the dog, Ueki and Domino (1961) found that application of various odors, as well as odorless air and isotonic sodium chloride solution, to the olfactory epithelium elicited rhythmic waves in the OB. Hara and Gorbman (1967) demonstrated that a sodium chloride solution applied to the olfactory epithelium induced a similar rhythmic wave in the OB of the goldfish.

Takagi *et al.* (1978a) studied whether various kinds of ions other than sodium can stimulate olfactory receptor cells in the bullfrog and the carp. Among isotonic monovalent cations, Rb and K induced the most striking rhythmic bulb waves in both animals. In contrast, TEA and choline showed hardly any stimulating effects. Powerful stimulating effects were also shown by some monovalent and divalent anions. Fe, Sr, Ca, Mg, and other divalent cations and some divalent anions also showed strong stimulating effects. Cd and other heavy metal ions suppressed responses induced by other ions, but tetrodotoxin did not. All of these effects were found to be similar in both the bullfrog and the carp.

Many ions were applied with increasing concentrations in the bullfrog (Fig. VI-7A). Very low concentrations elicited responses with long latencies ("delayed response"). Responses nearly disappeared at intermediate concentrations, but responses with short latencies appeared at higher concentrations ("initial responses"). In contrast, the carp showed the initial response, but not the the delayed response in most cases (Fig. VI-7 B). The dual effects found in the bullfrog, however, were remarkably coincident with findings in the palatal organ of the carp (Konishi, 1966).

(b) Effects of gustatory stimulants

Past theories have held that olfactory receptors in terrestrial animals are stimulated by chemical molecules in vapor, and gustatory receptors are stimulated by chemical molecules in water, thus implying that the mechanisms of the two sensations are clearly different. However, as Hasler (1957) once stated in regard to olfaction: "A smellable substance must pass into solution on the mucous film to be perceived by a terrestrial vertebrate; therefore, the view most generally accepted is that olfaction in all animals is aquatic in the final sense."

To evaluate the similarities between these two systems, Takagi *et al.*

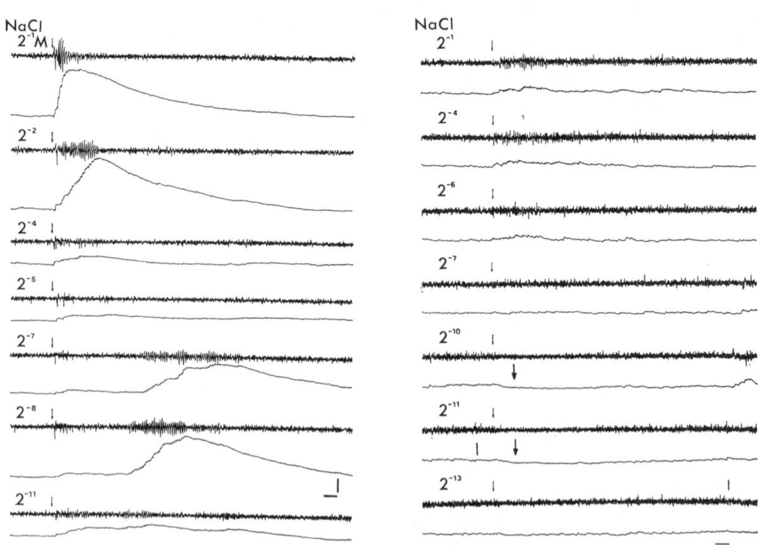

Fig. VI-7 A: Induced bulb waves elicited in the bullfrog by NaCl solutions. With NaCl solution of 2^{-1}, 2^{-2}, or 2^{-3} M, induced waves appeared immediately ("initial response"). With NaCl of 2^{-7}, 2^{-8}, or 2^{-9} M, induced waves appeared after a long latency ("delayed response"). With NaCl of 2^{-11} M, practically no response was induced. B: Induced bulb waves elicited in the carp, by NaCl solutions. With NaCl solutions of 2^{-1} to 2^{-8}, the initial responses appeared. However, with NaCl solutions of 2^{-10} to 2^{-13}, no special waves were elicited, but the intrinsic waves were suppressed, as indicated by thick arrow ("negative" delayed response). Thin arrows indicate dripping of NaCl solutions. In A and B, under each bulb wave, an integrated curve of the induced and intrinsic waves is shown. Calibrations at the right bottom show 1 sec and 100 μV. (from Takagi *et al.*, 1978a)

(1978b) studied the effects of gustatory stimulants on the olfactory epithelia of the bullfrog and the carp. Application of the four basic taste solutions revealed that salty, bitter, and acidic substances elicited responses in the OB of both animals, but the sweet substance had no effect. High alkali or acid solutions, and temperatures beyond 35°C or under 10°C, elicited similar responses in both animals. Many amino acids were effective stimuli in the bullfrog, but only betaine and 1-aspartic acid were effective in the carp. Mechanical stimuli elicited responses in the carp, but did not in the bullfrog. The "water response" occurred in the bullfrog, but did not in the carp. When various odorants were applied in the gaseous and aqueous phases, responses to odorants in both phases were found in the bullfrog, but only odorants in the aqueous phase induced responses in the carp (Iino and Takagi, 1978).

1.6. Water response

Studying the glossopharyngeal nerve activity, Pumphrey (1935) was the first to report a response elicited by dripping distilled water onto the tongue. Zotterman (1949) later named this phenomenon the "water response." Since then, the water response has been extensively studied in the frog tongue and in other organs of various animals (see references in a paper by Arito *et al.*, 1978). It was eventually discovered in the olfactory epithelium of the bullfrog (Takagi *et al.*, 1978b).

Arito *et al.* (1978) studied the effects of various electrolyte solutions on the generation of the water response by dripping distilled water on the olfactory epithelium after adaptation to each of these electrolyte solutions. The number of OB cells responding to distilled water increased when the charge of the adapting cations was increased and when the size of the cations was decreased, with a few exceptions. The magnitude of the water response increased with decreasing concentration of salt in the dripping solution after adaptation to the isotonic solution of the same salt. The water response was effectively depressed by an electrolyte solution, but not by a non-electrolyte solution.

The properties of the receptive sites involved in the water and odor responses were then studied using two group-specific reagents, uranyl ions and parachloromercuri-benzoate (PCMB). Uranyl ions are known to form a complex with phosphoryl ligands resulting in a high affinity for phospholipids of the cell membrane (Rothstein and Meier, 1951). PCMB is known to block sulfhydryl groups of proteins (Boyer, 1954). The water response was selectively blocked by treatment of the olfactory epithelium with the uranyl ions, but the response to an odorous (ENB) solution was not. Conversely, Arito and Takagi (1980) showed later that the response to the ENB solution was selectively blocked by treatment with PCMB, but the water response was not. From these results, it was considered highly probable that there are at least two different receptive sites: one composed of phospholipids and responsible for the generation of the water response, and another composed of protein macromolecules containing cysteine residues responsible for the odorant reception.

1.7. Mechanisms of odor discrimination——Neurophysiological studies
(a) Analysis of cellular responses

Numerous French investigators have recorded spike discharges from single olfactory cells and studied mechanisms of odor discrimination in the olfactory epithelium of the frog.

Duchamp *et al.* (1974) recorded extracellular single-unit activity in frog olfactory receptors. A total of 23 cells (28.4%) of 81 olfactory cells tested did not respond to any of the 20 odors applied. Among the remaining 58 cells, 12 cells (14.8%) responded to only one odor, and the rest of the cells responded to two to 14 odors. Not a single cell responded to more than 14 odors. In general, the number of cells responding decreased as the number of odors to which they responded increased. Following a total of 1,160 stimulations delivered to 58 cells, 241 activating and 59 inhibitory responses were recorded, leading to an overall selectivity of 25.8%.

To evaluate the similarities between chemicals as regards their action on the population of receptors, they calculated Pearson's 'r' correlation coefficients for the responses of the units to each pair of stimuli. The mean value of 'r' for the total of 190 pairs was found to be 0.252, well under the significance threshold. This implied that, considered as a whole, the stimuli were independent of each other. In contrast, highly significant values were found for some pairs, demonstrating definite similarities between some chemicals. No significant negative 'r' values were found. The names of the correlated compounds with 'r' values greater than 0.45 and the magnitude of their correlation are presented in Fig. VI-8. The six highest values ($0.7 <$ 'r' < 0.87) were found among six chemicals which were insignificantly correlated with the other 14, including benzene, naphthalene, anisole, acetophenone, and a benzaldehyde-nitrobenzene pair whose almond odor was judged qualitatively similar by human subjects. No other pairs reached as high a 'r' value as pairs found within this group. Among the primary aliphatic alcohols, only propanol and butanol displayed a close similarity, which could be expected from their chemical proximity. This and subsequent experiments used Benzecri's "analyse factorielle des correspondances" (Benzecri, 1969), and the above results were supported.

Since the first study by Duchamp *et al.* (1974), a series of five painstaking studies have been carried out in France. In the second study, Revial *et al.* (1978) selected four odors (BEN, ANI, CAM, and ABU) from the 20 odors used in the first study, and added another 16 odors. Thus, 20 odors were used as stimulants and the odor-induced responses of 76 olfactory cells were examined. A total of 317 excitatory responses (20.8%) and 33 inhibitory responses (2.2%) were obtained in 1520 odor applications. Sixteen cells did not respond to any of the 20 odors. Among the remaining 60 cells, the most odors to which cells responded was 7 out of 20. Some cells in the first study responded to 14 out of 20, and in the third study

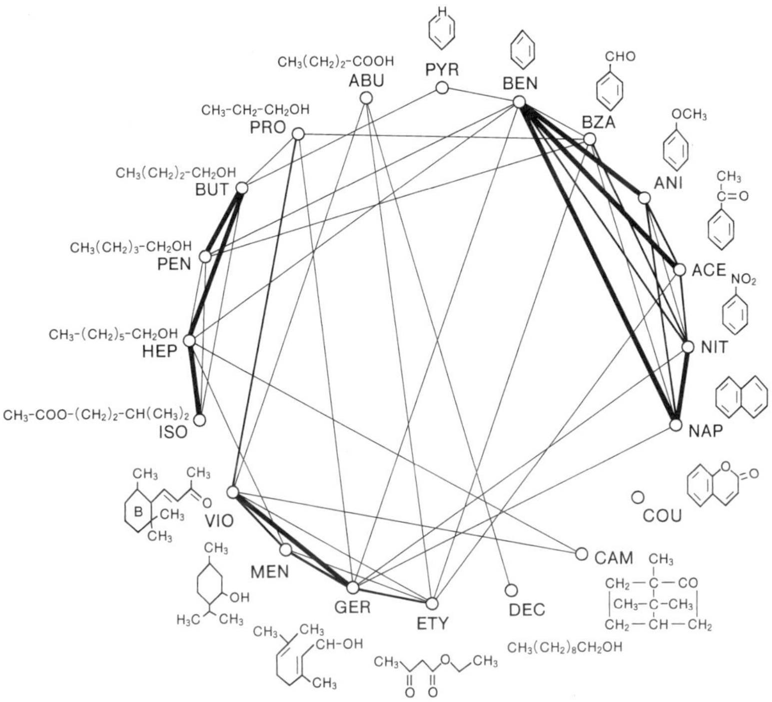

Fig. VI-8. Correlations between odorants according to the calculation of the Pearson's 'r' correlation coefficient. Three classes of significant 'r' values are represented. $0.45 < r < 0.59$: Light lines; $0.60 < r < 0.69$: medium lines; $r > 0.70$: heavier lines. Abbreviations are: propanol (P&O), butanol (BUT), pentanol (PEN), heptanol (HEP), decanol (DEC), benzene (BEN), benzaldehyde (BZA), nitrobenzene (NIT), anisole (ANI), acetophenone (ACE), naphthalene (NAP), β-ionone (VIO), ethylaceto-acetate (ETY), coumarin (COU), camphor (CAM), menthol (MEN), pyridine (PYR), geraniol (GER), isoamyl acetate (ISO), and butyric acid (ABU). (from Duchamp *et al.*, 1974)

(Revial *et al.*, 1982) four cells among 65 examined responded to all of the 20 odors applied. In the fourth study, three cells among 46 responded to 17 odors out of 19. In the fifth study, Siccard and Holley (1984) applied 20 odors to 74 cells and observed 514 excitatory responses (35%) and two inhibitory responses (0.13%). Sixty cells responded to more than one odor, two cells responded to only one odor, and 14 cells did not respond to any odors. Although the species and experimental method were consistent throughout the five experiments, the rates of responses to odors differed in each experiment. The differences may be due to the variety of odors applied. Throughout these experiments, odors were used only at one respective concentration. It is well known that odors change, not only in intensity, but also in quality when the concentration is altered. Seasonal changes in the physical conditions of the frog may also be a contributing factor. These difficulties may be inherent and incorrectable in this type of experiment.

The fifth experiment by Sicard and Holley (1984a) contained seven odors in common with the first study.

In separate studies, Duchamp and Sicard (1984a) found that frog olfactory receptors discriminate odorants better at low rather than at high concentrations. They also found a closer association between odorants of different concentrations than between odorants of different quality. As was briefly mentioned before (p. 149), Delaleu and Holley (1983) found that the photoactivable compound phenylazide interacts reversibly with the frog olfactory epithelium when delivered as an olfactory stimulus. Its application together with UV irradiation resulted in a differential reduction of the EOG to several odorants. Thus, the responses to chemicals representative of the aromatic group were clearly more reduced than those to compounds of the camphoraceous group. It was suggested that olfactory discrimination mechanisms can be approached by the use of this method.

As mentioned above (p. 189), examining odor-induced responses of single neurons by means of a microelectrode, Gesteland *et al.* (1980) found that stimulus sensitivity of olfactory receptor neurons develops in two stages: The cell first develops nonselective irritability to different stimuli, which occurs together with ciliogenesis. Later, the cell becomes selectively responsive to odors following synaptogenesis in the glomerulus of the olfactory bulb. At any given time, therefore, there will be a population of cells in the olfactory epithelium which is in the process of development and are nonresponsive to any odors. Another population of cells will be in an immature state, not in contact with the olfactory bulb. Still another population of cells will be mature, connected to the olfactory bulb, and contributing to the sensation of odors. The remaining cells are worn out and shrinking. Consequently, when recording neuronal responses from single cells, the experimenter should determine the stage of development of the observed cell. This point, however, has never been addressed in the past. It is essential to remember that the results of simple extracellular recordings of impulses in the olfactory epithelium cannot contribute to the solution of information processing problems in the olfactory nervous system.

Up to the present, there have been few reports on the properties of single mammalian olfactory receptors. Using the guinea pig, Ito (1968) may have been the first investigator to record extracellular responses to odors in single cells of the olfactory epithelium. However, he only obtained recordings of spike discharges in response to two odors, amyl acetate and menthone. Sicard (1985) recorded responses to odors in mice and indicated that the olfactory receptors in this mammal are more selective than those in the frogs.

(b) Spatio-temporal coding

Elucidation of the mechanisms underlying olfactory discrimination has been an important subject for research. Most attention has been focused upon the sensitivity of the single receptor cell as a basis for the discrimination of different odors. However, evidence has accumulated to suggest that a spatial and temporal analysis of odorants may be carried out across the extent of the olfactory epithelial sheet.

As early as 1950, Adrian (1950a, 1951) proposed a theory of regional analysis of odorants by the olfactory mucosa. He found that some chemicals would, at low concentrations, excite the units of the anterior part of the OB more readily than the units of the posterior part of the OB, whereas other chemicals excited the posterior units more readily than the anterior units. Several esters were included in the former group of chemicals and several hydrocarbons in the latter. In addition, Adrian reported a temporal differentiation, where multiunit discharges in response to esters had shorter durations and were more abrupt in their growth and decay than were responses to hydrocarbons.

(i) Imposed spatial patterning

Mozell and Pfaffmann (1954) demonstrated a spatial differentiation of odors at suprathreshold concentrations. Moulton (1965, 1967) later sampled simultaneous discharges from many regions with an array of implanted electrodes in the unanesthetized rabbit OB and was able to demonstrate well-defined spatial differentiations for different chemicals. Mozell (1967) simultaneously recorded the multiunit responses to odorants from the two most widely separated nerve branches on the dorsal aspect of the frog olfactory sac. These records clearly indicated spatial difference of coding.

Mozell (1964, 1966, 1967, 1970) and Mozell and Hornung (1981) proposed a chromatographic-like model for the differential adsorption between air and mucus phases to account, at least partially, for odor discrimination. Moulton (1976) termed this process "imposed spatial patterning." This process is dependent upon nasal patency, nasal air flow, and the relative air/mucus solubility of an odorant.

(ii) Inherent spatial patterning

Mustaparta (1971) studied spatial distribution of EOGs in the frog by stimulation with different odors. EOGs elicited by twenty-one odorants were recorded simultaneously from one electrode at a fixed position and from another placed different distances from the first. He found a non-homologous distribution of receptors with different specificities.

Kauer and Moulton (1974) devised a technique to deliver odors to small restricted areas on the olfactory epithelium. They then recorded neuronal responses to this punctuate stimulation in the salamander OB. The data suggested the presence of odor-specific restricted excitatory receptive fields in connection with some OB neurons. Moulton (1976) called this process "inherent spatial patterning." In fact, Thommesen and Döving (1977) recorded EOGs elicited by 31 odorants at 38 positions and also demonstrated a nonhomologous distribution of different receptors in the rat. Mackay-Sim and Shaman (1984) duplicated the previous work in the tiger salamander and reached the same conclusion.

Kubie *et al.* (1980), using the punctuate stimulation technique, recorded EOGs at one anterior and one posterior position, and/or at 18 to 30 different positions on each epithelium. They found that certain odorants, including limonene and camphor, consistently elicited larger EOGs from the posterior position ("posterior stimulators"), and that others, for instance butanol and geraniol, elicited larger responses from the anterior position ("anterior stimulators").

Moulton (1976) and his colleagues (Mackay-Sim and Kubie, 1981; Mackay-Sim *et al.*, 1982), proposed that a differential distribution of receptor neurons of similar responsiveness accounts at least partially for odor discrimination. Then, Mozell and Hornung (1981) proposed an experiment which would assess the relative role of each neuron in discrimination of an odorant.

Finally, it may be concluded that spatiotemporal analysis of odorants based upon a chromatographic effect across the mucosa and the selective sensitivity of the receptor cells both play essential roles in olfactory discrimination (Hoarung and Mozell, 1981; Kauer, 1980).

1.8. Drug effects on olfactory cell activity

Bouvet *et al.* (1984, 1988) studied the effects of acetylcholine and substance P on the frog olfactory mucosa. Stimulation with either of these chemicals elicited slow, low-threshold electrical potentials. In addition, prior application of substance P strongly depressed the electrical response of the mucosa to acetylcholine. These results suggest that acetylcholine and substance P could affect the functioning of olfactory neuroreceptors.

2. Odorous Stimulation of Trigeminal Nerves

The trigeminal nerve in the olfactory cavity was once believed to respond only to irritative and minty odors. However, many workers have shown that the trigeminal system responds to chemicals also known to affect the olfactory system (Bouvet *et al.*, 1987b; Cain, 1974a; Silver and Moulton, 1982; Silver *et al.*, 1985; Ito, 1968). Furthermore, Tucker (1963a, 1971), Beidler (1965), and others reported that some odorants were capable of stimulating trigeminal nerve endings at concentrations below those required to stimulate olfactory nerves.

2.1. Trigeminal responses to odors

Ito (1968) recorded responses to odors from a nerve twig of the trigeminal nerve in the guinea pig. Nineteen odors were selected as stimulants from Amoore's classification (Amoore, 1962a, b). A filter paper soaked in one of these odorous solutions was put in front of the nostrils (at a distance of 3–5 mm). Impulses were recorded in correspondence with each inspiration and, as anticipated, very strong responses were found in response to irritative odors, such as acetic acid and formic acid. Spike discharges continued even after the cessation of stimulation in these cases. The nerve responded fairly well to ethereal, pepperminty, putrid, and camphoraceous odors. However, the nerve only responded slightly to lemon odor, and hardly at all to flower and musk odors. Comparison of these responses to those observed in the olfactory nerve revealed similarities in that both responded well to ethereal and putrid odors, but both barely responded to flower and musk odors. Differences were seen with limonene, which induced strong responses in the olfactory nerve, but only slight responses in the trigeminal nerve. In addition, the trigeminal nerve responded much better to camphoraceous and pepperminty odors than the olfactory nerve. In particular, with the olfactory nerve response to amyl acetate designated as 100%, the responses of the olfactory nerve to cineole was calculated to be only $29 \pm 9\%$, while the responses of the trigeminal nerve were striking, calculated to be more than 100%. When the responses of the trigeminal nerve were examined to 19 odorants, the trigeminal nerve responded well to 16 of 19 odorants, with ionone, diphenyl ether, and musk having little effect. Thus, the trigeminal nerve appears to respond not only to irritating odors, but to a great number of odorants provided they are above certain concentrations.

2.2. Influences of trigeminal nerve stimulation on olfactory cell responses

Bouvet *et al.* (1987a) found that antidromic electrical stimulation of the ophthalmic branch of the trigeminal nerve (NV-ob) in the frog evoked slow potentials in the olfactory mucosa, modified the activity of receptor cells, and modulated responses to odors. They also found that substance P (SP) elicited similar electrical responses. These results implied that olfactory system function might be controlled at the receptor cell level, and suggested that the trigeminal system could modulate the activity of olfactory receptor cells via a local axon reflex inducing the release of SP. Thus, humoral control of olfactory cell activity was indicated.

E. Projection of the Olfactory Axons to the OB

1. Studies on the topographic relations

Olfactory cells project their axons to the OB through one, several, or many bundles of olfactory nerve fibers. The topographic relationships between the olfactory epithelium and the OB have been investigated anatomically using HRP, autoradiography, and other methods (Clark, 1951; Land, 1973; Land and Shepherd, 1974; Land *et al.*, 1970; Mackay-Sim and Nathan, 1984; Pinching and Doving, 1974; Sharp *et al.*, 1975; Stewart *et al.*, 1985). This same subject has also been studied electrophysiologically (Costanzo and Mozell, 1976; Costanzo and O' Connell, 1978; Freeman, 1974a; Kauer, 1980; Kauer and Moulton, 1974; Thommesen, 1978; Thommesen and Döving, 1977). Evidence indicates that the olfactory cells in the lateral portion of the epithelium mainly project their axons to the lateral portion of the OB, while those in the medial epithelium project to the medial OB. Costanzo and O'Connell (1980) demonstrated that hamster bulbar neurons receive input from receptors found in a restricted area of the sensory epithelium, displaying a more localized projection than in the frog (Costanzo and Mozell, 1976). These findings are consistent with recent anatomical evidence (Macrides *et al.*, 1985) and observations suggesting that odor stimuli seemed to activate restricted glomeruli in the OBs of mammals (Jourdan *et al.*, 1980; Skeen, 1977; Dubois-Dauphin *et al.*, 1981; Greer *et al.*, 1981).

Lancet *et al.* (1981) analyzed patterns of neural activity elicited by odor stimulation in the olfactory epithelium of the salamander and the mouse using a high-resolution modification of the 2-deoxyglucose (2DG) method. Autoradiograms taken from animals exposed to the odor of amyl acetate showed restricted regions of high (^{14}C)-2DG and (^{3}H)-2DG uptake. These studies suggest that autoradiography can be used to analyze odor-elicited cellular activity patterns in the periphery and correlate them with those observed centrally (pp. 274–276).

As stated previously, Fujita *et al.* (1985) identified a relatively precise topographical organization of the olfactory nerve projection in the rabbit using a monoclonal antibody (MAb R4B12). The spatial segregation of olfactory cell and axon subtypes defined in the epithelium and in the OB may be relevant to the spatial distribution of odorant-specific neuronal activity found in the olfactory epithelium (Mackay-Sim *et al.*, 1982; Mozell, 1966), to the observation that different odors elicit different spatial patterns of neuronal activity in the OB (Adrian, 1956; Moulton, 1963, 1976), and to the pattern of 2-deoxyglucose uptake in the OB (Jourdan *et al.*, 1980; Skeen, 1977; Stewart *et al.*, 1979).

Astic *et al,* (1986, 1987) studied the topographical organization of olfactory nerves in 15-day-old rats by tracing the retrograde transport of HRP and found that bulbar glomeruli receive input from a well-organized but diffuse region of the sensory epithelium. Their study represents a very extensive anatomical mapping of the neuroepithelial projection to the OB and the results cannot be fully described in the present account due to the complexity of the mammalian nasal cavity. Interested readers are advised to read their papers.

In a recent review, Holley (1986) claimed that evidence converges to suggest that the olfactory epithelium is a collection of narrow and elongated territories in a fan-like arrangement. If one follows any branch of the fan, i.e., any rostrocaudal axis, one finds a minimum of variation regarding the projection area in the OB. In contrast, moving away at a right angle to the rostrocaudal axis, one encounters maximal variation in projection sites. The same situation is present in the OB: There is a minimum of differentiation with respect to the origin of the projections along the rostrocaudal axis, while the dorsoventral and mediolateral directions bear maximum differentiation. Holley coined the term "somatotopy" to summarize this description, meaning that the projection from the olfactory epithelium to the OB appeared to be the projection of one set of axes onto another set of axes.

Clancy *et al.* (1985) provided evidence for two topographically organized central olfactory pathways in the hamster: (a) an interbulbar commissural system via the pars externa of the anterior olfactory regions of the left and right main OB (MOB). This projection system is topographically organized with respect to the mediolateral and dorsoventral axes of the MOB. (b) An intrabulbar association system exhibiting a true point-to-point topographic organization that interconnects opposing medial and lateral areas within each MOB.

In this connection, it is interesting that Pedersen *et al.* (1986a) mapped an olfactory population that projects to a specific region in the rat OB and suggested that olfactory receptor neurons form a functional mosaic within the olfactory epithelium.

2. Comparison of Neuronal Response Patterns in the Olfactory Epithelium and in the Olfactory Bulb

Mathews (1972a) studied neuronal response patterns to 27 odors between the olfactory epithelium and OB in the tortoise (*Gopherus polyphemus*). Some improvement in discriminative processing was observed from the olfactory epithelium to the OB (p. 365). Similar comparative studies were performed by Duchamp (1982), and Duchamp and Sicard (1984a, b). (For further information see p. 373.)

VII. The Olfactory Bulb

A. Neuroanatomy of the Olfactory Bulb

In general, vertebrates have both main and accessory olfactory bulbs (Fig. VII-1). The main OB (MOB) or simply the OB constitutes the primary olfactory area which receives nerve impulses originating in the olfactory epithelium and after information processing sends them to the secondary olfactory areas in the olfactory cortex via the lateral olfactory tract (LOT). The accessory olfactory bulb (AOB) receives nerve fibers from receptor cells in the vomeronasal organ. This olfactory system is wholly or partly lacking in humans and higher primates, including the old world monkey.

There is a large body of literature on the morphology of the OB stemming from research carried out long before Cajal (1911) published his famous book on the histology of the nervous system of the human and

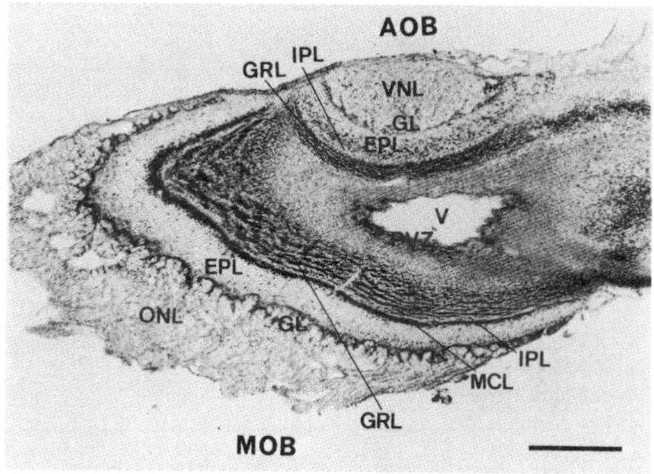

Fig. VII-1. Nissl-stained parasagittal section (50 μm thick) through the rabbit olfactory bulb. (from Mori, 1987a)

vertebrates. Many excellent reviews on the OB have also been written (e.g., Shepherd, 1972; Mori, 1987a).

The OB is remarkably constant in morphology throughout the subphylum of vertebrates. Although certain differences in the relative size and position occur, the OB is generally spherical in shape and, in cross section, has a similar concentric structure composed of six layers. A schematic diagram of the morphology of the OB is illustrated in Fig. VII-2.

1. Olfactory Glomerulus

In examining cross sections of the OB, one is struck by the many clearly defined, large spherical bodies (100–200 μm in diameter) arranged in a layer near the surface (Fig. VII-2). The "glomerulus" can be regarded as the most distinctive feature of the OB. According to Shepherd (1981),

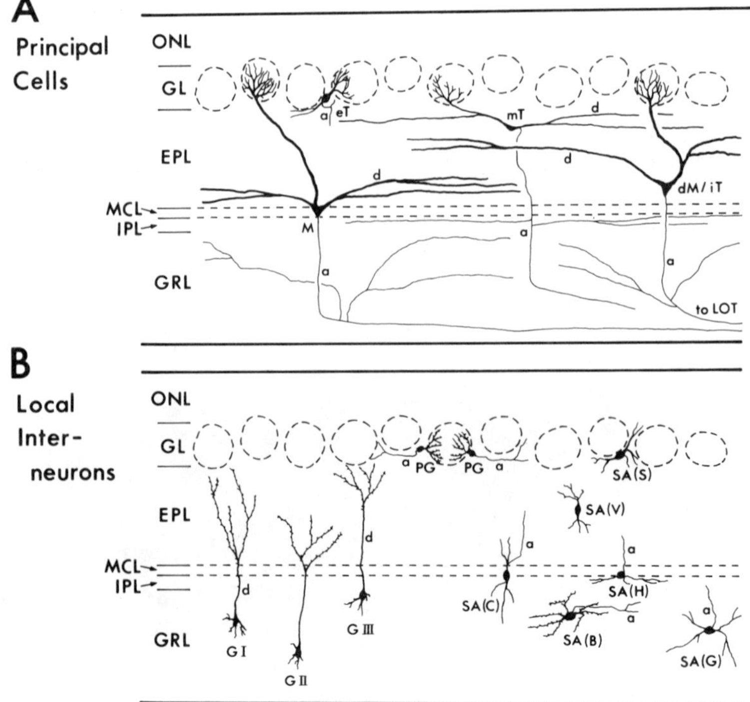

Fig. VII-2. Schematic diagrams of the structure of the OB. A: Locations and structures of principal cells in six layers. B: Locations and structures of granule cells and short axon cells. (from Mori, 1987a)

this term was first used in biology in 1681 by Malpighi in reference to the glomerulus of the kidney. In that context it meant simply a ball or tight cluster of tissue elements. Meynert, in 1878, was probably the first to apply the term to the nervous system when he described the small spherical regions of the OB within which the axons of olfactory cells ramify among the terminal branches of mitral cell dendrites (Akert and Steiger, 1967). The characteristic arrangement of glomeruli into a thin layer, one or two glomeruli thick, may be seen in a photomicrograph showing the striped pattern of a rabbit OB stained for cytochrome oxidase activity (Onoda and Imamura, 1984).

Several investigators have studied intraglomerular synaptic organization (Pinching and Powell, 1971a; White, 1972, 1973). The incoming olfactory nerve fibers do not divide before entering the glomeruli, but ramify freely after entering it. Hence, the axon of an olfactory receptor cell does not terminate in more than one glomerulus, and each glomerulus receives impulses from a segregated and independent collection of olfactory cells.

A characteristic feature seen in electron micrographs is the gathering of synaptic terminals into clusters, with partial enclosure by glial membranes. The main types of synaptic connections within an olfactory glomerulus consist of axodendritic synapses between olfactory axon terminals and the dendrites of mitral, tufted, and periglomerular (PG) cells. These synapses may either be single or arranged in reciprocal pairs (Figs. VII-7, -10, -11).

For further information, readers are advised to read the excellent reviews by Akert and Steiger (1967), Szentágothai (1970), Pinching and Powell (1971a, b), and Shepherd (1972, 1977).

2. Classification of Cells in the Olfactory Bulb (Fig. VII-2)

The characteristic morphological features of principal cells in the mammalian main olfactory bulb (MOB) have been described by studies using the Golgi methods (Cajal, 1911; Allison, 1953a; Valvelde, 1965). These cells were classified into mitral cells and tufted cells on the basis of their morphology and the laminar groupings of the cell bodies. Mitral cells have their cell bodies in a thin layer (mitral cell layer, MCL). In the rabbit OB, large principal cells which resemble mitral cells in shape were found in the deepest portion of the external plexiform layer (EPL) just superficial to the MCL. Their responses to LOT stimulation of the central portion of the pyriform cortex (PC) were similar to those of mitral cells. Because of this similarity they were called displaced mitral cells in intracellular horseradish peroxidase (HRP) injection studies (Kishi *et al.*, 1982a; Mori *et al.*, 1983a). Thus, the mitral cell type was subdivided into two groups: mitral and displaced mitral cells.

The tufted cells were found to be distributed in the EPL and the periglomerular (PG) region. Cajal (1911, 1955) classified tufted cells into three subgroups according to the position of the cell body: internal, middle, and external tufted cells. The internal tufted cells, with relatively large somata, and the middle tufted cells, with medium sized somata, are located in the deeper one-third and superficial two-thirds of the EPL, respectively. The internal tufted cells correspond to the displaced mitral cells.

The somata of external tufted cells are located in the periglomerular region, or at the border between the glomerular layer (GL) and the EPL. A recent study using retrograde HRP labeling (Schoenfeld and Macrides, 1984) indicated that only a subset of external tufted cells were principal (relay) neurons with axonal projections outside the MOB. Most of the external tufted cells located at the superficial part of the GL did not appear to have projections outside the MOB. Such external tufted cells may be classified into the category of local interneurons. Mori (1987a) tentatively called external tufted cells of this type L (local)-type external tufted cells, and those which project axons outside of the MOB P (projection) type external tufted cells.

The dendrites of mitral and tufted cells are classified into primary and secondary dendrites (Cajal, 1911). Both mitral and tufted cells usually project a single primary dendrite to a glomerulus which receives olfactory nerve inputs at the terminal tuft. They also give rise to from one to several secondary dendrites, which are distributed in the EPL.

Studies using the Golgi method have revealed the characteristic features of the dendritic distribution of mitral and tufted cells (Cajal, 1911; Allison, 1953a; Valverde, 1965; Price and Powell, 1970a; Macrides and Schneider, 1982). For example, Allison (1953a) noted that the secondary dendrites of tufted cells ramify mainly in the outer stratum of the EPL, whereas those of mitral cells divide primarily in the inner stratum of the EPL. However, it was difficult to reconstruct the whole dendritic arborization of mitral and tufted cells with the Golgi method, especially when the secondary dendrites extend for long distances in various directions. The recently developed technique of intracellular injection of HRP has made it possible to stain and reconstruct nearly all of the dendritic processes (Shepherd, 1985).

Six types of cells have been classified in the OB (Cajal, 1911; Golgi, 1875; Price and Powell, 1970a, b; Pinching and Powell, 1971a; Pinching, 1972). They are mitral, tufted (external, middle and internal), granule, and PG cells (Fig. VII-2).

Local interneurons in the mammalian MOB have been differentiated into three main types; granule cells, PG cells, and short-axon (SA) cells (Cajal, 1911, 1955). The granule cells and periglomerular cells exhibit a similar shape under Nissl staining, but their soma and processes are located in different laminae of the MOB.

The granule cell has no axon, and thus falls into the category of axonless, or amacrine cell, similar to the amacrine cell of the retina. Granule cells have two types of dendrites: peripheral processes and deep dendrites (Cajal, 1911; Valverde, 1965; Price and Powell, 1970a; Thamke *et al.*, 1973; Schneider and Macrides, 1978). Each granule cell soma gives rise to a single trunk of a peripheral process which projects into the superficial EPL.

The peripheral process ramifies within the EPL and a number of spinelike appendages called gemmules occur on the branches. The granule cell makes dendrodendritic reciprocal synapses with mitral and tufted cells at these gemmules (Fig. VII-6). Granule cells also extend a few finer dendrites (deep dendrites) laterally or toward the deeper portion of the granule cell layer (GCL). These deep dendrites have a number of spines and varicosities, and they project for a relatively limited distance. Recent studies have demonstrated that the granule cells can be classified into subtypes according to the laminar distribution of their peripheral processes in the EPL (GI, GII, and GIII in Fig. VII-2B) (Mori *et al.*, 1983 a, b; Orona *et al.*, 1983).

The PG cells have been found to have small cell bodies and to be situated in the periglomerular region of the GL. Each PG cell typically has spiny dendritic arborizations within a single glomerulus. Axons of PG cells are relatively short and are distributed predominantly in the periglomerular region of the GL (Figs. VII-2, -7, -10, -11) (Pinching and Powell, 1971c).

The third type of local interneuron has been termed SA cells. SA cells have been further classified into the following six subtypes based on the morphology and laminar distribution of their somata; in Fig. VII-2 B they are shown by abbreviations: SA(B), Blanes cells; SA(G), Golgi cells; SA(C), Cajal cells; SA(H), horizontal cells; SA(V), Van Gehuchten cells, and SA(S), superficial short-axon cells (Cajal, 1911, 1955; Price and Powell, 1970d; Pinching and Powell, 1971a; Schneider and Macrides, 1978). Most of the Blanes cells and Golgi cells have their somata in the GCL, but some of them have also been found in the internal plexiform layer (IPL) and the MCL. Somata of the Cajal cells and the horizontal cells are most commonly located in the IPL and the MCL, while those of the Van Gehuchten cells are found in the EPL. The superficial shortaxon cells have somata in the periglomerular region of the GL.

The dendrites of PG cells arborize within one (or occasionally two) glomerulus and receive synaptic input from olfactory nerve fibers. These dendrites also receive dendrodendritic synapses from the dendrites of principal cells (Fig. VII-9 C) (Pinching and Powell, 1971a). Thus, signals carried by olfactory nerve fibers are transmitted to PG cells via two different ways: by a direct synaptic input from olfactory nerve fibers, and by an indirect route via the primary dendrites of principal cells. PG cell axons are distributed in the GL and at the border between the GL and the EPL. Some of the PG axons are thought to synapse with the primary dendrites of principal cells, and with the somata and dendrites of PG and superficial short-axon cells (Pinching and Powell, 1971c). Of interest is the finding, by immunohistochemical studies, that a large proportion of PG cell populations show immunoreactivity for glutamate decarboxylase (GAD), a synthesizing enzyme of GABA (Fig. VII-11) (Riback et al., 1977; Mugnani et al., 1984; Kosaka et al., 1985).

Examination of principal cells and local interneurons in the main OB has resulted in the differentiation and classification of 15 types of cells (Fig.VII-2).

3. Laminar Structures in the Olfactory Bulb
3.1. Olfactory nerve layer (ONL)

Olfactory nerves, which consist of unmyelinated axons (from 0.1–0.4 μm in diameter in different species) of olfactory cells in the olfactory epithelium, enter the olfactory bulb at its surface and compose the most superficial layer. Because of the ongoing turnover of receptor cells, the axons in this layer are at different stages of growth, maturity, or degeneration (Chapter IV, p. 123).

3.2. Glomerular layer (GL)

The olfactory nerve fibers then enter a region where many glomeruli are found in a thin sheet called the glomerular layer. Incoming olfactory fibers divide only within glomeruli, and synapse on terminal dendritic tufts of secondary olfactory neurons, i.e., mitral cells, tufted cells, and PG cells. This is the intraglomerular neuropil. The layer surrounding the glomeruli contains cell bodies of small PG cells and slightly larger, but rare superficial short-axon cells, and the dendrites from mitral and tufted cells. This layer is called the extraglomerular neuropil. Within the glomeruli, the main types of connections consist of axodendritic synapses between the sensory axon terminals and dendrites of mitral, tufted, and PG cells, and dendrodendritic synapses between the dendrites of these cells (Pinching and Powell, 1971b).

Recent histochemical studies indicate that the olfactory nerve fibers are chemically heterogeneous and can thus be further classified into subtypes (Fujita et al., 1985; Mori et al., 1985; Allen and Akeson, 1985a, b). An example of two subtypes of olfactory nerve fibers with segregated projections to the MOB was described and illustrated in Chapter IV (p. 142).

3.3. External plexiform layer (EPL)

A bushy layer approximately 500 μm in depth can be distinguished under the glomerular layer. It is called the external plexiform layer and includes numerous dendrites, both principal and secondary, of the mitral and tufted cells. It is divided into two sublayers: the superficial and deep EPL. The superficial EPL consists of the somata of tufted cells, and smaller versions of the principal mitral cells, which project many secondary dendrites. The deep EPL contains many secondary dendrites of the mitral cells, and also includes the cell bodies and secondary dendrites of displaced mitral cells.

3.4. Mitral cell layer (MCL)

Under the external plexiform layer is the mitral cell layer (MCL), where the somata of large mitral cells lie in a thin sheet. In mammals, each mitral cell sends one principal dendrite to a glomerulus, many secondary dendrites to the EPL, and one axon toward the deeper layers of the OB. Some of the granule cell somata are also found in this layer.

3.5. Internal plexiform layer (IPL)

Again this is a bushy layer which is composed of axons, recurrent and deep collaterals of the mitral and tufted cells, and peripheral processes of the granule cells. Some centrifugal fibers from other olfactory structures and deep collaterals of the mitral cells terminate and synapse on the peripheral processes of granule cells. Some of the granule cell somata are also found in this layer.

3.6. Granule cell layer (GCL)

This layer is composed chiefly of numerous cell bodies of granule cells and their superficial and deep processes. Granule cells are the most numerous neurons in the MOB. There are about 3×10^6 granule cells in each MOB of the rat (Struble and Walters, 1982). The cell bodies are small (5–8 μm in diameter), and characteristically arranged in small, tightly-packed clusters. Some of the granule cell somata can be found in the IPL and the MCL. The entire synaptic output of this layer is through the spines to the mitral dendritic branches in the EPL. All of the output is inhibitory, and all of the output is directed to the mitral and tufted cell dendrites. This is, therefore, an example of an extremely powerful and specifically targeted inhibition.

In addition to the numerous granule cells, SA cells (deep short-axon cells) are also present. Furthermore, numerous nerve fibers pass through this layer: some are mitral cell axons directed to their projection areas in the olfactory cortex (OC), and others are centrifugal fibers coming from other olfactory areas. Centrifugal fibers from the anterior commissure (AC) and other higher olfactory areas synapse with the cell bodies and deep collaterals of the granule cells in this layer.

3.7. Periventricular or subependymal layer (PVL or SEL)

This is the deepest layer in the OB and surrounds the intrabulbar part of the ventricle (Allison, 1953a).

4. Recent Advances in Morphological Research

Recently, Mori and Kishi (1982), Kishi *et al.* (1982a), and Mori *et al.* (1983a, b) employed the HRP labeling technique to reconstruct all the dendrites and axons of mitral, displaced mitral, and tufted cells of the rabbit OB, and compared the spatial distributions of the dendritic field of tufted cells with those of mitral cells. The results revealed that tufted cells and mitral cells extend their secondary dendrites into different sublaminae in the EPL. It has been shown that the dendrites of these principal neurons from extensive dendrodendritic reciprocal synapses with the gemmules of the peripheral processes of granule cells in the EPL (Andres, 1965;

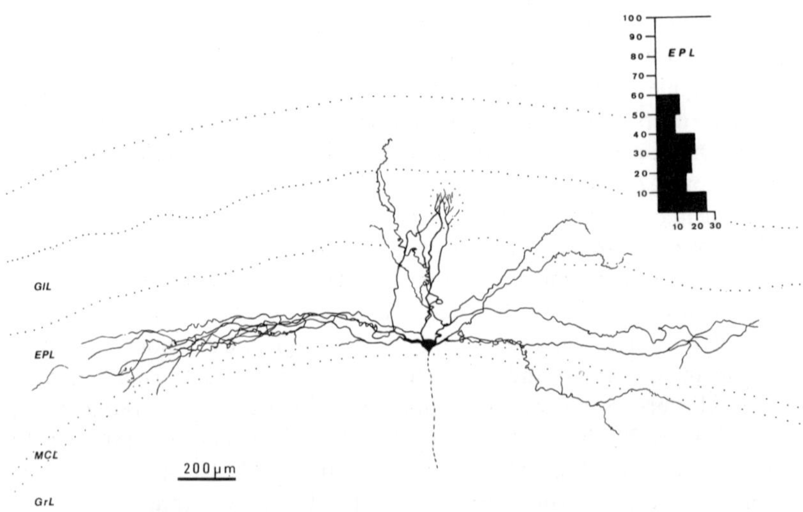

Fig. VII-3. Camera lucida drawings of mitral cells. Broken lines indicate the axon and the dotted lines show the boundaries of bulbar layers in the section containing the cell body. The graphs attached to each reconstruction show the distribution of the secondary dendrites in the sublayers of the EPL. The ordinate shows ten sublayers of EPL divided into ten percentile intervals from the MCL (zero) to the GL (100). The abscissa indicates the percentage of the secondary dendrites distributed in each sublayer of the EPL (from DAB preparations). (from Mori *et al.*, 1983a)

Rall *et al.*, 1966; Price and Powell, 1970a, b; Jackowski *et al.*, 1978). Therefore, Mori *et al.* (1982a, 1983) also attempted to elucidate the distribution pattern of the peripheral processes of granule cells of the rabbit OB in those sublaminae of the EPL (pp. 248–250).

4.1. Dendritic distribution of the mitral cells

Kishi *et al.* (1982, 1984) and Mori *et al.* (1983a, b) fully reconstructed 28 mitral cells of rabbits. These mitral cells responded to LOT stimulation with a prolonged inhibitory postsynaptic potential (IPSP) and, in about half the cases, this IPSP was accompanied by an antidromic spike (Mori and Takagi, 1978a). An example of the reconstruction of the dendritic distribution is shown in Fig. VII-3. Most mitral cells (21 of 28) had a single primary dendrite which passed through the EPL into the GL and arborized extensively in a single glomerulus. However, about 20% of the mitral cells (six of 28 cells) had two to three primary dendrites that ended in a single glomerulus (four cells had two primary dendrites and two cells had three primary dendrites). In one exceptional case, a single mitral cell was found to give rise to three primary dendrites that ended in two different glomeruli.

The lengths of the primary dendrites were measured from the soma to the point where they entered the glomerulus, ranging from 242 μm to 1,207 μm. The average diameter of the primary dendrites was 7.9 μm at the proximal portion near the cell body and 4.7 μm at the point where they entered the glomerulus.

Mitral cells showed extensive projections of their secondary dendrites in the EPL. A single mitral cell emitted two to five secondary dendrites in various directions parallel to the MCL in the rabbit. The tendency of the secondary dendrites to be distributed mostly in the inner half of the EPL was observed for all the mitral cells sampled in the various parts in the OB (Fig. VII-4 A).

The distribution of the secondary dendrites of a mitral cell projected in a plane tangential to the MCL was studied. The spatial range of the distribution was estimated by a line linking together the terminals of the secondary dendrites of each mitral cell. The secondary dendrites projected in virtually all directions within the plane tangential to the MCL and thus had a disk-like projection field. The average radius of the projection field was about 850 μm.

The secondary dendrites of a single mitral cell were observed to branch occasionally. They had an average of 18 branch points and 21 terminals. The length to the terminals ranged from 224 μm to 2,415 μm with a mean of 1,171 μm (for more information see p. 254).

Although the primary dendrites showed relatively small tapering, the secondary dendrites decreased considerably in diameter as they ran tan-

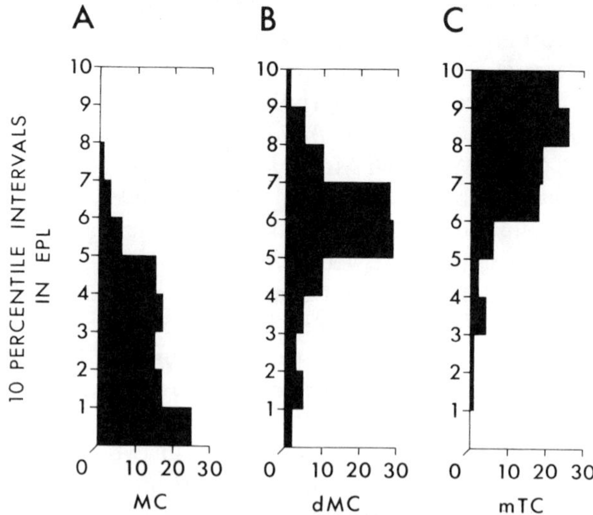

Fig. VII-4. A comparison of the distribution of secondary dendrites in the sublayers of the EPL in mitral cells (A), displaced mitral cells (B), and the middle tufted cells (C). Abscissa: percentage of the secondary dendrites distributed in each sublayer in the EPL. (from Mori *et al.*, 1983a)

gentially to the MCL. The diameter decreased rapidly immediately after the dendrites left the soma. However, the diameter suddenly increased just proximal to the branch points prior to the formation of the daughter dendrites. After the initial rapid decrease in diameter, the dendrites typically projected for a long distance with only a small tapering if no branching occurred. The diameter became 0.2 to 1.0 μm at the terminals.

The dendritic segment from the soma to the first branch point was called the first segment, that from the first to the second branch point, the second segment, and so on. The average length and diameter of the first segment of the mitral cell secondary dendrites was 82 μm and 5.8 μm, respectively. These values were 370 μm and 2.9 μm in the second segment, 460 μm and 2.0 μm in the third segment, 392 μm and 1.5 μm in the fourth segment, 400 μm and 1.3 μm in the fifth segment, 354 μm and 1.1 μm in the sixth segment, 188 μm and 0.9 μm in the seventh segment, and 228 μm and 0.7 μm in the eighth segment.

4.2. Dendritic distribution of the displaced mitral cells

Mori *et al.* (1983a, b) studied the cell bodies of the displaced mitral cells lying just superficial to the MCL. The diameters of cell bodies of HRP-injected displaced mitral cells (mean 33 μm) were similar to those of HRP-injected mitral cell bodies (mean 32 μm). As described below, in HRP-stained samples with background Nissl staining, the displaced mitral cells were clearly distinguished from mitral cells because of their characteristic dendritic projection pattern in the EPL as well as the superficial position of their somata. The responses of the displaced mitral cells to LOT stimulation showed an IPSP, with or without a preceding antidromic spike, similar to those of the mitral cells.

In fully reconstructed cells, a single primary dendrite projected from the somata and arborized dendrites in a single glomerulus. The average length of the primary dendrite of the displaced mitral cells was 401 μm and the mean diameters of the primary dendrites were 6.2 μm at the root and 5.3 μm at the point where they entered the glomeruli.

Secondary dendrites emerged from the soma of displaced mitral cells or from the proximal portion of the primary dendrites. Compared with the initial tangential direction of the mitral cell secondary dendrites, the direction of the displaced mitral cell secondary dendrites was more radially inclined toward the central portion of the EPL. At the middle or slightly superficial levels of the EPL, the secondary dendrites gradually changed their course to directions tangential to the MCL and projected for a long distance within those levels. The secondary dendrites were mainly distributed between the middle and superficial sublayers. In five displaced mitral cells, 77% of the secondary dendrites were distributed in the sublayers between the 40% and 80% level of the EPL (Fig. VII-4 B). The difference from the distribution of the mitral cell secondary dendrites is clearly seen: 73% of the secondary dendrites of the displaced mitral cells were found in the superficial half of the EPL.

A single secondary dendrite of a displaced mitral cell was found to branch up to six times. The average lengths and diameters of each segment of the secondary dendrites of displaced mitral cells were 63 μm and 5.2 μm (first segment), 483 μm and 3.2 μm (second segment), 626 μm and 1.9 μm (third segment), 466 μm and 1.4 μm (fifth segment), and 364 μm and 1.0 μm (sixth segment), respectively.

The spatial range of the dendritic field of the displaced mitral cells was studied in the plane tangential to the MCL. Compared with the relatively circular dendrodendritic fields of the mitral cells, the fields of the displaced mitral cells had slightly distorted shapes. The secondary dendrites of a single displaced mitral cell had an average of 11.2 branch points and 11.2 terminals. In the tangential plane, the terminals were distributed at distances between 213 μm and 1,760 μm. The total length of the secondary dendrites of displaced mitral cells ranged from 9,142 μm to 19,188 μm.

Mori *et al.* (1983) could not study the displaced mitral cells located in the granule cell layer (Haberly and Price, 1977) because they were not stained in their experiments.

4.3. Dendritic distribution of the middle tufted cells

Mori *et al.* (1983a, b) fully or partially reconstructed nine middle tufted cells. Under their experimental conditions, in which the stimulating electrodes were placed in the LOT bundles at the surface of the central or slightly rostral portion of the pyriform cortex, the responses of the middle tufted cells to the LOT stimulation varied from cell to cell: Some cells showed small IPSPs and others showed no response.

The somata of middle tufted cells were distributed in the superficial two-thirds of the EPL (Fig. VII-4 C). Middle tufted cells emitted radially from the somata a single primary dendrite that arborized in a single glomerulus. Superficially located middle tufted cells emitted their primary dendrites relatively tangentially. The length of the primary dendrite (excluding the terminal arborization) ranged from 189 μm to 340 μm.

Two to four secondary dendrites were found to emerge from the somata or the primary dendrite. In six middle tufted cells, 92% of the secondary dendrites were located in the superficial half of the EPL (Fig. VII-4 B).

The secondary dendrites showed infrequent branching and had tendency to project in particular directions, showing asymmetrical dendritic fields. For example, one middle tufted cell projected one secondary dendrite approximately 1100 μm in length in a predominantly rostrocaudal direction in the superficial portion of the EPL, and projected another secondary dendrite that emerged from the middle portion of the primary dendrite in a rostral direction similar to the projection of the primary dendrite. The rostrocaudal stretch of the secondary dendrites was approximately 2100 μm, and the lateromedial stretch was approximately 550 μm.

In tangential plane reconstructions, the terminals of the secondary dendrites of middle tufted cells were distributed at distances (from the soma) between 266 μm and 1464 μm. The mean length of the secondary dendrites was 829 μm. The total length of the secondary dendrites per cell was shorter (average 4050 μm) than that of the mitral and displaced mitral cells.

From the results presented in (a), (b), and (c) above distribution of secondary dendrites in sublayers of the EPL was compared among the mitral, displaced mitral, and tufted cells in Fig. VII-4.

Mori (1987a) has summarized data published in several papers (Macrides and Schneider, 1982; Mori *et al.*, 1983a, b; Kishi *et al.*, 1984; Orona *et al.*, 1984), and indicated a clear comparison of the morphology among mitral cells and three subtypes of tufted cells in the mammalian OB.

Hinds (1968) has shown that during the developmental stage, mitral, internal tufted, and middle tufted cells arise in that order. If the projection of the secondary dendrites also occurs in that order, the formation of the orderly laminar distribution of the secondary dendrites in the EPL might be explained by the sequential development of the dendrites of mitral, displaced mitral, and middle tufted cells. Mitral cell secondary dendrites presumably develop first and occupy the deeper portion of the EPL. The secondary dendrites of the displaced mitral and middle tufted cells may develop later, sequentially piling up in this order in the superficial portion of the EPL.

Greer and Shepherd (1982) reported that in the neurologically mutant, Purkinje cell-degenerated (PCD) mouse, there is selected loss of mitral cells with no apparent effects on tufted cells, and that because of the degeneration of mitral cells, reductions in width take place mainly in the deeper EPL. They therefore suggested that the genetic determinants of mitral and tufted cell differentiation and/or migration are distinct.

4.4. Distribution of local axon collaterals of mitral, displaced mitral, and tufted cells

To determine the projection fields of intrabulbar axon collaterals, Kishi *et al.* (1984) stained the mitral, displaced mitral, and middle tufted cells in the rabbit OB by intracellular injection of HRP. They found that the axon collaterals of mitral cells and displaced mitral cells were distributed exclusively within the GCL and that those of middle tufted cells were distributed predominantly in the GCL and on rare occasions in the mitral cell layer. None of these collaterals entered the EPL (Fig. VII-5).

Axon collaterals of mitral cells, emitted at the depth of the GCL, were widely distributed from the deep portion to the most superficial portion of the GCL. Displaced mitral cells also gave rise to collaterals in the deep portion, but they tended to be distributed more superficially in the GCL than mitral cells. Collaterals of middle tufted cells were released and distributed superficially in the GCL. The axon collaterals of these principal cells were typically found to extend for a longer distance than the secondary dendrites. They sometimes formed bush-like terminal arborizations.

These results indicated that the projection fields of the axon collaterals of principal cells are spatially separated from the dendritic projection fields. This suggested that the output of these principal cells via their collaterals has a functionally different role from the output via their dendrites. The observation that the three types of principal cells differ in the

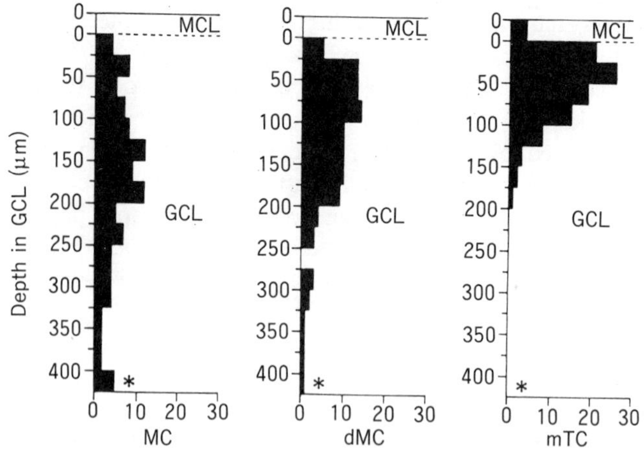

Fig. VII-5. A comparison of the distribution of axon collaterals in the MCL and sublayers of the GCL in mitral cells (MC), displaced mitral cells (dMC), and middle tufted cells (mTC). Abscissa: percentage of the axon collaterals distributed in each sublayer of the GCL and MCL. The deepest sublayer of the GCL is indicated by *, which includes all portions deeper than 400 μm from the MCL. (from Kishi *et al.*, 1984)

distribution pattern of their axon collaterals in the GCL suggested that the sublayers of the GCL are functionally distinct.

4.5. Dendritic distribution of the granule cell

In the EPL, the secondary dendrites of the mitral, displaced mitral, and middle tufted cells have been observed to synapse predominantly upon the gemmules of the peripheral processes of granule cells (Price and Powell, 1970a, b). Since the secondary dendrites of the mitral, displaced mitral, and middle tufted cells project to different sublayers (with partial overlapping) in the EPL, Mori *et al.* (1982a, 1983b) studied the projection pattern of the peripheral processes of the granule cells to the sublayers in the EPL. They examined 10 granule cells stained by intracellular injection of HRP, as well as 26 Golgi-impregnated granule cells. It was possible to check the distribution of the peripheral processes of granule cells with the Golgi method because of their relatively limited projection area.

The morphology of the HRP-stained granule cells was similar to that reported previously with the Golgi method (Cajal, 1911; Valverde, 1965; Price and Powell, 1970a; Thamake *et al.*, 1973; Schneider and Macrides, 1978).

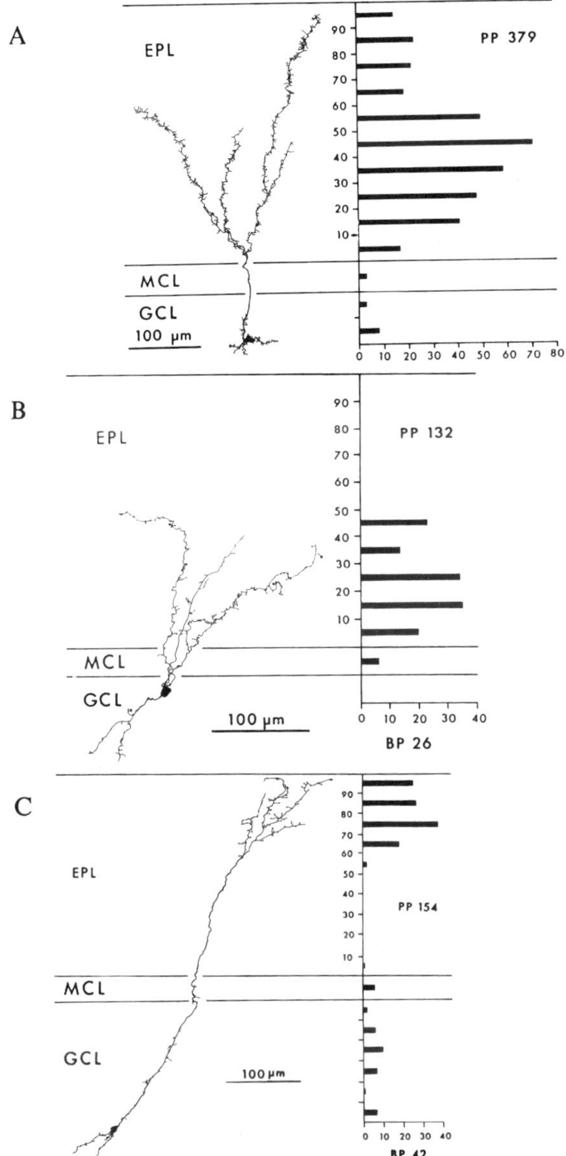

Fig. VII-6. Dendrite distributions of three types of granule cells. A, B, and C indicate types I, II, and III granule cells, respectively. PP and BP show numbers of gemmules on the peripheral and basal processes, respectively. (from Mori *et al.*, 1983b)

The distribution pattern of the peripheral processes of granule cells varied from cell to cell. Mori *et al.* (1983b) classified the granule cells into three conventional subgroups according to the distribution pattern of the peripheral processes. A type I granule cell is shown in Fig. VII-6 A. The peripheral process of this cell branched off at the border between the MCL and the EPL. One branch of the peripheral process projected to the most superficial portion of the EPL. A total of 379 gemmules were observed on the peripheral processes. The distribution of the number of the gemmules in the sublayers (with ten percentile intervals) of the EPL, MCL, and GCL is shown on the right of Fig. VII-6 A. In this cell, while the gemmules were distributed in all the sublayers of the EPL, they were mainly distributed in the deeper portion. Some other granule cells showed relatively uniform distribution of the gemmules in all the sublayers of the EPL. Mori *et al.* (1983a) classified those granule cells which had gemmules distributed in almost all the sublayers of the EPL as type I cells.

A type II granule cell is depicted in Fig. VII-6 B. The peripheral processes of this type of granule cells were limited to the inner half of the EPL. Consequently, the distribution of the gemmules found on the peripheral processes was also limited to the inner half of the EPL and thus suggested that they have dendrodendritic interactions predominantly with mitral cells.

A type III granule cell is shown in Fig. VII-6 C. The peripheral process of this granule cell ran radially without any branchings in the inner half of the EPL. The peripheral processes branched several times and had many gemmules in the superficial portion of the EPL. Consequently, these types were presumed to have dendrodendritic synapses with displaced mitral cells and/or superficial tufted cells.

4.6. Studies on mitral and tufted cells of other animals by other investigators

Macrides and Schneider (1982) morphometrically analyzed Golgi-impregnated mitral and tufted cells in the MOB of the hamster and indicated that these second-order neurons comprised several categories based on differences in the sizes and spatial organizations of their dendritic fields, and in their likely patterns of lateral and recurrent interactions with granule cells and/or periglomerular cells.

External tufted cells had their somata in the PG region, or at the superficial border of the EPL, and could be subdivided into three morphological categories: (i) Many external tufted cells lacked secondary dendrites and had only dendrites arborized within glomeruli. They were likely to receive interneuronal input predominantly from PG cells. (ii) External tufted cells in the second category were characterized by secondary dendrites which were highly branched and formed dense, spatially restricted fields in the EPL. (iii) In contrast, external tufted cells in the third category had sparsely branched secondary dendrites which extended tangentially in the EPL and tended to be asymmetrically distributed with respect to the soma. Most of the deeper-lying tufted cells, and the mitral cells, had sparsely branched secondary dendrites whose lengths and laminar distributions within the EPL were correlated with the depths of their parent somata.

Internal tufted cells, which had their somata in the deep one-third of the EPL, and mitral cells exhibited the largest secondary dendritic fields. These fields extended tangentially within the EPL in relatively symmetric radial patterns from their parent somata and covered a substantial proportion of the EPL. Middle tufted cells had their somata in the superficial two-thirds of the EPL and exhibited secondary dendritic morphologies which were intermediate between those of internal and external tufted cells.

These morphological differences in the tufted cells are likely to be associated with differences in lateral and recurrent inhibition mediated through synaptic contacts with granule cells. The morphological differences between mitral and tufted cells and the likely differences in their synaptic interactions within the OB, together with recent evidence that these neurons differ in their pharmacology and in their patterns of interconnections with more central regions of the forebrain, suggests that the mitral and tufted cells and the central olfactory circuits might be organized as functionally defined parallel pathways (Macrides and Schneider, 1982).

Orona *et al.* (1983) investigated relationships between the morphologies and positions of mitral, tufted, and granule cells in the OB of the rat, quite independently of the intracellular HRP studies by Mori *et al.* (1982a, 1983a, b) and by Kishi *et al.* (1982b, 1984). They injected small amounts of HRP extracellularly into superficial or deep parts of the EPL, or the GCL, in adult rats and found that iontophoretic injection of minute amounts of HRP labeled small numbers of neurons which permitted reconstruction of individual cells.

Comparison with the results obtained in the rabbit revealed several differences. For instance, in classification of principal cells, the displaced mitral cells in the rabbit seemed to correspond to either the type II mitral cells or to internal tufted cells in the rat. While the rabbit displaced mitral cells differed from the rat type II mitral cells in having cell bodies distinctly above the MCL, the secondary dendrites were similarly positioned and were somewhat shorter than those of other mitral cells.

Orona *et al.* (1983) also demonstrated the existence of subpopulations of granule cells in the OB, but found a slightly different correlation between the dendritic distribution pattern in the EPL and the laminar position of granule cell somata in the GCL. Thus, deep granule cells projected peripheral processes primarily in the deep EPL, where they branched extensively and had numerous gemmules. The peripheral processes of superficial granule cells, however, passed through the deeper half of the EPL without branching and reached the superficial part of the EPL where they made very extensive ramifications. Mori *et al.* (1983b) noted that subpopulations of granule cells in the rabbit MOB also show a tendency toward the relationship described previously (see section 4. 5) but they found it difficult to determine a particular subtype of granule cell based solely on the laminar position of the cell body in the rabbit.

Aside from some minor deviations, probably due to interspecific differences, we can say that the results obtained in the rat were very similar to those in the rabbit. Comparing these two series of HRP experiments, however, it is clear that the intracellular injection technique is far superior to the extracellular injection one in that only a single cell is labeled and the electrical activity of the labeled cell can be recorded simultaneously.

Based on the staining characteristics of cell bodies (stained with *o*-toluidine blue), Struble and Walters (1982) classified rat granule cells into two subtypes: light and dark granule cells. They reported that these two subtypes of granule cells also differ in nuclear diameter, spherical form of soma, basal dendritic arborization, and susceptibility to neonatal X-irradiation. Moreover, Skeen *et al.* (1985) showed the selective effect of early anosmia on dark granule cells. In the MOB, granule cell populations were composed of 85% dark granule cells and 15% light granule cells. Subpopulations of granule cells were also noted in immunohistochemical studies with antisera directed to enkephalin (Bogan *et al.*, 1982; Davis *et al.*, 1982; Shepherd, 1985).

4.7. Reciprocal dendrodendritic synapses

The presence of reciprocal synapses between the secondary dendrites of mitral cells and the peripheral processes of granule cells was first discovered morphologically by Hirata in 1964.

His and subsequent electron microscopic observations (Andres, 1965, 1970a, b; Rall *et al.*, 1966; Price and Powell, 1970a, b; Willey, 1973; Jackowski *et al.*, 1978) revealed that most synapses found on the secondary dendrites of these neurons are dendrodendritic reciprocal synapses with the gemmules of peripheral processes of granule cells. These synapses are widely distributed on mitral cells. They are found on secondary dendrites, proximal portions of the primary dendrites, the soma, and even on the initial segment of the axon. Reciprocal synapses have also been identified in the glomerulus (Pinching and Powell, 1971a) (Fig. VII-7 B). Mori (1987a) constructed a schematic diagram of the spatial distribution of input and output synapses of a mitral cell (Fig. VII-7). This figure depicts the

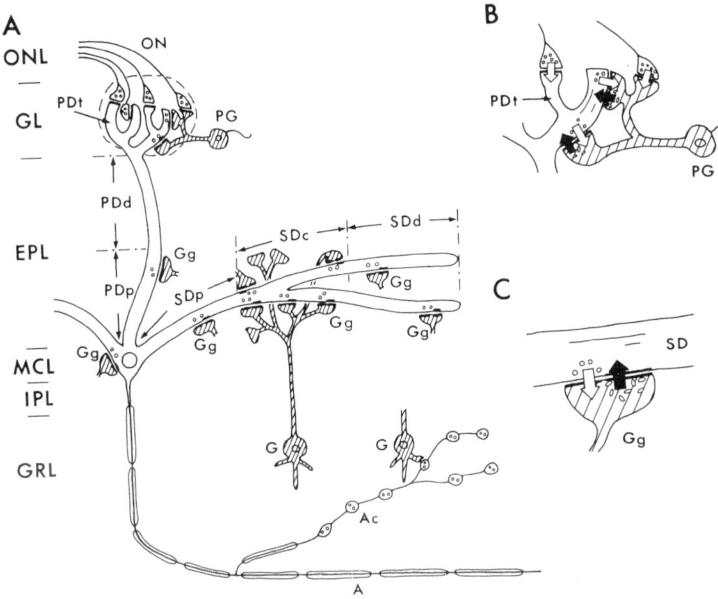

Fig. VII-7. Synaptic organization of the OB. A: A schematic diagram of the spatial distribution of input synapses on and output synapses from a mitral cell. ON, olfactory nerve; PDt, terminal portion of the primary dendrite; PG, periglomerular cell; PDd, distal portion of the primary dendrite in the EPL: PDp, proximal portion of the primary dendrite; SDd, SDc, and SDp, distal, central, and proximal portions of the secondary dendrite, respectively; G, granule cell; Gg, gemmule of the granule cell; Ac, axon collateral of the mitral cell. B: Synaptic input from olfactory nerve fibers to the PDt and reciprocal synapses with the dendrites with periglomerular cells. White arrows indicate synapses with round vesicles and asymmetrical membrane thickening, while black arrows show synapses with flattened vesicles and symmetrical membrane thickening. C: A dendrodendritic reciprocal synapse between the secondary dendrite of a mitral cell and the gemmule of a granule cell. (from Mori, 1987a)

positions and intercellular relations of all the serial and reciprocal synapses of a mitral cell (for function of this synapse, see p. 259). Electron microscopic study of mitral cells stained with intracellular HRP injection (Ojima *et al.*, unpublished) shows that the dendrodendritic synapses are not distributed to only a particular section of the secondary dendrites, but are found along the entire length of the secondary dendrites from the most proximal portion to the terminal. However, the density of the granule-to-mitral dendrodendritic synapses varies with distance from the cell body. The density is low at the proximal portion (0.17; number of synapses/μm^2), high at the central portion (0.30), and intermediate at the distal portion (0.20) of the secondary dendrites (Fig. VII-7 A). Dendrodendritic synapses are also found between the terminal portions of primary dendrites and dendrites of a PG cell (Fig. VII-7 B).

Mori *et al.* (1983b) revealed that there were differences in the distribution pattern of the gemmules of the peripheral processes of granule cells in the EPL. As already stated, they classified the granule cells into three main morphological subgroups (Fig. VII-6 A, B, and C).

The following relations between these granule cells and principal cells can be found: Mitral-to-granule dendrodendritic synapses are distributed mostly in the inner half (0 to 50% levels) of the EPL; displaced mitral (internal tufted cells)-to-granule dendrodendritic synapses are distributed mainly between the 40 to 80% levels (from the MCL) of the EPL; and middle tufted-to-granule dendrodendritic synapses are found in the sublayers superficial to the 60% level (from the MCL) of the EPL. Since different populations of granule cells extend their peripheral processes to superficial and deep sublaminae of the EPL, the above distributions suggest that each of the three principal cell types send dendrodendritic synaptic outputs to different subgroups of granule cells (Orona *et al.*, 1983; Mori *et al.*, 1983b).

4.8. Secondary dendrites of the principal neurons in various animals

It has been shown that a single mitral cell emits two to five secondary dendrites (mean 3.6 in rabbits and 4.5 in hamsters) which branch several times (p. 243; Macrides and Schneider, 1982; Mori *et al.*, 1983b; Orona *et al.*, 1984). Secondary dendrites of mitral cells extend laterally for a long distance (about 1 mm in the rabbit). The reported total length of the mitral cell secondary dendrites ranged from 6,559 μm to 26,653 μm (mean 15,016 μm) in the rabbit (Mori *et al.*, 1983b), and from 14,850 μm to 22,979 μm (mean 17,327 μm) in type I mitral cells, and from 5,325 μm to 11,175 μm in type II mitral cells in the rat (Orona *et al.*, 1984).

Remarkably long secondary dendrites of a mitral cell have been identified in the turtle. In *in vitro* preparations of the turtle OB (Nowycky *et al.*, 1978), Mori *et al.* (1981a), using intracellular injection of HRP, showed that turtle mitral cells had long secondary dendrites that projected up to 1800 μm from the cell body and extended around half of the bulbar circumference. They also showed that two primary dendrites characteristically supplied separate olfactory glomeruli (Fig. VII-8).

Displaced mitral cells also showed extensive ramifications of their secondary dendrites. The total length of the secondary dendrites per single cell (mean 12,189 μm) was slightly shorter than that of the mitral cells (p. 252). The total length of the secondary dendrites of the middle tufted cells (mean 4,050 μm; p. 253) was about one third of that of the displaced mitral cells. Although the total length is short, each secondary dendrite of the middle tufted cells extends for a considerable distance in a particular direction with few or no branchings. This suggests that dendrodendritic synapses of a single middle tufted cell are largely confined in a narrow band, or are distributed in a fan-like array.

Since all of these secondary dendrites form numerous reciprocal dendrodendritic synapses with granule cells (Rall *et al.*, 1966; Price and Powell, 1970a), the extensive lateral spread of the secondary dendrites has important functional implications for the spatial extent of the synaptic interactions.

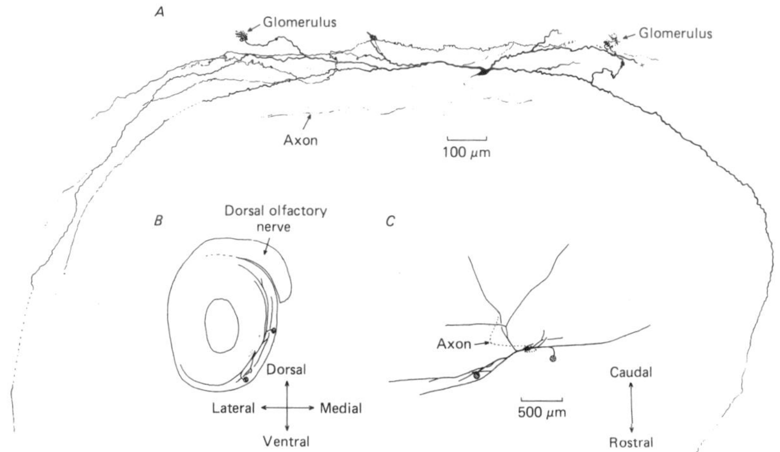

Fig. VII-8. A, camera lucida drawing of mitral cell injected intracellularly with HRP. B, outline drawing of transverse section of olfactory bulb showing position of injected cell. C, schematic tracing of injected cell, viewed from medial surface of the olfactory bulb. Electrical properties determined for this cell: input resistance (R_N), 88 MΩ; membrane time constant (τ_0), 62 msec; equalizing time constant (τ_1), 5.2 msec; electrotonic length (L), 0.95 (see text). (from Mori *et al.*, 1981a)

B. Neurophysiology of the Main Olfactory Bulb

1. Review of Electrophysiological Research (Adrian, 1953, 1954a)
1.1. Intrinsic and induced waves

Hasama (1934) may have been the first investigator to record electrical responses to odors in the OB, LOT, and hippocampal gyrus. He found the on- and off-responses to the onset and offset of odorous stimulation in the hippocampal gyrus.

Adrian (1950a, b) recorded brain waves from the surface of the OB and simultaneously recorded action potentials from internal layers. He (1950a) called the odor-elicited sinusoidal oscillatory potentials with frequencies of 50–60/sec the "induced wave," and the spontaneously occurring potentials with faster frequencies of 70–100/sec the "intrinsic wave." Similar brain waves have also been recorded with a deep electrode in the human OB (Sem-Jacobsen *et al.*, 1953, 1956).

The adaptation phenomenon in olfaction has long been known. When we keep sniffing the same odor, we soon lose olfactory sensitivity to the odor. Adrian (1950a) found in his experiment that when induced waves were elicited by odorous stimuli, the intrinsic waves were inhibited, but the latter waves soon grew in amplitude and masked the former waves. He concluded that this competition between the two waves may be a mechanism underlying olfactory adaptation, and that olfactory adaptation occurs in the OB rather than in the olfactory epithelium. This was an interesting explanation, but was later disproved by Ito's (1968) report of a striking and rapid decline in the olfactory nerve response to a sustained odorous stimulus in the guinea pig (see Fig. I-1). From this data, Ito concluded that about half of the adaptation occurs at the olfactory receptor level.

1.2. Evoked potential studies

Takagi (1962) studied the action of centrifugal nerve fibers on the electrical activity of the OB in the frog, bullfrog, and toad. When an electrical pulse was applied to the olfactory epithelium, an evoked potential was elicited in the OB (A-wave). Similarly, an evoked potential was elicited in the OB by electrical stimulation of the anterior part of the diencephalon (E-wave). When an E-wave preceded an A-wave, the A-wave disappeared, or decreased in amplitude for two to five sec (a strong inhibitory effect). This inhibition could be blocked by picrotoxin. In contrast, a preceding A-wave had a slight inhibitory effect upon the E-wave which lasted only 600 msec. The former inhibitory effect was attributed to a true inhibitory mechanism, while the latter effect was attributed to the refractory period of neurons in the OB.

In a later study on rabbits, Fujita *et al.* (1964) found a similar centrifugal inhibitory effect on OB activity by stimulating the amygdala and the anterior limb of the anterior commissure.

1.3. Electrical activity of the glomerulus

Ottoson (1959b) studied odo-relicited slow potential with superimposed induced waves in the frog OB. The slow bulb potential closely resembled the EOG, but the former was found to be sensitive to asphyxia, while the latter remained unaltered for hours after arrest of circulation. In addition, antidromic stimulation of secondary olfactory pathways blocked the superimposed induced wave but left the slow bulb potential unaffected. Ottoson concluded that the slow bulb potential originates in the glomeruli.

Leveteau and MacLeod (1966a, b) studied olfactory discrimination in the rabbit olfactory glomerulus. For this study, they designed a bipolar electrode consisting of two microelectrodes with tips set close together (about 150 μm apart) such that the size of the reception field would correspond to the mean volume of a single glomerulus. With this electrode they were able to record slow potentials evoked by odor stimulation in individual glomeruli of the OB, and could suggest a certain amount of discrimination.

They also found that glomerular selectivity increased slightly when the stimulus concentration was decreased. These results are reminiscent of the 2-DG study carried out by Stewart *et al.* (1979). A more recent study using high-resolution 2-DG autoradiography (Lancet *et al.*, 1982) confirmed the previous suggestions made by Stewart *et al.* (1979) and Jourdan *et al.* (1980), and provided evidence for regarding the individual glomerulus as a functional unit of activity. Details of these experiments will be mentioned toward the end of this section (2.3.).

In analogy with stereophonic hearing, von Békésy (1964) discovered a fairly well-developed bilateral reciprocal inhibition in the sense organs of taste and smell in man. Leveteau and MacLeod (1969a) and Leveteau, MacLeod, and Daval (1969) carried out experiments on rabbits to ascertain whether there was any electrophysiological support at the glomerular level for the reciprocal inhibition suggested by the above psychophysical findings. They recorded glomerular responses using a special bipolar glass micropipette, similar to the one described above. Odorous stimuli were presented at different onset times and at equal concentrations. When the contralateral stimulation was presented three msec before the ipsilateral one, the glomerular response fell to 50% of its normal amplitude. When both stimuli were adjusted for maximal temporal inhibition, the ipsilateral glomerular response was inhibited in proportion to the concentration of the contralateral stimulus, if it was weaker than the ipsilateral stimulus. For equal or superior contralateral stimulus concentrations, the inhibition reached a plateau at 50%. This interbulbar inhibition was suppressed by sectioning the anterior commissure (AC), but was unaffected by a retrocommissural bilateral transection of the forebrain.

1.4. Studies on an inhibitory mechanism in the OB

Green *et al.* (1962) and Mancia *et al.* (1962) found that spontaneous neuronal discharges were inhibited by a single antidromic volley applied to either the LOT or the anterior limb of the anterior commissure (Ochi, 1963; Phillips *et al.*, 1963; Nicoll, 1969). The short latency led Green *et al.* (1962) to suppose that a recurrent collateral of a mitral cell axon exerted an autogenic type of inhibitory action on the same mitral cell and negated a Renshaw type of inhibition due to the presence of an intercalated inhibitory neuron.

Yamamoto *et al.* (1962, 1963) were the first to succeed in intracellular recording of mitral and tufted cell activities in the rabbit. They found that a spike potential appeared accompanied by a prolonged hyperpolarization in response to an antidromic stimulation of the LOT. This hyperpolarization corresponded with inhibition of spontaneous discharges of the mitral cell. This and other facts led them to conclude that this hyperpolarization is an inhibitory postsynaptic potential (IPSP). Furthermore, since the latency of the hyperpolarization was never less than two msec, they had to recognize the presence of an inhibitory interneuron. They summarized the results obtained in their experiment in a schematic diagram depicted in Fig. VII-9. The interneuron A, very likely a granule cell, was included to explain the recurrent inhibition. Thus, their results

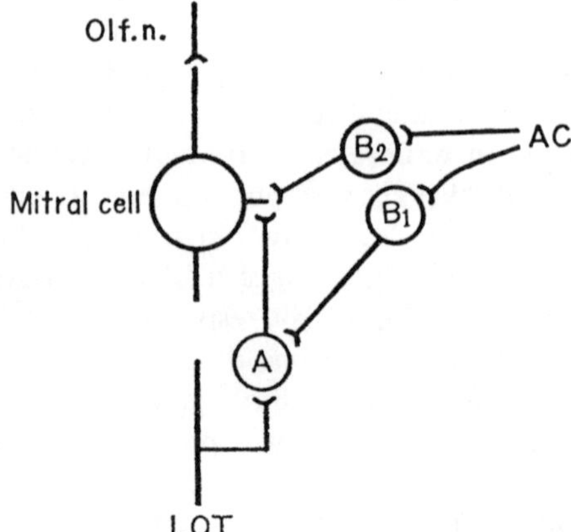

Fig. VII-9. Diagram summarizing the postulated pathways for generating and abolishing the IPSP's in the mitral cell. Details in text. Diagram to explain the roles of interneurons A, B₁, and B₂ in GCL. (from Yamamoto *et al.*, 1963)

supported the presence of Renshaw type inhibition. The interneurons B_1 and B_2 were included to account for the IPSP generated in the mitral cell and suppressed activity of the interneuron A following stimulation of the AC. In the present stage of experiments (pp. 266–268), the interneuron B_2 is thought to be a granule cell and the interneuron B_1 a short-axon cell. Their finding that the interneurons belonging to B_1 were few in number strongly suggests that some of the B neurons are short-axon cells. Similar findings have been obtained by Phillips *et al.*, (1963), Nicoll (1969), and Reese and Shepherd (1972).

2. Recent Findings on the Excitation and Inhibition of OB Neurons
2.1. Demonstration of the dendrodendritic synapse

Rall *et al.* (1966) and Rall and Shepherd (1968) postulated a new type of inhibitory pathway based on the theoretical analysis of field potentials in the OB elicited by LOT stimulation. It consisted of a dendrodendritic pathway via mitral cell secondary dendrites for synaptic excitation of granule cells and subsequent synaptic inhibition of the mitral cells via granule cell dendrites (for a review, see Shepherd, 1972). Electromicroscopic observations of reciprocal dendrodendritic synapses were already described (pp. 252–254).

Mori and Takagi (1978a) intracellularly recorded potentials from mitral cells and interneurons in the granular cell layer (GCL) following LOT stimulation. Most recordings from mitral cells showed large (5–21mV) and prolonged (60–650 msec) IPSPs subsequent to the antidromic spikes. Mori and Takagi then recorded EPSPs from cells in the GCL and found that the onset latency of the EPSP was approximately 0.6 msec shorter than that of mitral cell IPSPs. Comparison of the behavior of EPSPs in GCL cells and that of mitral cell IPSPs under various conditions of LOT stimulation suggested that these GCL cells are first excited through mitral-to-granule synapses and then elicit IPSPs in the mitral cells through granule-to-mitral synapses. It followed that GCL cells are the inhibitory interneurons mediating mitral cell inhibition. Electrophysiological studies have shown that the granule-to-mitral part of the dendrodendritic synapse is indeed inhibitory (Phillips *et al.*, 1963; Yamamoto *et al.*, 1963; Nicoll, 1969; Mori and Takagi, 1977a, b; Mori *et al.*, 1981b; Jahr and Nicoll, 1982). Thus, the hypothesis of dendrodendritic pathways for activation of granule cells and subsequent inhibition of mitral cells was supported.

The IPSPs decreased in amplitude and then reversed polarity as the intracellularly applied hyperpolarizing current was progressively increased. The IPSPs were accompanied by a prominent and long-lasting (up to 100 msec) conductance increase of the mitral cell membrane. The IPSPs reversed by hyperpolarizing current applied to the mitral somata had much shorter time courses compared with the original hyperpolarizing IPSPs. The asymmetrical reversal of the IPSP could be attributed to the wide-spread distribution of the granule-to-mitral inhibitory synapses on the mitral dendrites and somata. The early part of the IPSP might be primarily generated by the inhibitory synapses located at or in proximity to the soma, while those located on distal dendrites might be responsible for the later part of the IPSP.

Accummulating evidence suggests that the transmitter substance at the granule-to-mitral synapses is γ-aminobutyric acid (GABA; Felix and McLennan, 1971; McLennan, 1971; Nicoll, 1971a; Graham, 1973; Riback et al., 1977; Halász et al., 1979; Nowycky et al., 1980; Jahr and Nicoll, 1982), and that the action of GABA on the mitral cell membrane results in increased permeability to Cl ions (p. 272; Nowycky et al., 1981a; Jahr and Nicoll, 1982).

Greer et al. (1981) indicated a probability that combined use of bi-cucullin, a GABAergic antagonist and 2-DG would be useful for localization of synaptic responses in the OB.

2.2. Synaptic activity in the olfactory glomerulus

Olfactory nerve fibers from the olfactory receptor cells form one-way excitatory synapses with the terminal tufts of mitral cell dendrites in the glomerulus. Thus, a single volley in the olfactory nerves (ON) elicits an EPSP at the distal portion of the primary dendrite, which can be recorded as a clear EPSP from the mitral cell soma (Yamamoto et al., 1963; Mori and Takagi, 1975; Mori et al., 1981b, 1982b, 1984). Compared with the IPSPs elicited by antidromic LOT stimulation, the ON-elicited EPSPs are relatively insensitive to application of hyperpolarizing or depolarizing current into the soma. It is therefore clear that the EPSP generation sites are electrotonically more distant than the IPSP generation sites (Mori et al., 1981b).

In addition, dendrodendritic synapses are found between the dendrites of principal cells and dendrites of PG cells (Pinching and Powell, 1971a; Fig. VII-7 B). Thus, impulses carried by ON fibers are transmitted to PG cells via two different routes: i) a direct synaptic input from the ON fibers; and ii) an indirect route via the primary dendrites of principal cells (Fig. VII-11 C). A single volley in the ON strongly activates those neurons located in the GL; many of them are presumed to be PG cells, but they might also include external tufted cells and superficial short-axon cells (Shepherd, 1971; Freeman, 1974b; Getchell and Shepherd, 1975b). The axons of PG cells are distributed in the GL and at the border between the GL and the EPL. Some of the PG axons are thought to synapse on the primary dendrites of the principal cells, and on the somata and dendrites of PG cells and superficial short-axon cells (Pinching and Powell, 1971c; Fig. VII-11 C). To date, no intracellular recording from an identified PG cell synapse on principal cell dendrites has disclosed excitation or inhibition.

Elucidation of the transmitters which function at these synapses has also been a subject of interest. Thus, it is an important finding that a large proportion of PG cell populations showed immunoreactivity for glutamate decarboxylase (GAD), a synthesizing enzyme of GABA (p. 240; Riback *et al.*, 1977; Mugnani *et al.*, 1984; Kosaka *et al.*, 1985).

2.3. Fast prepotential (FPP) in the mitral cell dendrite

Electrical stimulation of the olfactory nerve bundles elicits spike potentials superimposed on EPSPs in mitral cells (Yamamoto *et al.*, 1963; Mori and Takagi, 1975, 1977; Getchell and Shepherd, 1975a; Mori and Shepherd, 1979; Mori *et al.*, 1981b, e; Jahr and Nicoll, 1982). A brief, depolarizing partial response, or FPP, has been observed on the rising phase of this spike potential both *in vivo* rabbit OB (Mori and Takagi, 1975) and *in vitro* turtle OB preparations (Mori *et al.*, 1982b; Jahr and Nicoll, 1982). In these studies, the amplitude and time course of the FPPs were found to be distinct from the IS-spike or the M-spike of the antidromically elicited action potential. Collision tests with antidromically elicited action potentials showed that the FPPs did not propagate into the axon. Compared to the SD-spike and the IS-spike, FPPs were relatively resistant to applications of hyperpolarizing currents. From results of these and other tests, it was considered highly probable that the active membrane responsible for the FPP is located within the glomeruli at the site of primary dendrite arborization (Mori and Takagi, 1975; Mori *et al.*, 1982b).

Generation of spike potentials in mitral cell dendrites far from the soma may be significant for the cell in several ways: (1) to serve as a booster so that EPSPs at distal dendritic sites can activate a full spike potential in the cell body (cf. Spencer and Kandel, 1961); (2) to convey the integrated outcome of complex synaptic interactions to the mitral cell body; (3) to convey a function related to the synaptic output to PG cell dendrites from the distal portion of the mitral cell primary dendrites (Mori and Takagi, 1975).

2.4. Spiking activity in the mitral cell dendrites

Impulse activity has been reported in a number of neuronal dendrites (Pearson, 1976; Schmitt and Worden, 1979; Roberts and Bush, 1981). Thus far, the only report of spiking activity in neuronal dendrites has been in mitral cells of the rabbit olfactory bulb (Mori and Takagi, 1975).

Mori *et al.* (1982b) studied spiking activity in the dendrites of mitral cells by means of intracellular recordings in an isolated turtle olfactory bulb preparation. Large somatic spike potentials and small FPPs were recorded in nearly all cells in response to an orthodromic volley applied to the olfactory nerves. However, the small FPPs were never found in antidromic responses from the LOT. Collision tests using antidromic and orthodromic volleys showed that the FPP did not propagate into the axon. Application of hyperpolarizing current resulted in delay and blocking of the soma spike potential with little effect on the FPP response. The results of these and other tests suggest that the FPP is localized in the dendrites and is distinct from injury potentials and from spikes originating at the axon hillock or initial axonal segment.

2.5. Excitatory and inhibitory interaction at distal dendritic synapses of mitral cells

Olfactory nerves terminate exclusively on the distal dendritic tufts of mitral cells in the OB, making the OB a favorable model for the analysis of synaptic responses in distal dendrites. Mori *et al.* (1984) performed intracellular recordings of responses to olfactory nerve volleys in the isolated turtle OB preparation. In this study, single mitral cells usually responded with similar EPSPs to volleys in two different ON bundles, indicating convergence from separate receptor neuron populations. An ON volley always elicited a spike potential arising from an EPSP (E_1 component). The spike was followed by a series of inhibitory and excitatory potentials (I_1, E_2, I_2, and I_3 components).

When paired ON volleys were applied, a conditioning volley resulted in a long-lasting inhibition of a test EPSP (Fig. VII-10, second wave in Aa). Two successively elicited waves in Aa are folded for comparison in Ab. In contrast, when a conditioning volley was applied to the LOT, a test ON-evoked EPSP was not suppressed (Fig. VII-10 Ba). The second wave in Ba is shown with a faster sweep speed in Bb. For comparison, evoked waves in Ab and Bb were combined in one diagram. Then it is clear that the wave in Bb shown by dotted line is even slightly increases in amplitude (Fig. VII-9 C). The lack of suppression of the EPSP by conditioning LOT in stimulation and, in contrast, the marked suppression of the EPSP by conditioning ON stimulation suggests that the inhibition of the EPSP occurred independently of the inhibition of mitral cells by granule cell interneurons in the deeper layers, and that the ON volley activated an additional inhibitory synapse at or near the distal portions of the primary dendrites of the mitral cells. Here, the activity of the PG cell may be quite important, because physiological studies have shown that PG cells are excited by ON stimulation and that PG cell dendrites are, in turn, inhibitory to the mitral and tufted cells at their terminal dendritic tufts in the glomerular layer (Shepherd, 1971; Getchell and Shepherd, 1975a).

These findings suggest that the local integration of excitatory and inhibitory synaptic interactions in the glomerular layer provides the basis

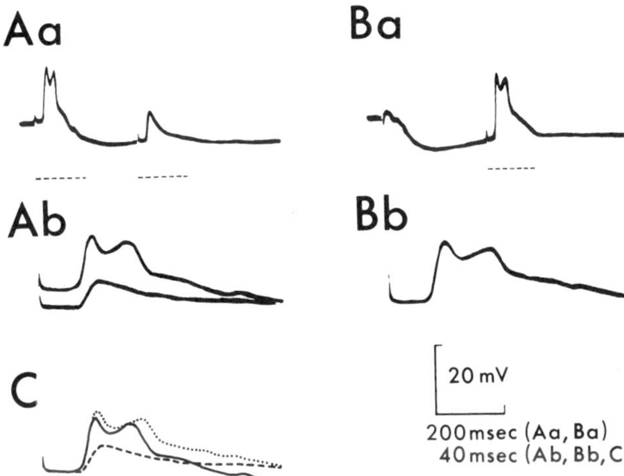

Fig. VII-10. Comparison of the effect of the conditioning LOT volley with the conditioning ON volley on the test ON-evoked response. C: The control ON-evoked response (solid line), the test ON-evoked response following conditioning LOT volley (dotted line), and the test ON-evoked response following the conditioning ON volley (dashed line) were superimposed for the comparison of the amplitude of the E_1 component. (from Mori *et al.*, 1984)

for the initial processing of olfactory input. In addition, the results demonstrate an exception to the classical doctrine that synaptic inhibition is preferably located near the cell body and axon hillock in order to control impulse generation at that site.

2.6. Na⁺- and Ca²⁺-spikes in the mitral cells

Mori *et al.* (1981a) analyzed the electrophysiological properties of mitral cells with intracellular recordings. Mitral cells were driven antidromically from the LOT, or activated directly by current injection. The *in vitro* turtle OB preparation provided a means for the analysis of ionic conductances responsible for spike generation in mitral cells (Jahr and Nicoll, 1980, 1982; Mori *et al.*, 1981a).

Applying tetrodotoxin (TTX) to the bathing medium only partially blocked the spikes generated by turtle mitral cells. Thus, in addition to the fast rising, TTX-sensitive Na⁺-spikes, mitral cells exhibited slow-rising, higher threshold, TTX-resistant spikes. The TTX-resistant spikes have been demonstrated in several other vertebrate central and peripheral neurons (Barrett and Barret, 1976; Llinas and Sugimori, 1980; Schwarzkroin and Slawsky, 1977; Matsuda *et al.*, 1978; Adams *et al.*, 1980).

Mori *et al.* (1981a) found that the TTX-resistant Ca^{2+}-spikes were enhanced in the presence of tetraethylammonium (TEA), but were completely blocked by cobalt, which is believed to be a specific inhibitor of regenerative calcium conductance (Hagiwara, 1975). Thus, these investigations considered it reasonable to conclude that the TTX-resistant spike component in the turtle mitral cells probably represents a substantial inward regenerative calcium current. The large size of the TTX-resistant component implies that the postulated calcium conductance is distributed in or near the soma, and might be in the more distal dendrites as well.

The distribution of synapses (pp. 252–254) and the well-known role of calcium in synaptic transmitter release (Katz and Miledi, 1969) led Mori *et al.* (1981a) to postulate that the inward movement of calcium during mitral cell impulse generation may be closely associated with the activation of mitral-to-granule dendrodendritic synapses.

2.7. Synaptic outputs of mitral cells

A single mitral cell has three different types of output synapses within the main OB. They are:

i) The mitral-to-PG dendrodendritic synapses located within the glomerulus (Pinching and Powell, 1971b), which physiological studies have shown to be excitatory (Shepherd, 1971; Getchel and Shepherd, 1975b).

ii) The numerous mitral-to-granule dendrodendritic synapses found along the secondary dendrites situated primarily in the deeper half of the EPL. These synapses are also present along the proximal portion of the primary dendrites and on the soma.

iii) The third type of output synapses is located along the axon collaterals of mitral cells which are confined to the GCL and IPL, as identified in the adult rabbit (Kishi *et al.*, 1984). Ojima *et al.* (personal communication) noted that mitral axon collaterals terminate, at least in part, on granule cell dendrites in the GCL and the IPL. These three output synapses have been found on spatially different parts of the mitral cells as well as in different laminae of the MOB (Fig. VIII-7).

Antidromic activation of mitral cells by LOT stimulation elicits monosynaptic EPSPs in granule cells (Yamamoto *et al.*, 1963; Nicoll, 1969; Mori and Takagi, 1978a; Mori and Kishi, 1982; Jahr and Nicoll, 1982). It is possible that these EPSPs are elicited (i) via the dendrodendritic synaptic pathway from the mitral cell dendrites to the peripheral processes of granule cells in the EPL and/or (ii) via a separate synaptic pathway from intrabulbar mitral axon terminals to granule cell collaterals. Theoretical analysis of the depth profile of the field potentials showed that the EPSP is generated predominantly in the deeper half of the EPL (Rall *et al.*, 1966; Rall and Shepherd, 1968) supporting the view that the major action is via the pathway described in (i) above. Neurophysiological studies have also supported this conclusion (Nicoll, 1969; Mori and Takagi, 1978a).

2.8. Electrical activity of tufted cells

Schneider and Scott (1983) carried out intracellular recordings of electrical activities of middle and external tufted cells in the rat OB which had been identified by intracellular HRP injection. They found that synaptic responses of these superficial tufted cells to a volley in the olfactory nerve consisted of an initial EPSP of short duration (about 15 msec) accompanied by an IPSP of much longer duration (200 to 400 msec). Thus, superficial tufted cells were basically similar to mitral cells with respect to these potentials. In contrast to the single spike response of mitral cells elicited by olfactory nerve volley, these superificial tufted cells showed a train of spike potentials (2–4) superimposed on the initial EPSP. They also showed that, when examined with an array of stimulating electrodes for activation of olfactory nerve fibers, the presumptive superficial tufted cells responded to stimulation of a greater number of sites across the ONL than mitral cells. From these results it is clear that the superficial tufted cells show more orthodromic excitability than mitral cells. This corresponded well to the finding by Onoda and Mori (1980; p. 276).

The somata of external tufted cells are closer to the glomerulus than those of mitral cells and have no, or relatively few, secondary dendrites. Thus, intracellular recordings from these cells would provide more direct data on the form of synaptic interactions between the olfactory nerve inputs and inputs from PG cells inside the glomerulus. However, no systematic intracellular studies on this problem have yet appeared.

The secondary dendrites of internal and middle tufted cells, like mitral cells, extend for long distances in the EPL, and make a number of dendrodendritic reciprocal synapses with granule cells. Mori *et al.* (1983a) found in the rabbit OB that the displaced mitral cells (internal tufted cells), and mitral cells, showed a large and prolonged IPSP in response to LOT stimulation at the central level of the anterior pyriform cortex. In contrast to this, stimulation of the LOT at the same site elicited no antidromic spike and no, or only a small, IPSP in the middle tufted cells in the same animal. This finding suggests that activation of the deep-lying mitral and internal tufted cells did not sufficiently activate those granule cells which form dendrodendritic inhibitory synapses on the middle tufted cells. In the rat, Schneider and Scott (1983) obtained a similar finding that stimulation of the LOT resulted in a barely detectable IPSP in the middle tufted cells, whereas olfactory nerve stimulation elicited a relatively large IPSP following the initial EPSP.

2.9. *Electrical activity of granule cells*

Identification of granule cells for the analysis of their membrane and synaptic properties was first performed by means of electrophysiological criteria (Mori and Takagi, 1978a, b). The validity of thesecriteria was confirmed by morphological identification of granule cells with intracellular HRP injection (Fig. VII-6; Mori and Kishi, 1982; Mori *et al.*, 1983b). The identified granule cells have been found to generate action potentials following supramaximal stimulation of LOT or AC, or during depolarizing current injection (Mori and Takagi, 1977c, 1978a, b; Mori and Kishi, 1982; Jahr and Nicoll, 1982; Mori *et al.*, 1983a). Furthermore, the spike activity in axonless granule cells in the OB resembles that observed in retinal amacrine cells (Miller and Dacheux, 1976; Murakami and Shimoda, 1977; Webline, 1977).

Granule cells in the rabbit OB frequently show two types (or more than two types, in some cases) of spike potentials: large-amplitude spikes and smaller spikes (Mori and Takagi, 1978a, b). The large-amplitude spikes are much more easily blocked by hyperpolarizing current than the smaller spikes. This suggests that granule cells have multiple sites of spike generation in the somatic and dendritic membrane, and that the large spikes might be elicited in the soma or proximal dendrites, while the smaller ones might be elicited in relatively remote dendrites.

Inhibition of mitral cell activity can be produced not only by LOT stimulation, but also by AC stimulation which originates in the contralateral anterior olfactory nucleus (Baumgarten *et al.*, 1962; Ochi, 1963; Yamamoto *et al.*, 1963). To extend these results, Mori and Takagi (1978b) studied the granule cell activity following AC stimulation. In contrast to the EPSPs generated via the dendrodendritic pathway, a train of volleys applied to the AC caused a temporal summation of the EPSP. With repetitive LOT stimulation, the EPSPs in granule cells did not summate, showing a depression, or an alteration, of their amplitude (Mori *et al.*, 1977; Mori and Takagi, 1978a). Thus, it was clearly demonstrated that the responses of granule cells elicited by AC stimulation are generated through a different mechanism from the responses elicited by LOT stimulation.

Nakashima *et al.* (1978) showed that stimulation of the deep portion (layer III) of the ipsilateral pyriform cortex elicits EPSPs in granule cells. The depth profiles of field potentials generated by these EPSPs indicated that the generation sites of these EPSPs were in the GCL. This finding agrees with studies demonstrating that centrifugal fibers from the ipsilateral pyriform cortex and the AC terminated predominantly in the GCL (Price and Powell, 1970c; Davis and Macrides, 1981). From these observations, it is clear that the IPSPs elicited in mitral cells by AC stimulation are generated by granule cells in the GCL. In contrast, theoretical analysis of the depth profile of the field potentials elicited by LOT volleys determined that the synaptic excitation of the granule cells occurs predominantly in the EPL (Rall *et al.*, 1966; Rall and Shepherd, 1968). This indicates that most of the synaptic input to granule cells is mediated through the dendrodendritic synaptic pathway. This theoretical conclusion was experimentally confirmed by means of intracellular recording and paired LOT stimulation (Nicoll, 1969; Mori and Takagi, 1978a).

Histological investigations (Cajal, 1955; Rall *et al.*, 1966; Price and Powell, 1970a, b, c) have shown the following: granule cells extend their peripheral processes to the EPL where they form reciprocal synapses with the secondary dendrites of mitral cells; they also extend, in an opposite direction, their deep dendrites to the depths of the OB; and centrifugal fibers from various sources terminate not only on the granule cell somata, but also on these deep dendrites. While the distance from the peripheral processes to the output synapses may be short, the deep processes are located far from the output synapses. This fact led to the question of how the EPSPs generated in those remote synapses could activate the output synapses in the periphery. Shepherd and Brayton (1979) used computer simulation to examine this problem and found that activation of the output synapse by passive spread of EPSPs from remote generation sites is not possible due to considerable damping. Mori and Takagi (1977c) found that in some granule cells, AC volleys elicited spike responses superimposed on the EPSP. Thus, the spike activity of the soma and peripheral processes of granule cells may serve as a booster for the action of the remote synapses on the output synapse.

According to histological studies of the rabbit OB two types of cells exist in GCL (Fig. VII-7). The type 1 category of cells includes numerous axonless cells and the type 2 category refers to the smaller number of short-axon cells (p. 241). Electrical stimulation of the AC elicits EPSP in type 1 GCL cells. The characteristics of these cells were found to be similar to those of the interneurons inferred from the analyses of mitral cell IPSPs. It has been suggested that these type 1 GCL cells are the common inhibitory interneurons (presumably granule cells) mediating both AC-evoked and LOT-evoked IPSPs in mitral cells. However, there are no anatomical reports of reciprocal synapses between the dendrites of mitral cells and short-axon cells. In addition, no electrophysiological studies of type 2 GCL cells have been published (Nakashima *et al.*, 1978). As mentioned before, test LOT-evoked responses in the type 1 cell are suppressed by conditioning LOT stimulation. Consequently, the type 2 GCL cells could be differentiated from type 1 cells, if the absence of such a depression were proven (Mori and Takagi, 1978a).

2.10. Electrical activity of local axon collaterals of mitral, displaced mitral, and tufted cells

Nicoll (1970b) showed that following a conditioning LOT volley, a testing LOT volley evoked a negative field potential in the GCL and a positive one in the EPL. This suggests that the conditioning and testing LOT volleys uncovered the synaptic input to the granule cells through the axon collateral pathway in the GCL. This corroborated the results of Kishi *et al.*'s (1984) HRP experiment demonstrating that the local axon collaterals of mitral cells are distributed within the GCL. Taken together, these results provide strong evidence for a second type of mitral to granule synaptic pathway, indicating the presence of both the lateral type and Renshaw type of inhibition.

Thus, two types of recurrent inhibitory pathways from mitral-to-granule-to-mitral cells have been elucidated. The possible functional differences between the two pathways have been discussed by Shepherd (1977), and more recently, Kishi *et al.* (1984) summarized the differences in the following four points:

(a) Distance between input and output synapses

Since the secondary dendrites are strictly confined within the EPL, it is highly probable that the mitral-to-granule dendrodendritic synapses terminate on the peripheral processes of granule cells next to, or near, the granule cell output synapses.

The mitral-to-granule synapses may therefore have quite powerful activating effects on the granule-to-mitral inhibitory synapses. In contrast, the synapses from the mitral cell axon collaterals terminate on the processes of granule cells exclusively within the GCL. Thus, these synapses are relatively distant from the output synapses of granule cells, suggesting that their influence on the output synapses may be relatively weak. However, AC volleys have been found to elicit spike responses superimposed on EPSPs in some granule cells (Mori and Takagi, 1978b), implying that spike activity may stand between the EPSP generated at deep dendritic sites and the activation of output synapses at the peripheral process. If true, the above discussion would be unfounded.

(b) Spatial range of the distribution of synapses from a mitral cell

A single mitral cell projects secondary dendrites in virtually all directions, having a disk-like projection field with a radius of about 850 μm in the deep portion of the EPL. In contrast, a single mitral cell projects axon collaterals up to about 3,500 μm from the soma. Thus, the influence of a single mitral cell may be more widespread through synaptic connections via the axon collateral pathway than through those via the dendrodendritic pathway.

(c) Directional and nondirectional projection fields

The secondary dendrites of mitral cells project relatively evenly in all directions, whereas the axon collaterals tend to project in particular directions, and in some cases, make dense arborizations in specific sites within the GCL.

(d) Synaptic connection with one cell type or multiple cell types

While secondary dendrites of mitral cells synapse with granule cells exclusively (Price and Powell, 1970b; Jackowski *et al.*, 1978) axon collaterals may synapse with granule cells as well as short axon cells in the GCL (Price and Powell, 1970d). As mentioned above in 2.8., an LOT volley elicits monosynaptic activation of the granule cell, and also results in an inhibitory influence on granule cells, possibly via a relay of short-axon cells.

Anatomical studies by Cajal (1911) and physiological investigations by Nicoll (1971b) led both to propose that recurrent axon collaterals of mitral and tufted cells (internal or middle tufted cells) synapse with other mitral or tufted cells to form a recurrent excitatory pathway. Kishi *et al.* (1984), however, found that all the examined axon collaterals of mitral, displaced mitral, and middle tufted cells are confined to the GCL (and MCL), and concluded that the proposed recurrent excitatory pathway to the principal cells is quite rare, at least in the adult rabbit OB, if it exists at all.

3. Further Analysis of Excitation and Inhibition in the Isolated Turtle OB
3.1. Passive membrane properties of turtle mitral cells

The stable conditions for intracellular recording in *in vitro* OB preparations permit the analysis of passive membrane properties of mitral cells (Mori *et al.*, 1981a). Mori (1987a) summarized the data on passive electrical properties of the turtle mitral cell, as shown in Table VII-1. He found that compared to rabbit mitral cells, or other mammalian central neurons, turtle mitral cells have a relatively high input resistance, and relatively large membrane and equalizing time constants. The electrotonic length (Rall, 1969) of turtle mitral cells ranges from 0.9 to 1.9. These data are important in estimating the spatial spread of the localized synaptic or spike potentials in mitral cell dendrites and soma, and also in providing a basis for computer simulation of the dendrodendritic synaptic interactions between mitral cells and local interneurons in the OB (Rall and Shepherd, 1968; Shepherd and Brayton, 1979).

3.2. Analysis of inhibitory responses of mitral cells

Mori *et al.* (1981b, d) showed that the responses of a mitral cell to a single

Table VII-1. Passive electrical properties of turtle mitral cells

Property	n	Range	Mean
Input resistance (R_N)	25	33–107 MΩ	60 MΩ
Membrane time constant (T_0)	7	24–93 msec	52 msec
Equalizing time constant (T_1)	7	2.5–11 msec	7.5 msec
Electrotonic length (L)	7	0.9–1.9	1.25

n: number of mitral cells examined. (from Mori *et al.*, 1981a)

volley in the LOT consisted of an antidromic spike and a complex hyperpolarizing potential that had the properties of an IPSP. They further showed that the inhibitory response consisted of two successive components, I_1 and I_2, followed by a prolonged hyperpolarization I_S. High-gain recordings revealed miniature hyperpolarizing potentials during the I_1 and I_2 responses. Both the miniature potentials and the I_1 and I_2 responses were increased in amplitude by depolarizing injected currents, and were decreased and reversed in polarity by hyperpolarizing currents. The input conductance was increased during the I_2 component.

Mori *et al.* (1981b) found that a single orthodromic volley in the olfactory nerve elicited a complex depolarizing-hyperpolarizing potential in mitral cells. The depolarization consisted of two successive components, E_1 and E_2. The hyperpolarization consisted of two successive components, I_1 and I_2, followed by a prolonged hyperpolarization, I_S. The depolarizing components had the properties of EPSPs. They decreased in amplitude with depolarizing current injection and increased with hyperpolarizing current injection. The hyperpolarizing components resembled the I_1 and I_2 components of the LOT-evoked responses in their timing and properties. They postulated that the E_1 component reflected the initial excitation of the mitral cell dendritic tufts induced by olfactory nerve terminals in the olfactory glomeruli. The I_1 component was postulated to arise from dendrodendritic synaptic input mediated by interneurons, primarily granule cells, indicating that it is the IPSP. The E_2 and I_2 components were thought to arise mainly from intrinsic synaptic circuits within the OB. While the two I components were seen only after LOT volley, the two E components were seen only after ON volley.

Relations between these components are summarized by the superimposed traces in Fig. VII-11 B. When the antidromic impulse is aligned with the orthodromic impulse, as in this figure, the timings of the subsequent I_1 and I_2 components can be seen to be nearly identical.

Mori *et al.* (1981b) summarized the provable pathways involved in eliciting the above-mentioned various components of mitral cell responses in Fig. VII-11 A and C. They believed that these reciprocal synapses provided for self- and lateral inhibition of the mitral cells, as well as the tufted cells. The similarity of the properties of I_1 and I_2 suggested to these workers that, to a large extent, I_2 is generated by similar synaptic mechanisms (IPSP) at similar sites on the mitral cell. It may be possible that an intrinsic property provides for delayed reactivation of the reciprocal synapse.

The E_2 component was found to be brief. When I_1 and I_2 were blocked pharmacologically, however, the E_2 component was found to be continuous with a prolonged depolarization that lasted through the period of E_2 and beyond. This finding will be further discussed below.

3.3. A long-duration inhibitory potential

Mori *et al.* (1981c) studied the I_S component, which appeared after the I_1 and I_2 components, by orthodromic or antidromic stimulation of the mitral cells. This slow I_S component was inhibitory, as shown by the interruption of spontaneous discharges and the blockage of responses to

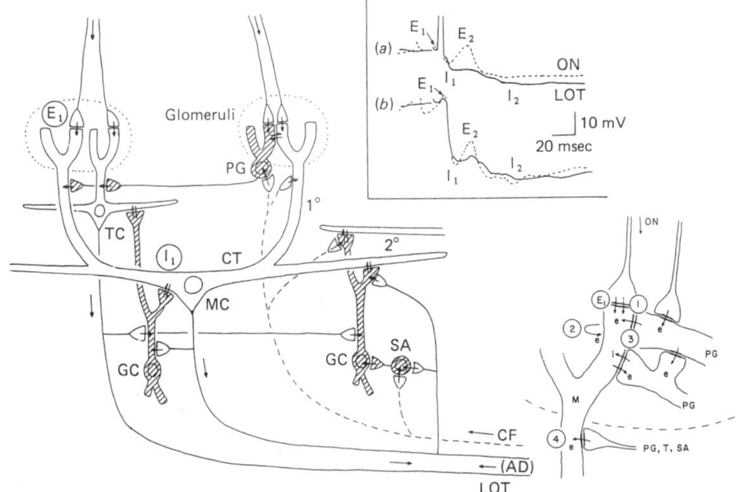

Fig. VII-11. Summary diagram illustrating possible sites of synaptic actions analysed in this study. AD, antidromic; CF, centrifugal fibers; CT, common trunk; GC, granule cell; LOT, lateral olfactory tract; MC, mitral cell; PG, periglomerular cell; SA, short-axon cell; TC, tufted cell, 1°, primary dendrite; 2°, secondary dendrite. Probable sites of generation of E_1 and I_1 components of synaptic potentials are indicated. Inset, comparison of mitral cell responses to ON and LOT volley. (a) Superimposed tracings of responses at normal resting membrane potential. (b) Responses during membrane depolarization by current injection. Note close alignment of corresponding components of LOT and ON responses. (A and B from Mori *et al.*, 1981b; C from Nowycky *et al.*, 1981b)

injected current and incoming volleys. The hyperpolarization following a single volley usually lasted up to five sec or longer and was robust, ranging up to 20 mV in amplitude. The I_S response decreased in amplitude with either depolarizing or hyperpolarizing current injection, without showing reversal potential. This contrasted with the earlier I_1 and I_2 components, which showed reversal potentials characteristic of IPSPs. The I_S response shares some characteristics with the slow potentials reported in certain other vertebrate neurons, for instance, sympathetic ganglion cells (Libet, 1970; Nishi, 1974; Weight, 1974), parasympathetic cardiac ganglion cells (Hartzell *et al.*, 1977), and cells of the rabbit olfactory cortex (Mori *et al.*, 1978). This type of response may be involved in longer-term neuromodulatory control of the excitability of the mitral cell.

3.4. GABAergic mechanisms of I_1 and I_2

Nowycky *et al.* (1981a) studied the effect of bicuculline, a GABAergic antagonist, on the responses of a mitral cell elicited by a single pulse applied to the LOT. They found that bicuculline initially blocked I_1 and I_2, and later blocked I_S. It also blocked the conductance increase and current sensitivity associated with the two early potentials. In addition, bicuculline inhibited the long-lasting inhibition of test volley induced by a preceding conditioning volley. In recordings with KCl electrodes, both I_1 and I_2 were found to reverse to a depolarizing response, but I_S was not reversed and was usually eliminated. Current injection revealed that I_1 and I_2, recorded with KCl electrodes, had a similar I-V relationship as when recorded with K acetate electrodes, but the curve was displaced by about 15 mV in the positive direction. When Cl in the bathing medium was replaced by the impermeable anions, propionate or glucose, the conductance increase and potential change during I_1 and I_2 were abolished. Responses in low Cl solution closely resembled those obtained with bicuculline.

These experiments provided evidence that, in the turtle OB, I_1 and I_2 represent the GABAergic, Cl-mediated inhibition of mitral cells by dendrodendritic synapses of granule cells, but that I_S is not GABAergic.

In another *in vitro* preparation, i.e., OB explants of fetal mice, Corrigal *et al.* (1976) demonstrated selective and reversible depression of slow wave potentials and spontaneous unit spikes after GABA administration. In contrast, bicuculline induced slow wave potentials and increased the frequency of unit discharges. Strychnine was unable to produce this effect, showing that GABA, not glycine, was the likely transmitter responsible for inhibition of mitral cells in this preparation as well.

3.5. Long-lasting synaptic excitation

Bicuculline blockage of I_1 and I_2, and then I_S, revealed a slow depolarizing excitatoty potential, termed E_S (Nowycky *et al.*, 1981b). This component has never been found during antidromic activation of mitral cells. E_S appeared after I_1, I_2, and I_S were abolished when Cl in the bathing solution was replaced by an impermeable anion, propionate or gluconate.

This slow depolarizing potential had an average amplitude of 10 mV and duration of 400–1,600 msec. The amplitude was increased by hyperpolarizing current injection and decreased by depolarizing current injection. In recordings with KCl electrodes, both I_1 and I_2 reversed to a depolarizing response, while I_S did not reverse. The properties of E_S were found to be similar to those of the E_2 component.

The sites of generation of excitatory response components in the mitral cell may be discussed in reference to the schematic diagram shown in Fig. VII-11 C. E_1 is generated by excitation of the glomerular dendritic branches of mitral cells by synapses of olfactory nerve terminals. Possible sites and mechanisms for the E_2 and E_S are indicated by ①–④ in the diagram. Full explanation of the diagram, however, is well beyond the scope of this summary and the reader is advised to read the original paper by Nowycky *et al.* (1981b).

4. Information Processing of Afferent Impulses in the OB
4.1. Patterns of afferent impulses recorded in principal OB cells

Since Walsh (1956) recorded single-cell spike discharges to odors in the rabbit OB, many others have studied them in various animals (Orrego, 1961, in the turtle; Mancia *et al.*, 1962, in the rabbit; Döving, 1964, 1966 a, b and Higashino *et al.*, 1969, in the frog). Shibuya *et al.* (1962) classified neuronal responses to odors in the frog OB into seven types: excitatory on, on-off, and off types; inhibition followed by excitation; excitation followed by inhibition; simple inhibition type; and no-response type.

4.2. Electrophysiological analyses of similarities and dissimilarities of odors

Döving (1966a) recorded extracellular single-unit responses in the frog OB to odorous vapors from a homologous series of normal aliphatic alcohols, acetates, and ketones. The results showed that the degree of similarity in olfactory stimulative properties between the odors was greatest among neighboring substances and gradually decreased as differences in chain length increased. These results were in accordance with psychophysical studies of odorous substances selected from a homologous series (Engen, 1962).

Döving (1966b) then recorded neuronal responses in the frog OB to vapors of 33 substances selected from five of Amoore's (1962a, b) seven primary odor groups. The results were again classified as excitatory or inhibitory, and were statistically processed for all pairs in given series. In cases of musky odors, he found that the chi-square values were consistent with the results of subjective judgments.

In these experiments, however, a third type of response, namely the no-response type, was not taken into consideration. Silent neurons are frequently encountered during experiments. These neurons often respond to the same odor at higher concentrations or to other kinds of odors. Thus, studies of neurons observing excitatory, no-response, or inhibitory responses are thought to be essential to the analysis of odor similarities.

In a similar experiment on bullfrogs, however, Higashino *et al.* (1969) attempted to reduce the three response types to two for simplicity. For the purpose, they diluted each of eight odorants so that it only elicited responses of the excitatory or no-response type, but not responses of the inhibitory type. The number of spikes occurring within the first second of the discharge were counted for different odorants, and correlation coefficients (r) were calculated for each pair of odorants. A fairly good correlation was found when the r values were compared with the similarities between the molecular profiles of the odorants (Sm). Thus, the importance of molecular shape in odor discrmination (Moncrieff-Amoore hypothesis) was generally supported

Discriminative mechanisms of odors were also studied in mammals. Mathews (1972b) studied neuronal responses to odors of the OB in the rat (Chapter VI, p. 233) and Tanabe *et al.* (1975b) investigated these responses in the old world monkey (Chapter IX, p. 364). Comparative studies of neuronal responses between the receptor and bulbar levels have been carried out in the amphibian olfactory system (Mathews, 1972a; Duchamp, 1982; Duchamp and Sicard, 1984a, b). The results also suggested that odor discrimination was slightly improved in the OB, but no sign of a novel, specific odor categorization could be found.

4.3. Studies on spatial information processing in the OB using the 2-DG and HRP methods

One of the main problems encountered in this research is that the olfactory system does not have an obvious spatial organization, which has been so critical in orienting research on visual or other sensory systems. In the previous chapter (pp. 225–233), we found that neuroanatomical and electrophysiological studies have already provided some hints of topographic relationships in the projection of the olfactory epithelium to the OB. However, they have been very limited in comparison with the advances of research in other sensory systems.

Since the development of the Sokoloff method (Kennedy *et al.*, 1975; Sokoloff, 1977), [^{14}C] 2-deoxy-D-glucose (2-DG) has been used in conjunction with X-ray autoradiography as a method for mapping experimentally induced alteratons in functional activity in the olfactory system. Using this method, Sharp *et al.* (1975, 1977) identified a pattern of increased density in parts of the OB, which peaked over the glomeruli.

Stewart *et al.* (1979), used the 2-DG method to demonstrate that the spatial patterns of activity for a given odor were scattered and punctate in the glomerular layer at low odor concentrations, defining an odor-specific spatial domain within the glomerular layer. The densities were found to grow and merge at higher concentrations. While the spatial domains for different odors overlapped, they remained distinguishable. Skeen (1977), Jourdan *et al.* (1980), and Greer *et al.* (1981) also demonstrated that the spatial patterns of areas showing intense 2-DG uptake in the OB are different for different stimulus odors. Recent advances in high-resolution 2-DG autoradiography made it possible to observe 2-DG uptake in individual cells and showed that selective populations of mitral and granule cells were labeled by a specific odor stimulation (Lancet *et al.*, 1982). It will be an important and interesting problem to elucidate which properties of odorant molecules can be represented spatially in the glomerular layer of the OB.

Especially important in Lancet *et al.*'s work (1982) were their findings in the glomerulus. First, individual glomeruli display different degrees of 2-DG uptake (glomerular selectivity). Second, when broadly examined, many glomeruli show a relatively uniform level of uptake throughout their neuropil (glomerular uniformity). Third, distinctly low levels of labeling are seen in some glomeruli (inactivated glomeruli). Glomerular selectivity appeared to be the fine-structural correlate of the odor-specific spatial patterns of activity in the glomerular layer observed in X-ray 2-DG autoradiography (Sharp *et al.*, 1977; Stewart *et al.*, 1979; Skeen, 1977; Jourdan *et al.*, 1980) or electrophysiological recordings of evoked potentials (Leveteau and MacLeod, 1966a, b). These workers came to regard the differential 2-DG uptake observed in individual glomeruli as related to odor specificity and thus concluded that the individual glomerulus works as a functional unit of activity.

Adamek *et al.* (1986) iontophoretically injected WGA*HRP (see p.323) into the olfactory epithelium and labeled glomeruli of the rat OB. In all cases, heavily labeled glomeruli were scattered among highly to faintly labeled glomeruli. Their results suggest that within any small, contiguous population of the primary olfactory neurons there are two subpopulations: one which projects diffusely throughout the OB and another which projects more focally to a few glomeruli. Since there is continual neuronal turnover in the olfactory epithelium, it was tempting for Adamek *et al.* to speculate that immature and relatively nonselective primary olfactory neurons generate the diffuse label and that more odor-specific, mature cells have a dense, focal projection.

In this connection, it is interesting that a subcomponent of the olfactory pathway was identified in the neonatal rat (Teicher *et al.*, 1980; Greer *et al.*, 1982). The subcomponent consisted of a small group of glomeruli within the OB, and was termed a modified glomerular complex (MGC). It is situated in the posterior part of the OB, at the medial border of the accessory OB, but anatomically distinct from it and from the glomeruli of the main OB. Studies with 2-deoxyglucose (2-DG) have shown that the MGC is active in suckling rat pups, suggesting that it is part of the information pathway for odor cues which mediate suckling (Teicher *et al.*, 1980; Greer *et al.*, 1982). Jastreboff *et al.* (1984) demonstrated that specific olfactory receptor populations (a matrix of receptor cells possessing functional and topographical specificity) project to these identified glomeruli in the rat OB.

As stated previously, Fujita *et al.* (1985) indicated, using a monoclonal antibody (MAb R4B12), a relatively precise topographical organization of the olfactory nerve projection in the rabbit. Spatial segregation of the subtypes of olfactory cells and axons thus defined in the epithelium and in the OB may be relevant to the following observations: the spatial distribution of odorant-specific neuronal activity in the olfactory epithelium (Mackay-Sim *et al.*, 1982; Mozell, 1966); different odors eliciting different spatial patterns of neuronal activity in the OB (Adrian, 1956; Moulton, 1963, 1976); and uptake of 2-deoxyglucose in the OB (Jourdan *et al.*, 1980; Skeen, 1977; Stewart *et al.*, 1979). Olfactory stimuli appear to activate selected glomeruli in the OB (Jourdan *et al.*, 1980; Skeen, 1977; Greer *et al.*, 1981).

4.4. Djfferences in response patterns of principal neurons in different OB layers

Onoda and Mori (1980) studied the depth distribution of temporal firing patterns of the OB neurons during the artificial intake of deodorized air in the rabbit. OB neurons were classified into single-peak units and multiple-peak units on the basis of the number of bursting activities during each air intake. The former showed a simple temporal firing pattern and were recorded mainly in the glomerular layer (GL), and superficial portion of the external plexiform layer (EPL). The latter showed a more complex temporal pattern and were distributed predominantly in the deep portion of the EPL and mitral cell layer. Thus, it was made clear that many units recorded in the superficial part of the EPL generated trains of impulses in response to an artificial sniff, whereas mitral cells typically did not respond without specific odorous stimulation. This difference in response patterns may also play an important role in the processing of olfactory information in the OB.

4.5. *Convergence of olfactory input on mitral cells*

Each mitral cell receives, via a glomerulus, input from approximately 1,000 neurons in the olfactory epithelium (Van Drongelen, 1978; Van Drongelen *et al.*, 1978). The spontaneous firing of the olfactory neurons is random, averaging about one discharge per sec, and interspike intervals can be described by a Poisson distribution. The same is true of firing in response to low concentrations of odorants. A typical olfactory discrimination can occur in 1 sec or less. Thus, it is frequently thought that a discharge frequency far above the spontaneous rate must occur in order for the information to be useful in such a short time. However, the fact that approximately 1,000 neurons converge on each mitral cell means that each mitral cell receives input of about 1,000 spontaneous discharges per sec. Thus, relatively small changes in the rate of firing of the olfactory neurons can provide significant information in a very short time (Van Drongelen, 1978; Van Drongelen *et al.*, 1978).

C. Neurochemistry of the Olfactory Bulb

In recent years, research on neurochemical aspects of the vertebrate brain has achieved remarkable progress. A variety of chemicals found in the brain have become candidates for neurotransmitters, neuromodulators, and neuroactive substances.

The OB is particularly rich in the quantity and variety of neuroactive substances it contains. Indeed, the OB has the reputation of being a virtual treasure trove of neuroactive substances and is known to contain the following: an extremely large, well-defined GABAergic system (p. 272); subpopulations of single neuronal types which use different putative neurotransmitters; a unique protein; it receives substantial contributions from brainstem transmitter systems; and it has the highest concentrations of taurine, carnosine, and thyroid-hormone releasing hormone (TRH) of any region in the brain (outside the hypothalamus, in the case of TRH; Halász and Shepherd, 1983). A number of research papers on these substances have appeared, often presenting various contradictory data.

The author recognized the extreme importance of neurochemical research on the OB and would like to review the neurochemistry of the OB in the present book. However, as a neurophysiologist, the author does not feel qualified and will only provide a few introductory remarks and brief summary of this subject which may serve as a reference for newcomers.

1. Candidates for Neuroactive Substances

Halász and Shepherd (1983) selected the following substances as candidates of neurotransmitters, neuromodulators, or neuroactive substances and reviewed many studies devoted to each of them. Prostaglandin D_2 studied by Watanabe *et al.* (1986) is added.

1.1. Acetylcholine (ACh)

1.2. Amino acids

(a) Gamma-aminobutyrate (GABA)

(b) Glutamate (Glu)

(c) Aspartate (Asp)

(d) Other amino acids

1.3. Biogenic amines

(a) Serotonin (5-hydroxytryptamine, 5-HT)

(b) Catecholamines

(c) Dopamine (DA)

(d) Noradrenaline (NA)

(e) Adrenaline

(f) Histamine

1.4. Peptides and hormones (Table I in Halász and Shepherd, 1983)

(a) Carnosine (β-alanyl-L-histidine) (Margolis. 1980b)

(b) Substance P (SP)

(c) Enkephalins (ENK)

(d) β-lipotropin, β-endorphin, vasopressin, and oxytocin

(e) Luteinizing hormone-releasing hormone (LHRH)

(f) Thyrotropin-releasing hormone (TRH)

(g) Somatostatin (SOM)

(h) Neurotensin, angiotensin II, and vasoactive intestinal polypeptide (VIP)

(i) Gastrin and cholecystokinin (CCK)

(j) Insulin

1.5. Specific proteins

(a) Olfactory marker protein (OMP)

(b) Parvalbumin

(c) Vitamin D-dependent calcium-binding protein

1.6. Prostaglandin D_2 (PGD_2)

2. Results

2.1. A survey of the results of research on various neuroactive substances

Abundant data from a number of reports indicate a rich diversity of neuroactive substances within the OB. However, these papers often present data which, though very extensive, are not always clear and are often mutually inconsistent.

A number of excellent reviews of neurochemical research on the OB and related areas have appeared in recent years (Macrides *et al.*, 1982; Margolis, 1980a, 1981; Halász and Shepherd, 1983). Interested readers are directed to these reviews for clear and full accounts of the present status of advances in neurochemistry of the OB.

Halász and Shepherd (1983) summarized the principal types of neurons and synaptic connections as well as the main putative neurotransmitter substances in the OB in a schematic diagram shown in Fig. VII-12. The identified neuronal elements are, at least in the case of PG cells, differentiated into subtypes on the basis of their different neuroactive substances. The unidentified terminals associated with several substances are meant to represent centrifugal fibers from the forebrain and midbrain, each type presumably containing a different substance.

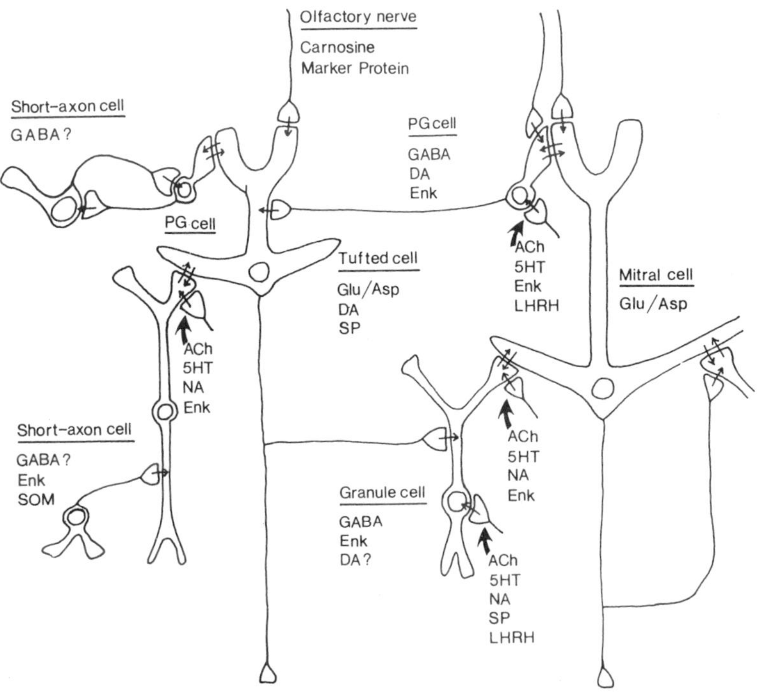

Fig. VII-12. Schematic diagram of the olfactory bulb summarizing the synaptic circuits and the main neurotransmitter and neuromodulator substances that have been proposed for the several types of neurons. (from Halász and Shepherd, 1983)

2.2. Recent studies on Prostaglandin D_2

Prostaglandin D_2 (PGD_2) is one of the major prostaglandins (PGs) in the brain of the rat (Narumiya *et al.*, 1982) and of man (Ogoroch *et al.*, 1984). The enzymes responsible for the biosynthesis (PGD synthetase; Shimizu *et al.*, 1979) and inactivation (NADP-linked 15-hydroxy-PG dehydrogenase; Tokumoto *et al.*, 1982; Watanabe *et al.*, 1980) of PGD_2 have been found and purified in Hayaishi's laboratory. PGD_2-binding protein was also identified in the synaptic membranes of the rat brain (Shimizu *et al.*, 1982) and in synaptosomes of the human brain (Watanabe *et al.*, 1985). PGD_2 exerts various physiological or pharmacological effects such as sleep induction (Ueno *et al.*, 1982, 1983), body temperature regulation (Ueno *et al.*, 1982), and analgesia (Horiguchi *et al.*, 1986). These effects are believed to be triggered by the binding of PGD_2 to its specific receptor.

Studies on regional distribution reveal that PGD_2, PGD synthetase, NADP-linked 15-hydroxy-PGD_2 dehydrogenase, and PGD_2-binding protein are concentrated in the hypothalamus and limbic system, and especially in the OB. These studies also show that the mitral cell layer is enriched with PGD_2-binding protein. Autoradiographic analyses indicate that the PGD_2 binding site is localized in or near the mitral cell layer of the rat OB (Yamashita *et al.*, 1983).

Watanabe *et al.* (1986) electrophysiologically studied the role of PGD_2 in the processing of odor signals in the rabbit OB. Iontophoretic, intra-arterial, and intravenous administration of PGD_2 enhanced and prolonged the responses of mitral cells to some of the olfactory stimuli tested. The extent and duration of granule cell inhibition of mitral cells was assessed by recording field potential responses in the OB following paired LOT volleys. The i.v. administration of indomethacin or diclophenac, both of which are inhibitors of PG biosynthesis, prolonged the granule cell inhibition of mitral cells without any significant change in the conditioning amplitudes and reduced the spike responses of mitral cells to olfactory stimuli. After treatment with indomethacin, i.v. administration of PGD_2 rapidly reduced the duration of the granule cell inhibition of mitral cells. These results indicate that PGD_2 modulates mitral cell responses to odor stimuli by suppressing the inhibitory synaptic inputs from granule cells to mitral cells.

3. Discussion and Summary

Despite the energetic and strenuous efforts of many researchers, biochemical aspects of neuronal and synaptic activities in the OB remain largely unknown.

The olfactory nervous system is interesting in that the receptor cells appear to have distinct and specific sensitivities to different odors, even at the molecular level. Thus individual olfactory axons may biochemically differ from each other in terms of the neurotransmitters present at their synapses with the glomeruli. Furthermore, these very biochemical differences might also prevail in the higher olfactory areas.

These problems alone should occupy olfactory researchers for a considerable time. More emphasis will hopefully be directed to these newer biochemical fields, which may hold answers to many basic problems in olfaction.

D. Effects of Odor Deprivation or Long Exposure to an Odor

1. Morphological Aspects

The effects of odor deprivation on the OB, or higer olfactory structures have been studied by many investigators. Margolis *et al.* (1974) found that intranasal irrigation with a zinc sulfate solution resulted in a marked decrease in olfactory bulb, weight, shrinkage of the glomerular layer of the olfactory bulb and specific disappearance of the unique olfactory marker protein from the OB. These changes were irreversible for several months. In contrast, no alterations were seen in the activities of 3 neurotransmitter-metabolizing enzymes, or in 16 high-affinity "neurotransmitter uptake" systems, or in the levels of cyclic nucleotides in the olfactory bulbs. Meisami (1976) electrically cauterized one naris of the rat one or three days after birth and showed that the growth of the OB was considerably retarded on the anosmic side. Furthermore, while the protein concentration of each OB remained essentially the same throughout the two months after the onset of unilateral anosmia, the total concentration of protein of the anosmic OB was always less than that of the normal OB (13% at day 10, 22% at day 25, and 30% at day 60). It is therefore apparent that odor deprivation reduced the overall protein synthesis in the developing OB of the operated side.

Meisami and Manoochehri (1977) destroyed the olfactory receptors of newborn rats by perfusion with zinc sulfate. Neonatal deafferentation caused a 32% and 41% retardation in OB growth at days 10 and 25, respectively. As previously mentioned, the mere absence of olfactory stimulation resulted in significant inhibition of ipsilateral OB growth (22% at day 25; Meisami, 1976). The more severe effect of neonatal anosmia on OB growth observed in this study (41% at day 25) could be attributed to the loss of olfactory afferent nerves and the associated transneuronal degenerative responses in the OB.

Benson *et al.* (1984) replicated Meisami's (1976) experiment in mice and confirmed again the retarding effect of odor deprivation on the developing OB. At postnatal day 30, odor-deprived OBs were smaller in volume than nondeprived OBs by 13% to 35%. Furthermore, mitral cells, mitral-to-granule synapses, and granule-to-mitral synapses were found to be smaller and less numerous in deprived OBs than in normal ones. In a similar experiment, Brunjes *et al.* (1985) found a 25% reduction in the size of the deprived OB, and also observed a significant reduction in dopamine concentration eight to 30 days after birth. However, no significant changes in norepinephrine concentration were found. Brunjes (1985) studied the time course of changes in OB size after surgical closure of one naris. He found that reliable differences between deprived and control OBs appeared by day 12 after surgery and that odor deprivation induced at a later age resulted in less severe alterations. Skeen *et al.* (1986) found that neonatal olfactory deprivation resulted in a substantial loss of tufted cells in mouse OBs. Since tufted cells are generated prenatally, their reduced number in the postnatally deprived OB is presumed to be a consequence of retarded migration or enhanced cell death.

The converse manipulation, prolonged exposure to an odor, was studied in rats by Döving and Pinching (1973) and Pinching and Döving (1974). They found for each of 44 different odors a specific pattern of mitral cell degeneration in the OB. Apfelbach and Weiler (1985) demonstrated that ferrets show a sensitive phase in their postnatal development, during which time they can become imprinted to food odors, and that at the same time the number of granule cell spines in the OB reaches a maximum, declining significantly thereafter. In ferrets exposed continuously to saturated levels of geraniol odor in the cage environment, the normal decline in spine numbers observed at days 60 and 90 was significantly enhanced. This late postnatal phase was further associated with a marked and significant decrease in total brain weight.

Panhuber *et al.* (1987) found that continuous exposure of adult rats to a single odor for two months causes a shrinkage of mitral cells in the OB which is more extensive and severe than that found in adult rats exposed to deodorized air or to normal rat colony odors.

The effects of odor deprivation or long exposure to an odor, described above in animals, are thought to have significant implications for humans living or working in odorous environments.

2. Neurochemical Aspects

Margolis *et al.* (1974) demonstrated that stringent intranasal irrigation with $ZnSO_4$ solution reversibly elicited a marked decrease in the OB weight, shrinkage of the glomerular layer of the OB, and specific disappearance of the OMP from the OB. In contrast, no alterations were observed in the activities of 3 neurotransmitter-metabolizing enzymes, acetylcholinesterase, choline acetyltransferase, and glutamic acid decarboxylase, or in 16 high-affinity neurotransmitter uptake systems, such as adenosine, GABA, etc., or in the levels of cyclic nucleotides in the OB. Margolis (1974) showed that a naturally occurring dipeptide of histidine, carnosine, decreases to less than 10% of normal levels, while other amino compounds are unaffected after similar deafferentation.

Harding *et al.* (1978) found that intranasal irrigation with $ZnSO_4$ resulted in the immediate and total loss of the ability to find a buried food pellet. This anosmia persisted for six weeks inmost of the treated mice, and was matched by a long-term reduction of the levels of carnosine synthesis and transport in the primary olfactory pathway. These biochemical parameters were virtually undetectable two weeks after treatment and did not exceed 5–10% of average control values even one year after treatment. Light microscopic observations indicated a substantial destruction of the epithelium with subsequent atrophy of the OB. At very long intervals after treatment, some receptor regeneration was apparent with accompanying reinnervation of the OB.

E. Removal of the Olfactory Bulb

1. Regeneration of Olfactory Axons into the Forebrain

In several vertebrates, from amphibians to mammals, the population of sensory neurons is reconstituted after degeneration due to olfactory nerve transection. The axons of these new neurons grow into their target, the OB, and establish morphological and functional connections with OB neurons (see p. 121). When the OB was removed, identical differentiation and maturation of the olfactory neurons were observed (Costanzo and Graziadei, 1983; Monti Graziadei, 1983). In this case, a morphological connection of olfactory axons with the forebrain was discovered.

After unilateral bulbectomy in neonatal mice, Graziadei *et al.* (1979) obtained the following findings: From five days of survival onward, there was a marked anterior displacement of the frontal cortex into the cavity previously occupied by the OB. By 20 postoperative days the axons of newly reconstituted olfactory cells had reached the level of the lamina cribrosa, and by 30 days the fibers had penetrated into the telencephalon and had formed typical glomerular structures within the host tissue. The identification of these fibers and glomerular structures as olfactory was confirmed by immunohistochemical techniques using antisera to the specific olfactory protein. Ultrastructural observations clearly indicated the typical glomerular pattern of the structures and demonstrated synaptic contacts between the sensory terminals (containing OMP) and dendritic processes, as yet unidentified, originating from the surrounding cerebral matrix (Graziadei *et al.*, 1978, 1979).

Their observations thus demonstrated that in the absence of their normal target, newly formed axons can penetrate a "foreign" environment, the cerebral cortex, and form typical glomerular structures and corresponding sensory synapses. Thus, they suggested a heretofore unsuspected degree of plasticity in the olfactory system as well as in the cerebral cortex. Other investigators have also shown similar findings (Small, 1977; Small and Leonard, 1983; Shafa, 1978).

Graziadei and Monti Graziadei (1986a, b) found that these ectopic glomerular structures in neonatal mice are formed by the commingling of the olfactory axon terminals and the dendrites of brain neurons that lie in their proximity, but that they are characterized by the absence of periglomerular cells. In the synaptic contacts between the sensory axon terminals and the dendrites of the brain neurons, they found large neurons, resembling mitral cells, that expanded their dendrites into the intracerebral glomeruli.

2. Effects of OB Removal on the Olfactory Cortex

Sandberg *et al.* (1984a, b) found that olfactory bulb removal and consequential degeneration of the lateral olfactory tract led to a decrease in the levels of glutaminase and malate dehydrogenase in the ipsilateral olfactory cortex. These changes in enzyme activity may account for the well-established decrease in the levels of aspartate and glutamate in the olfactory cortex following ipsilateral bulbectomy. Jancsar and Leonard (1984) found that bilateral OB removal caused marked changes in the "turnover" of several neurotransmitters in the amygdaloid cortex and the midbrain areas of the rat brain.

3. Effects of OB Removal on the Olfactory Epithelium

Margolis *et al.* (1974) found that surgical removal of the OB results in a rapid specific decrease of the level of the OMP in the olfactory epithelium. This treatment did not cause any alteration in the normally low levels of three neurotransmitter-metabolizing enzymes, acetylcholinesterase, choline acetyl transferase, and glutamic acid decarboxylase, in the olfactory epithelium.

4. Effects of OB Removal on Animal Behavior

A number of studies have been undertaken to clarify the effects of bulbectomy on behavior. Since, however, these experiments have been performed on animals, the author will leave a detailed survey of the extensive results to other appropriate reviewers. Here, only a short introductory comment is noted.

The bulbectomy experiments disclosed that the OB is not only a relay station of olfaction but also is considerably involved in the control of various aspects of animal behavior. In 1967, Heimer and Larsson reported that bulbectomy causes deficiencies in male mating behavior, and other investigators have found various abnormal types of behavior in rodents: speciesdependent increases or decreases in female mating behavior, a dramatic reduction in male aggressiveness, a general increase in irritability and exploratory activity, absence of maternal behavior, deficits in learning (evident in passive avoidance task and single-trial taste-illness tests), and several other physiological effects. Other abnormal behavior elicited after OB removal include muricides, loss of olfactory functions (anosmia), changes in neuroactive substances, changes in hormones, changes in estrous cycles, hyperphagia, and obesity (Cain, 1974b; Edwards, 1974 for reviews; Wenzel and Salzman, 1968).

Concerning the generative mechanism of these deficits, Cain (1974) concluded: "It is proposed that the generation of oscillatory and other electro-encephalogram patterns of feedback circuitry within the OB is important to normal limbic and basal forebrain functioning, and that such patterns constitute a nonspecific facilitatory and modulatory influence of the OB on these brain regions ..."

F. Projection of Afferent Fibers from the Olfactory Bulb to the Olfactory Cortex

1. Afferent Fibers from the OB

Mitral cells in the mammalian MOB project their axons to the superficial layer (layer Ia) of the ipsilateral olfactory cortex. The target areas of the olfactory cortex include all subdivisions of the anterior olfactory nucleus (AON), the dorsal peduncular cortex, and the entire lateral entorhinal cortex (ER) (Clark and Meyer, 1947; Lohman, 1963; Powell *et al.*, 1965; White, 1965; Scalia, 1966; Price, 1973; Scalia and Winans, 1975; Broadwell, 1975a; Devor, 1976a; Haberly and Price, 1977; Skeen and Hall, 1977; Heimer *et al.*, 1977; De Olmos *et al.*, 1978; Scott *et al.*, 1980, 1985; Davis *et al.*, 1978; Ojima *et al.*, 1984).

Turner *et al.* (1978) studied the locus and cytoarchitecture of the projection areas of the OB in a microsmatic animal, the rhesus monkey, (*Macaca mulatta*). In the monkey, the OB sends fibers entirely ipsilaterally to the AON, the olfactory tubercle (OT), the frontal and temporal prepiriform cortices (PPC), the oral, medial, and dorsal divisions of the anterior amygdaloid nucleus, and the polar and anterior ER. All of these structures were found to have a laminar organization throughout, except in the AON where it was found only in the anterior, peduncular portion. While the OB fibers projected to the entire extent and depth of the AON, the olfactory afferents to all other structures were confined to layer Ia of the plexiform layer. The rhesus monkey is different from other macrosmatic species in that it does not have a recognizable accessory OB, and has no projections to the nucleus of the stria terminalis, the taenia tecta, or the ventral division of the superficial amygdaloid nucleus. With these exceptions, the projections of the OB fibers in this old world monkey are similar to those in other macrosmatic species.

Ojima *et al.* (1984) studied the mitral cell axon and axon collateral projections to the olfactory cortex in the rabbit by intracellular staining with HRP. The stained mitral cell axons were reconstructed from the soma to the most caudal portion of the anterior piriform cortex (aPC).

Single mitral cells were found to project to cytoarchitectonically different areas of the olfactory cortex, i.e., the anterior olfactory nucleus (AON),

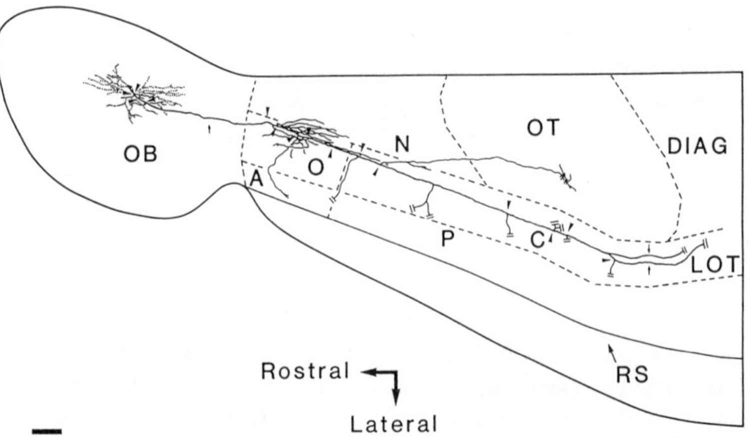

Fig. VII-13. Ventral view of a mitral cell and its innervation. This cell has five intensely stained axon collaterals projecting to the AON as well as four less intensely stained intrabulbar collaterals and six collaterals in the anterior PC. In addition, this cell has one more collateral which reaches the lateral portion of the OT. The main axon (arrow) itself bifurcates within the LOT at the caudal level of the anterior PC. Arrowheads indicate branching points of collaterals from the main axon. Symbols (=) indicate that due to the very weak staining, it is difficult to trace the collaterals precisely beyond this point. Dendrites are indicated by dotted lines. Scale: 1 mm. (from Ojima *et al.*, 1984)

the aPC, and the OT (Fig. VII-13). All the stained mitral cells projected to both the AON and the aPC, and about one-fourth of the mitral cells projected to the OT. At the surface of the AON and the aPC, the main axon running along the LOT gave rise to several thin collaterals at various intervals. The collaterals typically form in each area patchy terminal arborizations which tended to be elongated in an anteroposterior direction. In both the AON and the aPC, each mitral cell formed several terminal arborizations in layer Ia.

The axon terminals innervating the OT showed two types of projection patterns. One type of collateral was emitted from the main axon within the OB, coursed through the ventromedial portion of the olfactory peduncle without joining the main mass of the LOT, and usually terminated in the medial portion of the OT. The other type of collateral emerged from the main axon in the LOT, coursed medioposteriorly, and projected to the lateral portion of the OT.

Although individual mitral cells were found to project to several areas of the olfactory cortex, the fact that they formed dense terminal arborizations in specific locations of each area suggests that the bulbocortical connections are not diffuse but highly selective.

2. Morphologically, Neurochemically, and Functionally Distinct Fiber Projections from the OB

Parallel processing of olfactory information in the main and accessory olfactory systems have been well known (Broadwell, 1975a; Scalia and Winans, 1975; Davis *et al.*, 1978; Wysocki, 1979; and Meredith, 1983). Recently, parallel processing mechanisms of olfactory information were also demonstrated in the main olfactory pathways. Macrides *et al.* (1985) reviewed some of the evidence for heterogeneity among mitral and tufted cells: Some studies using retrograde axonal transport of HRP have shown that the axonal projections of tufted cells are different from those of mitral cells (Haberly and Price, 1977; Skeen and Hall, 1977; Scott, 1981; Schoenfeld and Macrides, 1984).

Projections of middle tufted cell axons are limited to the anterior portions of the olfactory cortex (AON, anterior PC, OT, etc.), and those of the external tufted cell (P-type) axons reach only as far as the AON at the olfactory peduncles. In contrast to these superficial principal neurons, deep principal neurons (i.e., mitral and deeply situated tufted cells) have efferent projections to the limbic system, such as amygdala and entorhinal cortex. This evidence indicates parallel processing of chemosensory information by deep versus superficial principal neurons of the MOB (Macrides *et al.*, 1985). Furthermore, it implies that deep principal neurons participate in broadly convergent and divergent patterns of intrabulbar and more central integrations, although these interconnections lack a prominent sector-to-sector topographic organization. In contrast, the superficially situated tufted cells were likely to function in ways that tend to preserve and possibly enhance spatially and temporally distributed information across populations of neurons.

The topographic organization of their intra- and interbulbar connections is complementary to the spatial patterns of 2-deoxyglucose labeling that have been reported in the MOB in response to stimulation with particular odorants or animal scents (Shepherd, 1985).

3. Reinnervation of Mitral Cells after the Olfactory Tract Transection

When projection tracts in the adult mammalian central nervous system are severed, the fibers distal to the cut degenerate and no longer innervate their terminal fields appropriately; specific functional deficits often accompany these lesions. When similar damage occurs in infant animals, the fibers degenerate, but there is sometimes partial or complete recovery from the functional deficit. A correlated regrowth of axons and appropriate reinnervation of their terminal regions has been demonstrated by many investigators (Nygren *et al.*, 1971; Devor, 1975, 1976b; Small, 1977; Small and Leonard, 1983; Kalil and Reh, 1979). Grafe (1983) proved that if the LOT of the golden hamster is transected in the first week of postnatal life, axons will grow back through the cut and reinnervate the terminal regions, and functional recovery occurs only when the terminal regions are reinnervated.

At first, the birth dates of the mitral cells were examined and found to be on gestational days 11 and 12 (E11 and E12); tufted cells were born on days 11 to 14 (E11 to E14). Secondly, the axons of early formed cells were found to reach the olfactory cortex before those of cells formed later. When a transection of the LOT was performed on postnatal day 3 (P3), Grafe found that cells that were formed early and sent out their axons early were able to reinnervate the olfactory cortex, whereas late formed cells did not. These results suggest that the factors which prevent the regrowth of axons when the LOT is cut after P7 may depend on the stage of development of the tissue into which the axons are growing, rather than the cells of origin and their axons.

G. Centrifugal Fibers to the OB

1. Anatomical Studies on the Centrifugal Fibers

Since the pioneer work by Kerr and Hagbarth (1955), many investigators have studied anatomy of centrifugal fibers to the OB, and have disclosed that the OB receives centrifugal afferent inputs from a number of

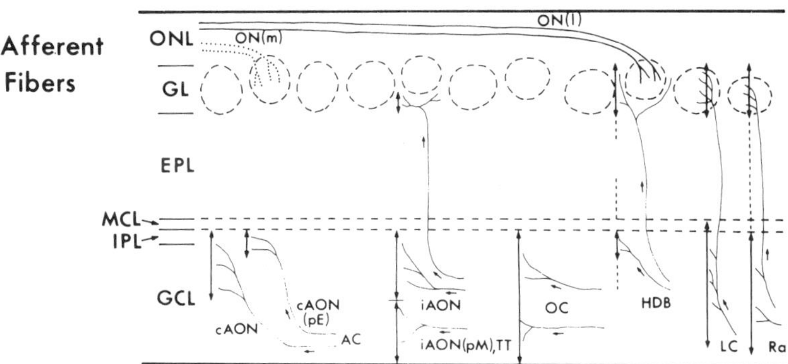

Fig. VII-14. Terminations of centrifugal fibers in the GCL. (from Mori, 1987a)

regions in the brain (Bennett, 1968; Broadwell, 1975b; Broadwell and Jacobowitz, 1976; Cajal, 1911; Cragg, 1962; Dahlström *et al.*, 1965; Davis *et al.*, 1978; Davis and Macrides, 1981; Dennis and Kerr, 1968, 1976; De Olmos *et al.*, 1978; Fallon and Moore, 1978a; Fujita *et al.*, 1964; Haberly and Price, 1978a, b; Heimer, 1968; Heimer *et al.*, 1977; Hernández-Peón *et al.*, 1960; Kerr, 1977; Lavin *et al.*, 1959; Lohman and Lammers, 1967; Macrides *et al.*, 1981; Nakashima *at al.*, 1978; Price, 1969; Price and Powell, 1970b, c, e; Powell *et al.*, 1965; Shafa and Meisami, 1977; Shipley *et al.*, 1985; Takagi, 1962; Yamamoto and Iwama, 1960). Centrifugal control mechanisms of these fibers were reviewed by Dennis (1985) and Kerr (1985). As shown in Fig. VII-14, each type of centrifugal afferent fiber shows a distinct laminar pattern of termination according to the origin of the fiber: one type consists of fibers arising in the ipsilateral olfactory cortex (OC), including the ipsilateral AON, OT, and PPC. A second type consists of commissural fibers arising in the contralateral AON. Fibers arising in the nucleus of the ipsilateral horizontal limb of the diagonal band (HDB) constitute a third type. A fourth type includes fibers from the locus coeruleus (LC) and the raphe nucleus (Ra) in the brain stem.

Many electron microscopic studies (Andres, 1970b; Hirata, 1964; Price, 1968; Price and Powell, 1970a, b, c, d) have shown that the peripheral processes of granule cells form reciprocal synapses with secondary dendrites of the mitral and tufted cells in the EPL and that granule cells receive various kinds of inputs from the telencephalon. Pinching and Powell (1972) and Price and Powell (1970a) have shown that these centrifugal afferents terminate predominantly on the periglomerular, granule, and short axon cells of the main OB and can thereby influence the synaptic integrations of mitral and tufted cells. Thus, it has been suggested that these cells function as interneurons between the extrinsic telencephalic fibers and the OB mitral cells.

Of related interest is the finding, in the hamster, that the centrifugal projections from a variety of cortical and subcortical structures have markedly different laminar patterns of termination within the main OB (Davis and Macrides, 1981; Macrides *et al.*, 1981).

2. Centrifugal Influence on OB Activity

Electrophysiological studies using the summed evoked potential method have demonstrated that centrifugal effects on the activity of the OB can be elicited from an extensive portion of the basal rhinencephalon in the cat (Dennis and Kerr, 1968), the ferret (Dennis and Kerr, 1975), and the rat (Callens, 1965). Furthermore, stimulation of centrifugal fibers originating in the basal rhinencephalon usually elicit a negative potential in the granule cell layer (GCL), and depress both unit and wave activity in the OB (Callens, 1965; Kerr and Hagbarth, 1955).

Nakashima *et al.* (1978) studied regions exerting centrifugal influence on OB activity. By delivering electrical stimuli to the AC, to the deeply located structures in the projection areas of the LOT, and to the medial forebrain bundle situated between the lateral hypothalamic area and the lateral preoptic area, negative field potentials were evoked in the GCL of the OB.

They then performed intracellular recordings from the mitral cells and the GCL neurons in the OB to clarify the modes of the centrifugal influences on the OB neurons. EPSPs were recorded in the GCL neurons by stimulating the deeply situated PPC as well as by stimulating the AC. Onset time and duration of the EPSPs corresponded well to those of the negative field potentials in the GCL. These negative potentials were thus thought to be caused by the EPSPs of a number of granule cells.

In almost all of the mitral cells, IPSPs were recorded by stimulating the AC and the deep structures of the LOT projection areas. The IPSPs came on several milliseconds after the negative field potentials in the GCL began. From the results of these experiments, Nakashima *et al.* postulated that the excitation of the centrifugal system tends for the most part to depress activity in the mitral cell, and that the GCL neuron (presumably the granule cell) appears to function as an inhibitory interneuron between extrinsic fibers from the telencephalon and the mitral cell.

3. Functions of the Centrifugal Fibers

The centrifugal fibers have their terminations in a strategic location in order to affect animals' responses to odorous stimuli. A good configuration was provided by Pager (1974). Recording multi-unit discharges from presumed mitral cells in rats that were hungry or satiated, she showed that responsiveness to food odors was significantly greater in hungry rats than in satiated rats, and that the effect was abolished by ipsilateral section of fibers passing to the OB in the AC. The effects may be presumed to indicate mechanisms by which central limbic structures, such as the hypothalamus, modulate or bias olfactory input activity.

VIII. Studies on the Olfactory Cortex and Diencephalic Olfactory Areas

A. A Survey on Multiple Functions of the Projecting Fibers from the Olfactory Bulb

The routes and terminal areas of fiber projections from the OB have been extensively studied anatomically and physiologically by a number of investigators. Most of them will be presently introduced in this chapter.

The olfactory pathway has often been regarded as concerned solely with transmission and processing of information about odors. Recent studies, however, have showed that functions of olfactory regions are not solely sensory in nature, but many of those areas are involved in the limbic system and social behavior.

Recently, Shepherd *et al.* (1981) wrote an excellent review on the diverse functions supported by multiple neuronal circuits within olfactory and basal forebrain systems. In it, they distinguished six circuits in the central olfactory pathways, as follows:

(a) A purely sensory circuit that convey olfactory information. This means pathways from the olfactory receptor through the OB to the olfactory cortex (OC) and then through the thalamus and other diencephalic areas to neocortical olfactory areas (see Chapter IX for old world monkeys and Chapter X for lower mammals). In addition, a pathway from the vomeronasal organ through the accessory OB to the amygdala is also included in this category.

(b) Circuits formed by axon collaterals of olfactory cortical neurons that terminate in various areas of the OC and in the OB. These circuits may perhaps generate the rhythmic brain waves often observed in the OB and pyriform cortex (PC).

(c) Circuits formed by sensory input pathways to the limbic system. Some of these central connections ultimately reach the level of the neocortex, and hence probably contribute affective qualities to odor perceptions.

(d) Circuits formed by neurons in the septal-diagonal band region, hypothalamus, hippocampus, and midbrain constitute the limbic system. Each of these areas has connections, directly or indirectly, with the OB and other regions. Thus, limbic system modulation occurs at all the main stages of synaptic transmission and integration in the sensory pathway. Release of different neuroactive peptides and the binding of steroid hormones at different sites in these circuits also helps limbic system modulation.

(e) Limbic circuits formed within the OB and basal forebrain independently of olfactory input. The reciprocal synapses between mitral and granule cells may be considered as one of circuit elements for controlling basal forebrain excitability and coordinating rhythmic activity. The circuits in (d) and (e) will remind the reader of the suggestions by Cain (1974b) that the OB can be viewed as a rostral extension of the limbic system.

(f) Circuits in the olfactory pathway that provide nonspecific arousal. Routtenberg (1971) postulated that there are two arousal systems in the brain. One, the ascending reticular activating system of Moruzzi and Magoun (1949), is primarily concerned with drive and behavior sequences (Cain, 1974). The second, the limbic-midbrain system of Nauta (1958), is concerned with incentive and reward; it is this system that is brought into play by self-stimulation techniques(Olds, 1956). The central olfactory areas appear to be integral parts of this second system. The olfactory pathway is the only sensory pathway with direct access to this forebrain arousal system. The effects of the OB removal may be largely due to depriving this system of its primary sensory input (Cain, 1974).

The above is a short digest and citation of Shepherd *et al.*'s review (1981). The reader is recommended to read the review for further information.

B. The Olfactory Cortex

1. The Anterior Olfactory Nucleus
1.1. Anatomical studies
The anterior olfactory nucleus (AON) is an assembly of several cell groups, situated in the olfactory peduncle between the OB and the other structures of the olfactory cortex. The name "anterior olfactory nucleus" was first used by Herrick (1910) in his study of amphibians and reptiles.

The AON was topographically divided into a minmum of six parts: pars rostralis, pars lateralis, pars dorsalis, pars medialis, pars ventralis, and pars externa (pE; Lohman and Lammers, 1967). The pars rostralis, pars lateralis, and pars dorsalis are situated within the OB; the remaining parts of this nucleus are located in the retrobulbar area. In addition, the pars posterior, pars caudalis, and pars rostralis are found in some animals, but not in others (Herrick, 1924; Rose, 1929; Crosby and Humphrey, 1939; Fox, 1940; Lohman, 1963; Lohman and Lammers, 1967; Dennis and Kerr, 1976).

The AON is a poorly laminated cortical structure consisting of a superficial plexiform layer (layer I), a small, uniformly distributed, compact pyramidal cell layer (layer II) and a loosely packed polymorphic cell layer (layer III). The main OB (MOB) efferents to the AON terminate exclusively in the superficial half of the superficial plexiform layer, layer Ia, while centrifugal efferents to the MOB arise throughout the layers II and III (Davis and Macrides, 1981).

Lohman (1963) studied fiber connections of the cell groups in the AON with other structures. Wysocki (1979) gathered data obtained by many workers and displayed them in a table (see Table 5 in his 1979 review paper). Some of the discrepancies may be accounted for by differences in experimental procedures or in the animal species (Davis and Macrides, 1981). Daval and Leveteau (1974) and Price (1973) provided evidence for internal organization within the subdivisions of the AON. Among the AON projections, Broadwell (1975b) noted the absence of a projection to the entorhinal area or the amygdala, sites which receive olfactory and/or vomeronasal inputs from the MOB and the accessory OB (AOB). He also noted the lack of an AON projection to the AOB. From these findings, an important conclusion may be drawn: that information from the AOB remains independent of the AON and very probably of its efferent projection sites, which was confirmed by a recent HRP study (Mori, personal communication).

Schoenfeld and Macrides (1984) indicated that both the afferent and efferent connections of the pE with the MOB are topographically organized and provide a short synaptic pathway between homotopic sectors of the two MOBs; this work provides an important base for neurophysiological studies.

In the mammalian OB, centrifugal fibers from the anterior limb of the anterior commissure (AC) terminate in the granule cells (Price and Powell, 1970c), and volleys in the AC fibers activate presumed granule cells which then elicit IPSPs in the mitral cells (Mori and Takagi, 1978b; Yamamoto et al., 1962). In regard to the origin of these AC fibers, many anatomists (e.g., Broadwell, 1975a; Dennis and Kerr, 1976; Haberly and Price, 1977; Shafa and Meisami, 1977), negating Cajal's view (1955), demonstrated that the AC fibers to the contralateral OB (COB) originate not from tufted cells but from cells in the AON. This conclusion was supported by HRP studies, which demonstrated peroxidase-positive somata in the contralateral AON, but not in the contralateral OB and also a large number of labeled cell bodies in the ipsilateral AON (Broadwell, 1975b; Dennis and Kerr, 1976; Shafa and Meisami, 1977). These findings confirmed earlier anatomical observations: the AON cells which project into the AC send one or sometimes two recurrent axonal collaterals to the ipsilateral OB (IOB) (Price and Powell, 1970e; Valverde, 1965). Tazawa *et al.* (1982b) analyzed electrophysiologically the axonal trajectories of many AON neurons to determine to what extent they project to the OB on both sides.

1.2. Physiological studies

Experimental data concerning birhinal interactions were first provided by Zwaardemaker (1895) and Elsberg (1935b), who observed a decreased threshold in response to birhinal stimulation. These views were also supported by later psychophysiological investigations (von Skramlik, 1925; von Békésy, 1964; Schneider and Schmidt, 1967). Bennett (1968) demonstrated that a rat previously conditioned for lateral discrimination of an olfactory stimulus loses that ability after bilateral section of the anterior limbs of the AC. Electrophysiological bases could, then, be found for the reciprocal inhibition suggested by these psychophysical and behavioral data: Leveteau and MacLeod (1969a), and Leveteau *et al.* (1969) reported the existence of an interbulbar lateral inhibition, the effect of which was maximum when the contralateral chemical stimulation was delivered 3 msec before the ipsilateral one. This very short latency indicates that no more than two synaptic relays are probably involved in the interbulbar route, a fact consistent with anatomical data from Lohman (1963) and Lohman and Lammers (1967). This interbulbar inhibition involved only the retrobulbar centers (i.e., AON; Leveteau *et al.*, 1969). Later, Leveteau *et al.* (1972) showed that the interval-dependent inhibitory effect in the AON responses was maximum when the contralateral chemical stimulation was given 2 msec before the ipsilateral one. Then, Daval and Leveteau (1974) studied the role of the AON in the interbulbar connections systematically exploring all possible pathways between the OB and the AON. Strong evidence was found for the presence of direct connections from mitral and tufted cells in the OB to AON cells and a synaptic relay of interbulbar fibers lying in the ventro-lateral and posterior regions of the AON.

In response to bilateral olfactory stimulation, Daval and Leveteau (1975) recorded extracellularly 40 units in the AON. Among them, 9 units (6%) recorded in the pars dorsalis of the AON responded with a short latency and faithfully up to high stimulation rates, demonstrating the existence of a direct pathway from the AON to the IOB (Valverde, 1965; Scalia, 1966 and Price and Powell, 1970 a, b). The other 31 units recorded in the pars ventralis of the AON had longer latencies and did not follow a high stimulation rate. These units were classified by their latencies into four statistically distinct groups.

Thus, two kinds of units were found in the AON. Since the units in the pars dorsalis could distinguish different time intervals, Daval and Leveteau (1975) thought that any given interval registered in the AON by a specific pattern of activation-inhibition provides the animal with a poweful tool for precisely discriminating the first-stimulated side. The units in the pars ventralis, in contrast, were found to be insensitive to time interval variations, but later they were found to be involved in odor quality coding (Boulet *et al.*, 1978). Thus, it seems that the AON is very well provided with immediate information concerning the position and the nature of an odor source.

These electrophysiological findings were in agreement with anatomical studies showing that the pars dorsalis and the pars ventralis of the AON are projection areas of the ipsilateral LOT (Broadwell, 1975b; Lohman, 1963; Scalia, 1966, 1968) but that only the pars dorsalis receives axonal projections from the contralateral AON (Broadwell, 1975b; Scalia, 1966).

The AON receives an afferent projection from the OB and projects to IOB, COB, and several olfactory cortices. In order to study the synaptic organization of the AON, Tazawa *et al.* (1981) recorded extracellular field potentials and intracellular synaptic potentials from the AON following stimulation of the LOT, IOB, and AC in the urethane-anesthetized rabbit. An LOT stimulation elicited an EPSP which was followed by two types of IPSP. The first IPSP was relatively brief (about 60 msec) and accompanied by an increase in the Cl conductance. In about 60% of AON neurons, this fast IPSP was followed by a hyperpolarizing potential (slow IPSP) with a large amplitude and a very long duration (several hundred msec). The OB stimulation sometimes elicited an antidromic spike superimposed on the same sequence of the synaptic potentials (EPSP–fast IPSP–slow IPSP). These synaptic potentials were quite similar to those reported in the pyriform cortex (PC) neurons (Mori *et al.*, 1978; Satou *et al.*, 1980, 1982a, b, 1983a, b, c; Fig. VIII-4).

Tazawa *et al.* (1982) demonstrated that the fast IPSP is generated mainly by an increase in chloride conductance, and that spatial facilitation of the amplitude occurs between the AC-evoked fast IPSPs and the LOT- or IOB-evoked fast IPSPs. The latter observation suggested that fast IPSPs are mediated by common interneurons.

Mori *et al.* (1979) inserted a glass capillary microelectrode into the AON neurons in the lateral portion of the pars dorsalis, the pars lateralis, and possibly also the pars externa, and analyzed the axonal projection of these AON neurons to the IOB and COBs and to the prepyriform cortex (PPC) in the rabbit. Of 117 AON neurons which sent their axons to the AC, 46 cells (39%) were activated antidromically by ipsilateral OB stimulation and 55 cells (47%) by contralateral OB stimulation; 22 AON neurons (19%) were activated from both, and 24 AON neurons (21%)

were activated by stimulation of the superficial layer of the PPC (SPC). Schematic drawings of axonal branchings and conduction times along the branches of two representative AON cells are shown in Fig. VIII-1.

The mean axonal conduction velocity of the AON neurons was found to be 2.8 m/sec in the AON-AC axonal segment, 1.6 m/s in the AON-COB segment, and 1.0 m/sec in the AON-IOB segment. These results and the collision tests between the antidromically evoked spikes indicated that a number of AON neurons send their axons to the COB via the AC and that the same neurons send thin axon collaterals to the IOB.

On the basis of Mori *et al.*'s (1979) clarification of the axonal trajectories of many AON neurons to the OBs and SPC, one can assume that the AON neurons are involved in a negative feedback pathway to the ipsilateral OB and in a negative feedforward pathway from the IOB to the COB.

Tazawa *et al.* (1982b) stained AON neurons (all of which showed an EPSP following LOT stimulation), injecting HRP intracellularly and demonstrated: (1) that layer II neurons in the AON (pars dorsalis and pars lateralis) extended apical dendrites into the most superficial portion of layer Ia. These apical dendrites had a tendency to project for a greater distance in the anteroposterior direction (Fig. VIII-2); and (2) that layer III neurons were classified into two types according to their dendritic patterns. One type of layer III neurons sent dendrites to layer Ia, while dendrites of the other type of layer III neurons were distributed only within layer III. Thus, LOT-evoked EPSPs in the second type of layer III neurons should be elicited polysynaptically. The axons of AON neurons were found typically to emit many axon collaterals within layer III of the AON.

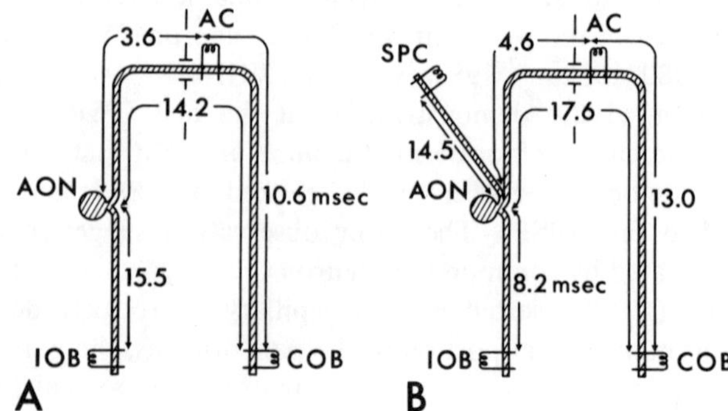

Fig. VIII-1. Schematic drawings of axonal branchings and conduction times along the branches of two representative AON cells. SPC: superficial layer of prepiriform cortex. (from Mori *et al.*, 1979)

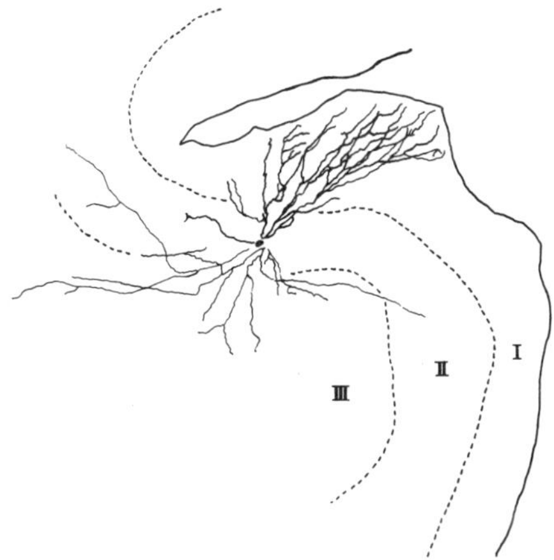

Fig. VIII-2. Dendritic and axonal arborizations of an AON cell stained by HRP. (unpublished data)

These were traced to the ipsilateral OB, to the neocortex dorsal of the rhinal sulcus, and to layers I, II, and III of the AON.

The temporal firing pattern of AON neurons together with PPC ones was investigated during the artificial intake of deodorized air in nine rabbits anesthetized with 30% urethane (Onoda, unpublished data; Onoda and Mori, 1980; Chapter VII, p. 276). Sixty units whose discharge rate changed consistently in relation to air intake pulse were isolated. According to the number of bursting activities during each air pulse, neurons were classified into 'single-peak' units (75%) and 'multiple-peak' units (25%). The result may be interesting when compared with the previous one in the OB that single-peak units were 56.4% while multiple-peak units were 43.6% (p. 276). Thus, the difference between the OB and the AON may be explained by the intervention of some sort of information processing. When the intensity of the air pulse was increased, units of 65% did not show an increase in the spike frequency, but only an increase in the duration of bursting activity.

2. The Prepyriform Cortex (PPC)

2.1. Anatomical studies

The pyriform lobe includes the PPC and the greater portion, at least, of the hippocampal gyrus. It is divided into several regions: the PPC, the periamygdaloid area, the entorhinal area (ER), the perirhinal area (PeR), the retrosubicular area, and the presubiculum (Peele, 1977).

The PPC extends along the LOT to the rostral amygdaloid region. Fibers in the LOT, derived from the OB and AON, end in this area. Consequently, the PPC is the major part of the olfactory cortex (OC) and is often called the primary olfactory cortex. The periamygdaloid area is a small region almost entirely included in the semilunar gyrus. This area also receives fibers from the LOT and hence is a part of the olfactory cortex. The most posterior part of the pyriform lobe is the entorhinal area (ER; area 28 in Brodmann's designation), which will be described in part 5 of this section. The perirhinal area (area 35 in Brodmann's designation) is found in the depths of the rhinal fissure, and serves as a transition area between allocortex and isocortex (Peele, 1977).

The PPC belongs to the paleocortex, and has rudimentary form of laminar structure (Brodmann, 1909). Although six layers can be differentiated, the three-layer classification has been preferred and is thought to be practical for the purpose of electrophysiological investigation (Peele, 1977). According to Price's (1973) nomenclature in the rat, the PPC consists of the following layers (Fig. VIII-3; Haberly, 1985; Switzer *et al.*, 1985):

Layer I: a superficial plexiform layer
Layer II: a superficial compact cell layer
Layer III: a deeper, more sparsely packed cell layer.

Layer I is subdivided into a superficial portion (Ia) receiving the LOT fibers from the OB and a deep portion (Ib) receiving association fibers from the olfactory cortical areas. Layer II is also subdivided into a superficial portion (IIa) containing semilunar cells and a deep portion (IIb) containing densely packed somata of the superficial pyramidal (SP) cells (Haberly, 1985). Somata of deep pyramidal (DP) cells and polymorphic cells are scattered throughout layer III. The LOT fibers terminate on the distal segments of the apical dendrites of the pyramidal cells, whereas the association fibers terminate both on the more proximal segments of the apical dendrites and on cells in layer III (Broadwell, 1975a, b; Haberly and Price, 1978a, b; Hartzell *et al.*, 1977; Heimer, 1968; Lohman, 1963; Powell *et al.*, 1965; Price, 1973; Price and Powell, 1971; Scalia, 1966; Stevens, 1969; Valverde, 1965). These structures are schematically illustrated in Fig. VIII-3 (Haberly, 1985).

Harberly and Shepherd (1973) separated layer IV, an endopyriform nucleus layer, from the remaining part of the layer III polymorph cell groups in the deep part of layer III.

Many investigators have studied the pyriform cortex using the Golgi

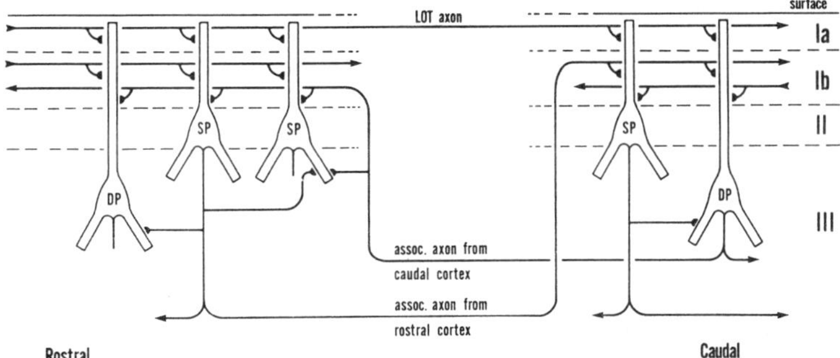

Fig. VIII-3. Model for excitatory inputs to pyramidal cells in pyriform cortex. Afferent fibers in the LOT excite distal apical segments in layer Ia. Local axon collaterals of pyramidal cells excite nearby pyramidal cells via their basal dendrites in layer III. Association fibers from the rostral part of the cortex excite distant pyramidal cells via intermediate apical segments in the outer part of layer Ib. Association fibers from the caudal pyriform cortex excite pyramidal cells via their proximalmost apical segments in the deep part of layer Ib and also probably via basal dendrites. (from Haberly, 1985)

method (Cajal, 1911, 1955; Calleja, 1893; O'Leary, 1973; Stevens, 1969; Valverde, 1965). Recently, Haberly (1983, 1985) classified nine types of PPC neurons in the adult opossum (Haberly, 1983, 1985 for further information).

Principal neurons   The principal neurons in the PPC are superficial pyramidal cells (SP) in layer II, and deep pyramidal cells (DP) in layer III. Pyramidal cells have several basal dendrites, which are primarily oriented to the deep layers, and a single apical dendrite, which almost invariably reaches the superficial part of the plexiform layer, where the terminal dendritic branches establish contact with the incoming OB fibers. They have axons directed toward the depth of the PPC and on their way give off many axon collaterals in the layer III. Some of them constitute a significant source of intracortical association fibers in layer Ib (Haberly and Bower, 1984; Haberly and Price, 1978a), and thus relay olfactory information to other parts of the brain (Fig. VIII-3).

Intrinsic neurons   In the superficial part of layer II pyramidal type cells lacking basal dendrites are found (Cajal, 1911). They are called "semilunar cells" (Valverde, 1965). In these cells, distal apical dendrites were found to give rise to very large, disk-shaped spines that receive synapses from afferent fibers. In contrast, proximal apical dendrites were found to have only a small number of diminutive spines (Haberly, 1983, 1985; Haberly and Behan, 1984).

Other types of intrinsic cells are found in the middle and deep layers. These nonpyramidal cells are mostly polymorphic or fusiform in shape. They are often referred to as stellate cells from the star-shaped form imported to the cell body by the several dendritic trunks.

One of these types is a large multipolar neuron concentrated in the deep part of layer III. They would appear to be responsible, at least in part, for the GABAergic feedback inhibition (Haberly, 1985) that follows LOT activation (Biedenbach and Stevens, 1969a, b; Haberly, 1973b; Satou *et al.*, 1982a). Their depth distribution corresponds to that of the neurons responsible for the feedback inhibition (Satou *et al.*, 1983a).

A second type of nonpyramidal cell is the large fusiform cell in layer I. These cells were also postulated to mediate a GABAergic inhibition, but, in contrast to layer III cells, they appeared to be in a feedforward rather than feedback configuration (Fig. VIII-3).

A third type of cell is the so-called neurogliaform cell (Haberly, 1983). These cells are found in all layers, accumulate [^{3}H] GABA (Haberly *et al.* 1987) and are GAD positive (Haberly, 1985), suggesting again that they mediate a GABAergic inhibition.

Recent studies (Haberly, 1985) of the afferent and association fiber systems in the olfactory cortex revealed (i) the segregation of afferent fibers to the superficial portion of layer I (Ia), and (ii) showed that the association fiber system that originates from cells within the pyriform cortex terminates primarily in the deep part of layer I (Ib), and in layer III.

The commissural projection between the two halves of the pyriform cortex shows that the anterior part receives a projection from the opposite AON and the posterior part a projection from the opposite anterior part (Haberly and Price, 1978a).

The afferent input to the PPC is from the OB through the mitral cell axons in the LOT. The branches of these axons terminate in the outer superficial layer (Ia); and none passes to the deeper layers. The PPC has another type of input through fibers entering the depths of the cortex. This "central input" (in Shepherd's (1979) term), comes from unidentified sources and terminates at various levels in the cortex, with the exception of the most superficial layer (Ia).

The pyriform cortex receives centrifugal inputs from the brainstem, thalamus, hypothalamus, and basal forebrain. They include inputs from presumed noradrenergic cells in the locus coeruleus, serotonergic cells in the dorsal raphe (Haberly and Price, 1978a), cholinergic cells in the magnocellular groups in the basal forebrain (Haberly and Price, 1978a; Luskin and Price, 1982), and possibly a light input from dopaminergic cells in the ventral tegumental area (Haberly and Price, 1978a).

The efferent fibers from the primary olfactory cortex are projected to the entorhinal area and other related structures (for details of these efferent fibers see Wysocki, 1979).

The endopyriform nucleus is located deep in the PPC. Krettek and Price (1978b) reported that projections from the PPC to the mediodorsal nucleus of the thalamus (MD) and to the central part of the putamen traverse the endopyriform nucleus. This nucleus was reported to project not only to the PPC but also to the medial and posterior cortical nuclei of the amygdala, the area which was said to receive accessory OB projections to the exclusion of the main OB projections. Thus, the endopyriform nucleus may be a very important relay nucleus between the PPC and the other olfactory areas. However, its olfactory function has not been studied.

2.2. Physiological studies (Imamura et al., 1984b)

Electrophysiological study of the PPC was commenced by Hasama (1934), who found that very regular rhythmic waves are elicited in the PPC by odorous stimulation. Recent series of studies were started by Freeman (1959, 1968a, b) and Biedenbach and Stevens (1969a, b). Neuronal properties and synaptic connections of PPC neurons were first examined using extracellular recording techniques. Then, with the success of intracellular recording from a pyramidal neuron in the cat and the opossum (Biedenbach and Stevens, 1969b; Haberly, 1973a), a series of potentials—EPSP with or without a superimposed spike potential followed by a long IPSP—was recorded in response to electric stimulation of the LOT. In olfactory cortex slice neurons of the guinea pig, Scholfield (1978a) showed that single shock applied to the LOT produces an EPSP usually generating a single spike, accompanied by a long (200–500 msec) afterdepolarization of 5–16 mV. It was concluded that this long afterdepolarization represented a Cl^--mediated IPSP, generated through deep-lying recurrent inhibitory loops (Scholfield, 1978b).

In the neurons of the rabbit PC, Mori *et al.* (1978b) found that OB stimulation elicits a slow hyperpolarizing potential accompanying the fast hyperpolarizing potential (Fig. VIII-4; Biedenbach and Stevens, 1969a, Haberly, 1973a, b; Scholfield, 1978b).

Mori *et al.* (1978) and Satou *et al.* (1982a) disclosed that these fast and slow hyperpolarizing potentials were elicited by electrical stimulation of either the LOT, OB, AC, or the deep-lying structure of the posterior part of the pyriform cortex (DPC). The hyperpolarizing potential

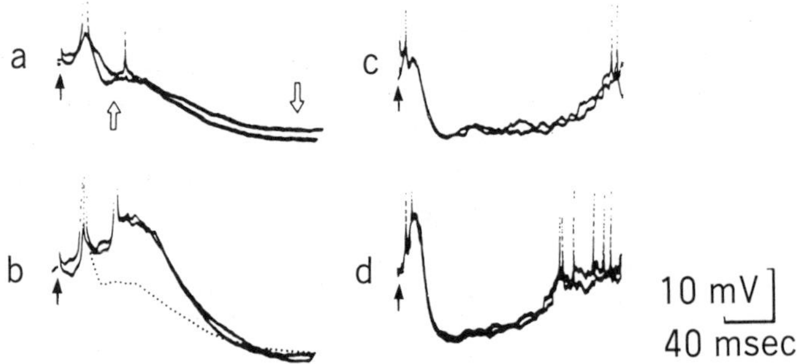

Fig. VIII-4. Intracellular records of fast and slow IPSPs from the neurons in the pyriform cortex. Responses to the OB stimulation recorded with K-Cl electrode. Photographs a, c, b, and d were taken in this order from the same cell in order to elucidate the effect of the changes of the Cl-ion concentrations inside the impaled cell upon the fast IPSP (upward white arrow) and slow IPSPs (downward white arrow). Note that c and d were recorded with a slower sweep speed. Trace a is superimposed on b with dotted lines. Arrows indicate the time of OB stimulation in this and next figures. Spikes were truncated. (from Mori *et al.*, 1978)

showed properties similar to those of well-known ordinary IPSPs (Eccles, 1964); it was quickly converted into a depolarizing potential when a Cl⁻-filled microelectrode was used or when a hyperpolarizing current was applied to the impaled cell; its amplitude was increased when a depolarizing current was applied. Consequently, the fast hyperpolarization was found to be an IPSP generated by a large increase in the membrane conductance for the Cl⁻, hence it was named the fast IPSP. The fast IPSP changed its amplitude in a direct linear relation to polarizing currents. In contrast, the peak membrane potential of the slow hyperpolarization was more negative than the reversal potential of the fast IPSP, and the diffusion of Cl⁻ into the impaled cell had little effect on the amplitude of the slow hyperpolarization. The inhibitory effect of the slow hyperpolarization on the excitability of the PC neuron was proven by the intracellular application of depolarizing current. Thus, the slow hyperpolarization was called a slow IPSP.

In contrast to the fast IPSPs, the slow IPSP showed an entirely different, nonlinear, relation in that both depolarizing and hyperpolarizing currents reduced the amplitude of the slow IPSP. On the basis of these observations, Satou *et al.* (1982a) proposed that a membrane conductance increase for K⁺ is involved in the generation of the slow IPSP.

Satou *et al.* (1982b) found spontaneous fluctuations of the membrane potential in the principal cells of the PC and showed that they were mainly composed of small, spontaneously occurring fast IPSPs. When any of the nearby structures—the OB, LOT, AC, and DPC—was electrically stimulated, the spontaneous fast IPSPs in principal cells were greatly suppressed for long periods with a time course similar to that of the slow IPSPs. Based on an analysis of the suppression of these spontaneous fast IPSPs, Satou *et al.* concluded that the slow IPSP was accompanied by a disinhibitory potential (for further details of this potential see Satou *et al.*, 1982b).

Satou *et al.* (1983a) studied the neuronal pathways responsible for the fast IPSPs elicited in principal cells in the pyramidal cortex by volleys from the OB, LOT, AC, and DPC. The relatively long latencies (5.1–5.5 msec) of the onset of the fast IPSPs elicited by the OB, LOT, AC, and DPC stimulation suggested that these fast IPSPs are not evoked monosynaptically, but that inhibitory interneurons exist.

Analysis of the effects of conditioning OB, LOT, AC, or DPC shocks upon the OB-, LOT-, AC-, or DPC-evoked fast IPSP suggested that the initial part of the OB-evoked fast IPSP was evoked by volleys through the LOT, and the later part through pathways other than the LOT, and that an excitatory synaptic transmission to the principal cells may not be involved in the major neuronal pathways responsible for the AC- and DPC-evoked IPSPs.

Spatial facilitation was observed among the fast IPSPs evoked by volleys from the OB, DPC and AC when shocks were applied at suitable intervals. Also, a temporal facilitation of fast IPSPs was observed when the OB, DPC, or AC shocks were applied repetitively at short intervals.

From a number of the results obtained by Satou *et al.* (1983a) and by others (Biedenbach and Stevens, 1969a, b; Haberly, 1973a, b; Haberly and Shepherd, 1973), Satou *et al.* (1983b) assumed the existence of interneurons mediating the fast IPSPs in the principal cells and determined 11 criteria for identifying the inhibitory interneurons (p. 90 in Satou *et al.*'s paper in 1983b)

According to these criteria, they identified 30 units as inhibitory interneurons. These interneurons were located mostly in the deeper part of layer III of the PC.

Intracellular recordings from the presumed inhibitory interneurons showed that OB stimulation elicited two successive EPSPs on which bursting discharges were superimposed. These EPSPs were followed by a long-lasting hyperpolarizing potential. A comparison of the latencies of the antidromic activation of the principal cells and the synaptic activation of the inhibitory interneurons following OB or DPC stimulation suggested that the inhibitory interneurons are activated at least partly through the axon collaterals of the principal cells, which project their main axons to the OB or DPC. Considering all these results, Satou *et al.* (1983b) proposed a circuit diagram for the neuronal pathways responsible for the fast IPSPs of principal cells in the PC (Fig. VIII-5).

As described above (pp. 299–303), principal cells of layers II and III extend their dendrites to layer Ia and have synaptic contacts with the LOT terminals.

The results of the electrophysiological analysis of the PC, using both *in vivo* (Biedenbach and Stevens, 1969a, b; Haberly, 1969, 1973a, b; Haberly and Shepherd, 1973) and *in vitro* (Richards and Sercombe, 1968; Scholfield, 1978a, b; Yamamoto and Matsui, 1976) preparations were in agreement that volleys in the LOT monosynaptically activate pyramidal cells through excitatory synapses on the distal portion of their apical dendrites. In addition, it was suggested that the activated superficial pyramidal cells (layer II cells) give rise to excitatory synaptic inputs by way of their axon collaterals to other cells in layer III (Fig. VIII-3; Biedenbach and Stevens, 1969b; Freeman, 1968b; Haberly and Shepherd, 1973).

Because of the importance of discrimination between the direct input from OB neurons and the indirect input through the relay from olfactory cortex neurons, Satou *et al.* (1983c) analyzed the monosynaptic and poly-synaptic excitation of PC neurons evoked by volleys in the LOT in the

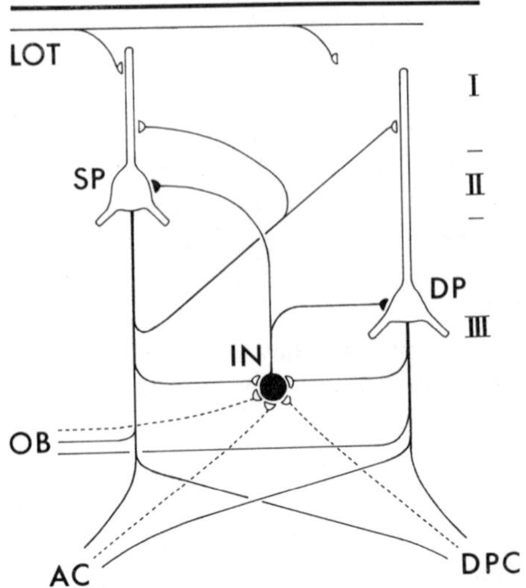

Fig. VIII-5. Schematic circuit diagram of the pyriform cortex. (from Satou *et al.*, 1983b)

rabbit. Based on the analysis of the onset latency of the EPSP and of the spike discharges evoked by the LOT volley, two types of principal cells could be discriminated (Satou *et al.*, 1982a; Shepherd, 1979); cells of the first type (type I) receive monosynaptic inputs from the LOT fibers, and those of the second type (type II) receive di- or polysynaptic inputs.

To elucidate the organization of synaptic inputs to PPC neurons, intracellular and extracellular responses of single cells to the LOT or OB shock were analyzed and two types of EPSP were revealed. The EPSP in the type I cells had shorter latencies (0.0 to 0.9 msec) and the EPSP in the type II cells had longer latencies (1.0 to 6.0 msec). A conditioning LOT or OB shock did not suppress the testing EPSP in the type I cells, whereas the conditioning stimulation greatly suppressed the test EPSP in most of the type II cells.

On the basis of these and other results, Satou *et al.* (1983c) concluded that type I cells are monosynaptically activated by LOT volleys, whereas type II cells are activated di- or polysynaptically by way of a relay from type I cells. The type I cells were recorded in both the superficial and the deep parts of the PC, although they were recorded more frequently in the superficial part. On the other hand, most of the type II cells were recorded in the deep part of the PC. These results supported and extended the previous model circuit shown in Fig. VIII-5, in which the monosynaptically activated superficial pyramidal cells give rise to excitatory inputs to other pyramidal cells and neurons in deep layers.

2.3. Role of the pyriform cortex in olfactory information processing

Due to the lack of highly specific feature detector cells in the olfactory system, many investigators postulated that odors are coded in the form of complex, spatially distributed patterns (Adrian, 1942; Leveteau and MacLeod, 1966; Kauer and Moulton, 1974; Holley and Döving, 1977). Recent studies of the OB using the 2-DG method have revealed spatially distributed, odor-related patterns that may represent a visualization of at least one aspect of the olfactory code (Sharp *et al.*, 1975; Skeen, 1977; Stewart *et al.*, 1979; Jourdan, 1982; Lancet *et al.*, 1982). Thus a higher olfactory area, the primary olfactory cortex, was expected to enhance these patterns in a topographically organized fashion by analogy with other sensory systems. Results with the 2-DG method, however, indicated that the opposite occurs (Sharp *et al.*, 1975, 1977; Astic and Cattarelli, 1982; Astic and Saucier, 1982). In contrast to the results of the above 2-DG experiments, Tanabe *et al.* (1975c) found a clear improvement in the rate of odor discrimination along the olfactory pathway from the OB through the PPC-MA to the LPOF, a neocortical olfactory area (p. 372).

In Tanabe *et al.*'s experiment in an unanesthetized monkey (1975c), Haberly (1985) calculated that each of the eight stimulating odors (Table

IX-3) activated an average of 45% of neurons in the OB and an average of 40.1% of neurons in the PPC, and finally that in the neocortical olfactory area, the LPOF, each odor activated an average of 22% of neurons, which was probably on the order of 10^6 neurons by Haberly's estimate. Haberly had an overall impression from Tanabe *et al.*'s results that information was still in an ensemble coded form, and that each odor would appear to be activating tens of thousands of cells and each cell would appear to be participating in the coding of a very large number of different qualities of odors.

Haberly (1985) examined the properties of neurons and synapses as well as of neuronal circuits in the PC and found that afferents to the PC are arranged in a broad and overlapping rather than point-to-point fashion. On the basis of these studies, Haberly (1985) postulated that olfactory information is in the form of a highly distributed ensemble code in the pyriform cortex and hypothesized that the olfactory cortex serves as a content-addressable memory (CAM) for association of odor stimuli with memory traces of previous odor stimuli (e.g., Anderson, 1970, 1972; Kohonen, 1977; Kohonen *et al.*, 1981; Hopfield, 1982). If most or all parts of the olfactory cortex have the CAM property, neuronal circuits with a CAM property would be able to explain various phenomena such as stimulus generalization, a high degree of damage immunity, and so on (for further details see his 1985 review paper).

3. The Olfactory Tubercle (OT)

The olfactory tubercle has been considered one of the constituents of the olfactory cortex and has been studied relatively well by anatomists. However, the physiological data on the OT have been fragmentary and remain unspecified.

3.1. Anatomical studies

In macrosmatic animals, such as the rat and the rabbit, the OT is a conspicuous eminence on the base of the brain just caudal to the olfactory peduncle. The laminar structures of the OT have been identified in macrosmatic mammals since the studies of early histologists (Ganser, 1882; Calleja, 1893; Cajal, 1955). The following three layers have been found in most animals:

(1) An outermost or superficial plexiform layer;
(2) An (intermediate) pyramidal cell (or cortical) layer;
(3) A (deep or innermost) polymorph cell layer.

Thus, lamination in the OT of macrosmatic mammals is clear-cut and is given as evidence that the OT is a part of the olfactory cortex. Danner and Pfister (1981) clearly differentiated OT neurons into seven types according to anatomical criteria.

In microsmatic animals, like the rhesus monkey (Heimer *et al.*, 1976), only a small anterolaterally situated area of the OT receives direct input from the OB (Allison, 1953a; Broadwell, 1975a; Devor, 1976a; Ferrer, 1969; Heimer, 1969; Price, 1973, 1984; Skeen and Hall, 1977; Turner *et al.*, 1978), and lamination is not apparent, except perhaps in the later-almost segment, which receives OB fibers. In many primate brains, a cortical organization has not been clearly demonstrated in the olfactory trigone (Heimer *et al.*, 1976).

In the human brain, the OT occupies that part of the basal hemisphere wall just posterior to the point of attachment of the olfactory tracts. The anterior perforated space or substance is the region of the human brain that corresponds to the olfactory tubercle of lower mammals (Calleja, 1893; Beccari, 1911; Cajal, 1911, 1955; Crosby and Humphrey, 1941; Humphrey, 1967; Takimoto *et al.*, 1962; Brodal, 1981; Björklund and Lindvall, 1978). The OT, which belongs to the paleocortex, can be divided into rostral, middle, and caudal regions (Peele, 1977).

The anterolateral portion of the OT receives olfactory tract fibers from both medial and lateral divisions. The olfactory fibers arise from the ipsilateral OB and AON. Fibers from the amygdaloid complex enter the OT, and fibers from the dorsomedial nucleus of the thalamus (DM) reach the OT via the inferior thalamic peduncle. Fibers from the dopaminergic cell bodies of the midbrain (A_{10} region in the rat) enter the OT. Fibers from the locus coeruleus reach the OT (Solano-Flores *et al.*, 1980).

Fibers arising largely in the nonolfactory part of the OT pass to the septal area and into the stria medullaris thalami and the medial forebrain bundle. Other fibers from the OT go to the entorhinal area, subiculum, hippocampus, periamygdaloid areas, and basal and lateral amygdaloid nuclei, and to certain neocortical regions (Peele, 1977). Price (1984) proved that OT fibers constitute the majority of the projections to the nucleus gemini of the LHA.

3.2. Biochemical studies

The laminar organization, the well-separated major cell types, and the high levels of transmitter-related substances make the OT very attractive for neurochemical studies. In fact, concentrations of dopamine, GABA, and acetylcholine within the OT were found to be among the highest in the brain. Krieger *et al.* (1983) showed that the granule cells in the rat OT accumulate ^{3}H-γ-amino-butyric acid. Fallon *et al.* (1983) studied the islands of the Calleja complex of the rat basal forebrain and found histochemical evidence for a striatopallidal system.

Ungerstedt (1971) demonstrated the presence of dopamine-containing fibers throughout this region. D'Ambrosio *et al.* (1982) labeled dopamine receptors in the OT by [^{3}H] haloperidol. Krieger *et al.* (1977) showed

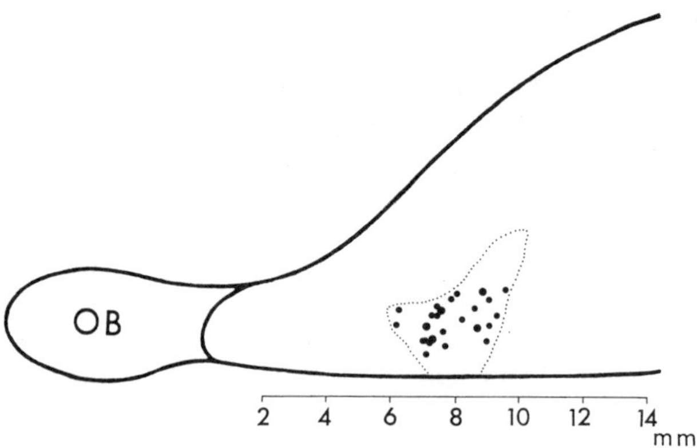

Fig. VIII-6. Sites of recorded OT neurons in the dorsal surface of the right hemisphere of the rabbit, as indicated by dots encircled by a dotted line. The abscissa show distances from the frontal pole. (Iino and Takagi, unpublished data)

that the adenylate cyclase was mainly present in the molecular layer of the OT, and Church *et al.* (1982) suggested that this enzyme occurs in the neurons. It was of interest that GABA and GAD had a different localization (Krieger and Heller, 1979). The locus ceruleus influences the different neural elements of OT (Solano-Flores *et al.*, 1980, 1981; Guevara-Aguilar *et al.*, 1982). Readers are referred to an excellent review written by Krieger (1981).

3.3. Physiological studies

In contrast to the recent advances in histological and biochemical studies, neurophysiological research on the OT has been lacking, and olfactory and limbic connections of the OT remain problematic.

Very probably Iino and Takagi (unpublished data) were the first investigators to examine responses of single OT neurons to odors. They inserted vertically a glass pipette microelectrode into the OT of rabbits (Fig. VIII-6). As stimuli, 13 odors were used, e.g., vapors of the eight chemicals (Table IX-2) and of five biologically significant odors, e.g., of urines and feces of the experiment and other rabbits and of a pellet (Fig. X-3). Neurons in the neocortical olfactory area in the orbitofrontal cortex of rabbits responded well to odors of animal products, but did so much less frequently to the odors of pure chemicals (Fig. X-3 and-6A; also compare with Fig. IX-15 B). Very different from the neocortical neurons, the OT ones responded relatively well not only to the above biological five odors, but also to the odors of pure chemicals. These findings proved that at least some of the OT neurons of a macrosmatic animal respond well to odorous stimuli.

3.4. Pathological studies

The above histological and neurochemical findings not only indicate that the OT is more than one of the olfactory structures, but also lend insight into the possible role of the basal forebrain in relationship to human behavioral disorders including schizophrenia (van Hoesen *et al.*, 1976), Huntington's and Parkinson's diseases (Klawans, 1973), and others (Krieger, 1981).

4. The Amygdala or the Amygdaloid Complex

The amygdala (AMG) is a large almond-like mass located above the tip of the inferior horn of the lateral ventricle within the rostral end of the temporal lobe. It is continuous medially with the cortex of the uncus and of the temporal lobe. The AMG is directly linked to the main and accessory OB, olfactory association areas, the thalamus, the hypothalamus, midbrain, hippocampus, neocortex, and more (Table 6 in Wysocki, 1979; Girgis, 1969; Motokizawa, 1985).

4.1. The structure of amygdaloid complex in humans

The AMG consists of a number of subdivisions which vary between investigators and depending upon animal species (Fox, 1940, 1943; Crosby and Humphrey, 1941, 1944; Brodal, 1947a; Koikegami, 1955, 1957; Winans and Scalia, 1970; Eleftherieu, 1972; Broadwell, 1975a, b; Scalia and Winans, 1975; Skeen and Hall, 1977; Krettek and Price, 1978b; Wysocki, 1979). Because of these subdivisions, the AMG is often called the amygdaloid complex.

The human amygdaloid complex has been subdivided into basolateral and corticomedial groups of nuclei, and anterior amygdaloid and corticoamygdaloid areas. In contrast to the increase in development of the basolateral group, the corticomedial group as a whole is relatively decreased in humans, and certain of its subdivisions are poorly differentiated (Peele, 1977). This may correspond with the fact that the vomeronasal system is lacking in humans (and higher primates and old world monkeys). In lower mammals (Chapter X, B), it has been histologically confirmed that the nuclei of the former group receive afferent fibers from the main OB, while nuclei of the latter receive efferent fibers from the accessory OB. Since the vomeronasal system has been known to play an essential role in sexual behavior and in feeding reactions (Winans and Scalia, 1970), and is very different in function from the main olfactory system, it may be wondered what kinds of roles the amygdaloid nuclei of the latter group play in olfaction and olfactory behavior in humans. It would be interesting to know whether in humans the main OB provides an afferent supply for the corticomedial amygdaloid field (which receives solely vomeronasal nerve fibers in lower mammals) or whether some other arrangements exist.

4.2. Olfactory afferents to the amygdaloid complex

The direct olfactory afferents lead through the LOT to the nucleus of the LOT, and to the cortical and medial amygdaloid nuclei. The indirect olfactory afferents from the PPC join the longitudinal association bundle and lead to the basal and lateral amygdaloid nuclei (Peele, 1977; Wysocki, 1979). Cellular responses to odors were recorded in the medial portion of monkey AMG (Tanabe *et al.*, 1975a).

4.3. Olfactory efferents from the amygdaloid complex

Many investigators have described projections from the amygdaloid complex to various subcortical and cortical regions, but disagreement has existed as to the identity of the paths leaving the complex. In the monkey, Nauta and his colleagues described two major amygdalofugal systems to subcortical loci—one diffuse, the second compact. The diffuse system connects with the substantia innominata (SI), lateral preoptic, and hypothalamic regions. One component of the system, bypassing the preoptic region, ends in the medial magnocellular division of the mediodorsal nucleus of the thalamus (MDmc). Other amygdalocortical components go to the rostral cingulate cortex and also to the superior prefrontal granular cortex.

Tanabe *et al.* (1973, 1975a, b) and Yarita *et al.* (1977, 1980) demonstrated the presence of two neocortical olfactory areas, in the orbitofrontal cortex (OFC) of the old world monkey (pp. 334 and 341). Porrino *et al.* (1981) proved direct and indirect pathways from the amygdala to the frontal lobe in old world monkeys. Then, Naito *et al.* (1984), using the HRP technique, showed the presence of a direct amygdalo-LPOF pathway and an indirect amygdalo-LPOF pathway via the SI. Since the amygdalo-MDmc pathway was shown above, and the MDmc-CPOF pathway was proved by Naito *et al.* (1984), it follows that the amygdalo-CPOF olfactory pathway via the MDmc exists. In this way olfactory information is sent to the two olfactory areas in the orbitofrontal cortex.

The second compact projection system is the stria terminalis. Although it has been shown that both the basolateral and corticomedial nuclei contribute to the stria, the further specific origin of the stria in the amygdaloid complex has not been agreed upon. The stria contains not only amygdalofugal fibers but also fibers coming to the amygdaloid complex.

Krettek and Price (1978a) showed amygdaloid projections to the bed nucleus of the stria terminalis, lateral hypothalamus, and other subcortical structures within the basal forebrain and brainstem. It is very probable that these amygdalofugal pathways play important roles in olfaction. The details of the roles, however, remain still to be studied in the future.

4.4. Physiological studies

Electrophysiological studies of amygdalofugal system have confirmed the data obtained by anatomic methods and provided more findings. Two groups of fibers were shown to exist in the stria terminalis (with conduction velocities of 1.8–2.3 m/sec and 0.6–0.75 m/sec) (De Molina and Sanchez, 1967). Neuronal activity was studied in three amygdaloid projection areas (anterior hypothalamic, and lateral and ventromedial hypothalamic areas) (De Molina and Marcos, 1967). Since the efferent fibers from the CPOF to the mediodorsal nucleus (MD) was found by Yarita *et al.* (1980), the reciprocal circuitry between MD and CPOF may involve modulation of olfactory information.

Machne and Segundo (1956) may be the first investigators who recorded unitary responses to odors and other afferent stimuli in the amygdaloid complex. They showed in the cat that olfactory and other sensory stimuli act upon single-cell discharges in the anterior, central, lateral, and basal nuclei of the amygdaloid complex. Their finding of these unitary responses shows that olfactory stimuli exert influences upon numerous somatic and visceral functions. Tanabe *et al.* (1975a) recorded single-cell responses to odors in the medial portion of the amygdala (MA) in the unanesthetized old world monkey (Chapter IX, 3; Figs. IX-15B and -18). Although the precise sites of neurons recorded were not located in the MA, their response patterns were found to be very similar to those of neurons in the PPC. Motokizawa *et al.* (1982) recorded neuronal responses in the anesthetized cat to electrical stimulation of the LOT and PPC. The responses were found in the deep amygdaloid nuclei, namely in the lateral, basolateral, basomedial and central amygdaloid nuclei.

4.5. Clinical studies

Chitanondh (1966) reported seven patients with olfactory seizures, hallucinations, or auras. Andy *et al.* (1975) reported three cases with complaints of olfactory auras. After amygdalotomy, there was no clinical recurrence of the olfactory disturbances in these patients. On the basis of these reports, Andy *et al.* (1975) took the view that the amygdaloid nucleus is essential for elaboration of an olfactory aura and for identifying odors (Peele, 1977; Brodal, 1981).

5. The Entorhinal Area

5.1. Anatomical studies

The posteriormost part of the PC is called the entorhinal area (ER); it was designated area 28 by Brodmann. The ER in humans is the largest component of the olfactory cortex (OC), occupying the posterior part of the ambient gyrus and the major portion of the hippocampal gyrus anteriorly (Peele, 1977). It is common to distinguish a medial and a lateral part (areas 28a and 28b; Blackstad, 1956; Haug, 1976; Krettek and Price, 1977c; Brodal, 1981).

The ER is composed of six laminas: lamina I, plexiform layer; II, layer of star cells; III, layer of superficial pyramids and, in the deeper portion, a dense plexus; IV, layer of deep pyramids; V, layer of small pyramids with recurrent axons; VI, layer of polymorph cells. All the specific afferents end in the external lamina (I, II, III), and the same cells there receive both these afferents and recurrent collaterals from the axons arising in the internal lamina (IV, V, VI).

The lateral part of the ER is the main recipient of the fibers from the OB and of those from the PPC and periamygdaloid cortex (e.g., Powell *et al.*, 1965; Price, 1973; Broadwell, 1975a; Scalia and Winans, 1975; Krettek and Price, 1977c; Skeen and Hall, 1977). There are fibers from neocortical temporal areas (Van Hoesen and Pandya, 1975a), from the prefrontal granular cortex (Van Hoesen *et al.*, 1975; Leichnetz and Astruc, 1976), from the amygdala (Krettek and Price, 1977c; De Olmos, 1972), from the medial septal nucleus, from the dorsal raphe nucleus, and from the nucleus locus coeruleus (Segal, 1977; Beckstead, 1978). Thus, the lateral part of the ER was supposed to be the seat of convergence and integration of impulses of olfactory origin with others.

The ER sends many fibers to the dentate gyrus and the hippocampus (HPC) (Van Hoesen and Pandya, 1975b), to the frontal lobe by way of the uncinate fasciculus, to neighboring cortical regions, including areas 21, 22, 36, 37, and 38, and also to midbrain tegumental areas (Peele, 1977).

5.2. Physiological studies

Studies on the distribution of evoked potentials elicited by odor application and electrical stimulation of the OB corresponded well with the anatomical findings (Dennis and Kerr, 1975). Resection of the ER bilaterally from the marsupial phalanger and from the baboon resulted in a reduction of aggressive behavior in response to provocation (Peele, 1977). From these observations, however, it would be difficult to presume any olfactory function which the ER may have.

5.3. Tanabe et al. *'s experiments on old world monkeys (1975c)*

Electrical stimulation of the ER elicited evoked potentials at various sites of OFC (Fig. VIII-7B, c-e and g-j; LPOF, see p. 334). ER stimulation, however, never elicited evoked potentials in areas 10 and 11, in which large-amplitude potentials were evoked by stimulation of the thalamic MD nucleus.

When the recording electrode was moved from the anterior end backward to the posterior end of the PC, electrical stimulation of the OB

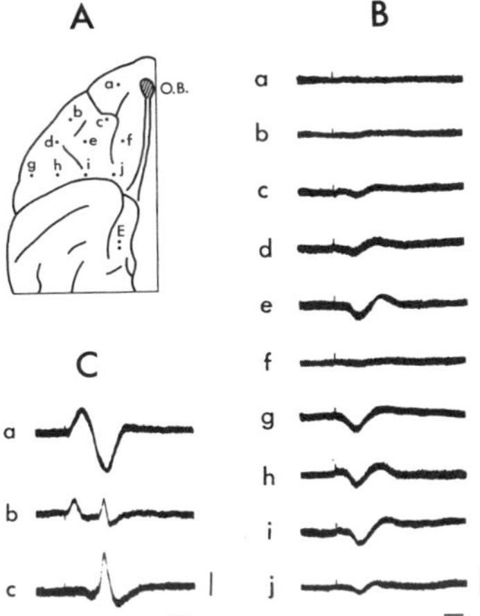

Fig. VIII-7. Evoked potentials in the LPOF due to stimulation of the ER in A and B. A: Records obtained at sites designated by letters on the orbitofrontal cortex in A. Evoked potentials are found from c to e and g to j. The site of stimulation is indicated by E. Calibrations are 20 msec and 30 μV. C: Evoked potentials in the anterior PC and the ER cortex due to stimulation of the OB. C, a: evoked potential recorded at AP. C, b: evoked potentials in the middle section of the PC C, c: Evoked potential in the ER. Calibrations are 20 msec and 30 μV. (from Tanabe *et al.*, 1975b)

elicited single-peak potentials with latencies of 3.5–7.5 msec initially (Fig. VIII-7 C, a). Such potentials continued to appear until the electrode was moved backward to a certain intermediate point, where the potential suddenly became "double-peaked" (Fig. VIII-7 C, b). When the recording electrode was moved farther backward and entered the posterior portion of the PC, the first peak disappeared and the second one remained (Fig. VIII-7 C, c). The latency of the second peak was as long as 25–40 msec. The recording sites at which only the second peak was obtained corresponded with the ER. The long latencies of the second peak made it quite clear that it was not elicited by direct fibers from the OB to the ER. This electrophysiological experiment does not necessarily negate the presence of direct OB fibers to the ER, although it would indicate that, if any, only a small number of such direct fibers exist.

When these latencies were compared with those of the evoked potentials in the LPOF (24.1 msec), the former latencies were found to be longer than the latter. In addition, the potentials evoked in the LPOF by OB stimulation countinued to appear even when the ER was ablated. Thus, it was made clear that the ER is not situated as a relay in the olfactory pathway from the OB to the LPOF.

Evoked potentials were recorded in the ER when only Walker's areas 12 and 13 were stimulated in the OFC. Then, a broad extent of the OFC, including areas 12 and 13, were aspirated. In spite of this ablation, an evoked potential in the ER elicited by OB stimulation survived (Fig. IX-8 C, c).

From these results it was concluded that reciprocal connections exist between the ER and Walker's areas 12 and 13, including the LPOF, and furthermore that the afferent pathway from the OB to the LPOF does not involve the ER, whereas impulse conduction from the OB to the ER is independent of the LPOF.

5.4. Connection of the LPOF with the ER

It is well known that the ER projects heavily to the HPC and the dentate gyrus. The anatomical studies of Van Hoesen *et al.* (1972) indicated three cortical areas as direct sources of afferents to the ER. In addition to afferents from the PPC and ventral parts of the temporal neocortex, a third afferent system was shown to originate in Walker's areas 12 and 13. Accordingly, Tanabe *et al.*'s experiments showed that stimulation of areas 12 and 13 evoked potentials in the ER, whereas stimulation of the other areas in the OFC did not.

The findings of Van Hoesen *et al.* (1972) also indicated that the ER is a final cortical link in the conduction routes from the sensory systems of the neocortex to the hippocampus and the dentate gyrus of the limbic system. Since the LPOF was proven to be an olfactory area (Tanabe *et al.*, 1975b, c), the ER may be a relay area which conveys information from a high-order olfactory area to the limbic system (Adey *et al.*, 1956). However, since the present experiments indicate that the ER is not involved in the main and specific olfactory pathway, it may be said that the ER does not play a primary role in olfaction.

C. The Diencephalic Olfactory Areas

1. The Thalamus

It has been established that the thalamus has various sensory relay nuclei. Such relay nuclei have also been sought in the olfactory nervous system. Powell *et al.* (1965) proved olfactory inputs to both the "mediodorsal" and the "medioventral" thalamic nuclei (MD and MV respectively) via the PC. Of these two projections, the one to the MD, has already been studied extensively anatomically and electrophysiologically.

1.1. Studies on the MD in old world monkeys

Anatomical investigations have disclosed that the MD of the monkey is composed of at least the following three cytoarchitectural subdivisions (Fig. IX-4 A; for reference see Yarita *et al.*, 1980c);

1. A pars magnocellularis (MDmc) is situated in the most medial portion of the MD, and receives afferent fibers from the PPC and the prefrontal cortex.

2. A pars multiformis (MDmf) is situated in the most lateral portion of the nucleus and has a relationship with vision (e.g., Scollo-Lavizzari and Akert, 1963).

3. A pars parvocellularis (MDpc) occupies a large intermediate region of the MD nucleus and projects to the dorsolateral convexity of the prefrontal cortex (area 9; Akert, 1964). This area may be related to the somatic sensorium (Benjamin and Jackson, 1974).

Among the above areas, Yarita *et al.* (1978, 1980) demonstrated in the old world monkey that olfactory signals are sent to the MDmc, supporting former physiological findings on other animals (Komisaruk and Beyer, 1972; Wedgwood, 1974; Jackson and Benjamin, 1974; Motokizawa, 1974a; Benjamin and Jackson, 1974 and a histological finding (Powell *et al.*, 1965).

Precise intracellular research on the activity of the MDmc neurons was first performed by Yarita *et al.* (1978, 1980; Fig. VIII-8). OB stimulation elicited a spike potential with a latency of about 35 msec and CPOF stimulation elicited a spike potential in the same neuron with a latency of about 2.3 msec and without an EPSP (Fig. VIII-8 A and B). The latter potential was concluded to be an antidromically elicited spike potential. It was immediately followed by a hyperpolarization which was later concluded to be an IPSP. Since it is well known that the OFC also projects efferent fibers to the MDmc (Nauta, 1971, 1972), the hyperpolarization could be due to stimulation of some corticothalamic inhibitory fibers, or of Renshaw type neurons, or both.

On these corticothalamic inhibitory fibers, repetitive stimulation of the CPOF with frequencies of 13, 40, and 55 Hz provided very important findings as shown in 1, 2, and 3 of Fig. VIII–8 D, In D_1, antidromic spike potentials faithfully followed the repetitive stimuli of low frequency (13 Hz) as indicated by the arrows. In D_2, however, repetitive stimuli of a higher frequency (40 Hz) rendered the membrane more hyperpolarized and often interrupted the appearance of antidromic spike potentials with the exception of the case when an abrupt depolarization occurred. In D_3, the membrane became more hyperpolarized and antidromic spike potentials never appeared when repetitive stimuli with a higher frequency (55 Hz) was applied. In a study on a transthalamic olfac-

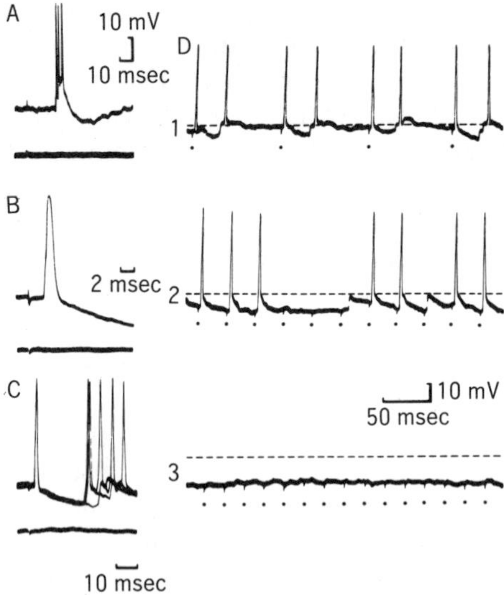

Fig. VIII-8. Intracellularly recorded potentials evoked by OB (A) and CPOF (B-D) stimulation. The lower traces in A–C are extracellular field potentials just outside the impaled cell. Three traces are superimposed in C. In D, repetitive stimulations of 13 (1), 40 (2), and 55 (3) HZ were applied to the CPOF. Times of stimulations are indicated by dots. The dashed lines indicated the resting membrane potentials. (from Yarita and Takagi, 1978)

tory pathway to the CPOF (Chapter IX), Yarita *et al.* (1980) found that 38 MDmc neurons of the 58 examined did not follow high frequency stimulation. The finding can now be explained by this hyperpolarization mechanism, which may be produced by the activation of concurrent inhibitory pathways to the MDmc neurons.

Another important finding in Fig. VIII-8 D_2 (Yarita *et al.*, 1978) was that a depolarization with a steep rise and often with superimposed spike potentials appeared abruptly in the middle of the hyperpolarization. This was presumed to be an EPSP elicited by simultaneous stimulation of some other excitatory efferent fibers from the CPOF to the MDmc.

Yarita *et al.* (1980) found a broad range of latencies (9–400 msec) in the responses of MDmc neurons to OB stimulation. Such a broad range of latencies had also been found in MD neurons of squirrel monkeys (Benjamin and Jackson, 1974), and in the PPC neurons (Biedenbach and Stevens, 1969a; Benjamin and Jackson, 1974; Mori *et al.*, 1978; Satou *et al.*, 1982a, b, 1983a, b, c). These broad ranges of latencies may reflect activation of a direct pathway as well as a number of indirect pathways of variable length and complexity from the OB to the thalamus. In fact, these pathways have been studied by many investigators in several animals (e.g., Benjamin and Jackson, 1974; Fox, 1949; Nauta, 1971, 1972). Tanabe *et al.* (1975b) and Yarita *et al.* (1980) did not study the indirect pathways in particular. In the light of numerous comparative studies on the secondary projections so far performed (e.g., Skeen and Hall, 1977; Turner *et al.*, 1978; for more information see Yarita *et al.*, 1980) it was thought quite probable that the MDmc in the old world monkey receives not only direct afferents from the PPC, but also indirect ones from the other olfactory-related structures found in macrosmatic animals.

1.2. Studies on the MD in lower mammals

Among the thalamic nuclei, Heimer (1972) and Benjamin and Jackson (1974) have established anatomically and neurophysiologically a prominent olfactory projection to the central segment (Krettek and Price, 1977b) or internal segment (Leonard, 1969) of the MD. Price and Slotnick (1983) demonstrated in the rat thalamus dual olfactory representation in the central segment of the MD and in the submedial nucleus (Fig. X-10).

Imamura *et al.* (1984a) studied odor response characteristics of rabbit MD neurons extracellularly by applying electric shocks to the LOT and to the olfactory projection area (OPA) in the neocortex. Of the 87 LOT-responsive MD neurons, 48 MD neurons responded well not only to eight chemical odors (Table IX-2), but also to five biologically significant odors (Fig. X-3). The number of odors that elicited neuronal responses in the MD were not much different from those in the OT.

The details of olfactory pathways from the OB to the MD have been studied by many investigators (Powell *et al.*, 1965; Sanders-Woudstra, 1971; Leonard, 1972; Lammers and Lohman, 1974; Price and Slotnick, 1983; Price, 1985). As described above, the central or internal segment of the MD receives afferent fibers from the PPC directly or indirectly through a part of claustrum, from the olfactory tubercle (e.g., Lammers and Lohman, 1974), from the AON (Ferrer, 1969), from the AMG (e.g., De Molina and Ispizua, 1972; Krettek and Price, 1977b), and from the septum (Guillery, 1959). However, a prominent olfactory input to the medial segment of the MD was not shown from the basal or basolateral nucleus of the AMG (Krettek and Price, 1977b; Porrino *et al.*, 1981; Benjamin *et al.*, 1982), although Motokizawa (1974a) asserted otherwise.

In accordance with these histological findings, many physiologists recorded responses of single MD cells to electrical stimulation of the OB or to odorous stimulation (Benjamin and Jackson, 1974; Jackson and Benjamin, 1974; Komisaruk and Beyer, 1972; Motokizawa, 1974a; Tanabe *et al.*, 1975b).

On the other hand, Krettek and Price (1978a) and Wiegand and Price (1980) proved that direct projections from the OC to the neocortical areas in the dorsal bank of the rhinal sulcus are interconnected with the MD and submedial nuclei. However, projection of olfactory impulses to the thalamic nuclei from the neocortical areas could be negated, because of the short latency effects: Responses to OB stimulation were recorded in the neocortex with about the same latencies (8–20 msec) as the response latencies obtained in the thalamus (Wiegand and Price, 1980).

Siegel *et al.* (1977) and Benjamin *et al.* (1982) identified a widespread population of cells in the deep polymorphic layer of the OC which projects to the thalamus. Price (1977) and Luskin and Price (1983) showed that neurons in the superficial layers of the OC project to the region of these deep cells, suggesting that they mediate the short-latency responses elicited in the MD and submedial nuclei by OB stimulation. Price and Slotnick (1983) found such cells not only throughout the deep layer of the posterior PC (the ventral endopyriform nucleus), but also in the polymorphic zone of the OT and in a zone deep to the periamygdaloid cortex. Benjamin *et al.* (1982), however, showed considerable differences among animal species in the origin of olfactory afferents to the thalamus.

1.3. Studies on other thalamic olfactory areas

The thalamic projection to the MV, described by Powell *et al.* (1965), does not appear to have been examined further.

Giachetti and MacLeod (1977) reported electrophysiological evidence for an olfactory input to the "ventroposterior medial" nucleus of the thalamus in the rat. But this nucleus appears to be well caudal to that described by Powell *et al.* (1965).

Komisaruk and Beyer (1972) reported responses to olfactory stimuli in both dorsal and ventral parts of the medial thalamus, but they were not precisely localized to specific nuclei.

Imamura *et al.* (1980) recorded electrical activity of 26 cells in the ventrobasal complex of the thalamus in the rabbit. Among them, 10 cells (39 %) responded to stimulation of the LOT, four cells (15%) to stimulation of the tympanic chordal nerve (TCN) or the tongue, and 12 cells (46%) to stimulation of both the LOT and TCN or tongue. In addition, they classified six patterns in the responses of these neurons.

To examine recording sites, they injected potamine sky blue through recording microelectrodes and found small blue spots in the nucleus ventralis posteromedialis (VPM). Since Imamura *et al.* (1984a) showed responses to odors of MD neurons, the above report by Imamura *et al.* (1980) indicated the presence of dual thalamic representations at the MD and VPM in the rabbit.

When Imamura *et al.* (1980) examined recording sites, however, they found a third group of neurons which responded to both LOT and TCN stimulation at a far lateral area to the VPM. Neuronal responses to odors at this area were entirely unexpected. Probable olfactory functions and fiber projections in this area and the VPM shown above remain to be studied.

Krettek and Price (1977b), Herkenham (1979), and Craig *et al.* (1982) found that a more rostral part of the ventromedial thalamic complex, the submedial nucleus, is interconnected with a cortical area situated between the PC and the cortical projection field of the MD. In the submedial nucleus, the olfactory input is concentrated in the ventral part, and the anterior part may also be related to olfaction. Later, the MD and submedial nuclei were proved to receive olfactory information in the rat (Price and Slotnick, 1983).

In summary, possibilities of multiple olfactory representation in the thalamus were thus indicated in different animals. Elucidation of whether or not any of those cells are olfactory in nature, and of what kind of role they would play in olfaction, remain important subjects for future research.

1.4. Thalamic pathways to the frontal cortex

In the old world monkey, Yarita *et al.* (1977, 1980) proved an olfactory pathway from the MDmc to the CPOF and concluded that this pathway plays an important role in the integration of odorous information (Chapter IX).

Leonard (1972), Krettek and Price (1977a), Wiegand and Price (1980), and others showed that the central segment of the MD projects to several prefrontal cortical areas including the lateral orbital cortex. Recently, Price and Slotnick (1983) histologically showed that afferent fibers from two thalamic nuclei (the MD and MV) project to at least three cytoarchitectonically distinct neocortical areas. They assumed that these neocortical areas are related to olfaction (Fig. X-10) and have provided a new subject for future research (Chapter X).

1.5. Olfactory behavior changes due to MD lesions

Slotnick and Kaneko (1981) found that lesions of the MD in the rat produced severe deficits in odor reversal learning. In the hamster, lesions of the MD or of the cortex in the dorsal bank of the rhinal sulcus produced changes in odor preference and in male sexual behavior (Eichenbaum *et al.*, 1980; Sapolsky and Eichenbaum, 1980). Consequently, it is probable that the thalamic and neocortical structures related to olfaction may not be involved in sensory processing as such, but in a variety of higher level functions involving olfactory-guided behavior in rats and other macrosmatic mammals.

2. The Hypothalamus

The hypothalamus has been studied extensively by numerous investigators (Morgane and Panksepp, 1979). In this section, research data related only to olfaction are described.

2.1. Anatomical studies

The hypothalamus is situated ventral to the thalamus and forms the floor and lower walls of the third ventricle. It is composed of many nuclei which are grouped and classified in various ways. A division of the hypothalamus into the three longitudinal zones—lateral, medial, and periventricular—is easily made in some species, but in the human only a longitudinal division into medial and lateral zones is better appreciated (Peele, 1977). In the listing of preoptic and hypothalamic nuclei, these longitudinal zones have been subdivided into the following four regions:

(1) Preoptic area (POA)
(2) Supraoptic region
    a. Anterior hypothalamic area
    b. Others
(3) Infundibular region
    a. Ventromedial hypothalamic nucleus (VMH)
    b. Lateral hypothalamic area (LHA) and tuberal nuclei
    c. Others
(4) Mamillary region

Among these nuclei, the lateral hypothalamic area (LHA) and ventromedial hypothalamic nucleus (VMH) seem to have closer relations with olfaction than the others. The preoptic area (POA) is not a part of the hypothalamus developmentally, but it was included in this classification because the area is functionally related to the hypothalamus. However, relationship between this area and olfaction has not been made clear.

There is much evidence that LHA neurons receive olfactory inputs. Neuroanatomists have demonstrated that cells projecting to LHA neurons are found in several regions of the basal forebrain such as the AON, OT, AMG, and PC (Ban and Zyo, 1962; Broadwell, 1975a, b; Heimer, 1972; Heimer and Nauta, 1969; Millhouse, 1969; Powell et al., 1965; Scott and Chafin, 1975; Scott and Leonard, 1971; Zyo et al., 1963). Recently, Price et al. (1984), injecting wheat germ agglutinin conjugated to horseradish peroxidase (WGA*HRP) into the LHA, demonstrated labeled cells in the above four regions. Corresponding to these facts, olfactory input to the LHA was electrophysiologically proven by many workers (pp. 324–325).

Heimer (1972) negated the presence of a direct monosynaptic connection between the PPC and hypothalamus, and found fibers from rostral olfactory areas, only traversing the preoptic area, terminate either in the nuclei gemini or in the MD of the thalamus. This supposition was later confirmed by Price *et al.* (1984) who electrophysiologically obtained short-latency responses consistently in the region of the nuclei gemini to electrical stimulation of the OB.

Price and Slotnick (1983) noted that the cells in the OT and PC that project to the LHA are in the same layers as those which project to the MD, and proved that at least some of the cells deep to the OT and PC have bifurcating axons which terminate in both the LHA and MD. The fact that the same cells convey the same messages simultaneously to the LHA and MD would be very important not only in olfactory research but also in other studies for the elucidation of experimental results in the future.

The periamygdaloid cortex and the amygdaloid complex send the following two fiber bundles to the hypothalamus:

(i) The stria terminalis (ST)

Some fibers of one (precommissural) component of the ST reach the VMH, and some other fibers of the same component may also terminate in the neighboring MD and LHA (Heimer and Nauta, 1969). These fibers reach the VMH from the anterior parts of the cortical and medial amygdaloid nuclei, which receive a direct projection from the main OB (Lohman, 1963; Heimer, 1968; Scalia, 1966; De Olmos, 1972).

The fibers of one other (postcommissural) component of the ST terminate roughly in the anterior hypothalamic nucleus (Leonard and Scott, 1971; De Olmos, 1972). Via this component, the neocortex may have access to the medial hypothalamus.

(ii) The ventral amygdalofugal pathway

The fibers of the OC may also reach the hypothalamus through this amygdalofugal pathway. De Olmos (1972) indicated that the central amygdaloid nucleus may also send long fibers to the LHA. It is not known, however, whether these long fibers actually terminate in the LHA or are fibers of passage en route to the the MD of the thalamus.

Since Nauta's (1962) study in the monkey, it has commonly been believed that a direct pathway exists from the medial division of the MD to the hypothalamus. Siegel *et al.* (1973), however, provided evidence that such a direct connection between the MD and the hypothalamus does not exist in all mammals, but only in the monkey (Nauta, 1964); in other mammals the connection is multineuronal, interrupted by at least two synapses.

Another connection from the MD to the hypothalamus is formed by the thalamocortical and corticohypothalamic pathways. The intermediary in this circuit in the monkey is the orbital surface of the prefrontal cortex; in the rat, it is the sulcal cortex lying along the dorsal bank of the rhinal fissure. In both species these cortical areas receive projections from the medial division of the MD, and are connected by efferent fibers to the lateral preoptico-hypothalamic region (Freeman and Watts, 1947; Akert, 1964; Nauta, 1962; Leonard, 1969).

The hypothalamo-hypophyseal fiber system is the best-known efferent pathway of the hypothalamus. Among them, projecting fibers from the mammillary nuclei to the anterior nuclei of the thalamus is called the mammillothalamic tract of Vicq d'Azyr. This tract is a part of a composite pathway, the so-called Papez circuit (Papez, 1937), which links the hippocampal formation, the mammillary nuclei of the hypothalamus, the anterior nuclei of the thalamus, the cingulate gyrus, and again the hippocampal formation. This famous circuit is presumed to play some important roles in olfaction, but those roles remain to be determined.

The medial forebrain bundle (MFB) is at present the best-established major efferent pathway which connects the lateral preoptic-hypothalamic zone with the septum and the mesencephalon in rats and cats (Guillery, 1957; Nauta, 1958; Wolf and Sutin, 1966; Chi and Flynn, 1971).

The LHA was also reported to emit fibers to the amygdaloid complex. These fibers, together with collaterals of the MFB fibers, course toward the thalamus (Nauta, 1958; Wakefield as cited by Hall, 1972; Milhouse, 1969).

2.2. Physiological studies

Neuronal responses to odorous stimulation of the olfactory epithelium as well as to electrical stimulation of the OB were recorded in the LHA (Scott and Pfaffmann, 1967, 1972; Komisaruk and Beyer, 1972). Thus, olfactory input to the LHA was demonstrated by many investigators in rats, mice, and rabbits (Barraclough and Gross, 1963; Pfaff and Gregory, 1971; Pfaff and Pfaffmann, 1969a, b; Scott and Pfaff, 1970).

In the LHA of unanesthetized old world monkeys, Tazawa *et al.* (1983, 1987) recorded spontaneous discharges of 287 neurons and found that only 22 neurons among them exhibited significant responses to no more than four of the eight odors (Table IX-2). The response profiles of the 22 neurons were shown as a matrix and a histogram (Fig. IX-16 B, and 18).

Kogure *et al.* (1981) extracellularly recorded responses of 103 neurons to eight odors (see Table IX-2) in the LHA of unanesthetized rabbits.

Among them, 90 units responded to one to eight odors (Fig. VIII-9). Neurons that responded to three odors were found to be most numerous (22%) and higher response probabilities were observed to some odors than to the others. 94% of the units recorded showed an excitatory type response to odors, while only 6% showed an inhibitory type response. These results suggested that the olfactory effect on the LHA is predominantly excitatory.

Compared with the response profiles of LHA neurons in old world monkeys (Figs. IX-16B and -18), it appeared that the LHA neurons in rabbits do not contribute much to discrimination of odors.

2.3. Behavioral studies

It is well known from our daily experience that our sensation of smell depends upon the condition of our body. For instance, we sense food odors differently depending upon whether we are hungry or not. When smokers catch colds, they often say that tobacco tastes unpleasant. When women become pregnant, their sense of smell undergoes a change. However, the mechanisms which produce such changes in sensation have not been clarified, and Kogure *et al.* (1980) felt that neurophysiological data which may be counterparts of these daily psychological experiences should be sought.

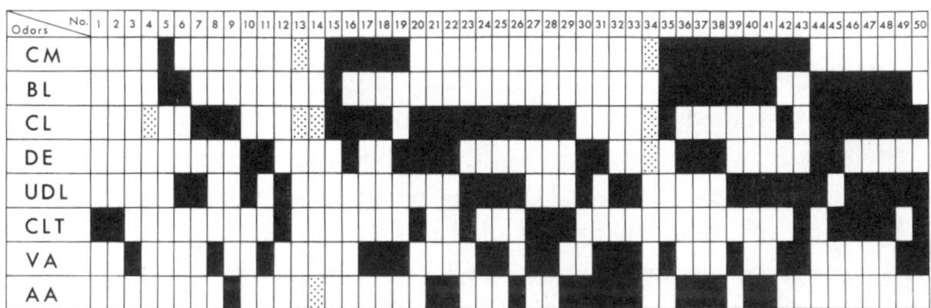

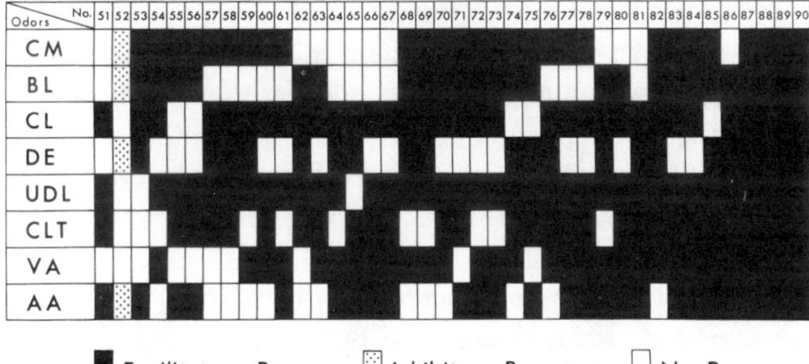

Fig. VIII-9. Response patterns of 90 LH neurons to eight odors in the rabbit. (from Kogure *et al.*, 1980)

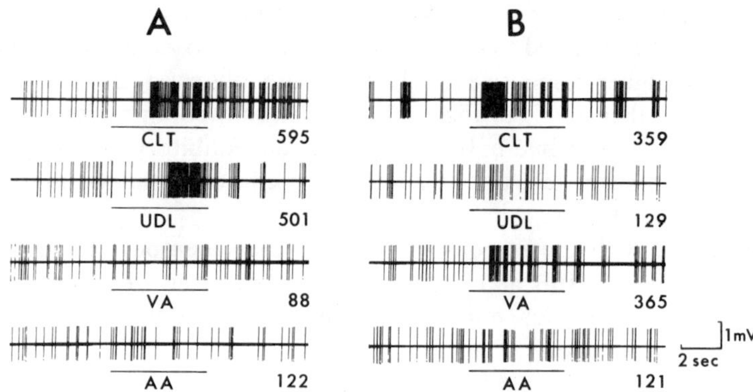

Fig. VIII-10. Effect of stomach distension on the response pattern of an LH neuron to odors. A: Response to 4 kinds of odors before stomach distension (A) and during (B) distension. (from Kogure *et al.*, 1980)

Thus, the influence of stomach conditions upon neuronal responses to odors were examined in the LHA of unanesthetized rabbits. After response patterns of all recorded neurons to eight odors (Table IX-2) were studied, the rabbits' stomachs were inflated and subsequent changes in their response patterns were recorded. At first, the inflation itself was found to increase the number of spontaneous discharges. However, the impulses first increased by inflation gradually decreased in number during sustained inflation and then increased again when the intragastric balloon was deflated. In addition, the kinds of odors that elicited neuronal discharges in the LHA also became different (Fig. VIII-10). For instance, a neuron that responded to CLT and UDL, but not to VA and AA, became responsive only to CLT and VA when the stomach was inflated. In addition, the number of odors that elicited neuronal responses before stomach distension in most cases changed with inflation.

It is still not entirely clear where in the brain the sensation of hunger or food satiety occurs. However, the finding that neuronal responses to odors in the feeding center (LHA) differ depending upon the condition of the stomach strongly suggests that odors which were appetite stimulating in the state of hunger become less and less attractive as the stomach is distended by food intake.

The hypothalamic areas have been commonly regarded as integrators of sensory and visceral cues contributing to food intake behavior. Kogure *et al.*'s study has proven that the idea is true.

Regarding the mechanism of Kogure *et al.*'s finding in LHA, anatomical findings on the olfactory centrifugal pathway must be considered (Broadwell and Jacobowitz, 1976; Price and Powell, 1970c 1971). This pathway has also been studied electrophysiologically by many investigators (Chapter VII, p. 290). Pager *et al.* (1972), recording multiunit discharges in the mitral cell layer of the OB, showed that a food odor elicited "positive (enhanced)" responses in unanesthetized food-deprived rats, whereas the same odor elicited "significantly different negative (inhibited)" responses in food-satiated rats. Pager (1974) also showed that the specific and selective activation elicited by a food odor in the mitral cell layer of the OBs of hungry rats disappeared when the anterior limb of the AC was sectioned on the recording side. These and Kogure *et al.*'s data strongly suggest that distension of the stomach influences mitral cell activity in the OB through the anterior limb of the AC, and modifies mitral cell responses to odors. Thus, Kogure *et al.*'s finding that inflation and deflation of the stomach modulated neural responses to eight odor s in the LHA indicates the possibility that olfactory information that has already been modulated in the OB is sent to the LHA directly or indirectly.

3. *Substantia Innominata*

The substantia innominata (SI) of Reichert is situated lateral to the LHA and beneath the lenticular nucleus and overlies part of the OT and the posterior orbital cortex. It is thought to receive fibers from the MD of the thalamus, via the inferior thalamic peduncle. Within the SI is a longitudinal collection of large polygonal cells with long dendrites, constituting the basal ganglion. The basal ganglion projects to the temporal lobe. Fiber systems traversing the SI include interconnections between the amygdaloid complex and the hypothalamus, and between the former and the MD of the thalamus. Fibers from the posterior orbital cortex and anterior cingulate gyrus also pass through it (Peele, 1977).

When Tanabe *et al.* (1973, 1975a, b, c) discovered an olfactory area in the LPOF of the old world monkey, they tried to find a pathway to the LPOF. Observing an evoked potential in the LPOF in response to electric stimulation of the OB, they first destroyed the thalamus bit by bit. After destruction of the entire thalmus in five monkeys, the evoked potential still remained. When, however, the investigators began to destroy a portion under the thalamus, the potential suddenly disappeared. The hypothalmus was at first suspected to be a relay area through which olfactory fibers pass to the LPOF from the PPC and/or the MA. In order to locate precisely the area where fibers to the LPOF are relayed, Naito *et al.* (1984) injected HRP into the LPOF, and searched for labeled cells in the LHA and its adjacent areas. A few labeled cells (seven in all) were found in the LHA, but many more labeled cells were found in the SI (Fig. IX-7). Thus, it was considered very probable that the SI is the main relay nucleus that receives olfactory afferents from the OB and sends efferent fibers to the LPOF, but that the LHA does not. However, electrophysiological studies on SI neurons have never been attempted. Consequently, the presence of relay cells in the SI awaits electrophysiological substantiation. At the present stage of research, it would be difficult to decide whether or not the SI contain fibers of passage en route to the LPOF from the PPC and MA.

4. The Septum (SPT)
4.1. Anatomical studies
The septum is the basal region of the hemisphere immediately rostral to the lamina terminalis. Ventral to it is the anterior perforated substance, superior to it is the subcallosal gyrus, and dorsal to it is the fornix. The septum is sometimes referred to as the parolfactory area. The septal nuclei are well developed in lower mammals (Stephan and Andy, 1962). In man, the upper part of the septal region forms the thin septum pellucidum, free of nerve cells; the lower part (the "precommissural" septum) is usually subdivided into lateral and medial septal nuclei (Stephan, 1975).

Some of the fibers in the descending medial forebrain bundle (MFB) from the septum terminate in the lateral preoptico-hypothalamic zone, and this is reciprocated by ascending MFB fibers originating in the mesencephalon and the LHA and ending, respectively, in the medial and lateral septal nuclei (Guillery, 1957; Wolf and Sutin, 1966). It appears that the lateral septal nucleus mainly receives afferents and that the medial septal nucleus sends most of the efferents (Brodal, 1981). Since the septal region has reciprocal connections with the hippocampal formation (e.g., Swanson and Cowan, 1977; Meibach and Siegel, 1977; and others; see Brodal, 1981), it represents a structural relay in the pathways that connect the limbic forebrain area with the limbic midbrain area and vice versa. The relationship of most of these fibers with olfaction, however, remains to be studied.

The axons of mitral cells in the OB form the LOT and the MOT, which send fibers to higher olfactory areas. According to Peele (1977), the MOT reaches the septum directly only in humans. The presence or absence of this pathway has not been studied in the old world monkey. In lower mammals, fiber connections from the OB to the septum and probably to the nucleus accumbens via the LOT and the OT have been found histologically (Huber et al., 1962; Peele, 1977). Again, such findings have yet to be made in the old world monkey.

Tazawa et al. (1983, 1987) studied in the old world monkey the septum and nucleus accumbens electrophysiologically and histologically as a composite body (Fig. IX-9). They recorded neuronal responses to odors in the septum (Fig. IX-17) and demonstrated olfactory input at least to the septum. They also proved septal projection to the LHA. Thus, an olfactory pathway through the septum to the LHA was demonstrated in the old world monkey (Chapter IX, pp.353–355).

4.2. Behavioral studies

In recent years much work has been devoted to the septal nuclei, since they seem to influence various behavioral patterns and autonomic function (De France, 1976).

Several functions have been found to be influenced by stimulation or damage to the septum. The name "septal syndrome" has been invented as a common denominater for the changes seen after destruction of the septal nuclei. In general, this syndrome is characterized by behavioral over-reaction to most environmental stimuli, in the form of sexual and reproductive behavior, feeding and drinking, and rage reactions and aggressive behavior. In view of the many interconnections between the septal nuclei, the hippocampus, the amygdala, and the hypothalamus, it is no wonder that damage to any one of them elicits a large number of rather similar symptoms. To elucidate olfactory functions which the septum may have, ablation and stimulation studies have been attempted (Grossman, 1976; Brodal, 1981). Several behavioral changes were observed. However, it was not possible to determine whether or not and to what degree those behavioral changes are related to olfaction.

D. The Hippocampus

This gross anatomical structure, which got its name from its resemblance to a seahorse, is sometimes called Ammon's horn. It is common to speak of a "hippocampal formation" or "hippocampal region" including, in addition to the hippocampus (HPC) itself, the dentate gyrus and the subiculum (Brodal, 1981).

1. Olfactory Input to the Hippocampus—Anatomical Studies

Among the afferent fibers to the HPC, projection from the entorhinal area (ER; area 28) is, at least quantitatively, most important. The medial and lateral perforant paths from the medial and lateral parts of the ER, respectively, have been found to supply the entire length of the HPC and the dentate gyrus. Correspondingly, the termination of these two fiber groups has a distinctly different laminar distribution (Hjorth-Simonsen, 1972; Stewart, 1976). Although the ER sends olfactory impulses to the HPC through these paths, as has been shown physiologically, the functional significance of the distribution of these two fibers in olfactory sensation remains unknown.

The septohippocampal connections, mostly from the medial nuclei of the septum, are presumed to be the second important input. Fibers from the hypothalamus (Segal and Landis, 1974; Pasquier and Reinoso-Suarez, 1976) are said to exert a strong inhibitory influence on the HPC (Segal, 1979). In addition, other fibers are known to lead from the anterior thalamic nucleus (Swanson, 1978) and from the raphe nuclei and the nucleus locus coeruleus (Segal and Landis, 1974; Pickel *et al.*, 1974; Moore and Halaris, 1975).

As efferents, it is generally agreed that most projections from the HPC pass through the fornix. Some of the fornix fibers descend anterior to the anterior commissure (precommissural fornix): others descend posterior to it (postcommissural fornix). Determination of the function of these fibers in olfaction also awaits future research.

2. Olfactory Influences on the Hippocampus——Physiological Studies

Histologically, it was proven that afferent fibers do not reach the HPC directly from the OB. Electrophysiological evidence, however, showed that the HPC receives olfactory information indirectly, particularly via the ER. Cragg (1960) recorded action potentials in the HPC following stimulation of the OB. That the entorhinal area markedly influences hippocampal activity was shown by Andersen *et al.* (1966). Yokota *et al.* (1970) obtained intracellular recordings from hippocampal neurons of conscious squirrel monkeys. In an attempt to identify hippocampal pyramids by antidromic volleys, fornix stimulation was applied and three prin cipal types of postsynaptic potentials were recorded: a) EPSPs with latencies ranging from 17 to 27 msec and amplitudes of 4.5–15 mV, associated with spike discharges; b) IPSPs resembling those elicited by fornix stimulation except for longer latencies ranging from 10–18 msec; and c) combined IPSP-EPSPs. In response to OB stimulation, EPSPs were obtained which showed variations in amplitude, rise time, and duration that were similar to those of septal-induced EPSPs in other neurons. Thus, responses with long latencies indicated the presence of olfactory input to the HPC through plural synapses.

3. The Hippocampus and the Sense of Smell

The HPC is generally included among the regions of the brain collectively referred to as the "rhinencephalon." For a long time it was thought that the HPC is related to olfactory function. Evidence to the contrary, however, was the fact that the size of the HPC has no relation to the development of the sense of smell in various animals. Anosmic whales, for example, have HPCs (Addison, 1915; Langworthy, 1932; Ries and Langworthy, 1937). Some human brains which lack OBs and LOTs were found to have well-developed HPCs. Microsmatic man has the largest HPC of all animals (Rose, 1927).

4. Animal Experiments and Clinical Observations

Decorticate cats suffered no marked reduction in olfactory capacity, and were still in possession of highly complex feeding reflexes (Dusser de Barrenne, 1933), even with rather extensive damage to the paleo- and archicortex (Bard and Rioch, 1937). It appeared that most of the olfactory reflexes are mediated through subcortical structures. According to Allen's (1940, 1941) studies, conditioned positive and negative reactions to agreeable and disagreeable smells were not cancelled after bilateral extirpation of the HPC, but the correct differential response was lost when the PC was also ablated, although the elementary reflexes persisted. The loss of correct response can be attributed to the evidence that the pyriform cortex is situated as an important relay station in the middle of the main olfactory pathway, and bilateral destruction of this cortex in practice deprives the main olfactory system of this principal function. The fact that elementary reflexes remained, however, may be ascribed to the function of the vomeronasal system.

In conscious human beings, stimulation of the OB was followed by olfactory sensation, whereas electrical stimulation of the HPC yielded no response at all (Penfield and Erickson, 1941). Olfactory sensations occurring in the so-called uncinate fits appeared to be associated with lesions of the uncus and the hippocampal gyrus, but not with lesions of the HPC. In patients with epileptic seizures, marked cell loss was frequently observed in the HPC, but there was no decisive evidence that these changes were responsible for the disturbances of smell which occasionally occurred in these patients (Meyer, 1957).

In conclusion, Brodal's view expressed in his 1947 review still seems to be valid: that any relation between the HPC and the sense of smell must be very remote (Brodal, 1947b; Allison, 1953b; Brodal, 1981).

IX. Search for Previously Undefined Olfactory Areas and Their Afferent Projections in the Neocortex of the Old World Monkey

When we embarked upon a series of studies on neural mechanisms of olfaction in our laboratory in 1971, we faced many important but unresolved questions regarding olfactory neural pathways.

We first sought to determine whether or not an olfactory area existed in the neocortex. Other sensory systems were known to have corresponding areas in the neocortex, and the neural pathways to these sensory areas had already been defined. In contrast, the projections of the olfactory system had only been characterized from the olfactory epithelium through the olfactory bulb (OB) to the olfactory cortex, a division of the paleocortex. A sensory pathway to the neocortical olfactory area had never been demonstrated. Although histological studies had shown that afferent fibers from the olfactory cortex projected to many areas of the brain, physiological investigations had not yet determined which specific fibers were olfactory in nature.

Our next goal was to establish whether an olfactory pathway passed through the thalamus. While transthalamic pathways had been identified in all other sensory systems, the presence of such a pathway in the olfactory system had not been demonstrated.

The third problem concerned the localizations of discriminative mechanisms within the olfactory nervous system. Although there had been several investigations of this question in the lower olfactory areas of the frog, the turtle, and the rat, no substantial progress had been made in solving this problem.

We will initially describe how we solved the first and second problems. The third problem will be discussed in the last half of this chapter.

In his famous book, *The Physiology of the Nervous System*, J.F. Fulton (1949) dedicated one chapter to olfaction and cited several studies, including a series of works by Allen. Using dogs as subjects in a conditioned-response paradigm Allen (1940, 1943a) discovered that ablation of a ventrolateral portion of the prefrontal cortex markedly impaired olfactory discrimination. In a follow-up study, Allen (1943b) applied electric shocks to the pyriform lobe in dogs and identified evoked potentials in the ventrolateral portion of the prefrontal cortex. Allen thus suggested that the ventrolateral portion of the prefrontal cortex was involved in olfactory processes.

Summarizing this work, Fulton stated: "Allen's studies represent the first systematic attempts to unravel the physiology of the olfactory system in higher animals, and it is earnestly hoped that this profitable line of investigation will be actively continued." However, since the publication of Allen's last papers in 1943, this problem had lain dormant and, to the best of our knowledge, we were the first workers to resume this line of research.

When I first delved into Allen's pioneering work, I immediately perceived gaps of knowledge that needed to be filled in. Remarkable as his efforts were, Allen's findings constituted no more than a few fragmentary pieces of the puzzle—hardly enough to form a conclusive picture. Furthermore, if the prefrontal cortex played an important role in olfactory processes, I realized that it would be more profitable to study a species having a far more developed prefrontal cortex than the dog. We therefore chose to study brain mechanisms of olfaction in crab-eating and Japanese monkeys. Thus, my long-cherished desire to study the neural mechanisms of monkey olfaction as a model of the human was completely fulfilled.

A. Location and Identification of an Olfactory Area in the Neocortex

1. Location and Identification of an Olfactory Area in the Lateroposterior Area of the Orbitofrontal Cortex

Tanabe *et al.* (1973, 1974, 1975a, b) applied pulses of electrical stimulation to the olfactory bulb (OB) of crab-eating monkeys (*Macaca irus*) anesthetized with nembutal and used a concentric electrode to search for evoked potentials throughout the prefrontal cortex. Evoked potentials with an average delay of 24.1 msec and amplitudes ranging from 30 to 110 μV were identified in the lateral and posterior area of the orbitofrontal cortex (LPOF) (Fig. IX-1 D).

Sites which displayed evoked potentials are shown with hatched lines in Fig. IX-1 B and C. They include the posterior region of Walker's (1940a) area 12, the lateroposterior portion of the area between the frontal and temporal lobes, and the opercular region of the frontal lobe. Evoked potentials were not recorded in any other frontal lobe areas. For the sake of simplicity, the LPOF will be defined as the area where these evoked potentials were recorded. Tanabe *et al.* (1975b) found that electrical stimulation of the OB produced evoked potentials only in an area (LPOF) where, by means of microelectrode recordings, cells were shown to respond exclusively to odors.

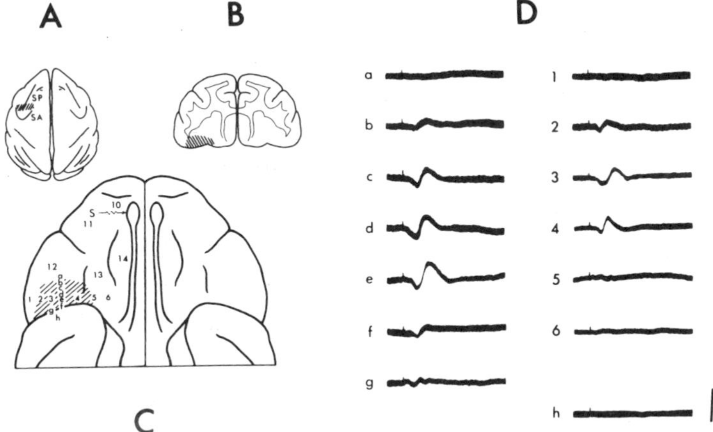

Fig. IX-1. OB-evoked potentials in the orbitofrontal cortex of the old world monkey A: Dorsal surface of the cerebral hemispheres of a monkey. SP: sulcus principalis; SA: sulcus arcuatus. The hatched area indicates where a recording electrode was inserted. B: Coronal section of a monkey brain at the level of electrode insertion. The area indicated by hatched lines is where evoked potentials were recorded. C: Ventral view of the orbitofrontal cortex. Numbers from 10 to 14 indicate areas classified by Walker (1940). Numbers from 1 to 6, and letters from a to h, represent recording sites. The hatched area (LPOF) is a portion in which evoked potentials were elicited by stimulation of the OB, as indicated by an arrow and S. D: Records obtained at sites indicated in C. Evoked potentials are seen in b to g, and in 2 to 4. Calibrations are 40 msec and 100 μV. (from Tanabe *et al.*, 1975c)

Stimulation of the prepyriform cortex (PPC) produced evoked potentials in the LPOF that were similar to those obtained with OB stimulation. However, these potentials had larger amplitudes and appeared with an average delay of 16.5 msec, which was shorter than that observed with OB stimulation.

2. *A Neural Pathway to the LPOF*

Having established a neocortical olfactory area, we wanted to identify the neural pathway projecting to the LPOF.

While stimulating the OB and recording evoked potentials in the LPOF, a third electrode was inserted into the PPC. Using the PPC electrode as an anode and a silver plate attached to the temporal muscles as a cath-

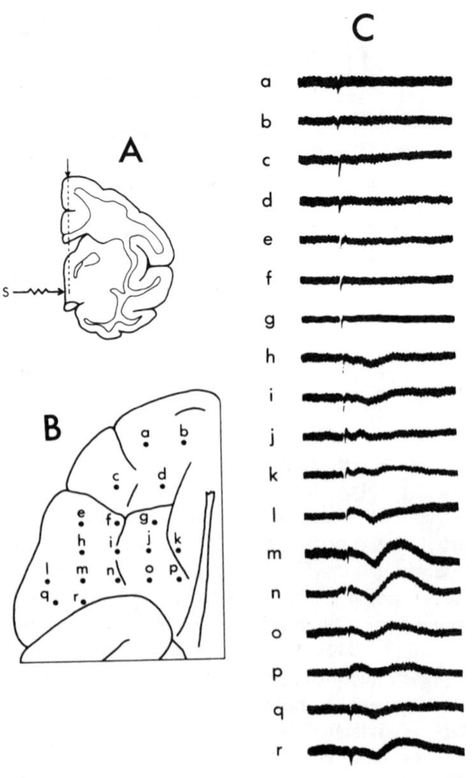

Fig. IX-2. Evoked potentials in the OFC due to stimulation of the hypothalamic area (at s in A). A indicates the site of stimulation in a coronal section of the hypothalamic area. B is a ventral view of the orbitofrontal cortex; the 18 letters represent recording sites. C shows records obtained at the sites shown in B. Evoked potentials are observed in 1 to r (especially m and n). Calibrations are 20 msec and 30 μV. (from Tanabe *et al.*, 1975b)

ode, an electric current was used to electrocoagulate the PPC (and probably part of the medial portion of the amygdala (MA) in some cases). After a pause, the OB was electrically stimulated again, but evoked potentials in the LPOF were no longer observed. To rule out the possibility that the loss of evoked potentials was the result of a shock effect, we stimulated the OB on either side of the brain, recording evoked potentials in the lateral olfactory tract on the electrocoagulated side and in the PPC on the intact side. Evoked potentials from either side of the brain appeared strikingly, indicating that olfactory fibers projecting from the OB to the LPOF pass through the PPC (and probably the MA as well).

Having thus demonstrated the presence of an olfactory pathway from the OB through the PPC-MA to the LPOF in the monkey, we sought to determine whether these fibers passed through the thalamus. This was an important issue, because its resolution determined whether the olfactory system was unique in being the only special sensory system which did not involve the thalamus.

We initiated a series of studies to define the entire route of this olfactory pathway from the OB to the LPOF. In acute experiments on three monkeys, pulses of electrical stimulation were applied to the OB at regular intervals, and the evoked potentials were recorded in the LPOF. While observing these potentials, a glass aspirating pipette with a tip diameter of approximately 1.5 mm was repeatedly inserted into the thalamus. Although a considerable area of the thalamus was gradually aspirated, the evoked potentials in the LPOF did not disappear in any of the subjects.

Following these and further lesion studies, each monkey was sacrificed and the extent of thalamic damage was determined. In the first monkey, the anterior half of the thalamus was removed bilaterally but the posterior half remained intact. The second monkey sustained extensive bilateral damage to the thalamus, but the medial and lateral geniculate bodies were spared. In addition, the entorhinal cortex was destroyed on the stimulated side (Fig. 5 in Tanabe *et al.*, 1975b). The third monkey also sustained extensive bilateral damage to the thalamus, with only small portions of the medial and lateral geniculate bodies left intact. In addition, a lesion was found in the hypothalamic region on the stimulated side. Electrolytic lesions of the thalamus were carried out in another two monkeys. A concentric electrode was repeatedly inserted into the thalamic region and was displaced by a distance of 1 mm in the anteroposterior, mediolateral, and vertical planes after each insertion. A current was passed through the electrode after each change in position, and we examined whether or not the potentials evoked in the LPOF by OB stimulation disappeared. The evoked potentials neither disappeared nor decreased in size during the process of electrocoagulation. After these and further lesion studies in these two monkeys, we confirmed that the thalamus was extensively damaged in each case, and also found that the corpus callosum sustained partial damage.

Even with nearly complete damage to the thalamus, the evoked potentials in the LPOF failed to disappear. From these results, we concluded that the fibers projecting from the OB to the LPOF do not appear to pass through the thalamus.

However, when areas ventral to the thalamus, including the anterolateral and posterodorsal regions of the hypothalamus, were damaged by aspiration or electrocoagulation, the evoked potentials suddenly disappeared. Furthermore, stimulation of these hypothalamic areas elicited evoked potentials in the LPOF. These results suggested that the olfactory pathway to the LPOF passes through the hypothalamus and/or its adjacent areas and not through the thalamus (pp. 347–351).

3. Potentials Evoked in the Orbitofrontal Cortex by Electrical Stimulation of the Hypothalamic Area

In order to obtain a more positive indication that the hypothalamus and/or its adjacent area lay along the pathway between the OB and the LPOF, we applied single pulses of electrical stimulation to the OB and used a concentric electrode to search for evoked potentials in the hypothalamic and adjacent area. Evoked potentials were identified in areas including the hypothalamus and its environs. A recording electrode was then inserted into the LPOF, and single pulses of stimulation were applied to hypothalamic areas through a stimulating electrode. Stimulus sites that would elicit evoked potentials in the LPOF were located by shifting the stimulating electrode in the anteroposterior and mediolateral planes. The clearest evoked potentials were obtained when the anterolateral or dorsal regions of the posterior hypothalamic area were stimulated. Furthermore, potentials could be recorded in either of these loci when the OB was stimulated.

A stimulating electrode was fixed at a site in the dorsoposterior hypothalamic area, because stimulation of this locus elicited a very clear evoked potential in the LPOF. While passing single pulses of electrical stimulation through the electrode, a concentric recording electrode was repeatedly inserted into the orbitofrontal cortex (OFC) and evoked potentials were sought. Sequential placements of the recording electrode throughout the OFC revealed that evoked potentials appeared mainly in the posterior regions of Walker's areas 12 and 13 and only slightly in the medial region of area 13 and the posterior region of area 14. In summary, the area in which the evoked potentials were elicited by stimulation of the hypothalamic area overlapped with the area in which evoked potentials were elicited by OB stimulation. From these experiments we concluded that an olfactory pathway originating in the OB passes through the PPC and the medial portion of the amygdala, and then projects to the LPOF through areas ventral to the thalamus. This olfactory pathway can thus be considered to be an "extrathalamic olfactory pathway." The exact routes of this pathway will be analyzed later (pp. 347–351).

4. *Neuroanatomical Studies on the Hypothalamic Projections to the Prefrontal Cortex*

Several investigators have performed histological studies on the olfactory projection to the hypothalamus. After lesioning the PPC in the rat, Powell *et al.* (1965) traced the degeneration of the afferent fibers to the entire anteroposterior extent of the hypothalamus via the medial forebrain bundle. Scott and Leonard (1971) also obtained a similar result in the rat. Electrophysiological studies have further demonstrated the existence of an olfactory projection to the hypothalamus; responses of single neurons in the hypothalamus were elicited by presenting odors to the nostrils or by applying pulses of electrical stimulation to the OB (Barraclough and Gross, 1963; Komisaruk and Beyer, 1972; Scott and Leonard, 1971; Scott and Pfaffmann, 1967, 1972; Kogure *et al.*, 1978, 1979, 1980). Heimer (1972), in a degeneration experiment in the rat, found no evidence of fibers projecting from the PPC to the hypothalamic region. His negative results, while arguing against a direct connection between these two areas, do not discount the possibility of multisynaptic olfactohypothalamic connections.

Since electrical stimulation of the OB in the monkey elicits evoked potentials which can be recorded from the anterior to the posterior hypothalamus and its adjacent areas, we must conclude that the OB and the hypothalamus and/or its adjacent areas are connected via the PPC.

It has long been known that fibers project from the hypothalamus to the thalamic MD nucleus (Clark and Boggon, 1933; Murphy and Gelhorn, 1945). Thus, we could have thought that the olfactory afferent fibers entered the hypothalamus or its adjacent area and proceeded to the LPOF via the MD. However, this was not found to be the case in our thalamic ablation studies, which ruled out a projection of fibers from the MD to the LPOF. However, other studies revealed that stimulation of the hypothalamic area elicited evoked potentials in the LPOF. Thus, we could only postulate that the hypothalamus and/or its adjacent areas have fiber connections with the LPOF that do not pass through the MD.

The "prefrontal cortex" has been defined as that part of the frontal lobe that receives direct projections from the MD (Nauta, 1971; Rose and Woolsey, 1948). Our studies, however, bring to light an area that would be an exception to this definition. In connection with this topic, it is interesting to note that negative findings in retrograde-degeneration studies have led several investigators to doubt that all regions of the prefrontal cortex receive MD projections (Akert, 1964; Bonin and Green, 1949).

B. Location and Identification of Another Olfactory Area in the Neocortex and Its Afferent Projections

1. Olfactory Input to the Thalamus

Using electrophysiological techniques, Tanabe *et al.* (1975b) demonstrated that the olfactory pathway to the LPOF passes through some areas ventral to the thalamus but not through the thalamus itself.

In contrast, several investigators have histologically demonstrated a projection from the PPC to the MD (Heimer, 1972; Leonard, 1972; MacLeod, 1971; Nauta, 1961, 1971, 1972; Powell, Cowan, and Raisman, 1965; Sanders-Woudstra, 1971; Scott and Leonard, 1971) and also a projection from the MD to the OFC in primates and carnivores, and its probable equivalent in rodents (Akert, 1964; Benjamin, Jackson, and Golden, 1978; Bonin and Green, 1949; Kievit and Kuypers, 1977; Krettek and Price, 1977b; Lammers and Lohman, 1974; Leonard, 1969, 1972; MacLeod, 1971; Nauta, 1962; Pribram, Chow, and Semmes, 1953; Rose and Woolsey, 1948; Rosene and Heimer, 1977; Tobias, 1975; Walker, 1940). These studies imply that the posterior region of the OFC is no more than a few synapses removed from the olfactory receptor cells. Therefore, a functional olfactory pathway from the OB to the PPC and thence through the MD to the OFC was postulated. An olfactory projection to the MD had already been demonstrated electrophysiologically in lower mammals (Motokizawa, 1974a). In anesthetized old world monkeys, Tanabe *et al.* (1975b) demonstrated afferent fiber connections from the OB to the MD by recording evoked potentials and spikes. Then, employing eight specific odors (Table IX-2; Tanabe *et al.*, 1975c), Yarita *et al.* (1980) recorded neural responses in the magnocellular subdivision of the MD (MDmc) in unanesthetized old world monkeys.

We thus demonstrated that the pathway from the OB to the MDmc is olfactory in nature, but the olfactory nature of the pathway from the MD to the OFC remains to be elucidated.

2. Projection Area of the MD

Since projection of fibers from the MD to the orbital, or orbitofrontal, cortex had been demonstrated histologically, Tanabe *et al.* (1975b) inserted a bipolar stimulating electrode into the MD and recorded evoked potentials in the OFC. Figure IX-3 shows the thalamic stimulation sites in A and the OFC recording sites in B. When the MD was stimulated,

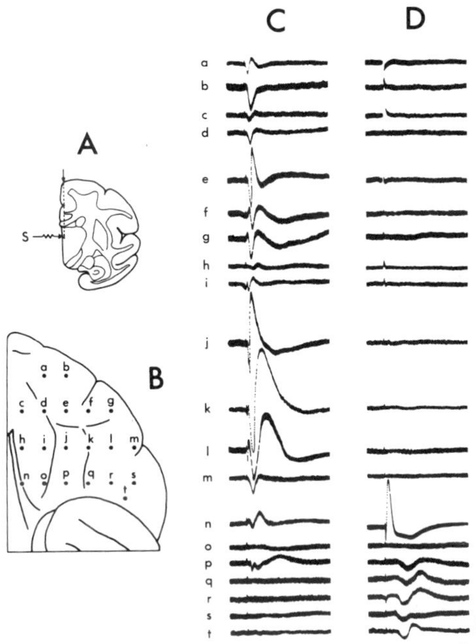

Fig. IX-3. Evoked potentials in the OFC due to stimulation of thalamic MD nucleus (at s in A). A: A frontal section of monkey brain. B: Sites of recording on the OFC as indicated by letters. C: Records from a to t are potentials due to stimulation of MD nucleus. D: Records are potentials at the same sites due to stimulation of OB. A large potential at n is the compound action potential of the lateral olfactory tract. Calibrations are 20 msec and 30 μV. (from Tanabe *et al.*, 1975b)

evoked potentials were found throughout the whole extent of Walker's areas 10 and 11 and in the anterior part of areas 12 and 13 (as seen in column C), but were not observed in those areas indicated in column D. This suggested that the MD projects fibers to that portion of the OFC which lies anterior and anteromedial to the LPOF. We thus found it important, when referring to the OFC, to distinguish clearly between the LPOF and this MD projection area.

Yarita *et al.* (1977, 1978, 1980) then initiated studies to determine whether these MD projections were olfactory in nature.

3. Finding Olfactory Representation in the Centroposterior Area of the Orbitofrontal Cortex (CPOF)

In acute experiments, Yarita *et al.* (1977, 1978, 1980) applied single pulses of electrical stimulation to the OB and inserted a low-resistance tungsten electrode into the thalamus to search for evoked potentials in the MD. A triphasic evoked potential with a latency of about 10 msec appeared in the magnocellular portion of the mediodorsal nucleus (MDmc). In some cases, a spike discharge was found to be superimposed on the first negative wave. Using the same recording electrode at the same position in the MDmc, Yarita *et al.* looked for evoked potentials while applying electrical stimuli to various sites in Walker's area 13 and surrounding regions, and usually obtained a large negative-field potential with a brief latency. They occasionally observed an initial positive potential, depending on the specific location of the recording electrode. They then applied two successive stimuli to the same OFC site. A second evoked potential, having the same amplitude as the first, was recorded until the interval between the two stimuli was reduced to 2.7 msec. This meant that the cells could respond to high-frequency stimuli up to 370 Hz. The small positive- and large negative-field potentials thus appeared to be caused by antidromic activation of neurons in the MDmc. In addition, a spike potential with a relatively fixed latency appeared sporadically, superimposed on the large negative potential. Thus, it appeared that at the same recording site in the MDmc, evoked potentials were elicited transsynaptically by OB stimulation and antidromically by stimulation of the OFC.

We repeated the same procedure while placing the recording electrodes at different depths of the MDmc and in other thalamic nuclei. Figure IX-4 presents sample results from some of the experiments. The thick vertical line in A represents the track of a recording electrode passing through the MDmc. Figure IX-4B and C displays the responses to stimulation of the OB and OFC, respectively. As shown in B and C, orthodromic and antidromic evoked potentials appeared preferentially at depths 4 and 5 along the track indicated in B. These evoked potentials could not be recorded in the lateral portion of the MD or its immediate surroundings, which suggested that impulses originating in the OB are transmitted to a region of the OFC through the MDmc. No other thalamic area contained units responding to either OB or OFC stimulation. We therefore concluded that the MDmc is one of the relay nuclei along the olfactory pathway to the OFC.

In order to delineate the MDmc projection area in the OFC, a stimulating electrode was placed at various sites in the OFC to elicit antidromic field potentials in the MDmc. From the results of this experiment in several monkeys, an area was delineated (stippling) in the central and posterior portion of the ventral surface of OFC (Fig. IX-5 A) in which stimulation effectively produced antidromic evoked potentials in the

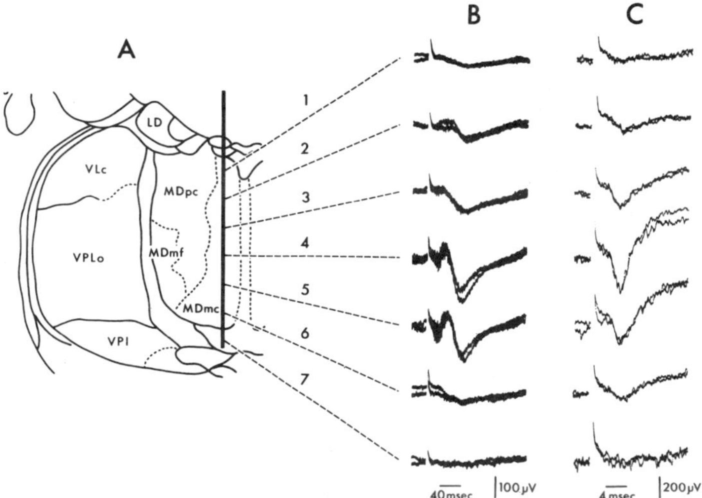

Fig. IX-4. Field potentials in the MDmc elicited by OB (B) and OFC (C) stimulation. A: Coronal section of the thalamus at A = +7.5 mm taken from the Olszewski (1952) stereotaxic atlas. A thick vertical line indicates the track of an inserted recording electrode. Dotted lines with numbers from 1 to 7 indicate evoked potentials elicited by stimulation of the OB and OFC, which were obtained at intervals of about 1 mm ventrally from the dorsal surface of the thalamus. Evoked potentials are clearest at the dorsoventral placements numbered 4 and 5 in B and C. (from Yarita *et al.*, 1980) (see p. 317)

MDmc (Fig. IX-5 B). This area, termed the CPOF, corresponds approximately to Walker's (1940a) area 13 and von Bonin and Bailey's (1947) area FF. Note that the stippled area lies medial and just anterior to another olfactory area, the LPOF (hatched area in Fig. IX-5 A) characterized by Tanabe *et al.* (1975c). We could not be certain, however, of the extent of overlap between these two areas, which is shown as only a small region in Fig. IX-5 A. In 3 of Fig. IX-5 C an electrode placed at site C in A is represented by a thick vertical line. As shown in Fig. IX-5 D, stimulation of this area at depths of 15 to 18 mm elicited antidromic evoked potentials in the MDmc. Stimulation was most effective in the deeper part of the gray matter and in the border area between the gray and white matter, whereas stimulation of the superficial part of the gray matter or the deeper part of the white matter failed consistently to produce antidromic evoked potentials in the MDmc. These experiments demon-

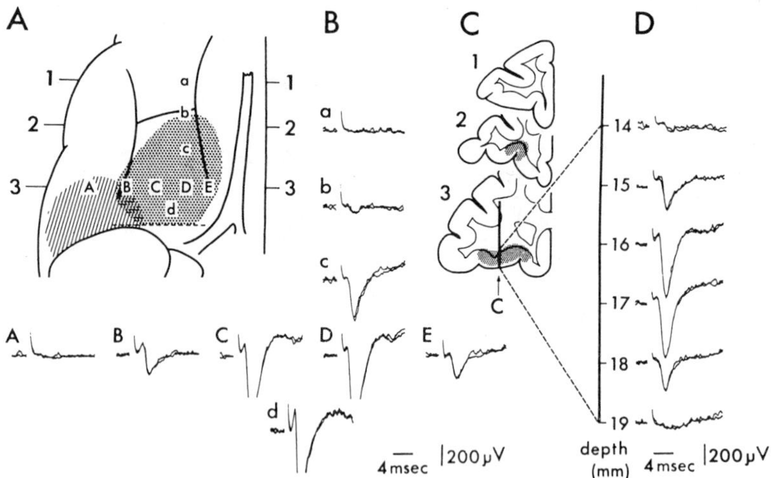

Fig. IX-5. Diagrammatic representation of the MDmc projection area. A: Ventral view of the monkey's frontal lobe. Letters A-E and a-d indicate points where a stimulating electrode was inserted vertically from the dorsal surface of the brain. Stippling indicates the area where stimulation evoked antidromic field potentials in the MDmc. This stippled area, abbreviated as CPOF, is the MDmc projection area. The area caudal to the horizontal dotted line was not extensively studied. The hatched area on the left side of the CPOF is the LPOF, identified by Tanabe *et al.* (1973, 1974, 1975a, b). B: Evoked potentials in the MDmc. C: Coronal sections of the frontal lobe at the levels of of 1–1, 2–3, and 3–3 in A. Stippled areas indicate regions in which stimulation evoked antidromic field potentials in the MDmc. D: Evoked potentials recorded at dorsoventral placements of 14 to 19 mm ventral to the dorsal surface of the brain, which are indicated by the two dotted lines between C and D. (from Yarita *et al.*, 1980)

strated that neurons in the MDmc project to the region corresponding approximately to Walker's (1940) area 13 and von Bonin and Bailey's (1947) area FF, or more precisely, to the CPOF, shown as the stippled area in Fig. IX-5 A. We could thus envision the possibility of olfactory information being conveyed to the OFC through the MDmc.

4. A Transthalamic Olfactory Pathway to the CPOF

To investigate the question of a transthalamic olfactory pathway, tungsten microelectrodes were used to make extracellular recordings of single-cell activity in the MDmc. Yarita *et al.* (1980) conducted studies to determine whether neurons in the MDmc exhibit orthodromic spike

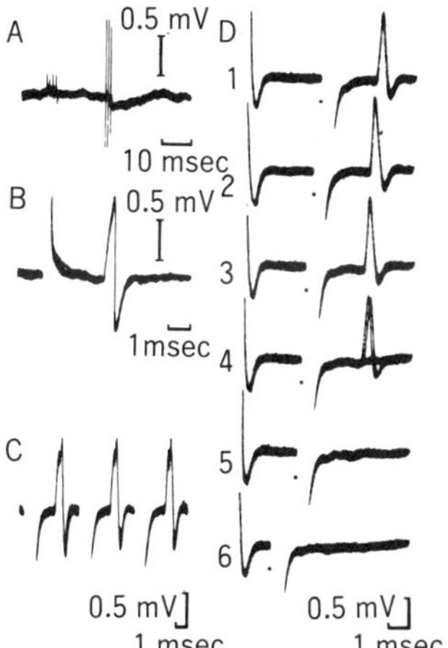

Fig. IX-6. Extracellular responses of a thalamocortical relay (TCR) neuron in the MDmc. A: Spike potentials elicited by OB stimulation. B: Spike potentials elicited by suprathreshold CPOF stimulation. Five traces were superimposed. Repetitive stimulations of 150 Hz were applied to the CPOF in C. D: A collision test. Stimulations were applied to the CPOF (indicated by dots) at various intervals after spontaneous spike discharges. Evoked spike discharges collided when the interval between a preceding spontaneous spike and a stimulation was shortened to 3.0 msec or less (5 and 6). In each case, four traces were superimposed. (from Yarita and Takagi, 1978)

potentials in response to OB stimulation, and antidromic spike potentials in response to CPOF stimulation.

Three standard neurophysiological criteria had to be met to establish that spike potentials elicited by stimulation of the CPOF are antidromic: i) The latency of the spike potential must be stable when a stimulus of threshold intensity is applied; ii) the spike potential must be able to respond faithfully to stimulation frequencies exceeding 100 Hz; and iii) the spike potential must pass a collision test with a spike potential originating in the *soma*.

Figure IX-6 shows the responses of a typical MDmc neuron that satisfied all three criteria. This neuron responded to OB stimulation with a latency of about 25 msec (Fig. IX-6 A), and produced a spike potential with a relatively stable short latency (2.8 msec) in response to threshold stimulation of the CPOF (Fig. IX-6 B). This spike potential continued to appear with the same latency in response to tetanic stimulation of up to about 150 Hz (Fig. IX-6 C). Furthermore, spike potentials elicited by CPOF stimulation ceased to appear when the CPOF was stimulated within about 3 msec after the occurrence of a spontaneous spike potential originating in the soma of this MDmc neuron (Fig. IX-6 D). This cessation was attributed to a collision of the elicited and spontaneous spike potentials along the axon between the soma and the site of stimulation. This convinced us of the antidromic origin of the spike potential elicited by CPOF stimulation. Among 58 neurons studied, 20 neurons completely satisfied the above three criteria. The remaining 38 neurons did not respond consistently to high-frequency tetanic stimulation of the CPOF, but did satisfy the other two criteria.

To clarify the effect of high-frequency stimulation, intracellular recording of MDmc neuronal activity was carried out. The results obtained revealed the presence of inhibitory input to the MDmc neurons (Chapter VIII, pp. 317–318 and Fig. VIII-8). This concurrent inhibitory mechanism was believed to suppress the consistent appearance of repetitive antidromic spike potentials in these 38 neurons, rendering them incapable of responding to high-frequency stimulation. Thus, these 38 MDmc neurons may have been thalamocortical relay neurons to the CPOF.

From the results of this study we could conclude that neurons in the MDmc receive olfactory afferents from the OB and send efferent projections to the CPOF or, more specifically for our purposes, that the MDmc neurons are in a position to function as a thalamocortical relay in the olfactory nerve pathway to the CPOF.

5. Olfactory Thalamic Projection to the Prefrontal Cortex

Yarita *et al.* (1977, 1978, 1980) demonstrated that MDmc neurons receive olfactory input and project to the CPOF. This region corresponds approximately with the MDmc projection area reported in anatomical studies (Akert, 1964; Kievit and Kuypers, 1977; Nauta, 1962, 1971, 1972; Pribram *et al.*, 1953; Tanaka, 1976; Tobias, 1975; Walker, 1940). In a previous study, Tanabe *et al.* (1975b) applied electrical stimuli to the medial portion of the MD and recorded evoked potentials throughout the whole extent of Walker's areas 10 and 11 in addition to the anterior part of areas 12 and 13. These areas are broader than the region delineated by Yarita *et al.* (1980), but this disparity may be attributable to the methodological differences between the two experiments. Whereas Tanabe *et al.* (1975b) stimulated the MD to record evoked potentials in the OFC, Yarita *et al.* (1978, 1980) stimulated the OFC to find antidromic evoked potentials in the MD. The directional differences of the two methods and the spread of current from electrical stimuli could result in such a discrepancy. Consequently, this discrepancy does not weaken our conclusion that a transthalamic olfactory pathway terminating in the CPOF exists in the old world monkey.

C. Histological Examination of the Olfactory Neural Pathways to the Two Neocortical Olfactory Areas

As mentioned before, Tanabe *et al.* (1973, 1974, 1975a, b, c) demonstrated an olfactory area in the LPOF of the old world monkey, and Yarita *et al.* (1977, 1978, 1980) identified another olfactory area in the CPOF. The studies of Yarita *et al.* (1978, 1980) also indicated the presence of a partial overlap between the LPOF and the CPOF.

With this new information on old world monkeys, the author proposed a multiple system model of olfactory sensory processing in the higher primates (Takagi, 1979a, b, 1980, 1981). It is very likely that man has a similar complex of systems for olfaction. The findings of neuroanatomical (Allison, 1953a; Raisman, 1972; Keverne, 1978; Wysocki, 1979) and neurophysiological (Motokizawa and Ino, 1981a, b; Onoda and Iino, 1980; Onoda *et al.*, 1981a, b, 1982) studies in lower mammals make it clear that the olfactory neural pathways of the old world monkey differ markedly from those in the lower mammals (Takagi, 1981), which will be discussed in Chapter X.

Naito *et al.* (1984) used horseradish peroxidase (HRP) to focus on differences in the routes and possible relay nuclei between the two main olfactory pathways in the old world monkey. The distribution of labeled cells was evaluated after injecting HRP into the LPOF in three monkeys and into the CPOF in two monkeys. The results obtained in the LPOF experiments agreed in general with those of Potter and Nauta (1979) except for some points to be mentioned later. Brain structures found to contribute projections to both the LPOF and the CPOF included some of the thalamic nuclei (including, of course, the MDmc), the amygdaloid complex (BM and BL), the SI, and the cerebral cortex.

We also identified the expected hypothalamic projections to both the LPOF and the CPOF. However, we counted only eight labeled cells in the hypothalamic area in each of two monkeys. Potter and Nauta (1979) found only nine such neurons in their study. While the two HRP experiments did reveal hypothalamic projections to the LPOF and the CPOF, we were puzzled by the dearth of labeled cells identified in the hypothalamus. In contrast, a fair number (25–50) of large neurons (up to 35 μm in diameter), located in the basal nucleus of Meynert, were also shown to project fibers to both the LPOF and the CPOF. It appeared that no clear-cut differences between the pathways to the two olfactory

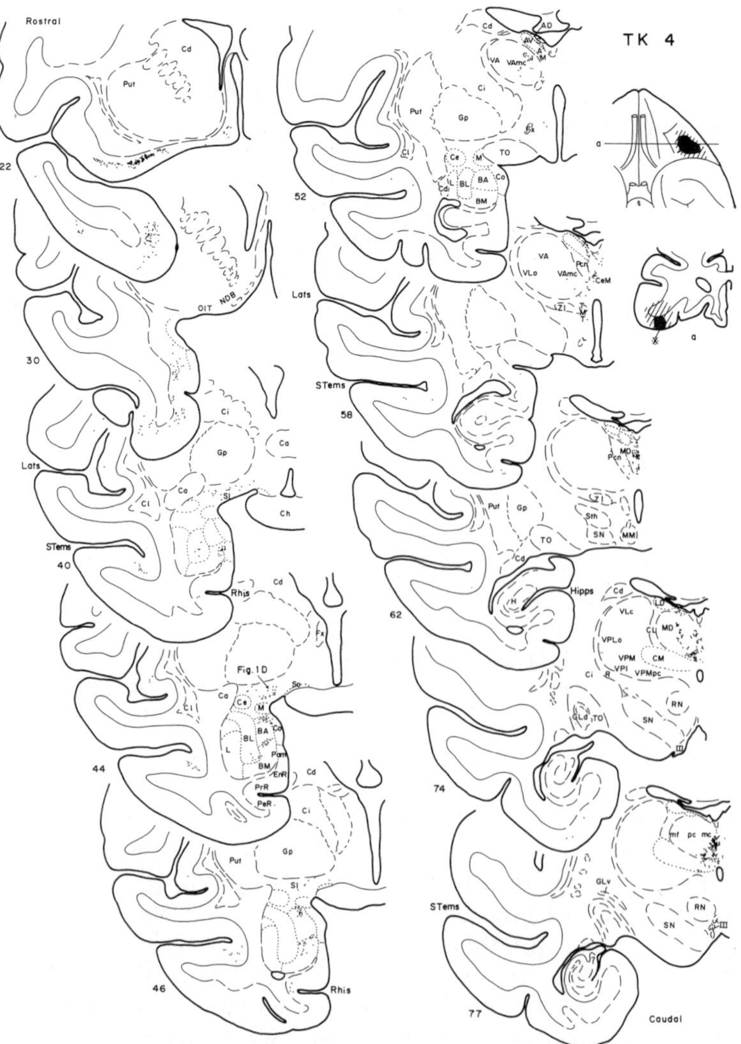

Fig. IX-7. Diagrammatic representation of the HRP study findings in monkey TK4. A series of coronal sections through the telencephalon and diencephalon shows the distribution of retrogradely labeled cells following injection of HRP into the OFC, as seen on the ventral surface of the cerebrum in the top right figure. Areas stained heavily with HRP are indicated in black and those with lighter staining are hatched. One dot corresponds approximately to one labeled cell. (from Naito *et al.*, 1984)

areas could be demonstrated. Common sites of origin for labeled cells projecting to the LPOF and to the CPOF, observed in Naito *et al.*'s (1984) and in Potter and Nauta's (1979) experiments, may be attributed, at least in part, to the slight overlapping of the LPOF and the CPOF noted by Yarita *et al.* (1978, 1980).

After examining the projections to the LPOF and the CPOF in greater detail, however, some differences began to emerge. In the CPOF study, labeled cells were densely clustered in the MDmc, whereas the LPOF experiment revealed labeled cells scattered throughout the thalamus. Most cells in the parvocellular portion of MD (MDpc) and magnocellular portion of ventroanterior thalamic nucleus (VAmc) were found to project fibers to the LPOF (sections 52 and 74 of Fig. IX-7). While cells in the lateral region of the SI tended to send their axons to the LPOF (sections 40, 44, and 46 in Fig. IX-7), cells in the medial region sent more projections to the CPOF (sections 39 and 41 in Fig. IX-8). Many fibers projected from the amygdala, especially the basal nuclei (BA, BM, BL), to the LPOF (sections 40, 44, and 46 in Fig. IX-12), while amygdaloid projections (BM, BL) to the CPOF were much less common (section 43 in Fig. IX-8). The cortical wall in the region of the superior temporal sulcus appeared to send a fair number of fibers to the LPOF (sections 40–46 in Fig. IX-8), whereas the cortical area near the lateral sulcus sent a greater number of fibers to the CPOF (sections 30–43 in Fig. IX-8).

These studies revealed organizational differences in the distribution of afferent projections to the two orbitofrontal areas. Were all these differences in fiber distribution supposed to provide some important anatomical basis for entirely different modes of processing olfactory information? We had previously speculated on the significance of the differences between the pathways to the LPOF and the CPOF areas (Tanabe *et al.*, 1975b; Yarita *et al.*, 1980; Takagi, 1979a, b, 1980, 1981). When we found several relay nuclei in the LPOF experiment, we began thinking of multiple extrathalamic olfactory pathways to the LPOF.

Corticocortical pathways described by Potter and Nauta (1979) and by Naito *et al.* (1984) forced us to reconsider such deep structures as the prorhinal and entorhinal cortices, which had not always been destroyed in the Tanabe *et al.* (1975b) study. As long as some of the neurons in these deep structures maintained fiber connections to the LPOF via areas deeper or more lateral than the hypothalamic areas, evoked potentials in the LPOF elicited by OB stimulation would have survived in the Tanabe *et al.* (1975b) study. In this set of experiments, the central diencephalic areas ventral to the thalamus, including the SI and hypothalamic areas, were extensively destroyed. However, no such evoked potentials survived the experimental ablations, which clearly indicated that projec-

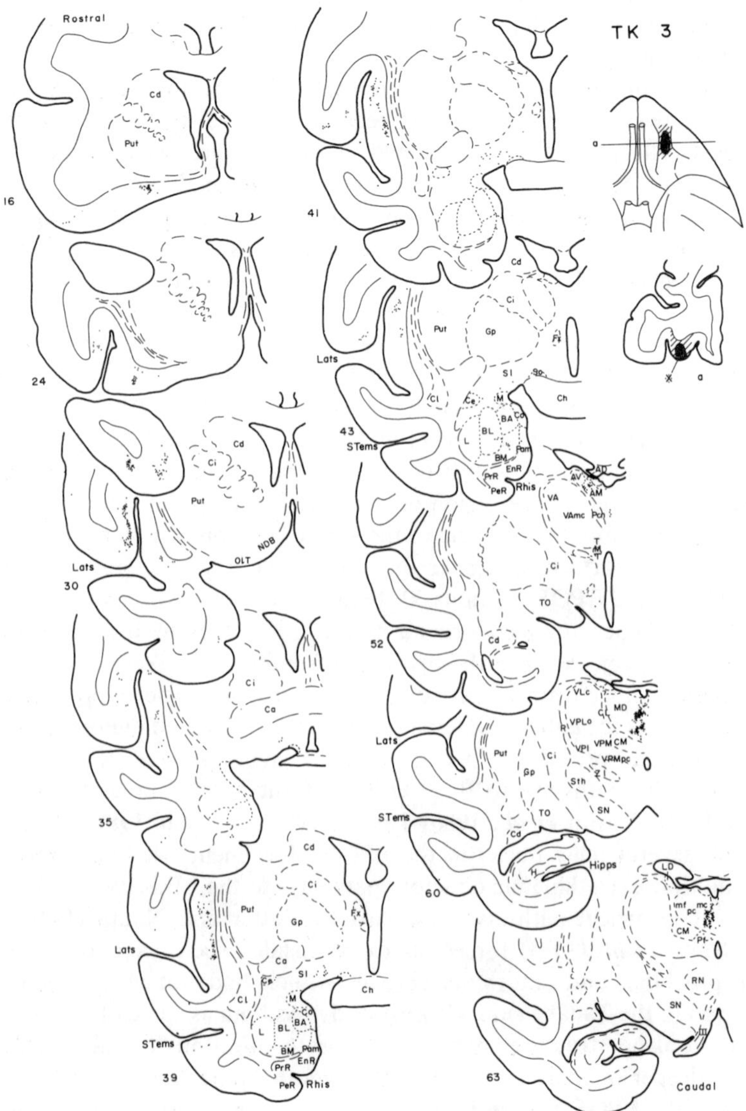

Fig. IX-8. Diagrams showing the HRP-stained areas and the distribution of labeled cells in monkey TK3. (from Naito *et al.*, 1984)

tions to the LPOF traversing the SI and the hypothalamus as well as those originating in the prorhinal and entorhinal cortices, had to pass through areas immediately ventral to the thalamus, and certainly not through any areas deeper or more lateral than that.

Although a neuroanatomical study in general cannot identify which specific neurons are olfactory in nature, I could not help but speculate upon the significance of two different olfactory nerve pathways—one to the LPOF described by Tanabe *et al.* (1975b) and the other to the CPOF outlined by Yarita *et al.* (1980). The postulation of two pathways provided a solid framework for the results of the Naito *et al.* (1984) HRP study. The extrathalamic olfactory pathways to the LPOF contained relay neurons primarily in the amygdala and the lateral region of the SI, and secondarily in the hypothalamus and the prorhinal cortex. This stood in contrast to the transthalamic olfactory pathway to the CPOF, which contained relay neurons mainly in the MDmc.

D. Search for an Olfactory Pathway to the Lateral Hypothalamic Area (LHA)—A Subcortical Olfactory Area

1. An Evoked Potential Study

Tazawa and co-workers (Tazawa, 1983; Tazawa, Onoda, and Takagi, 1983, 1987) applied electrical stimulation to the OB and recorded field potentials evoked in the LHA (Fig. IX-9 D). Then, while stimulating the OB, evoked potentials were sought in many other areas, and were identified in a region that was principally the septum (Spt) but probably also included the nucleus accumbens (Acc) (Fig. IX-9 A).

While observing field potentials evoked in the LHA by OB stimulation, electrolytic lesions of the Spt were performed (1mA for 5 min), resulting in the disappearance of these evoked potentials (Fig. IX-9 E). In these procedures, a large part of the Acc bordering on the Spt was probably involved in the destruction. It was therefore concluded that the neural pathway mediating the evoked potentials in the LHA elicited by OB stimulation passed through the Spt, and probably also the Acc.

2. An HRP Study

HRP studies were carried out in an attempt to characterize the pathway to the LHA. After injecting HRP into the LHA, HRP-labeled neurons were identified in several cortical and subcortical structures in the ipsilateral hemisphere. The sites of labeled neurons found in two monkeys are represented in sketches of their brain sections (Fig. IX-10). Labeled neurons were found primarily in the Spt and Acc, which suggested

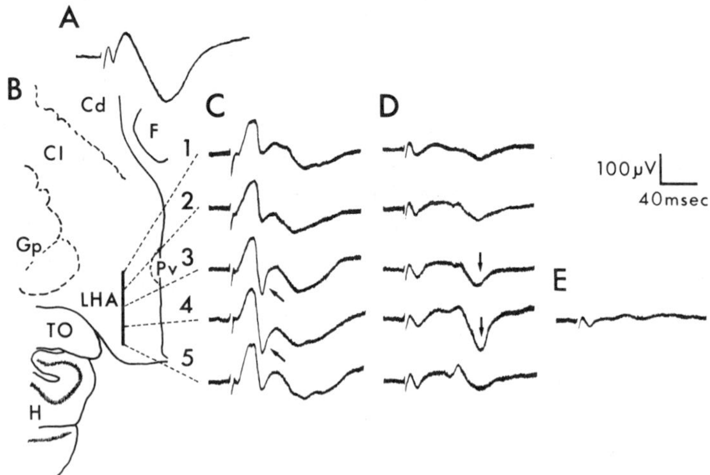

Fig. IX-9. Evoked potentials in the Spt and LHA. A: Evoked potentials in the Spt elicited by electrical stimulation of the OB. Such potentials were recorded with the same type of stainless steel electrode used for stimulation of the OB. When a potential of this shape was obtained, the recording electrode was fixed at the location and was then used to stimulate electrically the Spt. B: A sketch of the frontal section of the brain. The thick vertical line indicates the path of a recording electrode through the LHA. Dotted lines numbered 1 to 5 represent different dorsoventral sites of recording. Evoked potentials recorded at sites 1 to 5 in the LHA during electrical stimulation of the Spt (C) and the OB (D) (F=fornix). The other anatomical abbreviations are defined in the List of Abbreviations. E: Disappearance of the evoked potential elicited by stimulation of the OB and recorded at site 4 in the LHA before electrolytic destruction of the Spt. (from Tazawa, 1983)

the presence of fibers extending from the Spt and Acc to the LHA. Although the number of labeled neurons was small, they were located in the dorsal, medial, and ventral portions of the frontal cortex, about 5 mm posterior to the frontal pole, thus suggesting the existence of corticohypothalamic pathways, as already shown by Nauta (1971, 1972).

The identification of labeled neurons in the LPOF and CPOF was particularly important, because this implied that these areas had inputs to the LHA. Such inputs would allow transfer of highly processed information from the LPOF to the LHA. Furthermore, the close similarity of response patterns between the LPOF and LHA which is shown in Fig. IX-18 can be easily explained. The possibility of transfer of olfactory

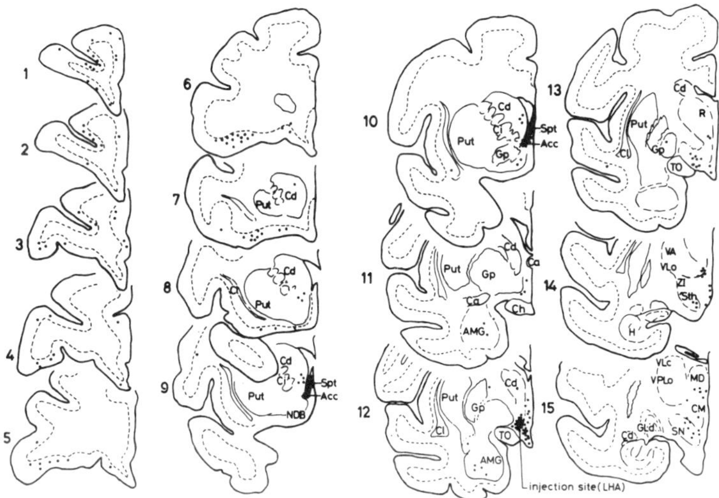

Fig. IX-10. Diagrammatic representation of the distribution of cells labeled by the retrograde action of HRP injected into the LHA. Each dot represents one labeled cell. The first section is 5 mm caudal to the frontal pole of the hemisphere. These 15 sketches represent sections taken at intervals of 1400 μm. Abbreviations: see the List of Abbreviations. Further explanation in the text. (from Tazawa, 1983)

information from the LPOF to the LHA will be discussed further in the next section.

Particularly impressive was the apparent absence of labeled neurons in the PPC, entorhinal area, or olfactory tubercle (Fig. IX-10, sections nos. 9–12). This finding suggests that fiber projections to the LHA from these areas do not exist in the old world monkey, which is significantly different from the histological findings so far obtained in lower mammals (de Groot, 1966).

The histological findings and the above electrophysiological data together indicate the existence of an olfactory pathway from the OB through the Spt-Acc to the LHA.

3. Studies of Olfactory Pathways to the LHA in Old World Monkeys

After injecting HRP into the LHA, Tazawa *et al.* (1983) identified labeled cells in the orbitofrontal area, including the LPOF and CPOF (Fig. IX-10). These orbitofrontal cortices corresponded well to a structure in the old world monkey described by Oomura *et al.* (1980), in which antidromic spikes and field potentials were detected during stimulation of the LHA.

If olfactory fibers project to the LHA from these neocortical areas, especially from the LPOF, then a particular response pattern from the LPOF may be transferred to the LHA. However, if the olfactory pathway from the LPOF to the LHA does exist, we must explain the disappearance of the field potential evoked in the LHA by OB stimulation when the Spt-Acc area was destroyed. Such field potentials could still have been evoked in the LHA by way of an olfactory pathway from the OB through the PPC-MA to the LPOF (Tanabe *et al.*, 1975b), and leads us to speculate that, even if a corticohypothalamic pathway from the LPOF to the LHA does exist, it may not necessarily be olfactory in nature. Thus, the presence of a specifically olfactory pathway to the LHA through the LPOF could be disputed. Indeed, the studies of Potter and Nauta (1979) and Naito *et al.* (1984) suggested that only a minor pathway connects the LHA to the LPOF and/or CPOF, if one exists at all. These findings provide little support for Tanabe *et al.*'s (1975c) assumption of a transhypothalamic olfactory pathway to the LPOF.

Analysis of information processing in the other sensory systems reveals a progression from simple to more complex response patterns as the analysis is carried from lower to higher sensory areas. Tanabe *et al.* (1975b) found a similar progression in the olfactory pathway from the OB through the PPC-AM to the LPOF. However, there was an exception to this rule between the LHA and LPOF, as these two olfactory areas displayed nearly identical response patterns (Fig. IX-19). From this perspective, Tanabe *et al.*'s (1975b) original assumption of a transhypothalamic olfactory pathway to the LPOF was not supported.

In summary, we can be confident in proposing that specifically olfactory activity in either the corticohypothalamic pathway to the LHA or the hypothalamocortical pathway to the LPOF is quite improbable, and may thus conclude that olfactory information is probably conveyed from the OB through the Spt-Acc to the LHA.

To complete this study, the routes of nerve pathways from the OB to the Spt-Acc must be determined. The efferent axons of OB neurons allegedly give rise to the lateral (LOT) and medial (MOT) olfactory tracts. According to Peele (1977), the MOT projects to the Spt directly only in humans. Unfortunately, the presence of this pathway has not been studied in the old world monkey. In lower mammals, an olfactory pathway has been shown to reach the Spt, and probably the Acc, via the lateral olfactory tract (LOT) and the olfactory tubercle (Huber *et al.*, 1962; Peele, 1977). In the case of the old world monkey, Tazawa *et al.* (1983, 1987) demonstrated the presence of an olfactory pathway from the OB through the Spt-Acc to the LHA, but researchers have yet to determine whether olfactory information is conveyed to the Spt-Acc through the LOT or the MOT in the old world monkey.

As shown in Fig. IX-19, our electrophysiological and HRP studies in the old world monkey have elucidated multiple olfactory pathways. The old world monkey has two neocortical olfactory areas, but no functional vomeronasal system. This fact strongly suggests that the neural mechanisms of olfaction in the old world monkey are very different from those in lower mammals which have functional vomeronasal organs (Chapter X, Figs. X-9 and -13).

E. Studies on Neuronal Responses to Odors in Various Olfactory Areas

1. Functional Studies of Neurons in the Neocortical Olfactory Areas

Having identified apparent olfactory areas in the neocortex, we next wanted to unequivocally establish whether or not these "olfactory" areas were indeed olfactory in function. We began by performing a number of behavioral experiments.

2. Behavioral Studies

Many years ago, I noticed while visiting the local zoo that monkeys have a habit of smelling food before eating it, and that they dislike food with a bitter taste. Taking advantage of these habits, we (Tanabe *et al.*, 1975b) conditioned monkeys to avoid odors associated with small pieces of bitter-tasting bread. To give the pieces of bread particular odors, we dipped each piece in one of the odorous solutions shown in Table IX-1. The original compounds or solutions were diluted to the appropriate concentrations in an odorless mineral oil, Nujol (Plough, Inc., Memphis, Tennessee). The concentrations and abbreviations used are listed in Table IX-1.

We prepared many pieces of bread with the five odors and trained monkeys to associate camphor odor with a bitter taste and to discriminate a piece with camphor odor from the other scented pieces. All solutions were colorless and bread pieces were made as equal in size (1-cm cubes) as possible. Using these pieces of bread with different odors, two kinds of tests were performed. In the simultaneous discrimination test, each trial consisted of presenting a monkey with two pieces of bread on a tray simultaneously. The right and left positions of the two odorous pieces on the tray were varied according to the Gellerman series. In the successive discrimination test, each trial consisted of presenting a monkey with only one odorous piece of bread on the tray. Again we used the Gellerman series to determine what odor to choose. After a period of training

Table IX-1. Five kinds of odorous solutions used to condition old world monkeys. Bitter taste (3 % quinine) was added to a camphor solution.

Odorous solution	Concentration	Abbreviation
dl-Camphor solution with 3 % quinine	(10^{-3})	(CQ)
Iso-valeric acid solution	(10^{-4})	(V)
γ-Undecalactone solution	(10^{-4})	(U)
Methyl cyclopentenolone solution	(10^{-3})	(C)
Skatole solution	(10^{-5})	(S)

Table IX-2. Classifications of eight chemical odors used for stimulation. Methyl cyclopentenolone is also known as cyclotene (Dow Chemical, Inc.). Classifications of these odors are shown on the left and an abbreviation of each odor is shown on the extreme right. (from Tanabe *et al.*, 1975c)

Class	Chemical	Abbreviation
Camphoraceous odors	*dl*-Camphor	CM
	Cineole	CL
	Borneol	BL
Ethereal odor	1,2-Dichloroethane (Ethylene dichloride)	DE
Burnt odor	Methyl cyclopentenolone	CLT
Fruity odor	γ-Undecalactone	UDL
Pungent odor	Isovaleric acid	VA
Miscellaneous odor	Isoamyl acetate	AA

(usually 7–10 days) the monkeys were able to discard bitter pieces of bread by sniffing alone, and put the nonbitter ones into their mouths. Seven monkeys (M1 to M7) underwent discrimination training in these behavioral studies. In the simultaneous discrimination test, the tray with two pieces was presented 10 times a day for 10 days for each odor pair, for a total of 100 trials. The number of successful responses was used as a score, for each pair. The successive discrimination test was carried out for five days, with 10 trials each day, for a total of 50 trials. The number of successful responses was used as a score.

Following the behavioral tests for the four odor pairs, various regions of the frontal cortex were ablated bilaterally. Approximately two weeks after surgery, discrimination training was administered to each monkey for a period longer than the time required for the first training. Discrimination tests were then begun, using the above two methods.

Upon completion of all these tests, the brains were removed and the extent of frontal lobe damage was determined. Among the seven monkeys, only in M4 and M5 was the LPOF ablated bilaterally.

The results of the behavioral tests were compared before and after ablation of the frontal lobe areas. The preoperative and postoperative scores were calculated (Tanabe *et al.*, 1975b). The differences between these two values for each odor pair represent the change in discriminative ability.

In monkeys M4 and M5, in which the LPOF was successfully ablated bilaterally, the rates of failure were much higher than those of the other monkeys, in which parts of the LPOF had been spared in one or both hemispheres.

We concluded that lesions of the frontal cortex markedly impair olfactory discrimination only when the lesions destroy the LPOF bilaterally. Even in M4 and M5, however, discrimination between *dl*-camphor and skatole was much less impaired than between other odor pairs.

In trying to assess the significance of these results, we should recall Allen's (1940) experiment. He reported the interesting observation that dogs with bilateral ablation of the frontal cortex, despite a severely impaired capacity to discriminate the odor of cloves from that of asafetida, retained the ability to locate meat by smell. Such selective retention of olfactory recognition of an odor associated with the animal's needs should also bring to mind. Pfaff and Gregory's (1971) finding that a high proportion of cells in the olfactory bulb and preoptic area of the rat respond more readily to urine odors than to artificial nonurine odors. Tanabe *et al.* (1975b) noticed a similar selectiveness. Among the odors applied, skatole can be reasonably assumed to be the most common and familiar to monkeys. Our findings in monkeys with bilateral LPOF ablation showed that the ability to discriminate skatole from *dl*-camphor was relatively well preserved. We thus speculated that odors immediately related to food-gathering or sexual behavior could be discriminated in lower olfactory areas, whereas odors not so immediately related to these behaviors needed further processing and could be differentiated only if the LPOF were intact.

Although we knew that the LPOF was an olfactory area in the OFC, we had yet to determine the extent to which the LPOF specifically mediated olfactory function. Allen (1943a) demonstrated that the discriminative abilities of dogs trained to discriminate auditory stimuli as well as general cutaneous stimuli remained unchanged after prefrontal lobectomy. Tanabe *et al.* (1975c) showed that cells in the LPOF which responded specifically to odors did not respond to visual or auditory stimuli. They also showed that the afferent pathway to the LPOF did not involve nonspecific thalamic nuclei; it passed through the areas immediately ventral to the thalamus. These data suggested that the LPOF is a projection area specific to the olfactory modality.

The posteromedial region, bordering the LPOF on the medial side, was known to be an inhibitory area for respiratory movement and a control area for blood pressure (Bailey and Sweet, 1940; Kaada, 1951, 1960; Kaada *et al.*, 1949; Sachs *et al.*, 1949; Spencer, 1894). We found it interesting that the inhibitory respiratory area in the orbitofrontal lobe lay just adjacent to the olfactory area, because we knew that smell sensations could modify respiratory movement and affect blood pressure in animals and humans.

3. Electrophysiological Studies (Tanabe et al., 1975a)

In acute experiments, animals anesthetized with Nembutal (35 mg/kg) were fixed in the head holder of a stereotaxic apparatus. Through a round hole opened in the skull a microelectrode was inserted into the brain. The loci of recordings were determined by referring to Snider and Lee's (1961) stereotaxic atlas of the monkey.

Applying eight odors (Table IX-2) as stimuli, we at first tried to record the responses of cells in the LPOF in monkeys anesthetized with Nembutal (Tanabe *et al.*, 1973, 1975c). Spontaneous discharges of 277 cells were recorded, but only 29 cells (10.5 %) responded to odors (Table IX-3). This finding coincided well with those of Haberly (1969) and Cain and Bindra (1972). Figure IX-11 presents examples of the excitation and inhibition types. We could hardly regard these response patterns recorded in anesthetized monkeys as natural, and attempted instead to record neuronal discharges in unanesthetized monkeys.

In chronic experiments, we preliminarily attached a cylinder to the skull. Following the completion of surgery and postoperative recovery from

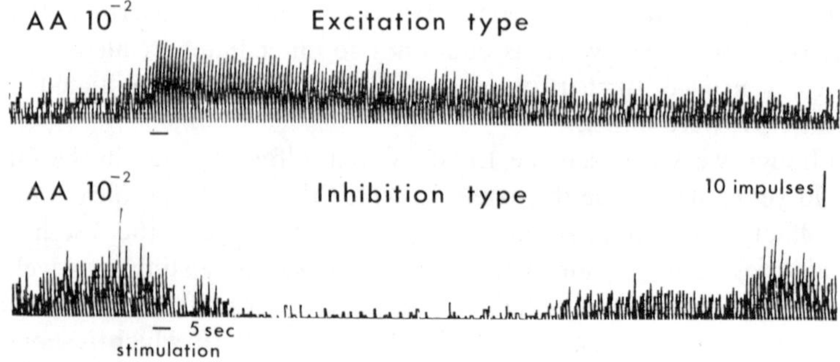

Fig. IX-11. Neuronal responses to odors in the LPOF of an anesthetized old world monkey. Once a neuron responded to an odor (amyl acetate in this case), an unusually long excitatory or inhibitory response was observed. AA: amyl acetate. (from Tanabe *et al.*, 1975a)

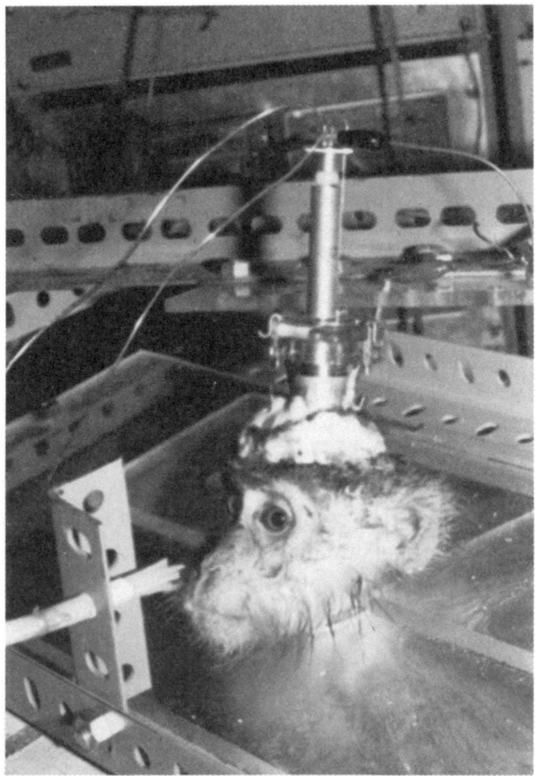

Fig. IX-12. An unanesthetized monkey sitting on a specially designed chair. Odors were presented through teflon tubes on the left. Responses to odors were recorded by means of an Evarts type remote control micromanipulator fixed on the head. (from Takagi, 1974)

the anesthesia, the monkey was placed in a restraining chair so that responses could now be recorded from the animal in an unanesthetized state. Thus, we tried to record neuronal responses to odors, using an Evarts (1968a, b) type micromanipulator (Fig. IX-12). We used 22 male monkeys (*Macaca irus*) in groups of five, 11, and six for single-cell recordings in the LPOF, the PPC-MA, and the OB, respectively.

Out of the 119 single cells recorded, 44 cells reacted to one or more of the eight odors (37%) (Table IX-3). Nearly four times as many cells as in the acute experiments responded to one or more odors in the chronic experiments (Table IX-3). This was one of the most important findings of our brain studies and the details will be discussed in the following section.

4. Comparison of the Neuronal Responses in Acute Anesthetized and Chronic Unanesthetized Monkeys

When the response types were examined in the acute experiments, half of the cells showed increases in the discharges (+ type) and the other half revealed inhibition (− type) (Table IX-3). In contrast, in the chronic experiments, 70.5% of cells showed the + type, 26.9% the − type, and 2.6% a mixed type, which exhibited a decrease followed by an increase in the firing rate (Table IX-3). Thus, the response types of unanesthetized monkeys were more varied and complex.

When responses to each of the eight odors were examined in the acute experiments, many cells were observed to respond to AA, VA, and DE (9, 8, and 6 cells, respectively), whereas few cells responded to UDL, CM, BL, and CLT (3, 2, 2, and 1 cell, respectively). Thus, the eight odors were divided into two classes according to the number of cells responding. In contrast, in the chronic experiments, the 44 cells responded almost uniformly to any of the eight odors. These results show very important differences, and indicate that responses to odors of cells in the neocortical olfactory area should be examined in unanesthetized animals.

In the acute experiments, approximately 80% of the cells maintained spike discharges for 5 to 20 sec once they were stimulated; 10% of the cells continued their discharges for longer than 60 sec, while only 10% of the cells discontinued their discharges within 5 sec. In contrast, in the chronic experiments, approximately 50% of the cells discontinued their discharges within 5 sec and approximately 80% of the cells did so in less than 10 sec. This important finding must also be considered in olfactory research.

From these comparisons, it is clear that cellular responses to odors can vary greatly, depending upon whether the animals are anesthetized or not.

Table IX-3. Comparison of data obtained from anesthetized and unanesthetized monkeys.

	Total No. of units tested	No. and % of units responding to an odor or odors		Type of response		
		No.	%	No. of + type	No. of − type	No. of + − or − + type
Acute anesthetized monkeys	277	29	10.7	51.6%	48.4%	0
Chronic unanesthetized monkeys	119	44	37	70.5%	26.9%	2.6%

(data based on Tanabe *et al.*, 1975a)

Animals were anesthetized in most of the previous olfactory experiments performed (Cain and Bindra, 1972; Haberly, 1969; Mathews, 1972a, b; Shibuya and Tucker, 1967). In fact, Giachetti and MacLeod (1975) compared response rates of pyriform neurons to odors in lightly anesthetized rats with those in deeply anesthetized ones (Haberly, 1969) and found a great discrepancy. They attributed it to a differential action of the anesthesia. We now believe, of course, that unanesthetized animals should always be used when the responses of neurons to odors are studied in the olfactory areas of the brain, and that results obtained in anesthetized animals are invalid. The same may be true of experiments investigating responses to odors of olfactory receptor cells in the olfactory epithelium.

5. Responses to Odors in LPOF Neurons

On the basis of the results of their preliminary experiments in anesthetized and unanesthetized monkeys, Tanabe *et al.* (1975c) initiated studies to examine the responses of LPOF neurons to odors in unanesthetized monkeys. They found that 44 cells among the 119 examined responded to at least one of the eight odors (Table IX-3). A cell that responded to four odors showed a particularly marked increase in impulses to BL odor over the spontaneous discharge rate and marked decreases to CL, UDL, and VA odors. Responses to odors recorded with a rate meter clearly showed that repeated application of the same odors elicit responses with nearly identical patterns. Since none of the 44 cells responded to visual or auditory stimuli, they were considered to be specific for olfactory stimuli.

These cells showed five types of response: (a) an increase in the rate of spontaneous discharge (+ type); (b) a decrease in the rate of spontaneous discharge (− type); (c) a decrease followed by an increase in the rate of spontaneous discharge, or vice versa; (d) different response patterns to different odors; and (e) no change in the rate of spontaneous discharge. The type (c) response, or the mixed response, was identified in the PPC and MA. The type (d) response was designated as a combined response. Seventy-five cells among the 119 did not respond to any of the eight odors.

Among the 44 cells with positive responses, 29 cells (65.9%) showed exclusively + type responses, eight cells (18.2%) only − type responses, and six cells (13.6%) combined type responses and one cell (2.3%) a combined and mixed type response. These 44 cells were arranged according to the increasing number of odors to which they responded and are shown in a matrix (Fig. IX-13 A) which includes the response types. We found that 22 (50%) of the cells responded to only one odor, 13

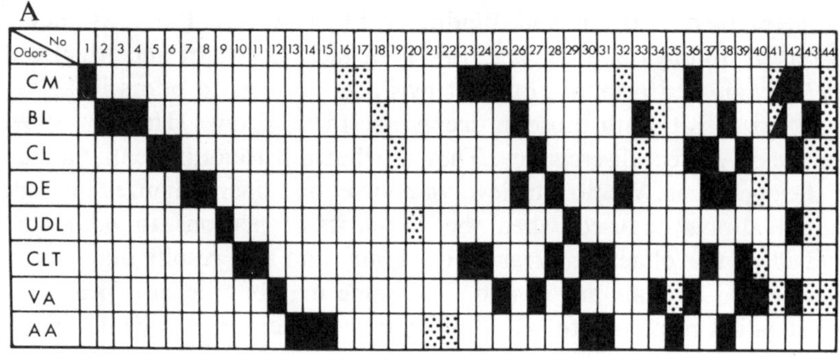

Fig. IX-13. A: Responses of LPOF neurons to odors. Horizontal column at top indicates cell number. The response types are indicated by five differently shaded blocks below the matrix. (from Tanabe *et al.*, 1975c). B: Responses of 23 cells in the LPOF. The eight odors shown by abbreviations at the top were applied at four different concentrations, as shown in the extreme left column. "No. of cells" in the extreme left column represents the number of cells that responded to the respective odor. While only one cell was found to respond to CM or VA, four or five cells responded to the other odors. Black, stippled, or white squares indicate excitatory, inhibitory, or no response, respectively. (Iino, unpublished data)

cells (29.6%) to two odors, six (13.6%) to three odors, and three (6.8%) to four odors. Thus, the number of cells responding was negatively correlated with the increase in the number of odors which elicited a response. No cell was found to respond to more than five odors. A histogram of these results appears in Fig. IX-18.

For comparison, recording was also performed in the dorsal portion of the prefrontal cortex. Among the 18 cells recorded, not a single cell responded to any of the eight odors applied.

6. *Response Patterns of LPOF Neurons to Odors of Different Concentrations*

In the previous studies on unanesthetized monkeys, the eight odors were applied consistently at a concentration of 10^{-2}, and the number of neurons responding to each odor was examined in the LPOF (Tanabe *et al.*, 1975c).

Later, Iino (unpublished data; Takagi, 1986a) attempted to study neuronal responses to the eight odors at four different concentrations (vapors from solutions of concentrations of 10^{-4}, 10^{-3}, 10^{-2}, and 10^{-1} of saturated odorous solutions). Thus, a total of 32 olfactory stimuli were applied to three unanesthetized monkeys under the same experimental conditions as in Tanabe *et al.*'s (1975c) experiment.

First, she presented the eight odors one after another at a concentration of 10^{-2}. If she obtained neuronal responses to one of the eight odors, she applied the odor at the other three concentrations. Some neurons responded only at the one concentration. Other neurons displayed higher rates of responding with higher concentrations. Still other neurons responded quite readily to the concentration of 10^{-2}, but were less active or did not respond at all at the other concentrations (Fig. IX-13 B). In other cases, neurons responded to more than one odor at the concentration of 10^{-2}, and showed responses to odors at the other concentrations (Takagi, 1986a).

In the next stage of her experiment, Iino applied 32 odors at random and found that several neurons did not respond to odors at the 10^{-2} concentration, but that they did respond to odors at one of the other concentrations. In one unusual case, she succeeded in recording responses to all 32 olfactory stimuli, but in most cases she found a response to VA only at the 10^{-2} concentration. Among the neurons examined, 20 neurons responded to at least one concentration of one or more odors. The results are summarized in a matrix in Fig. IX-13 B.

7. *Responses of Neurons in the CPOF*

Yarita *et al.* (1980) studied neuronal responses to the eight odors in the CPOF of unanesthetized monkeys. Altogether, responses of 12 CPOF cells were obtained in two monkeys (Fig. IX-14). Two cells (17 %) showed responses of a combined type; the other 10 cells (83 %) responded exclusively with the + type or − type. The + type cells (seven cells) were found more than the − type cells (three cells). A matrix and then a

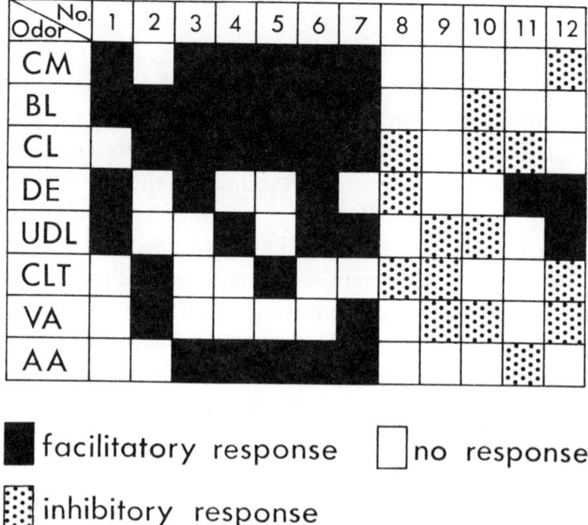

facilitatory response   □ no response

inhibitory response

Fig. IX-14. A matrix indicating response patterns of 12 neurons in the CPOF. Numerals in the top row represent the number of neurons that responded to the respective odors. The response types are indicated by three differently shaded blocks at the bottom. (from Yarita *et al.*, 1980)

histogram were made from these data (Figs. V-14 and 18). The number of odors that elicited neuronal responses ranged from three to six and no neuron responded to one, two, seven, or eight odors. These histograms will be compared later with those obtained for other olfactory areas described in the last section (Fig. V-18).

F. Responses of Neurons in the OB and Intermediate Areas of the Olfactory Pathways to the LPOF and CPOF

1. *Responses of Neurons in the OB*

From a total of 67 OB cells tested with the eight odors (Table IX-2), 40 cells (60%) responded to one or more odors. These 40 cells were arranged in a matrix according to the increasing number of odors which elicited neuronal responses (Fig. IX-15). Cells responding to five odors were the most numerous. Twenty-six cells (65%) showed the + type response, five cells (12.5%) the − type, and the remaining nine cells (22.5%)

the combined type response. None of the OB cells displayed mixed (+ or −) type responses which were found in higher olfactory areas such as the LPOF, PPC, and MA.

2. Comparisons with Other Studies in the OB

Walsh (1956) recorded single-cell responses to odorous stimuli in the OB of the anesthetized rabbit and concluded that odor specificity is a fundamental characteristic of only a small number of cells, his class III cells. Döving (1964) recorded responses of 177 cells: 29.4% of the cells showed the + type and 70.6% the − type response. He also found that some of the cells changed their response patterns from the + type to the − type when the stimulus intensities were increased. This is an interesting finding, because in the curarized bullfrog, Higashino *et al.* (1969) recorded only two patterns, the + type response and no change in activity. They found that responses of the − type never appeared when low concentrations of odors were used as stimuli. In contrast, Shibuya *et al.* (1962) found six types of responses, the + type, − type, two mixed types (+ and −), on-off type, and off-type, in the OB of the decapitated frog when high concentrations of odors were applied. A similar finding was made by Takagi and Omura (1963) in the olfactory epithelia of unanesthetized decapitated toads. In fact, in spontaneously active olfactory cells, responses of the − type were found in frogs when odors at high concentrations were presented (K. Aoki, unpublished data; T. Shibuya, personal communication). Thus, it is clear that the response types of OB cells vary depending upon the intensity of odors applied.

Mathews (1972a) anesthetized tortoises with urethane and examined the responses of 40 cells in the olfactory epithelium to 27 odors. He found that 18 of 40 cells responded to odors with a + type response. In the OB of this species he recorded only two response types, the + type (44.4%) and the combined type (56.6%). Although a slight increase of complexity in the response type was found when progressing from the receptor level to the next OB level, these findings in the OB of unanesthetized tortoises differed markedly from Mathews' (1972b) findings in the OB of rats unanesthetized but immobilized with Flaxedil. In that study, Mathews found that 42% of the OB cells displayed + type responses, 12% − type, and 46% combined type responses. The results of Mathews' second experiment on rats corroborate the findings of our studies on frogs and toads. Both sets of experiments were performed on unanesthetized animals, and both yielded three response types. Whether the different findings between the tortoise and the rat represent a true species difference or simply a difference in technique, such as anesthesia or the quality and concentration of odors applied, remains to be determined.

Human Olfaction

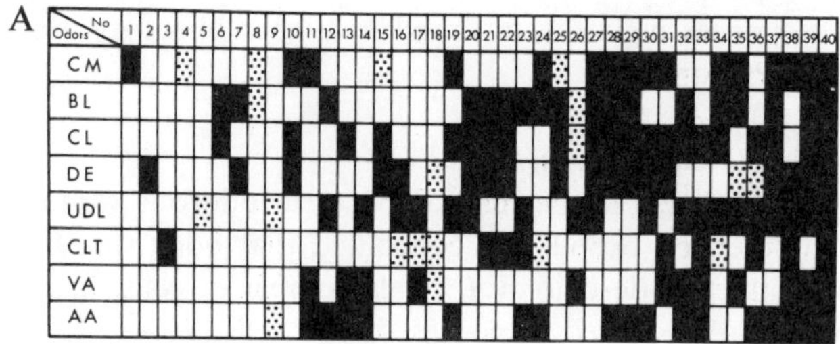

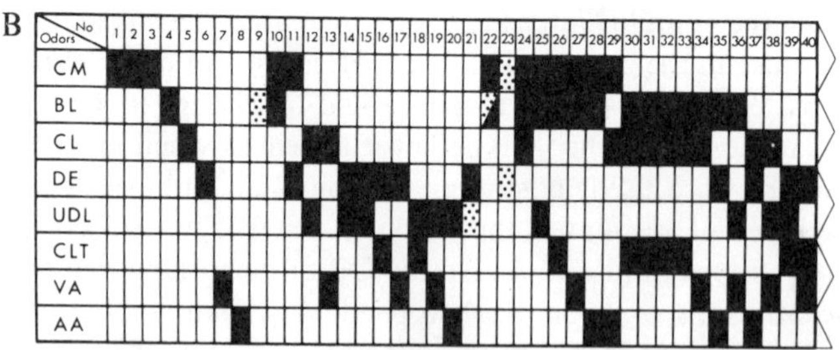

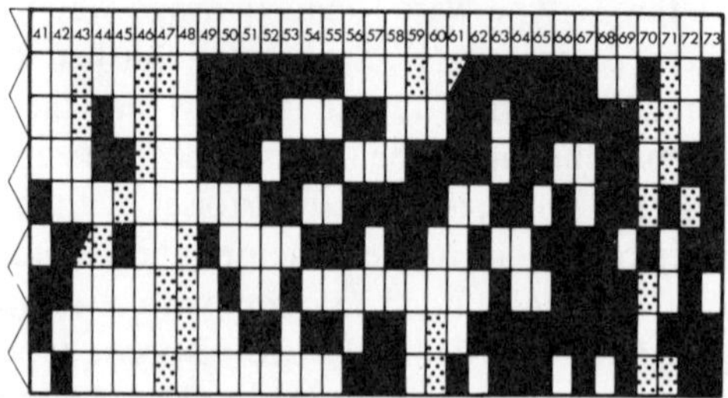

Fig. IX-15. Responses of OB cells (A) and PPC-MA cells (B). A: Responses of 40 cell in the OB. The matrix above indicates that cells Nos. 1–5 responded to only one odor, cells Nos. 6–9 to two odors, cells Nos. 10–18 to three odors, cells Nos. 19–26 to four odors, cells Nos. 27–36 to five odors, cells Nos. 37 and 38 to six odors, and cells Nos. 39 and 40 to seven and eight odors, respectively. Response types are indicated by the five differently shaded blocks (from Tanabe *et al.*, 1975c). B: Responses of 73 cells in an area extending from the PPC to the MA. Cells Nos. 1–9 responded to only one odor, cells Nos. 10–23 to two odors, cells Nos. 24–48 to three odors, cells Nos. 49–61 to four odors, cells Nos. 62 and 63 to five odors, cells Nos. 64–72 to six odors, and cell Nos. 73 to seven odors. (from Tanabe *et al.*, 1975c)

The number of odors that elicit responses in a single cell appears to be related to the cell's role in odor discrimination. Pfaff and Gregory (1971) could not find a single rat cell which responded to only one odor. However, Mathews (1972b) found two such cells among the 33 cells examined in the OBs of rats immobilized with Flaxedil. In Tanabe *et al.*'s (1975c) experiment, five of the 40 OB cells examined (12.5%) responded to only one odor. The small number of such cells suggests that most of the cells in the OB apparently belong to a "generalist type" of category. The results of Tanabe *et al.* coincide well with those of other investigators (Leveteau and MacLeod, 1966a, b; Mathews, 1972b; Pfaff and Gregory, 1971; Shibuya and Tucker, 1967).

3. Responses of Cells in the PPC and the MA

In the PPC and MA, spontaneous discharges of 145 cells were recorded and their responses to eight odors were examined. Seventy-three cells (50.3%) responded to one or more odors. The 73 cells are arranged in a matrix according to the increasing number of odors that elicited neuronal responses (Fig. IX-15 B). A histogram of the data in this matrix is presented in Fig. IX-18. Cells responding to three odors were the most numerous (25 cells, 34.3%). No cell responded to all eight odors.

Among the 73 cells, 57 showed responses exclusively of the +type (78.1%), and five cells of the −type (6.8%). The remaining 11 cells (15.1%) showed responses of the combined type (Nos. 21, 22, 43–45, 59–61, and 70–72 cells in Fig. IX-15 B), including three responses of the mixed type (Nos. 22, 43, and 61). Among them, three cells (Nos. 22, 43, and 61) also showed mixed-type responses.

In each of these experiments, we calculated as accurately as possible the loci at which the activities of 73 cells were recorded. We then sectioned the brain in a plane parallel to that of the microelectrode insertion, and examined the recording sites histologically. The 73 cells were classified into three groups: (a) cells in the PPC (18 cells), (b) cells in the MA (40 cells), and (c) cells in the border zone between the PPC and amygdala (15 cells).

The responses of the 18 cells in the PPC were classified according to the number of odors to which they responded. One of the 18 cells responded to seven odors, but not a single cell responded to all eight odors. Cells responding to three odors were the most numerous (33.3%). Thirteen of the cells (72.2%) showed responses exclusively of the +type, only two cells (1.1%) gave −type responses. The remaining three cells exhibited combined type responses.

Next, the responses of 40 cells in the MA were examined. In this region, cells responding to three odors were the most numerous. Among these 40, cells responded to as many as six odors, but no cell responded to seven or eight odors. Analysis of these responses indicated that 33 cells (82.5%) showed the +type response exclusively, only two cells (5%) the −type exclusively, and five cells (12.5%) the combined type. Comparison of the responses of the PPC and MA neurons showed that the response patterns in the two areas resembled each other in general, having common peaks at the three-odor level.

A few responsive cells recorded in the border zone between the two areas were also examined. In this region cells responding to three odors were also the most numerous. Comparison of response patterns in the three areas revealed a peak at three odors and a valley at five odors. We can say, therefore, that the neuronal responses in the three anatomical areas are generally similar.

4. Comparison with Other Studies in the PPC and MA

Responses of single cells in the PPC were studied by Haberly (1969). In rats deeply anesthetized with Nembutal, responses to odors of seven organic compounds and of solid rat pellets were examined. Among several hundred cells recorded, only 21 responded to odors. This coincides with our observation that very few cells respond to odors in the LPOF of monkeys anesthetized with Nembutal (Tanabe *et al.*, 1975a). Comparing these findings with those made in studies of olfactory receptor cells (Gesteland *et al.*, 1963) and cells in the OB (Döving, 1964), Haberly (1969) concluded, "It appears that the degree of unit specificity in the PPC is less than in the OB and the olfactory receptors."

The responses of cells in the amygdala of the anesthetized rat were studied by Cain and Bindra (1972). They examined single-cell responses to odors and found evidence suggesting a higher degree of discrimination in the cortical and basal nuclei than in the medial nucleus. Moreover, they recorded responses of the +type, −type, and mixed type. These three response types were also noted in the PPC, the MA, and the LPOF in our experiments. Since no single cell was found to respond to only one odor, Cain and Bindra stated: ". . . . the search for single units in the vertebrate brain which respond exclusively to individual odors would be futile, and (that) efforts should concentrate, as a first step, on correlating the similarities between across-fiber firing patterns (Erickson, 1968) to odors and some index of odor similarity. . . ." We previously discussed (pp. 360–361) the strong influences of anesthetics on the responses to odors in the LPOF. The results obtained by Haberly (1969) and by Cain and Bindra (1972) may thus be attributed to the effects of anesthesia. In our experiments on unanesthetized monkeys, 10% of the MA cells responded to only one of eight odors. Such cells were also found in the PPC (Tanabe *et al.*, 1975c) and in the OB (Mathews, 1972a; Tanabe *et al.*, 1974, 1975c; Ariki *et al.*, 1982).

5. *Responses of Cells in the Magnocellular Portion of the Thalamic Mediodorsal Nucleus (MDmc)*

In chronic experiments, responses of MDmc neurons to eight odors were examined (Yarita *et al.*, 1980). In seven monkeys, a total of 41 neurons showed responses to more than one odor. A matrix was made based on the differences in the response types of these neurons (Fig. IX-16 A). A histogram showing the numbers of cells responding to from one

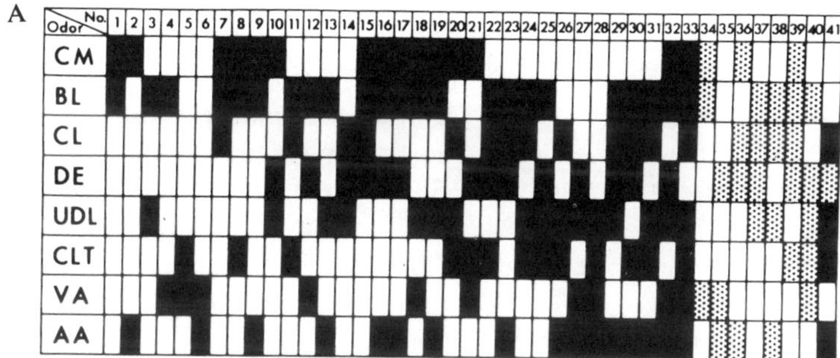

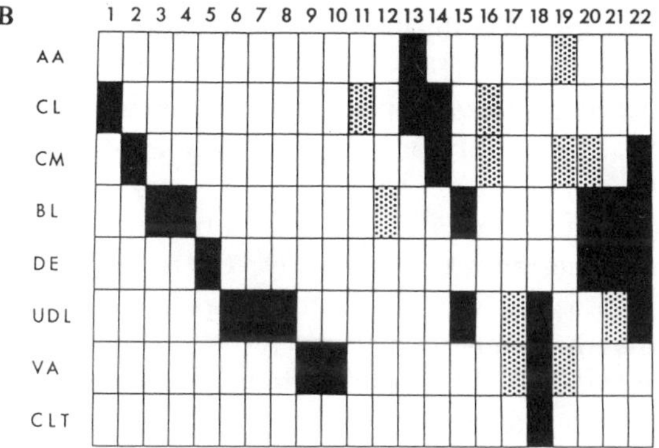

Fig. IX-16. Response profiles of neurons in MDmc (A) and in LHA (B). A: Response patterns of 41 neurons. The top row indicates the number of neurons studied. The response patterns are indicated by three differently shaded blocks (see Fig. IX-14). (from Yarita *et al.*, 1980). B: Response profiles of 22 LHA neurons examined. Numbers at the top of the matrix refer to individual neurons. (from Tazawa, 1983)

to eight odors is shown in Fig. IX-18. In Fig. IX-16 A, neuron No. 9 displayed +type responses to three odors, CM, BL, and AA, but did not respond to the other odors. One of the other neurons (No. 35) exhibited −type responses to three odors, but did not respond to the other odors. Forty of the 41 neurons showed +type or −type responses exclusively to two to seven of the eight odors. Only one neuron, No. 41, showed a combined type response pattern, that is, +type responses to CL, UDL, CLT, and AA, and a −type response to DE. None of the neurons responded to visual or auditory stimuli.

The finding that 98% (40/41) of MDmc neurons showed responses of either the +type or −type exclusively (Fig. IX-16 A) stands in sharp contrast to the responses observed in the other five olfactory areas; the LPOF, CPOF, PPC-MA, and OB (Figs. IX-13, 14, 15).

6. Responses of Cells in the LHA

Pfaff and Pfaffmann (1969a) recorded neuronal responses to odors in the LHA of anesthetized rats. Shortly thereafter, several investigators studied responses to odors of LHA neurons in rats or mice (Komisaruk and Beyer, 1972; Scott and Pfaff, 1970; Scott and Pfaffmann, 1972), but LHA neuronal responses to odors had not been studied in the monkey until Tazawa *et al.* (1983, 1987) examined them in unanesthetized old world monkeys. These investigators identified 287 neurons that displayed spontaneous discharges, but among them only 22 neurons exhibited significant responses to no more than four of eight odors. The sites of these neurons were later located (Fig. 1 in Tazawa *et al.*, 1987). Their responses were classified into two types: (a) increase in the rate of spontaneous discharges (facilitatory or +type response), and (b) decrease in the rate of spontaneous discharges (inhibitory or −type response).

These 22 olfactory neurons were arranged in a matrix from left to right according to the increasing number of effective odors (Fig. IX-16 B). From these data, a histogram was also made (Fig. IX-18). The histogram clearly shows that the number of neurons responding was negatively correlated with the increase in the number of odors which elicited a response. No neuron responded to five or more odors. None of these 22 neurons responded to either deodorized air, or visual or auditory stimuli.

When Tazawa *et al.* (1983) compared the histograms of response patterns between the LHA neurons (Fig. IX-16 B) and the LPOF neurons (Fig. IX-13 A), they were struck by the close similarity, because, until then, they had previously found entirely different response patterns among all the other areas investigated (Fig. IX-18). Such a close similarity in the response pattern initially caused them to wonder if the same processed information was simply transferred from the LHA to the LPOF, or vice versa (compare histograms of LPOF and LHA in Fig. IX-18). This problem was examined in an HRP study which was discussed previously (pp. 347–354).

Next, when the response patterns of the hypothalamic neurons were compared between the old world monkey (Fig. IX-16 B) and the rabbit (Fig. VIII-9), they were again struck by the difference. It is remarkable that the same structure shows such a different response pattern to the same odors among different animal species. This finding may also support the author's hypothesis that the animal world can be divided into two groups (Takagi, 1983, 1984c, 1986b; p. 390).

7. Responses of Cells in the Septum

Tazawa *et al.* (1987) recorded neuronal responses to eight odors (Table IX-2) in the Spt. Among 18 neurons examined, three neurons responded to odors, all showing facilitatory type responses, thereby demonstrating the presence of neurons responsive to odors in the Spt (Fig. IX-17). In the Acc, however, we failed to detect any neuronal responses to these odors.

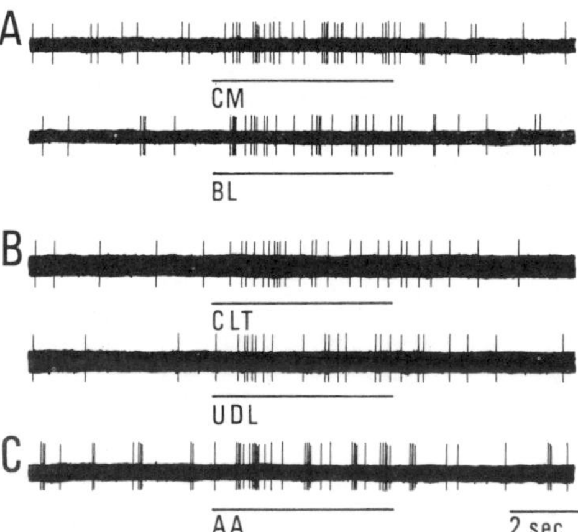

Fig. IX-17. Responses to odors of three neurons in the Spt. Neurons A and B responded to two odors, but not to the other odors. Neuron C responded to only one odor. (from Tazawa *et al.*, 1987)

G. Summary and Conclusions

The response patterns to odors in various olfactory areas have been clarified in our studies. Response patterns were found to be different in

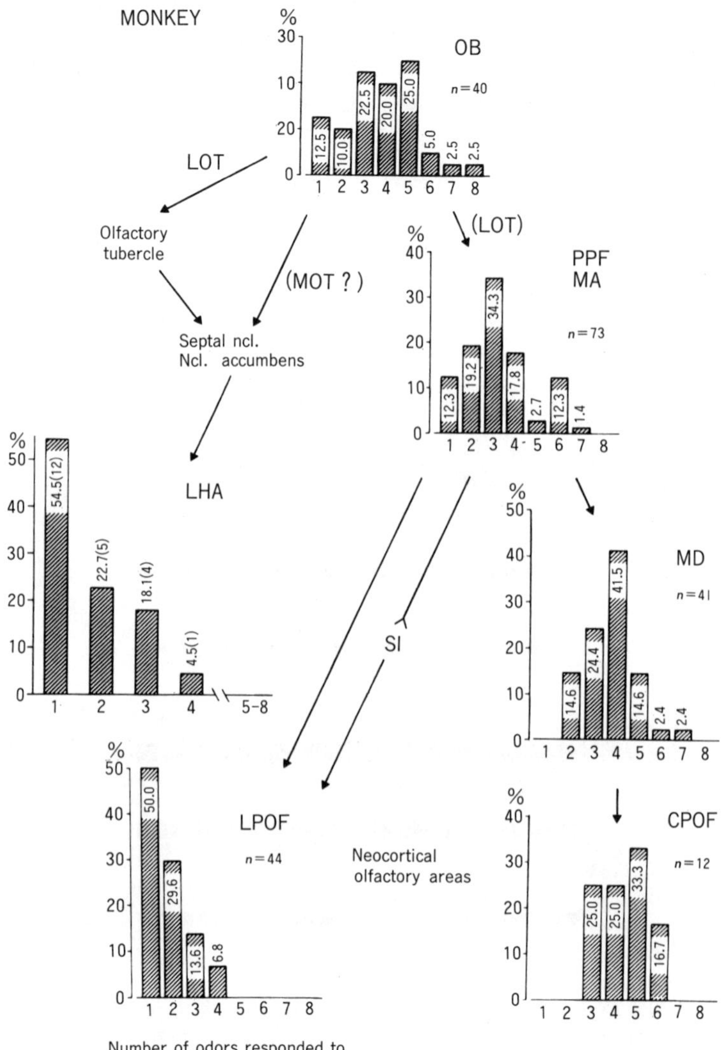

Fig. IX-18. Information processing in various olfactory areas of the old world monkey. (from Takagi, 1984a, b)

different olfactory areas except in the LPOF and the LHA. These results are summarized in Fig. IX-18.

When the response histograms in six olfactory areas are compared in Fig. IX-18, the following two points emerge:

(1) As one ascends the olfactory pathway from the OB through the PPF-MA to the LPOF, the number of odors which elicit cellular responses decreases. Finally in the LPOF, the number of cells responding to only one odor becomes the most numerous and no cell responds to more than five odors. Consequently, we may be able to conclude that this pathway to the LPOF plays the most important role in discrimination of odors.

(2) As one ascends from the OB through the PPF-MA and the MD to the CPOF, numbers of odors which elicit cellular responses become fewer. Finally in the CPOF, the number is limited to three to six. Considering the analytical role of the LPOF, we may presume that the CPOF plays an important role in the integration of odorous information. The author is tempted to say that our appreciation of perfumes and other pleasant odors may occur in the CPOF.

From the results of our electrophysiological and HRP studies in the old world monkey, multiple olfactory pathways have been elucidated (Fig. IX-19) as shown already in Fig. III-1. The old world monkey has two neural pathways to two neocortical olfactory areas, and one pathway to subcortical olfactory areas, but it has no functional vomeronasal system.

As was previously stated, the author has long wanted to clarify the neural mechanisms of olfaction in humans. If the above-mentioned results of our experiments on old world monkeys can be extrapolated, the "human olfactory pathways to the two neocortical olfactory areas" will be as shown in the frontispiece.

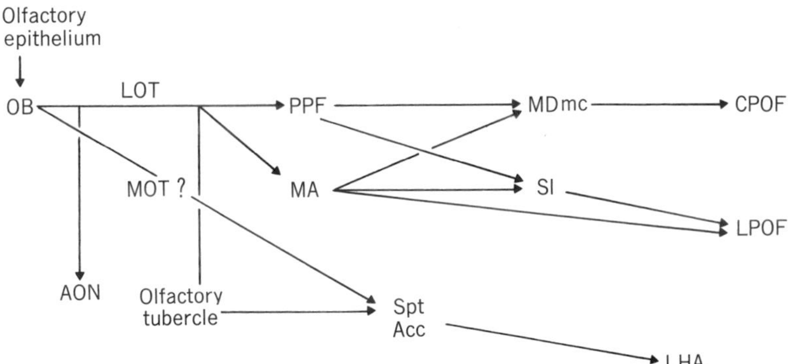

Fig. IX-19. Multiple olfactory pathways in the old world monkey. (from Takagi, 1984b)

X. Studies on Olfactory Nerve Pathways in Mammals Lower than Old World Monkey

In an attempt to elucidate the mechanisms of human olfaction, the author has collaborated with his associates to advance physiological research in the old world monkey, which has an olfactory nervous system akin to that in humans. The results of our experiments on monkeys have been discussed in Chapters VIII, IX, and X. The author has often had to admit that experimental data from the old world monkey are not always coincident with those obtained by other investigators in lower mammals. To explain such differences in the olfactory mechanism between old world monkeys and other lower mammals, therefore, he, with his collaborators, embarked upon olfactory research in lower mammals as well. The author's research question at first was: Do these mammals also have two olfactory areas in the neocortex, as do old world monkeys?

This chapter was added next to Chapter IX to contrast the neural mechanism of human olfaction with that of lower mammals.

A. Neocortical Olfactory Areas and Neural Pathways to Them in Rabbits, Dogs, and Other Mammals

1. Search for an Olfactory Projection Area in the Neocortex (NOPA) of the Rabbit

When Onoda and Iino (1980) started their research on the NOPA in the rabbit, two olfactory pathways had already been identified in lower mammals, mainly in rats, by many neuroanatomists (Broadwell, 1975a; Heimer, 1972; Powell *et al.*, 1965; Raisman, 1972; Scalia and Winans, 1975). They were the main and vomeronasal olfactory pathways: one arising from the olfactory cells, passing through the OB, PPC and MD, and terminating in the OFC; and a pathway from the receptors of the VNO, passing through the accessory OB into the corticomedial nuclei of the amygdala, and terminating in limbic areas such as the hypothalamic, preoptic, and septal nuclei (Fig. X-1). The first pathway may be called a "neocortical olfactory system" and the second a "vomeronasal olfactory system." Allison (1953a) and Keverne (1978) have written provocative reviews on this dual arrangement. It has been shown that old world monkeys and higher primates including man are devoid of the vomeronasal system, whereas new world monkeys and lower mammals have this system (Stephan, 1965; B, 1.3. in this chapter). Consequently, the latter group of animals are often called (lower) mammals with the vomeronasal system (VNS) in this book and separated from the former group of animals.

1.1. Identification of an olfactory projection area in the neocortex

The idea of an olfactory area in the OFC that receives afferent fibers from the mediodorsal nucleus of the thalamus (MD) had been generated by some neuroanatomical studies (Allison 1953a; Leonard, 1969; Powell, Cowan and Raisman, 1965) and one physiological investigation (Allen, 1943). Tanabe *et al.* (1973, 1974, 1975a, b, c) later demonstrated such an area in the old world monkey physiologically, and many morphological and physiological studies have further supported this idea (Benjamin *et al.*, 1978; Leonard, 1972; Motokizawa and Ino, 1981a, b; Naito *et al.*, 1984; Onoda *et al.*, 1981a, b, 1982a, b, 1984; Potter and Nauta, 1979; Powell *et al.*, 1965; Yarita *et al.*, 1978, 1980). Encouraged by neuroanatomical results (Allison, 1953a; Powell *et al.*, 1965) as well as by physiological findings (Tanabe *et al.*, 1973, 1974, 1975a, b, c), Onoda and Iino

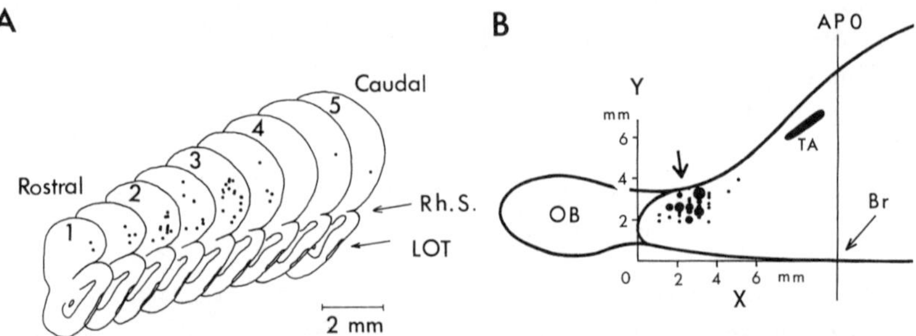

Fig. X-1. Locations of cells responding to odors in the rabbit neocortex. A: Three dimensional distribution of 51 neurons on frontal sections. Each unit is indicated by a small dot. The number on the top of each section gives the distance (mm) from the frontal pole. B: The dorsal surface of the anterior hemisphere. The dotted area indicated by a thick arrow shows the extent of 51 neurons analyzed. The size of dots expresses the density of neurons. The solid area in black shows a cortical taste area (TA) designated by Yamamoto and Kawamura (1975). X shows the distance from the frontal pole along the antero-posterior axis in mm. Y shows the distance from the sinus vein along the medio-lateral axis. Abbreviations: APO means 0 mm in the antero-posterior dimension of the rabbit brain atlas. (from Onoda *et al.*, 1984)

(1980) searched for neuronal responses to odors in the OFC of the unanesthetized rabbit. Initially, they applied the odors of the eight chemicals that were successfully used in monkey experiments (Table IX-2) by Tanabe *et al.* (1975c). In rabbit experiments, however, they could only rarely find neurons responding to these odors. Other kinds of odors were then used. They attempted to apply biologically meaningful odors, namely, the animal's own urine and fecal odors and those of conspecifics. The odor of dry laboratory food pellets for rabbits was also used. At least one of these five biologically significant odors elicited responses in 51 neurons from among more than 1000 neurons examined. The site of the neuron was marked after each recording. The sites of 51 neurons were later checked on coronal sections of the frontal lobe (Fig. X-1 A).

They identified only one olfactory projection area in the orbitofrontal cortex of the rabbit, occupying the lateral half of the OFC one to five mm caudal to the frontal pole (Fig. X-1 B).

1.2. Neural pathways to the NOPA

An electric stimulus was applied to the magnocellular portion of the MD (MDmc) and responses were recorded in the NOPA at depths of 1.2 and 1.5 mm, corresponding to the dorsal bank of the rhinal sulcus, to see if thalamic fibers terminate in the NOPA. This study demonstrated that the MDmc projects fibers to the NOPA.

This thalamocortical connection was then studied morphologically by injecting HRP into the NOPA (Onoda *et al.*, in preparation). Labeled cells were found predominantly in the MD (39.2%), with fair numbers of labeled cells in the PC (18.3%), VM (16.8%), and FC (14.8%; Fig. X-2). Smaller numbers of labeled cells were found in the preoptic area (3.5%), in the AON (about 2%), and in the frontal gyrus of the contralateral hemisphere (Fig. X-2). This indicated the presence of not only a transthalamic pathway but also a direct input to the NOPA from the PC. Motokizawa and Ino (1981b) obtained similar data in the cat.

These physiological and neuroanatomical experiments demonstrated transthalamic and extrathalamic olfactory pathways to the NOPA in the rabbit. In the hamster, projections of fibers to the NOPA from the OT and ER were similarly found by Reep and Winans (1982).

1.3. Responses of NOPA neurons to odors

Onoda and Iino (1980) and Onoda *et al.* (1984) searched for neuronal responses to eight chemicals (Table IX-2) and five biologically significant odors in the NOPA. The latter odors were vapors from the animal's own fresh urine and feces, conspecific's fresh urine and feces, and dry food pellets for the animal (Fig. X-3). Facilitatory responses were found to be more than twice as numerous as inhibitory responses. A marked

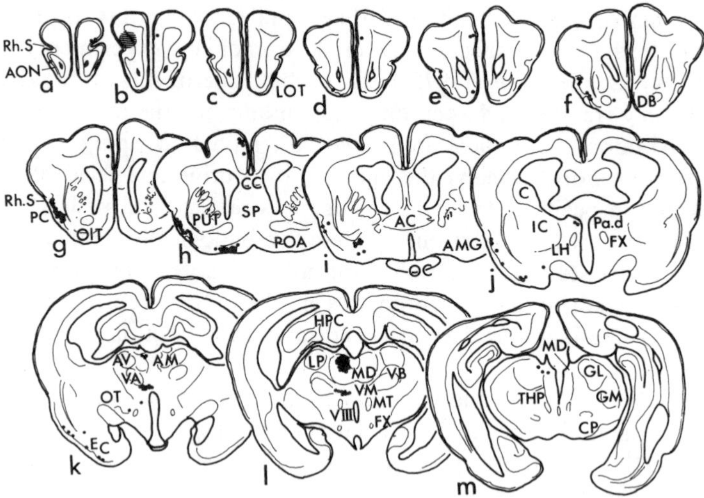

Fig. X-2. A series of representative frontal sections illustrating the distribution of HRP-labeled neurons obtained from six rabbits. The hatched area in b indicates the HRP deposit site. (from Onoda *et al.*, unpublished data)

tendency for these cortical neurons to respond readily to the biologically significant odors, but only rarely to the odors of the eight synthetic chemicals, was found when the responses of these 51 cells were examined (Fig. X-3). Facilitatory, inhibitory, and mixed response patterns were found, as shown in Fig. X-3. To evaluate statistically whether the response was significantly facilitatory or inhibitory, a cumulative sum analysis test (cusum test) was applied. For details of this statistical method, the reader should consult articles by Imamura and Onoda (1983), Imamura *et al.* (1984a), and Onoda *et al.* (1984).

Figure X-3 also shows other important findings. In some rabbits, the accessory olfactory nerves were cut bilaterally to remove the effect of the vomeronasal olfactory system, but no particularly different response patterns were obtained after nerve section (indicated by * in Fig. X-3). In other rabbits, the nasociliary nerves, branches of trigeminal nerves, were cut bilaterally. Again, no particularly different patterns were observed after the trigeminal effect was removed (indicated by #). Six neurons also responded to electric stimulation of the MD (indicated by ∂). Thus, these cells were concluded to receive a transthalamic olfactory input.

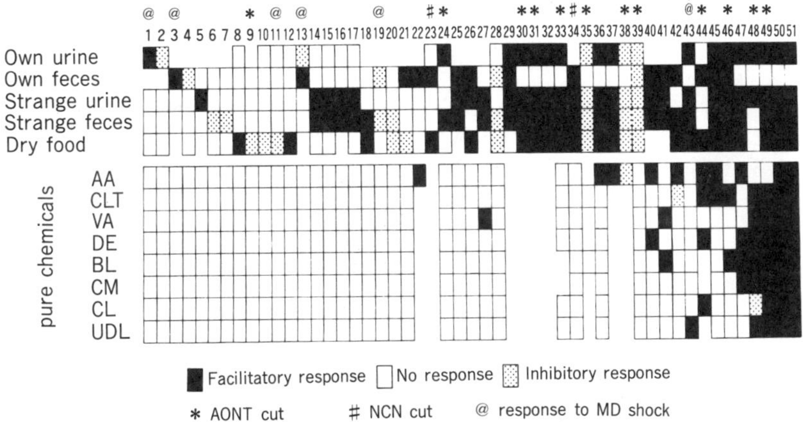

Fig. X-3. A panel histogram of neocortical responses obtained from 51 odor-sensitive rabbit neurons to odors. The number over the histogram indicates the unit number arranged in order of the number of odors responded to. AON (accessory olfactory nerve) cut (*): Units recorded from rabbits whose accessory olfactory nerves were cut. NCN (nasociliary nerves) cut (#): Units recorded from animals whose NCN were severed. @: Units that responded to MD shocks. (from Onoda *et al.*, 1984)

2. Search for a Neocortical Olfactory Area in the Dog

Allen (1940) studied the effects of ablating various areas of the neocortex on the positive and negative olfactory conditioned reaction in the dog. He found that the frontal lobe is concerned with such higher order responses as discrimination. He then wanted to locate an olfactory area in the frontal lobe, so he applied electrical pulses to the OB or the PPC and found an evoked potential in the orbital cortex (Allen, 1943b). This appears to have been the first report suggesting the presence of an olfactory area in the neocortex.

2.1. Finding an olfactory projection area in the orbital gyrus

Inspired by Allen's (1943) work and encouraged by Tanabe *et al.*'s (1973, 1974, 1975a, b, c) and Yarita *et al.*'s (1978, 1980) findings in the monkey, Onoda *et al.* (1981a, b) explored an olfactory area in the cortex. Applying eight chemicals and five biologically significant odors, they sought neuronal responses to odors in unanesthetized dogs. They recorded responses of 24 neurons and later marked the sites of recording, thereby delineating an olfactory area in the orbital cortex (Fig. X-4). A typical example of their recording sites is shown in Fig. X-5. It is a drawing of the frontal section and shows that the electrode track penetrated the

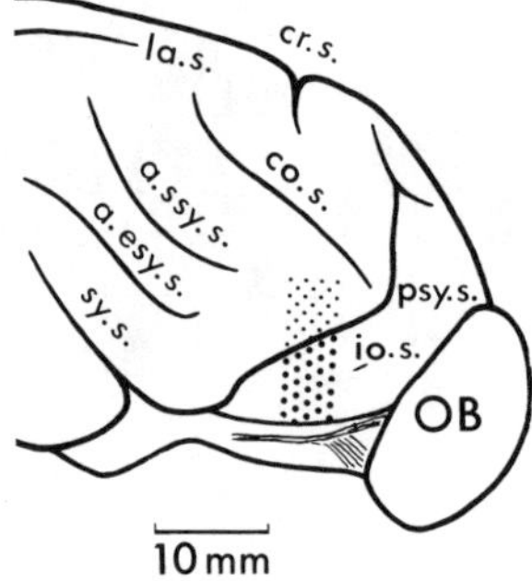

Fig. X-4. A lateral view of the dog hemisphere. The areas indicated by small and large dots show the location of units which responded to odors. The small dotted area indicates the dorsal bank of the ORG which is hidden by the anterior compositus gyrus. Abbreviations: a.esy.s., anterior ectosylvian sulcus; a.ssy.s., anterior suprasylvian sulcus; co.s., coronal sulcus; cr.s., cruciate sulcus; io.s., intraorbital sulcus; la.s., lateral sulcus; psy.s., presylvian sulcus; sy.s., sylvian sulcus. (from Onoda *et al.*, 1981a)

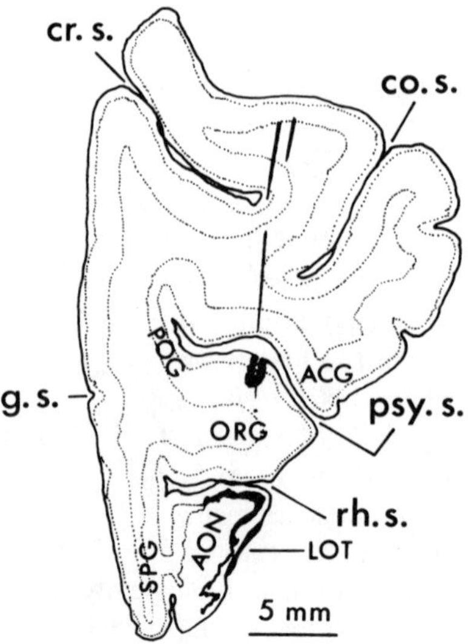

Fig. X-5. A typical example of unit location in a frontal section of the dog ORG. A whole view of a frontal section. Abbreviations: co.s., coronal sulcus; cr.s., cruciate sulcus; g.s., genual sulcus; psy.s., presylvian sulcus; ACG: anterior composite gyrus, POG: paraorbital gyrus, SPG: subprorean gyrus. For other abbreviations, see Table 1. (from Onoda *et al.*, 1982a)

anterior composite gyrus and reached the dorsal bank of the orbital gyrus (ORG) where a spot of blue dye (Hellon, 1971) beneath the end of the track was located in the fifth layer of the ORG. These histological examinations confirmed that the neurons were confined to a restricted region in the ORG (Fig. X-4) that corresponded to the 'ORB II a and b' according to Kreiner's (1961) classification. This region is anatomically comparable to area 13 in the monkey and to the sulcal cortex in subprimates (Benjamin *et al.*, 1978; Krettek and Price, 1977a; Leonard, 1969; Rose and Woolsey, 1948).

2.2. Responses of NOPA neurons to odors

In dog experiments, vapors from eight chemicals and five biologically significant odors were applied in the same way as in rabbit experiments.

Various patterns were observed in the neuronal responses to these odors. Odor responses obtained from 24 neocortical neurons are summarized in Fig. X-6 A. In this figure, for instance, neuron no. 1 responded solely to the self-urine odor, and showed a different facilitatory response to every application of this odor. Responses of neuron no. 10 showed a facilitatory response to two self-odors, feces and urine, but no response to other odors. Many units (29%) showed facilitatory responses. An inhibitory response and a mixed response were also found. Although not all of the odor stimuli were tested on each unit, the majority of units responded to only one of the odors (38%). These results suggest that

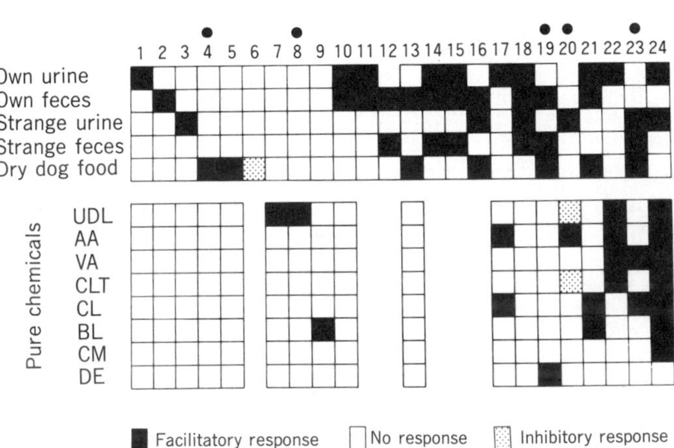

Fig. X-6 A. A panel histogram of neuronal responses to odors in the ORG of the dog. They are arranged in the order of the number of effective odors. Dots on the neuron numbers indicate that the nasopalatine ducts were closed on both sides. (from Onoda *et al.*, 1981a)

the dog OFC contains odor-specific neurons, such as that of urine or of feces, in a manner similar to neurons found in the rabbit neocortex. In the dog, a tendency for responses to odors of eight chemicals to be few in number was again found. In total, ten neurons (42%) out of 24 responded to these chemical odors, whereas 21 neurons (87.9%) responded to biologically significant odors. Three units (13%) showed a facilitatory response to only one synthetic odor but not to urine or feces.

In the rabbit, 29% of the neocortical neurons responded exclusively to biologically significant odors and 31% responded to these odors as well as to synthetic chemical odors. No neuron responded solely to synthetic odors (pp. 377–378 and Fig. X-3). In the dog, however, 42% of the recorded neurons responded to both kinds of odor. Furthermore, three units showed a facilitatory response to only one synthetic odor. Thus, the above evidence show that results obtained in the dog are not similar to responses in the monkey or in the rabbit, but lie somewhere in between.

Odors of animal waste and dry food pellets have been considered to stimulate the vomeronasal organ. However, some neocortical neurons responded to these odors even in preparations with the accessory olfactory nerve cut or with the nasopalatine ducts closed on both sides. Odors of animal waste and pellets thus appear to stimulate not only the vomeronasal organ but also the primary olfactory mucosa. Although the odors of animal waste and food pellets have not been adequately analyzed, we can at least say that biologically significant odors selectively elicit neocortical responses in the dog as well as in the rabbit.

2.3. Responses to odors of sex steroid hormones

Four steroids, progesterone, estradiol, testosterone, and androsterone, were used as odor stimuli together with the five biologically significant odors.

Neural responses to odor stimuli of more than several hundred units were studied. Most of the neurons failed to respond to any of the odors. Twenty-four units, however, showed significant changes in discharge rate during the application of at least one of the specific odors.

Out of those 24 ORG neurons, 14 (58.3%) responded to at least one of the steroids. Neocortical responses obtained from those 14 neurons to steroids are shown in Fig. X-6 B. All of those neurons showed facilitatory responses to odors of steroids. The number of neurons responding to only one steroid was the largest (eight neurons).

As reported in four previous studies (Onoda and Iino 1980; Onoda *et al.*, 1981a, b; Tanabe *et al.*, 1975c; Yarita *et al.*, 1980), individual neurons in the OFC of the rabbit, dog, and monkey showed highly selective responses to odors. Recent behavioral studies have demonstrated that lesions of the OFC disrupted odor preferences and impaired sexual be-

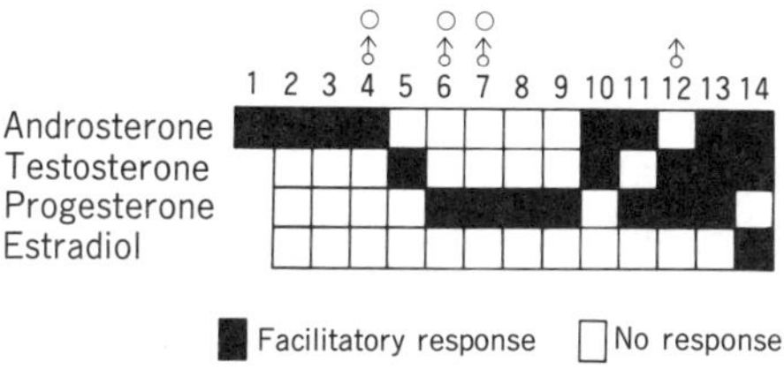

Fig. X-6 B. A panel histogram of responses to odors of sex steroid hormones of 14 dog ORG neurons. Open circles over the neuron number indicate neurons that responded to one steroid odor only. ♂ denotes a male dog. The neuron number was arranged in the order of the number of effective odors. (from Onoda *et al.*, 1982)

havior (Eichenbaum *et al.*, 1980; Sapolsky and Eichenbaum, 1980). Onoda *et al.* (1982a, b) also found that particular ORG neurons can re spond exclusively to steroid odors. Thus, in the dog, steroids would work as pheromones which could communicate olfactory messages concerning sexual differences or the sexual state of the animal. In fact, Kloek (1961) reported that the dog was capable of recognizing the odors of simple steroids, and that steroids can be classified as pheromonal substances.

Putative pheromones identified as steroids have been isolated from the male pig (Patterson, 1968), and 5-androsterone and testosterone have been shown to alter the action potential activity in the pig OB (MacLeod *et al.*, 1979). Since the metabolites of steroids are excreted from the animal, the use of highly active simple steroids is not ideal. It is interesting, however, to consider the effect of steroid-activated olfactory receptors on neuroendocrine structures. Considerable evidence indicates that olfaction affects mammalian reproductive behavior. Considering also the presence of anatomical connections between the lateral hypothalamic area and the ORG (Oomura *et al.*, 1980), the ORG neurons responding to steroid odors could presumably modulate the neuroendocrine system.

3. Search for a Neocortical Olfactory Area in the Cat

Fox *et al.* (1944), Berry *et al.* (1952), and other investigators applied electrical pulses to the OB and recorded evoked potentials in various areas of the cat brain. These investigators were looking for olfactory areas in the brain, but apparently such potentials were not found in the cat neocortex until Motokizawa (1974b) successfully recorded them.

3.1. An olfactory projection area in the neocortex

Motokizawa (1974b, 1976, 1977) and Motokizawa *et al.* (1977) applied electrical pulses to the OB and found evoked potentials in many areas of the cat brain. Eliminating artifactual potentials due to volume conduction, they located a true olfactory projection area (in his parlance an "olfactory receiving area") only in the orbital gyrus. To support this finding, Motokizawa and Ino (1981a) applied three odors, amyl acetate (AA), clove oil, or xylene (XY), and recorded a facilitatory response to AA and an inhibitory response to XY. This definitively demonstrated the presence of cells that respond to odors in this area (Motokizawa, 1986).

Although they found responses of cells both to odors and to electrical stimulation of the OB in this area, they also found that the same cells responded to stimulation of the trigeminal and vagus nerves. Thus, they showed that neurons in the olfactory area identified in the orbital gyrus are polysensory. However, they did not look for cells in this neocortical area that might respond specifically to odors. Motokizawa (personal communication) have sharply criticized the conventional method of applying odors, and have chosen instead to use mainly electrical stimulation in their olfac tory research, applying odors only in exceptional cases (Motokizawa and Ino, 1981a). In consequence, it is open to question whether or not their olfactory area contained cells that respond specifically to odors.

3.2. Neural pathways to the NOPA

Motokizawa *et al.* (1977) stimulated the OB electrically, and found that potentials evoked in the mediodorsal nucleus of the thalamus (MD) had latencies of 10 to 30 msec, whereas potentials evoked in the pyriform cortex and in the orbital gyrus had latencies as short as four and five msec, respectively. From these observations, they inferred the presence of a direct olfactory pathway to the orbital gyrus from the PPC. To confirm this electrophysiological finding, Motokizawa and Ino (1981b) injected HRP in the dorsal bank of the rhinal sulcus and found labeled cells in the ventral bank of the rhinal sulcus, lateral aspect of the PPC, olfactory tuberecle, and cortical amygdaloid nucleus but not in the MD.

Powell *et al.* (1965), in a histological study of the rat, demonstrated a projection of fibers from the PPC to the MD. Rose and Woolsey (1948) showed fiber connections between the MD and the OFC in rabbits, sheep and cats. This prompted neuroanatomists to assume the presence of an olfactory pathway from the PPC to the OFC through the MD. Motokizawa *et al.* (1977), however, found no labeled cells in the cat MD. Thus, while they could confirm the presence of a direct olfactory pathway from the PPC to an olfactory area in the orbital gyrus in the cat, they could not support the idea of the transthalamic olfactory pathway already demonstrated in monkeys and rabbits. Further studies are needed, however, to definitively establish the absence of this pathway in the cat.

3.3. Responses of NOPA neurons to odors

Motokizawa and his collaborators seldom studied responses of neurons to odors in the neocortical olfactory area. They only showed a facilitatory response to amyl acetate and an inhibitory response to xylene (Motokizawa and Ino, 1981a). Hence, this subject awaits future research.

4. Search for a Neocortical Olfactory Area in the Rat

Most neuroanatomical research on olfactory nerve pathways has been performed in the rat (Powell *et al.*, 1965; Heimer, 1972; Raisman, 1972; Scalia and Winans, 1975). A dual system of olfactory pathways (cf. Raisman, 1972; Keverne, 1979) has been suggested. The main pathway starts from the main OB, passes through the PC and MD, and terminates in the OFC. Electrophysiological studies on these olfactory pathways in the rat, however, have been rare and fragmentary.

4.1. An olfactory projection area in the neocortex

Although histological studies indicate a projection of olfactory fibers to the OFC (Powell *et al.*, 1965), this has yet to be studied electrophysiologically. Giachetti and MacLeod (1975, 1977a, b) demonstrated an olfactory input to a neocortical taste area, situated close to the somatic I area of the tongue (Fig. X-7). Convergence of olfactory and gustatory sensations has long been postulated, but has yet to be unequivocally demonstrated. Identification of this new olfactory sensory area may provide a physical basis for the heretofore "psychological" phenomenon of smell influencing taste.

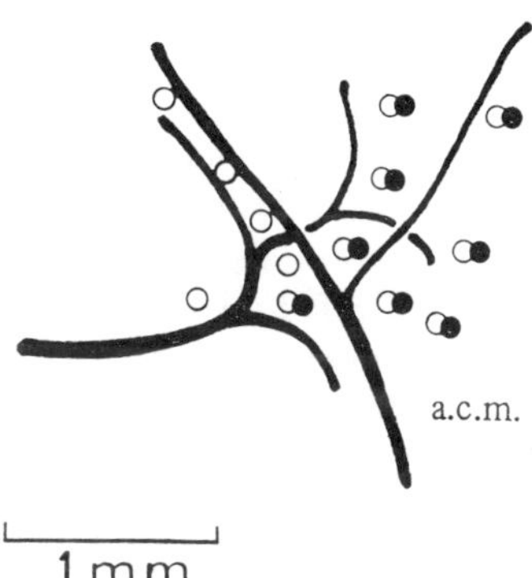

Fig. X-7. Sites where evoked potentials were recorded on the surface of rat brains. ○ and ● indicate loci of olfactory evoked potentials and somatic evoked potentials, respectively. a.c. m., arteria cerebri media. (from Giachetti and MacLeod, 1975)

4.2. A neural pathway to the NOPA

Giachetti (1977) and Giachetti and MacLeod (1975) found that an olfactory pathway to this newly defined olfactory area passes through the MV nucleus of the thalamus, but not through the MD. An olfactory connection between the PPC and the MV and MD was demonstrated by Powell *et al.* (1965). Olfactory projections to the MV were later confirmed and projections to other thalamic areas were also found in the rabbit (Imamura *et al.*, 1980).

4.3. Responses of NOPA neurons to odors

Giachetti (1977) and Giachett and MacLeod (1975) examined responses of cells in this olfactory area to their five odors. They found the cells to exhibit excitatory responses to some odors and inhibitory responses to others.

5. Possible Olfactory Pathways to a Neocortical Olfactory Area in Mammals Lower than Old World Monkeys

Regarding the olfactory nerve pathway to the neocortical olfactory area, Powell *et al.* (1965) and other histologists (Broadwell, 1975a; Heimer, 1972; Raisman, 1972; Scalia and Winans, 1975) have indicated in rats and other mammals that the main olfactory pathway starts from the olfactory cell, passes through the OB, PPC, and MD, in this sequence, finally terminating in the OFC. Electrophysiological experiments on rabbits, dogs, and cats mentioned above demonstrated that an olfactory area is located in the orbitofrontal or orbital cortex. Another important electrophysiological finding was the short latency of the evoked potential in the OFC elicited by OB stimulation. Latencies as short as four to five msec suggests the presence of direct nerve fibers from the OB to the OFC (Motokizawa *et al.*, 1977). On the other hand, Onoda and Imamura (1983a) injected HRP into the NOPA and found many labeled cells in the AON, pyriform cortex, and other areas including the MD. More precisely, Tazawa *et al.* (1982b) injected HRP intracellularly into AON neurons in the rabbit and traced stained axons to the neocortex dorsal to the rhinal sulcus (Fig. X-8), i.e., the NOPA. These two experiments indicate the presence of direct fibers from the AON to the OFC. In contrast, no histological paper has been found which demonstrates the presence of direct fibers from the OB to the OFC. Correspondingly, in the cat, Motokizawa (1974b, 1976) and Motokizawa and Ino (1981b) showed direct projection of fibers from the olfactory cortex to the orbital gyrus.

From these findings, olfactory pathways in the lower mammals were delineated as shown in Fig. X-9 (cf. Fig. IX-19).

Recently, Price and Slotnick (1983) histologically showed in the rat the presence of a substantial thalamocortical mechanism available for

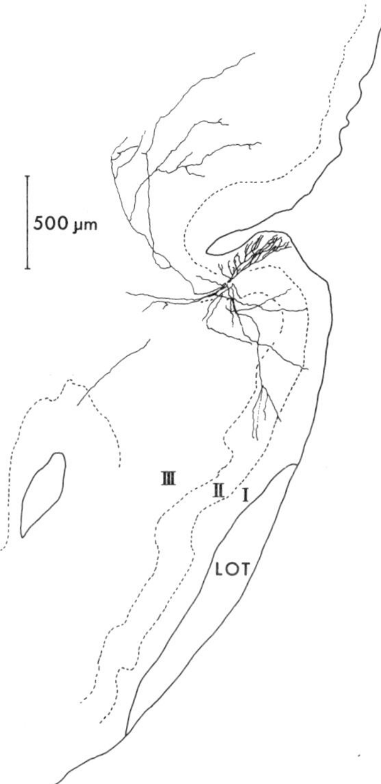

Fig. X-8. A camera lucida drawing of an AON celi in layer II. An arrow indicates axonal projection into the dorsal bank of the rhinal sulcus, i.e., the NOPA.

processing olfactory information.

The cells in layer Ia of the primary olfactory cortex receives afferent fibers from the OB and sends association fibers which terminate primarily in layer Ib of the cortex itself. The same cells also project to the prefrontal cortical areas in the dorsal bank of the rhinal sulcus, and to the large polymorphic zone of the olfactory tubercle (En/PZ). These cells send axons to the mediodorsal (MD) and submedial (SM) thalamic nuclei, which are themselves interconnected with the same areas of the prefrontal cortex. Three cytoarchitectonically distinct neocortical areas which are possibly related to olfaction are ventrolateral orbital area (VLO), lateral orbital area (LO), and ventral agranular insular area (AIv) (Fig.

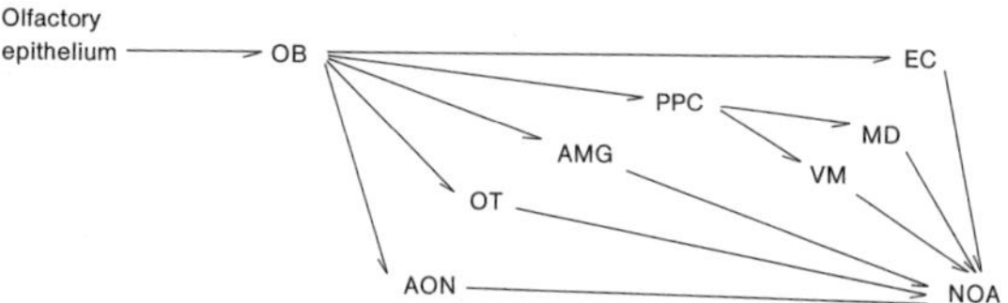

Fig. X-9. Multiple olfactory pathways in mammals with the VNS. (Onoda, personal communication)

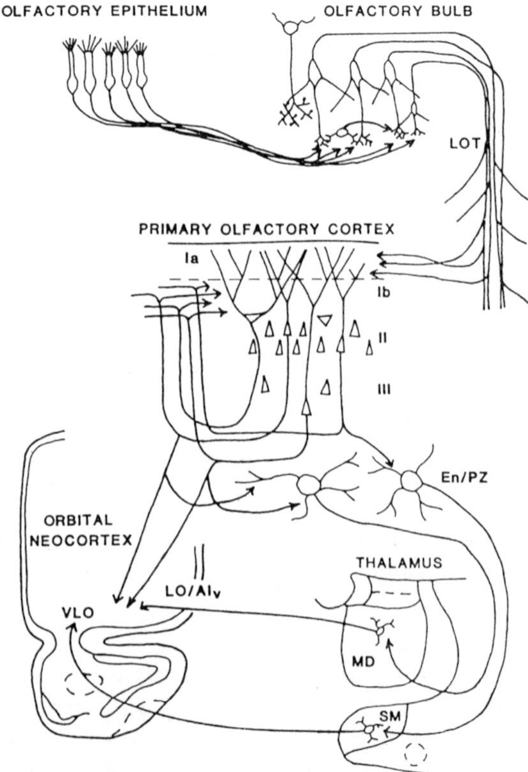

Fig. X-10. A summary diagram of some principal connections within the olfactory system. Abbreviations: En/PZ, entorhinal area/polymorphic zone of olfactory tubercle; Lo/Alv, lateral orbital area/ventral agranular insular area; VLO, ventrolateral olfactory tubercle; Lo/Alv, lateral orbital area/ventral agranular insular area. VLO, ventrolateral orbital area. (from Price and Slotnick, 1983)

X-10). As in other sensory systems, they assumed that each of these structures is involved in a different aspect of olfactory processing. This finding has provided olfactory neurophysiologists with new subjects for research. So far, however, no evidence which supports the presence of more than one neocortical olfactory area has been obtained in the rabbit and dog.

6. *Information Processing in the Olfactory Pathway of the Rabbit*

As already shown in Fig. VIII-6, Iino and Takagi (1983) examined neuronal responses of rabbit OTs to eight chemicals and five biologically significant odors and showed them in a matrix (Fig. X-11 A). Onoda and Iino (1980) and Imamura *et al.* (1984a) recorded neuronal responses of rabbit MDs to the same 13 odors and showed them in a matrix (Fig. X-11 B). When these two histograms are compared, it is apparent that neurons in these areas respond well not only to the five biologically significant odors but also to the eight chemical odors, and that neurons that respond to all 13 odors are more numerous in the OT, a lower level olfactory area, than in the MD, a higher level.

Imamura *et al.* (1984) compared the response pattern of MD neurons (Fig. X-11 B) with that of NOPA neurons (Fig. X-3). While in the old world monkey the number of odors to which single cells responded decreased along the olfactory pathway from the OB to the LPOF (Tanabe *et al.*, 1975c) and along the pathway from the MDmc to the CPOF (Yari-

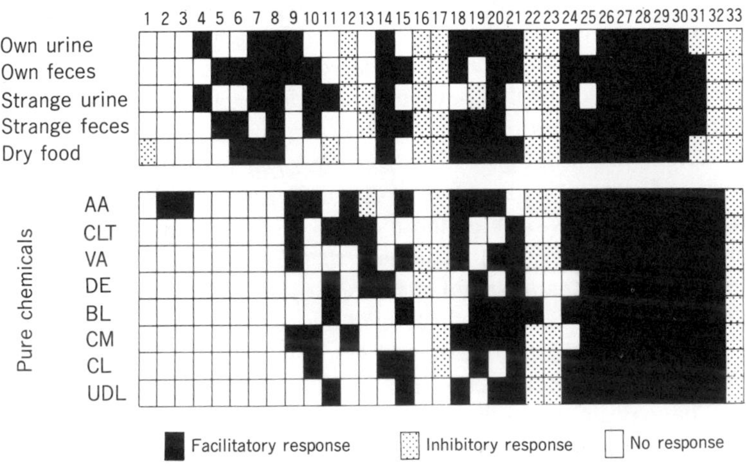

Fig. X-11 A. A panel histogram of responses to odors obtained from 33 odor-sensitive olfactory tubercle (OT) neurons. They are arranged in order of the number of odors responded to. (Iino and Takagi, 1983)

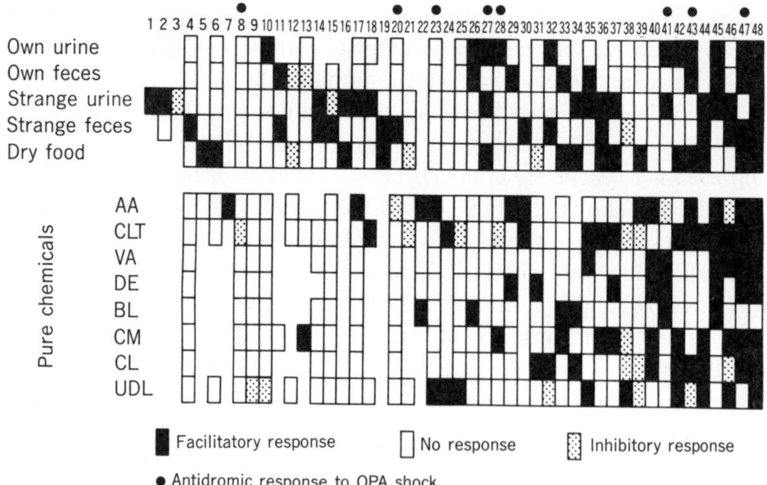

Fig. X-11 B. A panel histogram of responses of odors obtained from 48 odor-sensitive MD neurons of rabbits. Numbers over the histogram indicate the number of units recorded from. * Thalamocortical relay neurons which responded antidromically to OPA shocks (see text). Abbreviations of odors shown in the extreme left are listed in Table IX-2. (Imamura *et al.*, 1984a)

ta *et al.*, 1980), an entirely different tendency was observed in the rabbit. The number of neurons that responded to the eight chemical odors strikingly decreased in the NOPA compared with that in the MD. In fact, while over 60% of NOPA neurons responded exclusively to the five biologically significant odors (A), no NOPA neurons responded exclusively to the eight chemical odors (P). Most (76%) of the MD neurons, on the other hand, responded to odors of A and P, and neurons that responded only to either odors of A or those of P were few in number (Fig. X-12).

Imamura *et al.* (1984a) suggested, therefore, that most of the highly selective neurons that respond characteristically to biologically significant odors only appear finally at the level of the neocortical olfactory area, the NOPA. Very different from the old world monkeys and higher primates including man, in the lower animals it is very probable that information about biologically insignificant odors is suppressed in subcortical olfactory areas and only biologically important information is conveyed to the cortical olfactory area (NOPA).

7. Conclusions

At the present stage of physiological research on rabbits and dogs,

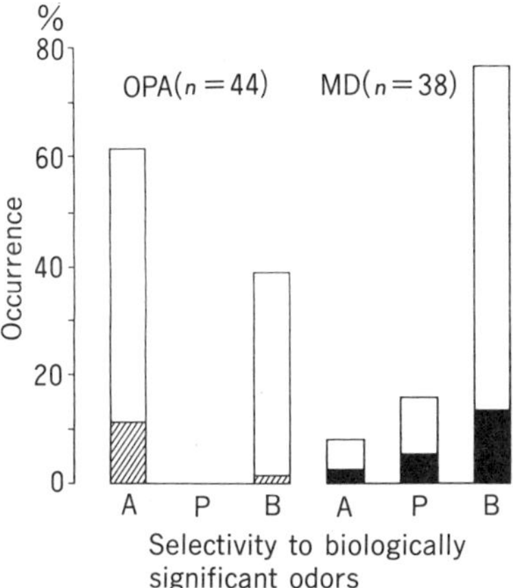

Fig. X-12. Comparison of odor responses of OPA neurons with those of MD neurons. Histograms show percentages of neurons responding to biologically significant and pure chemical odors. A: Units which responded exclusively to odors of animal products and/or dry food pellets; P: units which responded to odors of pure chemicals alone; B: units which responded to both odor groups, A and P. Striped and dotted bars indicate unit populations which responded orthodromically to MD shocks and antidromically to OPA shocks, respectively. (from Imamura *et al.*, 1984a)

the experiments of Onoda *et al.* (1981a, 1982a, 1984) reliably indicate that there exists a single olfactory area in the neocortices of these animals. The fact that only one neocortical olfactory area (NOPA) is present in rabbits and dogs, while two neocortical olfactory areas (LPOF and CPOF) exist in old world monkeys, may be extrapolated to a hypothesis that the animal world can be divided into two groups: old world monkeys and higher primates including man, and new world monkeys and lower mammals. The former group has two neocortical olfactory areas but has no VNS, while the latter group has only one neocortical olfactory area, but has the VNS.

B. The Vomeronasal System (VNS)

New world monkeys, lower mammals, and other nonmammalian vertebrates mostly have another kind of olfactory organ, named the vomeronasal or Jacobson's organ. This organ plays a very important, and often an essential, role in their daily lives. As will be mentioned in the later part of this section, sexual, chemoinvestigatory, and other functions are

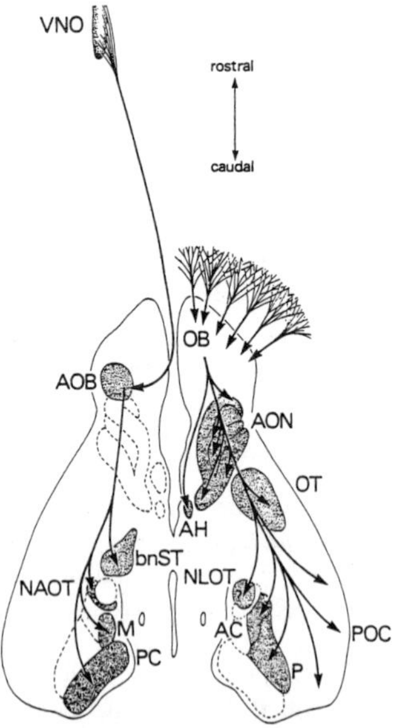

Fig. X-13. A scheme of a horizontal view of the rat brain comparing the independent primary and secondary projections (ipsilateral only) of the olfactory and vomeronasal systems (right and left halves, respectively). For clarity, secondary contralateral projections and many olfactory bulbar projections to rostral midline structures are not shown. AC: Cortical nucleus of the amygdala, AH: anterior hippocampus, bnST: bed nucleus of stria terminalis, M: medial nucleus of the amygdala, NAOT: nucleus of the accessory olfactory tract, NLOT: nucleus of the lateral olfactory tract, P: periamygdaloid region, PC: posterior cortical nucleus of the amygdala, POC: primary olfactory cortex. (from Wysocki, 1979)

mediated by this organ. The fact that human olfaction is inferior in many respects to that of animals may be partly or wholly explained by the absence of the VNO in humans.

In the author's 30 years of research into the olfactory nervous system, he has never had a chance to extend his study into this field. In this sense, he is again not well qualified to review the numerous experimental results on this subject. Readers interested in the VNS should consult two excellent review papers, in particular, published by Halpern (1987) and Wysocki (1979), which greatly facilitated the completion of the following review.

Based upon experimental data, Wysocki (1979) depicted neural pathways of the main olfactory and the vomeronasal systems in a diagram (Fig. X-13). Thus, differences between the two systems were clearly indicated (Licht and Meredith, 1986).

1. The Vomeronasal Organ (VNO) or Jacobson's Organ
1.1. Morphology of the VNO

The vomeronasal (Jacobson's) organ is situated in either a region, usually ventral, contiguous with the nasal cavity, or a separate pouch, a nasal diverticulum that in embryogenesis may have become totally separated from the nasal cavity, opening only into the mouth (Fig. IV-2).

The size and presence of the VNO varies considerably across mammalian orders but is probably best developed in pouched and egg-laying, meta- and protheterians. Among eutherians, the organ is best developed in lagomorphs and rodents (Wysocki, 1979).

In most mammals, the VNO is a cigar-shaped tube contained in a bony capsule located along the ventral edge of the nasal septum. Posteriorly the tube ends blindly and anteriorly it frequently communicates via its own duct with the nasopalatine duct, thus permitting communication with the nasal cavity, oral cavity, or both. The VNO has a lumen that is typically crescentric in cross section with a convex lateral wall lined with ciliated pseudostratified nonsensory epithelium and a concave medial wall lined with a specialized neurosensory epithelium (Barber and Raisman, 1974; Bertmar, 1981a, b; Cooper and Bhatnagar, 1976; Hunter *et al.*, 1984; Loo, 1977b; Maier, 1980)

1.2. Origin of the VNO

Some investigators including the author have thought that the VNO is phylogenetically older than the olfactory organ, and that the VNO appeared first in the most primitive form of vertebrates, then the olfactory organ appeared in the next stage of evolution, working together with the VNO and finally in the present stage of evolution, the VNO has disappeared or became vestigial and the olfactory organ has become well developed in old world monkeys and higher primates including humans, while the VNO, and olfactory organ are working together in lower mammals.

Parsons (1979) seemed to answer this question most appropriately and elegantly. Hence, his paper is cited with his permission:

Almost nothing is known about the origin or early history of the VNO. Since the nose is almost always surrounded by cartilage, fossils do not show its structure. One suggestion is that the VNO is the "original" nasal organ of the fish and that the olfactory epithelium of tetrapods is a new development adapted to function in an aerial rather than aquatic environment; the VNO is retained to smell substances dissolved in fluid, especially that in the mouth. The absence of Bowman's glands in VN epithelium and the ventral position of the latter support this theory, but the absence of the VN cilia does not. Also, the postulated functional difference is not substantiated by more recent work.

The differentiation between olfactory and VN epithelia presumably arose in the primitive amphibians or possibly their immediate ancestors among the fish. It seems likely that this differentiation occurred in relation to some factor important in terrestrial as opposed to an aquatic environment. The VNO is only weakly differentiated in many modern amphibians and turtles; possibly this is retention of a primitive condition, but modern amphibians are often secondarily simplified, so this is not a safe assumption. Neither are most characters of turtles primitive. In all higher forms the VNO develops as a ventromedial pocket of the nasal pouch in the embryo and, if it persists in the adult, become well, even completely, separated from the nasal cavity.

The fairly general retention of the VNO implies some important function, but its distribution among living tetrapods does not suggest any particular special function for it. Rather, it would appear simply to be part of the regular olfactory system. In certain groups, especially birds and crocodilians, the VNOs are greatly reduced or absent in adults, whereas they are best developed in snakes and lizards. In general, the VN epithelium is reduced or lost whenever and to much the same extent as is the olfactory epithelium. Thus, in many arboreal (chameleons and primates), aerial (birds and bats), and marine (sea snakes and whales) forms, the VNOs are greatly reduced or absent; that organ, like the olfactory system as a whole, is well developed in most pond turtles, terrestrial snakes and lizards, rodents, and carnivores among others. Such a distribution makes it most unlikely that the VNO possesses one distinct and unique function, and thus extensive extrapolation from studies of its use in one form is dangerous.

Thus, the evolution of the vomeronasal apparatus is very poorly understood. Palaeontology tells us almost nothing and comparative anatomy very little. Perhaps more physiological and behavioral information will give us a better idea of its history, but at present we can make few definite statements (Parsons, 1970).

*1.3. Distribution of the VNO throughout the mammalian and nonmam-
malian orders*

This problem was recently reviewed by Wysocki (1979), Bertmar (1981a), and Halpern (1987). In general, VN organs appear in terrestrial vertebrates, are absent in some aquatic vertebrates (e.g., fish), and are absent or vestigial in some adult forms of flying and arboreal species (e.g., birds). The VN system probably can be found in all orders of mammals with the possible exception of the Cetacea. There appears to be considerable interspecific variability within many orders in the extent of development of the VN system.

The absence of functional VNO system in humans was already mentioned in Chapter IV (p. 106). In nonhuman, old world primates, reports on the existence of the organ are also inconsistent (Wysocki, 1979). Many species appear to lack an accessory OB, but possess a Jacobson's organ, and some of them have intact sensory epithelium and vomeronasal nerves. In a comprehensive neuroanatomical investigation, however, Stephan (1965) was unable to locate the accessory OBs, or any rudimentary structure thereof, in any of seven old world monkey or ape species, (viz., *Macaca mulatta* (rhesus), *Cercopithecus ascanius* (black-cheeked, white-nosed monkey), *Cercopithecus mitis* (diadem monkey), *Colobus badius* (red colobus), *Papio doguera* (anubis baboon), *Pan troglodytes* (chimpanzee), and *Gorilla gorilla*). Thus, Stephan concluded that well-defined accessory OBs were absent in old world monkeys and apes and doubted the existence of remains or rudiments of accessory OBs.

On the other hand, Stephan's (1965) investigations were consistent with the view that Jacobson's organ is present in new world anthropoids. Prominent accessory OBs were localized in each of six new world monkey species [viz., *Aotes trivirgatus* (douroucouli), *Alouatta* sp. (howler monkey), *Cabus capucinus* (white-throated capuchin), *Saimiri sciurea* (common squirrel monkey), *Lagothrix lagothricha* (Humboldt's woolly monkey), and *Ateles ater* (spider monkey)]. In prosimians, the evidence was consistent and convincing. Stephan (1965) was impressed with the degree of development of the accessory OBs. Indeed, in these animals, Jacobson's organ possesses olfactory-like sensory epithelium (Wollard, 1925) and associated vomeronasal nerves (Stephan, 1965).

In conclusion, with some exceptions (i.e., Cetaceans, which have no functional olfactory and VN systems) one may be able to say that the old world monkeys and higher primates including man have no functional VN system, but it is present in the new world monkeys and lower mammals.

1.4. Histology of the sensory epithelium in the VNO

The fine structure of the VN epithelium has been described recently in representatives of virtually every major vertebrate group at the light microscopic and ultrastructural levels (Adams and Weikamp, 1984; Bhatnagar *et al.*, 1982; Ciges *et al.*, 1977; Kolnberger, 1971; Kolnberger and Altner, 1971; Kratzing, 1971a, b; Loo and Kanagasuntheram, 1972; Miragall *et al.*, 1979; Naguro and Breipohl, 1982; Taniguchi and Mikami, 1985; Taniguchi and Mochizuki, 1982; Vaccarezza *et al.*, 1981; Wang and Halpern, 1980a, b). There is a remarkable conservation of structure through widely separated groups. The VN epithelium lines the lumen and consists of three types of cells: receptor, supporting, and basal cells. Since these cells show much the same characteristics in most mammals, the findings of Amemori *et al.*'s (1981) study on the cat VNO are presented here:

The receptor cells are bipolar neurons with distal and proximal processes. In the cat, the distal dendritic process extends from the perikaryon which has a diameter of 4.5 to 5.5 μm and the proximal process, which is an axon with a diameter of 0.3 to 0.6 μm. The perikarya of the receptor neurons occupy the lower two-thirds of the epithelial layer. The distal end of the dendrite is slightly expanded, but does not form a terminal swelling comparable with the olfactory vesicle of the olfactory cell. The distal end has a characteristic tuft of microvilli, but not olfactory cilia.

The supporting cell possesses a perinuclear portion 3.0 to 3.5 μm in diameter in the upper third of the epithelium, and also has numerous microvilli at the distal end. The basal cell is found in the basal zone of the epithelium.

The organization of these cells into recognizable layers varies with different species, but Wang and Halpern (1980a, b) differentiated two types, the mouse type and snake type.

In many mammals, the VN epithelium has been demonstrated to differ from olfactory epithelium in several ways. It lacks Bowman's glands, which usually occur in the olfactory epithelium of all vertebrates excepting fish. The sensory cell in the VNO lack cilia, which occur on those of the olfactory epithelium in all vertebrates including fish. Axons from the olfactory cells project to the OB, while those from VN cells extend to the accessory OB.

1.5. Degeneration, regeneration, and turnover of receptor cells

Evidence for the turnover of VN receptor cells have been obtained in a number of animal species, and is entirely similar to findings obtained in the main olfactory system (e.g., Graziadei, 1973). Wang and Halpern (1980a) described the turnover process best in snakes.

Undifferentiated basal cells at the base of the epithelium have all the cytological features of stem cells. These cells have been observed in various stages of mitosis. Receptor cells differ in cytological characteristics, depending on their relative apical position in the epithelium. More deeply situated receptor cells appear less mature than more superficial receptor cells. The most superficially located bipolar neurons had characteristics suggestive of aging or degeneration. On rare occasions it was possible to observe a receptor cell lying between adjacent supporting cells, apparently in the process of being extruded into the lumen of the VNO (Wang and Halpern, 1980a, b).

The basal cells of the VNO of snakes (Wang *et al.*, 1979) and mice (Barber and Raisman, 1978a; Wilson and Raisman, 1980) incorporate ^{3}H-thymidine. In snakes, labeled cells are found closer to the surface, with long (21 days) survival times (Wang *et al.*, 1979). In mice, labeled columns of receptor cells are found further away from the neurosensory-nonsensory epithelial boundary region with increasing survival time (Barber and Raisman, 1978a). Wilson and Raisman (1980) estimated, based on counts of numbers of labeled cells, that the entire VN epithelium of mice turns over approximately every two to three months.

The turnover process in the VNO was studied by sectioning the VN nerve, similar to studies carried out in the olfactory nerve (Barber, 1981b; Barber and Raisman, 1978b; Kubie *et al.*, 1978; Wang and Halpern, 1982a). In mice, Barber (1981b) and Barber and Raisman (1978b) observed that sensory cells disappeared and frequent mitosis occurred eight days after the sectioning and the entire epithelium recovered to normal levels by 32 days. A similar process was found in garter snakes by Wang and Halpern (1982a, b).

2. The Vomeronasal Nerve (VNN)

2.1. Anatomy

The unmyelinated nerve fibers from the VN receptor cells show a consistent pattern across species. They pierce the basal lamina, and form compact, relatively long, vomeronasal nerves which penetrate the cribriform plate, and terminate in the glomerular layer of the accessory OB (Barber and Raisman, 1974; Barber and Field, 1975; Barber *et al.*, 1978; Wang and Halpern, 1982b; Meredith, 1982). It is now well known that there are two types of VN nerve fibers, as will be shown in the next section (2.2.).

The VNO (and the main olfactory mucosa) appear to be innervated by the terminal nerve (Bojsen-Møller, 1975), which travels in close proximity to the VNN (Graziadei, 1977, p. 141).

2.2. Immunohistochemical study on the vomeronasal nerve pathway

Onoda and Fujita (1988) sought to find a difference between the main olfactory organ and the VNO using monoclonal antibodies, MAb 2D5 and MAb 4G12 (Chapter IV, p. 144). In the adult rabbit, MAb 2D5

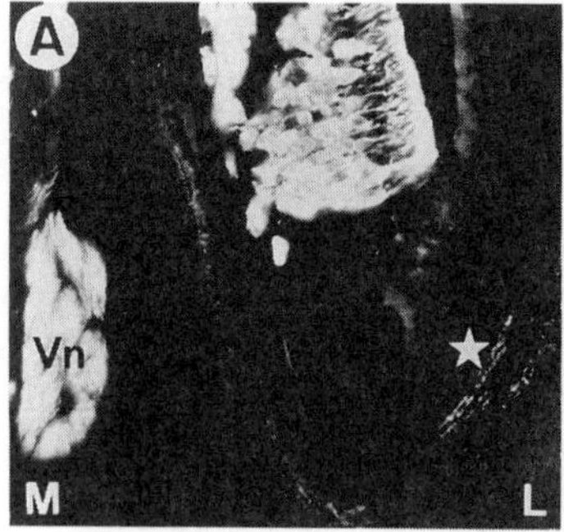

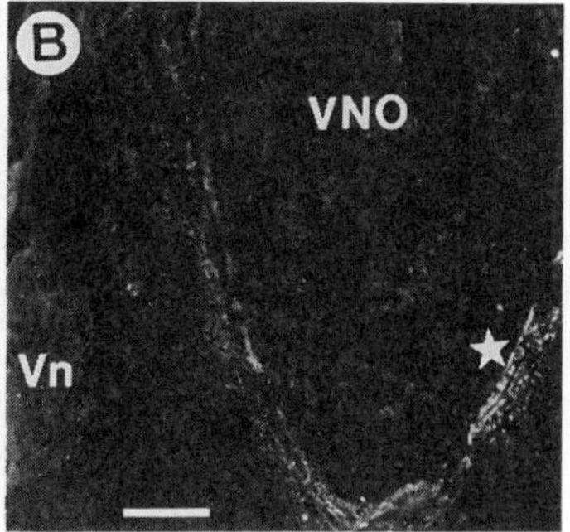

Fig. X-14. Immunostaining for MAbs 2D5 (A) and 4G12 (B) in the vomeronasal organ of the adult rabbit. (A) and (B) are adjacent sections; receptor cells in the vomeronasal organ (VNO) and their nerve axons (Vn) are 2D5 positive (A), but 4G12 negative (B); M is medial direction and L is lateral direction; stars show autofluorescent tissues. Autofluorescent Bowman's glands (*) can be seen. Scale: 50 μm. (from Onoda and Fujita, 1988)

stained the VNO or VNN, whereas MAb 4G12 stained neither the VNO nor VNN (Fig. X-14). Thus, the olfactory epithelium and nerve were clearly discriminated from the VNO and VNN. Onoda (1988a) found that MAb 2D5 stained the VNO and VNN at embryonic day 14 but that MAb 4G12 did not: throughout development and in childhood: Receptor cells of a characteristic premature stage were observed with thick dendrites, superficially located cell bodies, and protrusion of dendrites from the surface.

Imamura *et al.* (1985a, b) traced vomeronasal nerve (VNN) fibers and their terminations in the accessory olfactory bulb (AOB) immunohistochemically using three monoclonal antibodies (MAbs; Fig. X-15). One MAb (R2D5) labeled all VNN fibers, another MAb (R4B12) labeled a subgroup of the VNN fibers which terminated in the rostrolateral glomeruli in the AOB, and the third MAb (R5A10) recognized a complementary subgroup of the VNN which terminated in the mediocaudal portion of the AOB. These results showed for the first time occurrence of subtypes in the VNN axons with segregated terminations in the AOB.

Mori *et al.* (1987) analyzed the organization of the projections of subclasses of VNN fibers to the AOB using monoclonal antibodies generated against a homogenate of the rabbit OB. Monoclonal antibody R2D5 labeled all the somata of VN receptor cells in the VNO as well as all their axons, VNN fibers. Another monoclonal antibody (R4B12) which has been shown to bind selectively and thus identify a subclass of olfactory nerve fibers (Fujita *et al.*, 1985; Mori *et al.*, 1985), also labeled a subclass

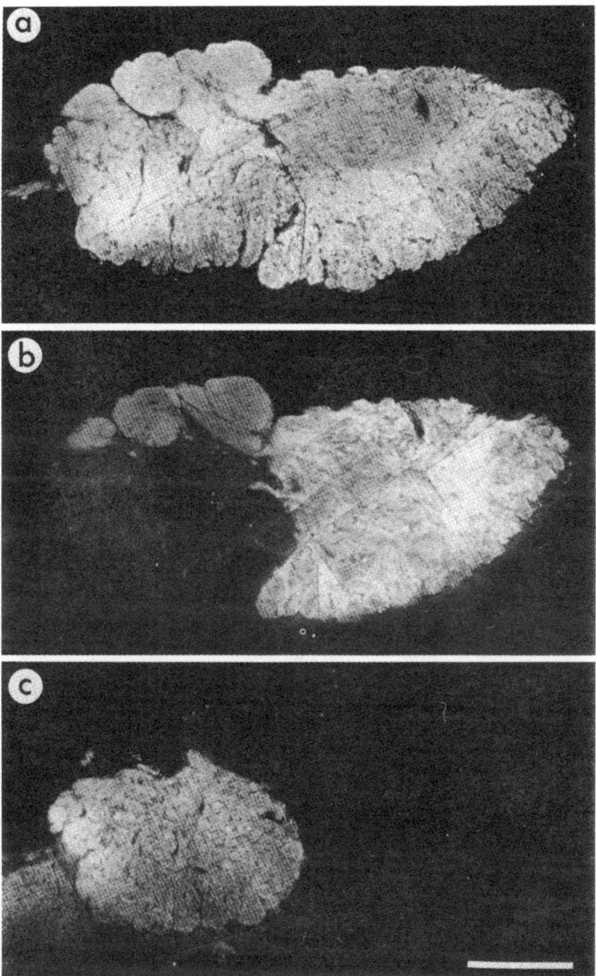

Fig. X-15. Indirect immunofluorescence on coronal sections of the rabbit AOB stained with MAb R2D5 (a), MAb R4B12 (b), or MAb R5A10(c). Neighboring sections (13.5 μm in thickness) of the middle portion of the AOB were stained with each antibody. In (a) MAb R2D5 stained the VNN layer and glomerular layer but did not stain other layers. MAbs R4B12 (b) and R5A 10 (c) stained different portions of the nerve layer and glomerular layer. These two MAbs weakly stained the external plexiform layer and granule cell layer (lower part of photomicrographs b and c.) Lateral is to the right. Top of figure is dorsal. Scale bar = 500 μm. (from Imamura *et al.*, 1985a)

of VNN fibers. The R4B12-positive subclass of VNN fibers projected to the glomeruli in the rostrolateral region of the AOB. The third monoclonal antibody (R5A10) recognized a complementary subclass of VNN fibers projecting to the glomeruli in the mediocaudal region of the AOB. In contrast to the clearly segregated terminations in the AOB, the two subclasses of VNN fibers intermingled with each other in the VNN bundless (Fig. X-16). Retrograde labeling of VN receptor cell somata following injection of HRP into the rostrolateral (R4B12-positive) region of the AOB indicated that VN receptor cells of this subtype were widely distributed in the VN sensory epithelium (Fig. X-16).

These results demonstrated the heterogeneity of VN receptor cells and the specificity of projections arising from subclasses of VN nerve fibers to the AOB.

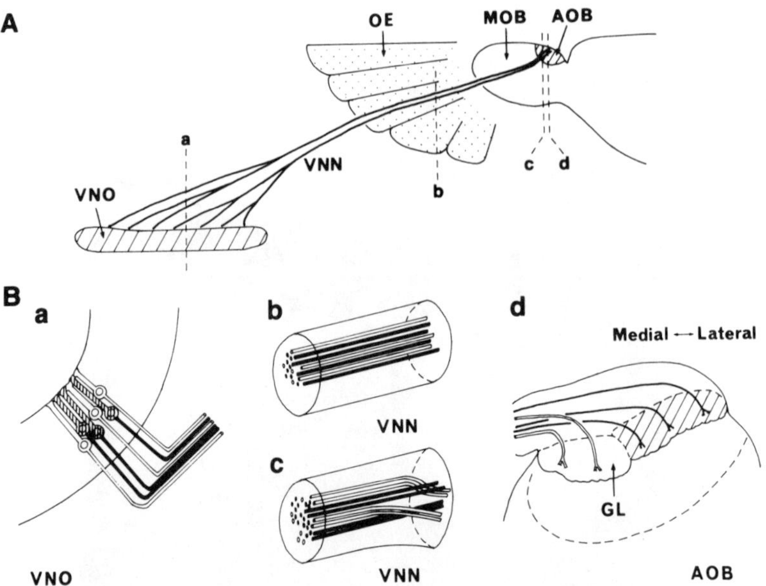

Fig. X-16. Schematic diagrams depicting the organization of the axonal projections from the two subtypes of vomeronasal receptor cells. A: Course of the projection of VNN from the VNO to the AOB. OE, the olfactory epithelium. B: Distribution of R4 B12-positive (filled) and -negative (open) VN fibers in transverse sections of the VNO in a, VNN bundles in b and c, and the AOB in d. Positions of each section (a–d) are shown by broken lines with corresponding letters in A. The receptor cell bodies which give rise to R4B12-positive axons are shown shaded in a. The R4B12-positive area in the glomerular layer (GL) is also shown shaded in d. (from Mori *et al.*, 1987)

Several MABs which bound to globo-series or lacto-series carbo-hydrates recognized types or subtypes of receptor cells in the VN systems. For example, MAbs anti-SSEA-3 and anti-SSEA-4 both stained the VN epithelium and VNN fibers. In contrast, they stained neither the main olfactory epithelium nor the olfactory nerve fibers. MAb 2C5, which labeled a subset of olfactory receptor cells, recognizes the Gal α 1→3, Gal β 1→4 GlcNAc-R structure (Mori, 1987a, b). MAb 4C9, which reacts with fucosyl-poly-N-acetyllactosamine, recognized a subset of vomeronasal receptor cells and their axons. These results indicate that different types or subtypes of receptor cells in olfactory systems expressed distinct car-bohydrates on their cell surfaces. The above results were compared with systematic investigations by Dodd *et al.* (1984, 1985) on carbohydrate expressions of various subsets of cutaneous primary afferent fibers. A striking similarity was noted between olfactory and somatosensory systems in the carbohydrate expressions of receptor cells and their axons. From these results, it is possible to imagine that the future may see the use of MAbs which define subsets of neurons in the MOB, AOB, and the olfactory cortex as well as further, undefined subsets of olfactory and VN receptor cells (Mori, 1987a, b, c).

3. The Accessory Olfactory Bulb
3.1. Anatomy

The AOB is the first relay station in the accessory olfactory system. In mammals, the AOB is a small structure, hemispherical in cross section. It is situated in the posterodorsal region of the main OB in most mam-mals (Fig. VII-1), but it is absent or vestigial in higher primates and hu-mans. Their laminar patterns are essentially the same but much simpler than those of the main OB (Allison, 1953a). The most superficial layer is composed of VNN fibers originating from receptor cells and passing into the glomerular layer. The glomeruli are relatively small and poorly defined, mitral cells are scattered throughout the external plexiform layer, and tufted cells are absent (Allison, 1953a). Compared with the corresponding layers in the OB, the external plexiform layer is smaller, and the principal cell layer is thicker in the AOB (Switzer and Johnson, 1977). For further information on the morphology of the AOB see the excellent review by Mori (1987a).

In contrast to the clear topographical organization of the olfactory receptor cell subclasses (MAb R4B12-positive and -negative) in the olfac-tory epithelium and their projections to the MOB, the accessory olfac-tory system shows little evidence of topographical projection. Thus, in spite of the segregated terminations of subclasses (MAb R4B12-positive and MAb R5A10-positive) of VNN fibers in the AOB, somata of these

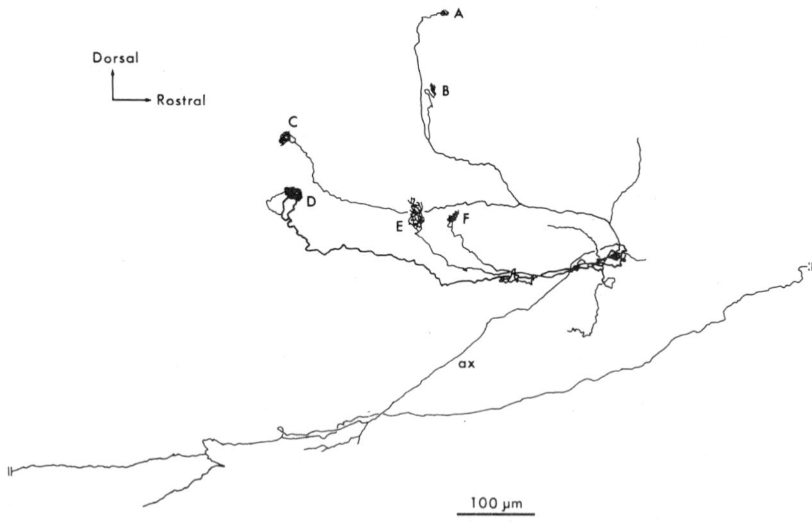

Fig. X-17. Camera lucida drawing of an HRP-injected mitral/
tufted cell in the rabbit AOB reconstructed in a parasagittal
plane. ax: axon; (A)–(F): glomerular tufts of primary dendrites.
(from Mori, 1987a)

subclasses of VN receptor cells are intermingled in the sensory epithelium
of the VNO (Mori *et al.*, 1986). The functional meanings of the presence
or the absence of the topographical projection is not yet clear.

Golgi studies indicate that principal cells in the mammalian AOB show
different morphological characteristics from the mitral or the tufted cells
in the MOB (Cajal, 1911; Allison, 1953a). A recent study using intracellular injections of HRP (Mori, 1983) made it possible to reconstruct the
whole dendritic tree of single cells in the rabbit AOB. These cells were found
to be very different from the mitral and tufted cells in the main OB in
terms of the number of primary dendrites, the size of terminal arborization,
and the length of secondary dendrites. Consequently, Mori (1983, 1987)
termed the principal cell the mitral/tufted cell. For instance, a single
mitral/tufted cell emits two or four stem dendrites from which primary
and secondary dendrites emerge (Fig. X-17). The primary dendrites branch
several times, project to the GL, and make terminal arborizations in
various shapes within the glomeruli (Mori, 1987a). The terminal tufts of

each mitral/tufted cell are distributed in widely disparate portions of the glomerular layer, which contrasts sharply with the contribution of a single primary dendrite to a single glomerulus in the case of the mitral and tufted cells seen in the MOB. The secondary dendrites of the AOB are relatively undeveloped.

The axons of the mitral/tufted cells project to the GCL and emit very thin axon collaterals in the IPL and superficial portion of the GCL. The main axons of the mitral/tufted cells run caudally in these laminae of the AOB and join the LOT at the dorsal portion of the anterior olfactory nucleus without giving off axon collaterals at the surface of this nucleus.

3.2. Development

Developmental studies of the VN organ were conducted in the mouse by Cuschieri and Bannister (1975) and in the hamster by Taniguchi *et al.* (1982a, b). Interesting findings demonstrated that the VN organ appears to develop later than the main olfactory epithelium, but that neurogenesis of the AOB precedes that of the main OB in the rat (Bayer, 1983) and the mouse (Hinds, 1967). For further information see Halpern (1987).

3.3. Sex differences

The VN system is sexually dimorphic in rats, and this dimorphism can be altered with perinatal manipulations of the endocrine system. The VN organ and neurosensory epithelial volume is greater in male rats at six months of age than in females. The volume of the AOB and the number of mitral cells in the AOB of three-month-old rats show sex differences; males have larger AOBs and more mitral cells than females. These sex differences can be eliminated by orchidectomizing males and androgenizing females on the day of birth (Segovia *et al.*, 1984a; Velencia *et al.*, 1986).

In postpubertal rats, gonadectomy results in reduction in the height of the VN epithelium and a decrease in the nuclear size of bipolar neurons in males and females (Segovia *et al.*, 1984b). For further review see Halpern (1987).

3.4. Physiology

There is a dearth of electrophysiological data on receptor activity, synaptic interactions, and information processing of odors in the VN system (Adrian, 1954b, 1955; Altner and Müller, 1968).

Tucker (1960, 1963b, 1971) may have been the first investigator to carry out in responses. Müller (1971) recorded electro-olfactograms (EOGs) in the VN organ of lizards (*Lacerta*) in response to a homologous series of alcohols, aldehydes, and fatty acids. These EOGs were not different from those recorded in the main olfactory epithelium.

Meredith and Burghardt (1978) recorded discharges of single neurons in the snake AOB and found increased discharges when swabs coated with prey extracts were pressed to the roof of the mouth. Meredith and O'Connell (1979) observed activation of AOB neuronal activity by simultaneously stimulating the nasopalatine nerve to activate the VN pump and applying continuous flow of dimethyl disulfide vapor over the VN pore.

Intracellular recordings from VNN cells have revealed (Reinhardt *et al.*, 1983) that the synaptic responses to VN nerve stimulation consist of three components: an initial EPSP, accompanied by a large IPSP, and a subsequent second EPSP with a long latency (about 60 msec). Action potentials are usually superimposed upon these EPSPs. The initial EPSP was interpreted as the one monosynaptically elicited at the excitatory synapses of VN nerve terminals on the glomerular tufts of the mitral/ tufted cell primary dendrites. The second EPSP occurred simultaneously with the late negative component of the VN nerve-evoked field potential (MacLeod and Rheinhart, 1983; Rheinhardt *et al.*, 1983). This second EPSP resembled the E_2 component of the olfactory nerve-evoked excitatory synaptic potential in mitral cells of the turtle MOB (Mori *et al.*, 1981b), although the origin of the second EPSP has not been clarified.

Comparisons of the latency of the LOT-evoked antidromic spikes between mitral/tufted cells in the AOB and mitral cells in the MOB revealed that the axonal conduction velocity of the AOB mitral/tufted cell is much slower and has much larger variation (0.4 to 3.1 m/sec) than that of the MOB mitral cells (about 10 m/sec; Reinhardt *et al.*, 1983; Mori, 1983).

The antidromic spikes of the AOB mitral/tufted cells were followed by a prolonged IPSP with relatively slow time-to-peak (Reinhardt *et al.*, 1983; Mori, 1983). By analogy with the main OB, this IPSP was considered to be generated mostly by dendrodendritic reciprocal synapses with granule cells in the AOB. Electron microscopic studies have demonstrated that nearly all the dendrites of mitral/tufted cells make dendrodendritic reciprocal synapses with peripheral processes of granule cells in the EPL of the mouse AOB (Barber *et al.*, 1978). The ultrastructural features of these reciprocal synapses are similar to those observed in the MOB, thus suggesting the presence of mitral/tufted-to-granule and granule-to-mitral/ tufted dendrodendritic pathways.

3.5. Neurochemistry

Macrides and Davis (1983) reviewed the chemical cytoarchitecture of the AOB, and noted the presence of several hypothalamic hormones in the AOB. For instance, luteinizing hormone-releasing hormone (LHRH) immunoreactivity was found in the rat (Witkin and Silverman, 1983), hamster (Jennes and Stumpf, 1980; Phillips *et al.*, 1980, 1982), and in new world primates (Witkin, 1985). Thyrotropin-releasing hormone

(TRH) was found in the AOB of the rat (Manaker *et al.*, 1985; Mantyh and Hunt, 1985). Somatostatin-positive mitral cells and somatostatin-positive fibers were also observed in the rat AOB (Vincent *et al.*, 1985).

Dulzen and Ramirez (1983) exposed male mice to male or female conspecifics and found that concentrations of LHRH and norepinephrine were higher in the region including the AOB, whereas those of TRH and DA did not differ between the VN and AOB regions. LHRH and DA increased selectively in the region including the AOB after exposure to male or female mice, whereas TRH and NA did not show significant differences between the two regions. These results imply that the VN system mediates endocrinological and behavioral responses to sex-related odors, and that chemical changes occur in the region of the AOB in parallel with behavioral changes (Halpern, 1987).

4. Projection of Vomeronasal Fibers to Higher Olfactory Areas

An important finding shows how afferents and efferents of the accessory OB appear to maintain independent pathways from the main OB and its subsequent projection areas (Fig. X-13). Many investigators have demonstrated that the AOB in mammals projects to the bed nucleus of the accessory olfactory tract, the medial nucleus and the posteromedial cortical nucleus of the amygdala, and the bed nucleus of the stria terminalis (e.g., Broadwell, 1975b; Davis *et al.*, 1978; de Olmos *et al.*, 1978; Devor, 1976a; Price, 1973; Scalia and Winans, 1975, 1976; Shammah-Lagnado and Negrao, 1981; Skeen and Hall, 1977; Winans and Scalia, 1970).

However, the major targets of mammalian main OB projections are the AON, the olfactory cortex, the olfactory tubercle, the antero-lateral cortical amygdaloid nucleus, and the lateral entorhinal cortex (e.g., Scalia and Winans, 1975). The pyriform cortex projects to the endopyriform nucleus, which in turn projects to the medial and posteromedial cortical amygdaloid nuclei (Kretteck and Price, 1978a, b). Hence, these areas seems to be the sites of interaction of the main and accessory olfactory pathways.

The AOB receives afferent fibers from the VNO; in addition, it receives centrifugal fibers, primarily from AOB target areas (Barber and Field, 1975; Broadwell and Jacobowitz, 1976; Conrad and Pfaff, 1976; Davis *et al.*, 1978; Halpern, 1980b; Kevetter and Winans, 1981; Kretteck and Price, 1977b, 1978a; Raisman, 1972), the locus coeruleus (adrenergic), and perhaps the raphe nuclei (serotonergic; Broadwell and Jacobowitz, 1976).

The tertiary projections of the VN system primarily innervate the preoptic-anterior hypothalamic junction, the ventromedial hypothalamic nucleus, and the ventral premammillary nucleus (deOlmos and Ingram, 1972; Halpern, 1980b; Kevetter and Winans, 1981; Kretteck and Price, 1977b, 1978a). These projections are the bases for the endocrinological and behavioral manifestations of the VNS. They are in contrast to the principal tertiary projections of the main olfactory system: the hippocampal complex, the frontal cortex, the dorsomedial thalamus, and the hypothalamus (e.g., Kretteck and Price, 1977, 1978a; Shepherd, 1985).

5. Some Functions of the Vomeronasal System (VNS)

The role of the VNS in vertebrate behavior and/or reproductive physiology has been reviewed by several investigators since 1970: Burghardt (1970) on reptilian VNSs and functional correlates; Madison (1977) on chemical communication in amphibians and reptiles; Halpern (1980a) on allelochemics in nonaquatic vertebrates; Halpern (1983; 1987) and Halpern and Kubie (1984) on nasal chemical sense in snakes.

On the mammalian VNS, Johns (1980), Meredith (1980, 1983), Johnston (1983, 1985), and Halpern (1987) have published excellent reviews. Especially on the role of the VNS in reproductive behavior of mammals, Wysocki (1979) and Keverne (1979, 1982) wrote also excellent reviews.

The critical role of the VNO and VNS in mediating neuroendocrine responses to chemical signals has already been demonstrated in mammals (Bellinger *et al.*, 1980; Kaneko *et al.*, 1980; Reynolds and Keverne, 1979; Wysocki *et al.*, 1983).

5.1. Primer pheromones

Various effects of primer pheromones that have been disclosed are: suppression of estrus in group-housed female mice (Lee-Boot effect; Van der Lee and Boot, 1956), induction of estrus and estrus synchrony in female mice produced by male odors (Whitten effect; Whitten, 1956), acceleration of puberty in female mice produced by male odors (Vandenbergh effect; Vandenbergh *et al.*, 1975), and pregnancy block caused by odor of a "strange" male mouse (Bruce effect; Bruce, 1964). It is now apparent that all these effects are mediated by the VNS that starts from the VN organ and goes to the AOB, to the corticomedial amygdala and to the neuroendocrine hypothalamus (Fig. X-13; Keverne, 1982; Halpern, 1987).

5.2. Releaser or signaling pheromones

These pheromones are chemical substances that play essential roles in releasing mating sequences, behaviors associated with aggregation, aggression, maternal care, and infant-mother recognition. Among them, sexual behavior has been studied. (Powers and Winans, 1975; Winans and Powers, 1977; Johnston and Rasmussen, 1984; Keverne *et al.*, 1986;

Meredith, 1980, 1983, 1986; Murphy, 1980; Steel, 1984; Steel and Keverne, 1985; Halpern, 1987).

Among nonsexual behavior, aggregation behavior was studied by Burghardt (1980) and Heller and Halpern (1982); agonistic behavior by Andren (1982), Bean (1982b), Clancy *et al.* (1984), Ingersoll (1981), and Wysocki *et al.* (1986); maternal behavior by Fleming *et al.* (1979) and Marques (1979); and newborn behavior by Teicher *et al.* (1984).

5.3. Chemoinvestigatory behavior (sniffing/licking, flehmen)

By sniffing a tiny drop of urine or sample of gland secretion, animals can discriminate species, subspecies, sex, sexual state, dominance status, and current diet, as well as information on individuality. The VNS palys a critical role in these behaviors (Beauchamp *et al.*, 1983).

Since the above functions are possessed by animals with the VNS, but not by humans, they are only briefly described in this book. The next function, however, seems to be closely related not only with lower animals but also with humans. Hence, it is summarized in detail.

5.4. Chemosensory recognition of genetic individuality

One of the attributes of the major histocompatibility complex (MHC) of genes is precise recognition of the cells of one individual by cells of another individual of the same species. Until recently, knowledge of such MHC-determined communication was confined to immunology settings. Yamazaki *et al.* (1976, 1978) at the Monell Chemical Senses Center, demonstrated that the recognition function of the MHC includes a chemosensory system that enables mice to identify one another according to their MHC type by olfaction.

Early studies indicated that the MHC of the mouse (H-2) affects mating choices in this species (Thomas, 1975). Generally, mating is more frequent between mice of different MHC types. This is the first example of reproductive behavior and mate selection in mice that has been traced to the variation of a particular gene or gene complex.

To further elucidate MHC-associated sensory recognition, and to provide a basis for chemical analysis of MHC-associated sensory signals, Yamazaki *et al.* (1979, 1982) demonstrated that they could train mice to distinguish arms of a Y maze scented by odors from MHC congenics or their urine. Further studies revealed that the mutation of a single known gene in this region was sufficient to mark odors of mice. Mice could be trained to discriminate between the urine of mutant and nonmutant mice. New data indicated that radiation chimeras, made by restoring lethally irradiated inbred mice with bone marrow cells of F1 hybrid donors, acquire a scent indicative of the reconstituting donors' H-2 type (Yamazaki *et al.*, 1985). Thus, cells of the hematopoietic system contribute to the MHC-related odorant profile. Studies designed to determine the chemical nature of the odors are underway.

Odors may also influence reproduction through their action on the neuroendocrine system. For example, a pregnant mouse may abort her litter before the embryos implant if she is exposed to an unfamiliar male (or his urine) within a critical period after copulation (Bruce effect, see p. 26). Yamazaki *et al.* (1983b) demonstrated that the incidence of pregnancy block was greater when the blocking male differs from the stud male at the MHC locus than when the blocking male has the same MHC type as the stud male. They suspected that pregnancy block represented a neuroendocrine response to the genetic (individual) identity of the blocking male generally reinforced by male pheromones.

As a first step in the study of interspecies chemosensory recognition of individuals by their MHC types, the ability of trained rats to distinguish the MHC types of MHC congenic mice by the scent of their urine was investigated. This was demonstrated with an automated testing apparatus designed for the rat.

It was remarkable that subsequent studies have provided substantial indications that humans can make similar distinctions. These studies have revealed that the same set of genes that is critical for immune response is also involved in the provision of individual scent. This concept is thought to be applicable not only to mice but to higher mammals as well. Therefore, it seems highly likely that the human MHC may also be involved in production of individual human odors. This establishes a link between immunology and olfactory communication. Further studies, now ongoing, will reveal the extent of this link (Beauchamp *et al.*, 1985a, b, c, 1986; Gilbert *et al.*, 1986; Yamazaki *et al.*, 1984, 1986a, b).

This line of research has also been conducted elsewhere. Singh *et al.* (1987) at the University of Cambridge studied MHC antigens in urine as olfactory recognition cues. They disclosed that classical polymorphic class I molecules in normal rats are excreted as urinary constituents. They observed that untrained rats can distinguish the smell of urine samples taken from normal donors that differ only at the class I MHC locus, and therefore excrete a different allelomorph of class I molecules in their urine. It would be interesting to know whether or not the same phenomenon exists in humans.

References

Adamek, G.R., Gesteland, R.C., Mair, R.G. and Oakley, B. (1984). *Brain Res.* **310**: 87–97.

Adamek, G.D. Nickell, W.T. and Shipley, M.T. (1986). *Chem. Sens.* **11**: 575–576.

Adams, R.G. and Crabtree, N. (1961). *Br. J. Indus. Med.* **18**: 216–221.

Adams, D.R. and McFarland, L.Z. (1971). *Comp. Biochem. Physiol.* **40A**: 971–974.

Adams. D.R. and Weikamp, M.D. (1984). *J. Anat.* **138**: 771–787.

Adams. D.J., Smith, S.J. and Thompson, S.H. (1980). *A. Rev. Neurosci.* **3**: 141–167.

Addison W.H.F. (1915). *J. Comp. Neurol.* **25**: 497–522.

Adey, W.R., Merrillees, N.C.R. and Sunderland, S. (1956). *Brain* **79**: 414–439.

Adrian, E.D. (1942). *J. Physiol.* **100**: 459–473.

Adrian, E.D. (1950a). *Electroence. Neurophysiol.* **2**: 377–388.

Adrian, E.D. (1950b). *Br. Med. Bull.* **6**: 330–331.

Adrian, E.D. (1951). *Année. Psychol.* **50**: 107–113.

Adrian, E.D. (1953). *Acta Physiol. Scand.* **29**: 5–14.

Adrian, E.D. (1954a). *Br. Med. J.* **1**: 287–290.

Adrian, E.D. (1954b). *J. Physiol.* **126**: 28–29.

Adrian, E.D. (1955). *Pflügers Arch.* **260**: 188–192.

Adrian, E.D. (1956). *J. Laryng. Otol.* **70**: 1–14.

Adrian, E.D. (1957). *J. Physiol.* **136**: 29.

Ai, N. and Takagi, S.F. (1963). *Jpn. J. Physiol.* **13**: 454–465.

Aiuchi, T., Tanabe, H., Kurihara, K. and Kobatake, Y. (1980). *Biochem. Biophys. Acta* **628**: 355–364.

Akert, K. (1964). The Frontal Granular Cortex and Behavior, ed. by Warren, M. and Akert, K., McGraw Hill, New York, pp. 372–396.

Akert, K. and Steiger, V. (1967). *Arch. Neurol. Neuroch. Psychiat.* **100**: 321–337.

Alberts, J.R. (1974). *Physiol. Behav.* **12**: 657–670.

Alberts, J.R. (1981). The Development of Perception: Psychobiological Perspectives. Vol. 1, ed. by Aslin, R.N., Alberts, J.R. and Petersen, M.R. Acad. Press, New York, pp. 322–357.

Alber, K. (1971). *Arch. Klin. Exp. Ohr. Nas. Kehlk. Heilk.* **199**: 687–692.

Allen, W.F. (1940). *Am. J. Physiol.* **128**: 754–771.

Allen, W.F. (1941). *Am. J. Physiol.* **132**: 81–91.

Allen, W.F. (1943a). *Am. J. Physiol.* **139**: 525–531.

Allen, W.F. (1943b). *Am. J. Physiol.* **139**: 553–555.

Allen, W.K. and Akeson, R. (1985a). *J. Neurosci.* **5**: 284–296.

Allen, W.K. and Akeson, R. (1985b). *Develop. Biol.* **109**: 393–401.

Alliez, J. and Noshida, M. (1945). *Ann. Med. Psychol.* **103**: 134–141. Cited by Doty, R.L. (1979).

Allison, A.C. (1953a) *J. Comp. Neurol.* **98**: 309–353.

Allison, A.C. (1953b). *Biol. Rev. (Cambridge)* **28**: 195–244.

Allison, A.C. and Warwick, R.T. (1949). *Brain* **72**: 186–197.

Allison, T. and Goff, W.R. (1967). *Electroence. Clin. Neurophysiol.* **23**: 558–560.

Altner, H. and Müller, W. (1968). *Z. Vergleich. Physiol.* **60**: 151–155.

Alter, M. and Seltzer, A.P. (1974). *Patient Care* **8**: 76–88. Cited by Amoore, J.E. and Ollman, B.G. (1983).

Amemori, T., Koide, N. and Okano, M. (1981). *JASTS* **15**: 36–39.

Amoore, J.E. (1962a). *Proc. Sci. Sec. Toilet Goods. Assoc.* **137**: Suppl. 1–12.

Amoore, J.E. (1962b). *Proc. Sci. Sec. Toilet Goods Assoc.* **37**: Suppl. 13–23.

Amoore, J.E. (1966). Symposium on Foods: The Chemistry and Physiology of Flavor. AVI Publishing Co., Westport.

Amoore, J.E. (1967). *Nature* **214**: 1095–1098.

Amoore, J.E. (1968). Theories of Odor and Odor Measurement. ed. by Tanyloaç, N. Robert College, Istanbul. pp. 71–85.

Amoore, J.E. (1969). Olfaction and Taste III. ed. by Pfaffmann, C. Rockefeller Univ. Press, New York, pp. 158–171.

Amoore, J.E. (1970). Molecular Basis of Odor. Charles. C. Thomas, Publishers, Spring-field, Ill.

Amoore, J.E. (1971). Handbook of Sensory Physiology. Vol. IV. Chemical Senses, Part I, ed. by Beidler, L.M. Springer-Verlag, Berlin, New York, pp. 145–156.

Amoore, J.E. (1975). Olfaction and Taste V. ed. by Denton, D.A. and Coghlan, J.P. Acad. Press, New York, London, pp. 283–289.

Amoore, J.E. (1977a). *Chem. Sens. Flav.* **2**: 267–281.

Amoore, J.E. (1977b). Food Intake and Chemical Senses. ed. by Katsuki, Y., Sato, M., Takagi, S.F. and Oomura, Y. University of Tokyo Press, Tokyo, pp. 89–99.

Amoore, J.E. (1979). *Chem. Sens. Flav.* **4**: 153–161.

Amoore, J.E. and Buttery, R.G. (1978). *Chem. Sens. Flav.* **3**: 57–71.

Amoore, J.E. and Forrester, L.J. (1976a). *J. Chem. Ecol.* **2**: 49–56.

Amoore, J.E. and Ollman, B.G. (1983). *Rhinology* **21**: 49–54.

Amoore, J.E. and O'Neill, R.S. (1986). *Chem. Sens.* **11**: 576–577.

Amoore, J.E. and Venstrom, D. (1966). *J. Food Sci.* **31**: 118–128.

Amoore, J.E., Venstrom, D. and Davis, A.R. (1968). *Percept. Motor Skills* **26**: 143–164.

Amoore, J.E., Popplewell, J.R. and Whissell-Buechy, D. (1975a). *J. Chem. Ecol.* **1**: 291–297.

Amoore, J.E., Forrester, L.J. and Buttery, R.G. (1975b). *J. Chem. Ecol.* **1**: 299–310.

Amoore, J.E., Forrester, L.J. and Pelosi, P. (1976b). *Chem. Sens. Flav.* **2**: 17–25.

Amoore, J.E., Pelosi, P. and Forrester, L.J. (1977). *Chem. Sens. Flav.* **2**: 401–425.

Andersen, I.B., Lundqvist, G.R. and Proctor, D.F. (1972). *Am. Rev. Resp. Dis.* **106**: 438–439.

Andersen, P.T. Blackstad, W. and Lømo, T. (1966). *Exp. Brain Res.* **1**: 236–248.

Anderson, J.A. (1970). *Math. Biosci.* **8**: 137–160.

Anderson, J.A. (1972). *Math. Biosci.* **13**: 197–220.

Anderson, P.A. and Ache, B.W. (1985). *Brain Res.* **338**: 273–280.

Anderson, P.A.V. and Hamilton, K.A. (1987). *Neurosci.* **21**: 167–173.

Andren, C. (1982). *Copeia.* 148–157.

Andres, K.H. (1965). *Naturwissenschaften* **52**: 500.

Andres, K.H. (1966). *Z. Zellforsch.* **96**: 250–274.

Andres, K.H. (1968). *J. Ultrastruct. Res.* **25**: 163.

Andres, K.H. (1969). *Z. Zellforsch.* **96**: 250–274.

Andres, K.H. (1970a). Taste and Smell in Vertebrates. ed. by Wolstenholme, G.E.W. Churchill, London, pp. 175–176.

Andres, K.H. (1970b). Taste and Smell in Vertebrates. ed. by Wolstenholme, G.E.W. and Knight, J. Ciba Found. Symp. Churchill, London, pp. 177–194.

Andy, O.J., Jurko, M.F. and Hughes, J.R. (1975). *Confin. Neurol.* **37**: 215–222.

Änggärd, A. and Denset, O. (1974). *Acta Otolaryngol.* **78**: 232–241.

Anholt, R.R., Aebi, U. and Snyder, S.H. (1986). *J. Neurosci.* **6**: 1962–1991.

Ansari, K.A. (1976). *Europ. Neurol.* **14**: 138–145.

Ansari, K.A. and Johnson, A. (1975). *J. Chron. Dis.* **28**: 493–497.

Anton, W. (1895). *Verh. d. Oeut. Otol. Gesellsch.* (Jena), **4**: 55–57.

Aoki, K. and Takagi, S.F. (1968). *Proc. Jpn. Acad.* **44** (8): 856–858.

Apfelbach, R. and Weiler, E. (1985). *Neurosci. Lett.* **62**: 169–173.

Araki, T., Ito, M. and Oscarsson, O. (1961). *J. Physiol.* **159**: 410–435.

Ariki, T., Onoda, N. and Takagi, S.F. (1982). *J. Physiol. Soc. Jpn.* **44**: 390.

Arito, H. and Takagi, S.F. (1980). *Proc. Jpn. Acad.* **56** (8): 189–193.

Arito, H., Iino, M. and Takagi, S.F. (1978). *J. Physiol.* **79**: 605–619.

Arstilla, A. and Wersäll, J. (1967). *Acta Otolaryngol. (Stockh.)* **64**: 187–204

Arvanitaki, A., Takeuchi, H. and Chalazonitis, N. (1967). Olfaction and Taste II. ed. by Hayashi, T. Oxford, London, Pergamon Press, pp. 573–598.

Asada, Y. (1963). *Jpn. J. Physiol.* **13**: 583–589.

Asaka, H. (1981a). *Otorhinolaryngology* (Tokyo) **53**: 767–772.

Asaka, H. (1981b). *Otorhinolaryngology* (Tokyo) **53**: 692–693.

Asaka, H. (1984). Recent Advances in Otorhinolaryngology. Igaku-kyoiku Publishing Co., Tokyo, pp. 225–234.

Asaka, H. and Umeda, R. (1978). *J. Jpn. Rhnol. Soc.* **17**: 79–81.

Asaka, H., Nagano, H., Fukushima, Y., Sanada, S. and Oka, T. (1978). Olfactory Disorders—Olfactometry and -therapy. ed. by Toyota, B., Kitamura, T. and Takagi, S.F. Igaku-Shoin, pp. 35–38.

Asaka, H., Fukushima, T., Oyaizu, K., Oka, T., Kuroishi, T. *et al.* (1979). *JASTS* **13**: 143–146.

Asaka, H., Morita, M., Fujii, S., Kuroishi, T., Oka, T. *et al.* (1980). *JASTS* **14**: 161–164.

Asaka, H., Okabe, E., Morita, M., Fujii, S., Oka, S. *et al.* (1983). *J. Otolaryng. Jpn.* **86**: 1181–1182.

Asaka, H., Morita, M., Sirakura, M., Fujii, S., Oyaizu, S. *et al.* (1984). *JASTS* **18**: 77–80; *Chem. Sens.* **10**: 134, 1985.

Asaka, H., (1985). Unpublished data.

Asaka, H., Fukushima, Y., Fujii, S., Morita, M. and Okabe, E. (1986). *JASTS* **20**: 61–64; *Chem. Sens.* **12**: 510, 1987.

Ash, K.O., Branseford, J.E. and Koch, R.B. (1966). *J. Cell Biol.* **29**: 554–561.

Ashton, E.Y., Eayrs, J.T. and Moulton, D.G. (1957). *Nature* **179**: 1069–1070.

Astic, L. and Cattarelli, M. (1982). *Brain Res.* **245**: 17–23.

Astic. L. and Saucier, D. (1982). *Der. Brain Res.* **2**: 141–156.

Astic, L. and Saucier, D. (1986a). *Brain Res. Bull.* **16**: 445–454.

Astic, L., Saucier, D. and Holley, A. (1987). *Brain Res.* **424**: 144–152.

Axelrod, F.B., Branom, N., Becher, M., Nachtigall, R. and Dancis, J. (1972). *J. Pediatr.* **81**: 573–578.

Baginsky, B. (1894). *Virchows Arch. Path. Anat.* **137**: 389–404.

Bailey, E.H.S. and Nichols, E.L. (1884). *New York Medicial Journal* **40**: 325.

Bailey, E.H.S. and Powell, L.M. (1885). *Trans. Kansas Acad. Sci.* **9**: 100–101.

Bailey, P. and Sweet, W.H. (1940). *J. Neurophysiol.* **3**: 276–281.

Bakay, L. and Cares, H.L. (1972). *Acta Neuro-chir.* **26**: 1–12.

Ban, T. and Zyo, K. (1962). *Med. J. Osaka Univ.* **12**: 385–424.

Bannister, L.H. (1965). *Quart. J. Micr. Sci.* **106**: 333–342.

Bannister, L.H. (1968). *Nature* **217**: 275–276.

Bannister, L.H. (1974). Transduction Mechanisms in Chemoreception. ed. by Poynder, T.M. IRL Press, London, pp. 39–46.

Baradi, A.F. and Bourne, G.H. (1951). *Nature* **168**: 977–979.

Barber, P.C. (1981a). *Neurosci. Lett. Suppl.* **7**: 577.

Barber, P.C. (1981b). *Brain Res.* **216**: 229–237.

Barber, V.C. and Boyde, A. (1968). *Z. Zellforsch.* **84**: 269–284.

Barber, P.C. and Field, P.M. (1975). *Brain Res.* **85**: 201–203.

Barber, P.C. and Lindsay, R.M. (1982). *J. Neurosci.* **7**: 3077–3090.

Barber, P.C., Parry, D.M., Field, P.M. and Raisman, G. (1978). *Brain Res.* **152**: 283–302.

Barber, P.C. and Raisman, G. (1974). *Brain Res.* **81**: 21–30.

Barber, P.C. and Raisman, G. (1978a). *Brain Res.* **141**: 57–66.

Barber, P.C. and Raisman, G. (1978b). *Brain Res.* **147**: 297–313.

Barnstable, C.J. (1980). *Nature* **286**: 231–235.

Bard, P. and Rioch, D. Mck. (1937). *Bull. J. Johns Hopkins Hosp.* **60**: 73–147.

Barraclough, C.A. and Gross, B.A. (1963). *J. Endocrinol.* **26**: 339–359.

Barraquer-Berre, L. and Fargas, J.A. (1950). *Rev. Esp. Otoneuroftalmol. Neurochir.* **52**: 5. Cited by Doty, R.L. (1979).

Barrett, E.F. and Barrett, J.N. (1976). *J. Physiol.* **255**: 737–774.

Barton, R.P.E. (1974). *J. Laryngol. Otol.* **88**: 355–361.

Baumgarten, R. von., Green, J.D. and Mancia, M. (1962). *J. Neurophysiol.* **25**: 489–500.

Baumgarten, R., Bloom, F.E., Oliver, A.P. and Salmoiraghi, G.C. (1963). *Pflügers Arch. Ges. Physiol.* **277**: 125–140.

Bayer, S.A. (1983). *Exp. Brain Res.* **50**: 329–340.

Bean, N.J. (1982). *Physiol. Behav.* **29**: 433–437.

Beauchamp, G.K., Wellington, J.L., Wysocki, C.J., Brand, J.G., Kubie, J.L. and Smith, A.B. III. (1980). Chemical Signals in Vertebrates and Aquatic Invertebrates. ed. by Muller-Schwarze, D., Silverstein, M. Plenum Press, New York, pp. 327–329.

Beauchamp, G.K., Martin, I.G., Wellington, J.L. and Wysocki, C.J. (1983). Chemical Signals in Vertebrates. ed. by Muller-Schwarze, D. and Silverstein, R.M. Plenum Press, New York, London, pp. 73–86.

Beauchamp, G.K., Yamazaki, K. and Boyse, E.A. (1985a). *Sci. Am.* **253**: 86.

Beauchamp, G.K., Yamazaki, K., Wysocki, C.J., Slotnick, B.M., Thomas, L. and Boyse, E.A. (1985b). *Proc. Natl. Acad. Sci. USA.* **82**: 4186–4188.

Beauchamp, G.K., Wysocki, C.J. and Wellington, J.L. (1985c). *Behav. Neurosc.* **99**: 950–955.

Beauchamp, G.K., Gilbert, A.N., Yamazaki, K. and Boyse, E.A. (1986). Chemical Signals in Vertebrates, IV: Ecology, Evolution and Comparative Biology. ed. by Duvall, D., Muller-Schwarze, D. and Silverstein, R.M. Plenum Press, NY, pp. 413–422.

Beccari, N. (1911). *Arch. Ital. Anat. Embriol.* **10**: 261–328. Cited by Krieger, N.R. (1981).

Beckstead, R.M. (1978). *Brain Res.* **152**: 249–264.

Bednár, M. and Langfelder, O. (1930). *Mschr. Ohrenh. U. Laryngo-Rhin.* **64**: 1133–1139.

Beidler, L.M. (1954). *J. Gen. Physiol.* **38**: 133–139.

Beidler, L.M. (1958). *Federation Proc.* **17**: 12 (Abstr.).

Beidler, L.M. (1965). *Cold Spring Harb. Symp. Quant. Biol. XXX*: 191–200.

Békésy, C. von. (1964). *J. Appl. Physiol.* **19**: 369–373.

Bellak, L. and Benedict, P. K. (1958). Schizophrenia. A Review of the Syndrome. Logos Press, New York.

Bellinger, J.F., Pratt, H.P.M. and Keverne, E.B. (1980). *J. Reprod. Fertil.* **59**: 223–228.

Benjamin, R.M. and Jackson, J.C. (1974). *Brain Res.* **75**: 181–191.

Benjamin, R.M., Jackson, J.C. and Golden, G.T. (1978). *Brain Res.* **141**: 251–265.

Benjamin, R.M., Jackson, J.C., Golden, G.T. and Wests, C.H.K. (1982). *J. Comp. Neurol.* **207**: 358–368.

Bennett, M.H. (1968). *Physiol. Behav.* **3**: 507–515.

Benson, T.E., Ryugo, D.K. and Hinds, J.W. (1984). *J. Neurosci.* **4**: 638–653.

Benzecri, J.P. (1969). Analyse factorielle des correspondence. Laboratoire de Statistiques, Université Paris VI, Paris

Berg, H.W., Pangborn, R.M., Poessler, E.B. and Webb, A.D. (1973). *Nature* **197**: 108.

Bernstein, J. (1912). Elektrobiologie. Friedr, Vieweg and Sohn, Braunschweig, Germany.

Berry, C.M., Hagamen, W.D. and Hinsey, J.C. (1952). *J. Neurophysiol.* **15**: 139–148.

Bertmar, G. (1981a). *Evolution* **35**: 359–366.

Bertmar, G. (1981b). *Arch. L. Biol. Bruxelles.* **92**: 343–366.

Bhatnager, K.P., Matulionis, D.H. and Breipohl, W. (1982). *Acta Anatomica* **112**: 158–177.

Bickerstaff, E.R. (1968). Neurological Examination in Clinical Practice. Blackwell, Oxford, Cited by Cain, W.S. and Gent, F. (1986).

Biedenbach, M.A. and Stevens, C.F. (1969a). *J. Neurophysiol.* **32**: 193–203.

Biedenbach, M.A. and Stevens, C.F. (1969b). *J. Neurophysiol.* **32**: 204–214.

Bignami, A., Eng, L.F., Dahl, D. and Uyeda, C.T. (1972). *Brain Res.* **43**: 429–435.

Bird, S. and Gower, D. (1980). *Endocr. J.* **85**: 8–9.

Bitensky, M.W., Gorman, R.E. and Miller, W.H. (1971). *Proc. Natl. Acad. Sci. USA.* **68**: 561–562.

Björklund, A. and Lindvall, O. (1978). *Limbic Mechanisms.* ed. by Livingston, K.E. and Hornykiewicz, O. Plenum Press, New York, pp. 307–331.

Blackstad, T.W. (1956). *J. Comp. Neurol.* **105**: 417–538.

Blakeslee, A.F. (1918). *Science* **48**: 298–299.

Blakeslee, A.F. and Salmon, M.R. (1931). *Eugen. News* **16**: 105–108. Cited in Chemical Senses in Health and Disease, ed. by Kalmus, H. and Hubbard, S.J. (1960).

Bloom, G. (1954). *Z. Zellforsch, Mikroskop. Anat.* **41**: 89–100.

Bloom, G. and Engström, H. (1952). *Exp. Cell Research* **3**: 699–701.

Bodian, D. and Howe, H.A. (1941a). *Bull. Johns Hopkins Hosp.* **68**: 248–267. Cited by Lorenzo, A.J.D. (1970).

Bodian, D. and Howe, H.A. (1941b). *Bull. Johns Hopkins Hosp.* **69**: 79–85, and other papers (pp. 86–215). Cited by De Lorenzo, A.J.D. (1970).

Boeckh, J. (1967). Olfaction and Taste II. ed. by Hayashi, T. Pergamon Press, Ltd., pp. 721–735.

Bogan, N., Brecha, N., Gall, C. and Karten, H.J. (1982). *Neuroscience* **7**: 895–906.

Boistel, J. and Fatt, P. (1958). *J. Physiol.* **144**: 176–191.

Boisvieux-Ulrich, E., Sandoz, D. and Chailly, B. (1977). *Biol. Cellulaire* **30**: 245–252.

Boisvieux-Ulrich, E., Sandoz, D. and Chailly, B. (1980). *Biol. Cellulaire* **37**: 261–268.

Bojsen-Møller, F. (1975). *J. Comp. Neurol.* **159**: 245–256.

Bonin, G. von and Bailey, P. (1947). The Neocortex of *Macaca mulatta*, Univ. Illinois Press, Urbana.

Bonin, G. von and Green, J.R. (1949). *J. Comp. Neurol.* **90**: 243–254.

Bonsall, R.W. and Michael, R.P. (1971). *J. Reprod. Fertil.* **27**: 478–479.

Booth, W.D., Baldwin, B.A., Poynder, P.M., Bannister, L.H. and Gower, D.B. (1981). *Q.J. Exp. Physiol. Cogn. Med. Sci.* **66**: 533–540.

Bostock, H., (1974). Transduction Mechanisms in Chemoreception. ed. by Poynder, T.M. IRL Press, London, pp. 27–36.

Boulet, M., Daval, F. and Leveteau, J. (1978). *Brain Res.* **142**: 123–134.

Bourne, G.H. (1948). *Nature* **161**: 445–446.

Bouvet, J.F., Dalaleu, J.C. and Holley, A. (1984). *ECRO VI*, Lyon, France.

Bouvet, J.F., Delaleu, J.C. and Holley, A. (1987a). *Neurosci. Lett.* **77**: 181–186.

Bouvet, J.F., Gordinot, F., Croze, S. and Delaleu, o.-C. (1987b). *Chem. Sens.* **12**: 499–505.

Bouvet, J.F. Delaleu, J.C. and Holley, A. (1988). *Neurosci. Res.* **5**: 214–223.

Boyer, P.D. (1954). *J. Am. Chem. Soc.* **76**: 4331–4337.

Boyse, E.A., Beauchamp, G.K., Yamazaki, K., Bard, J. and Thomas, L. (1982). *Oncodevel. Biol. Med.* **4**: 101–116.

Brain, W.R.B. (1955). Diseases of the Nervous System. Cited by Sumner, D. (1962). *Lancet* **ii**: 895–897.

Brain, W.R.B. (1960). Clinical Neurology. Cited by Sumner, D. (1962). *Lancet* **ii**: 895–897.

Brain, W.R.B. (1969). Diseases of the Nervous System. 7th ed. Revised by Brain, W.R.B. and Walton, J.M. Oxford University Press, London. Cited by Cain, W.S. and Krause, R.J. (1979).

Brandl, U., Kobal, G. and Plattig, K-H. (1980). Olfaction and Taste VII. ed. by van der Starre, H., IRL Press, London, pp. 401–404.

Breipohl, W. (1986). Ontogeny of Olfaction. Springer-Verlag, Berlin, New York, London.

Breipohl, W. and Fernandez, M. (1977). *Cell Tiss. Res.* **183**: 105–114.

Breipohl, W., Naguro, T. and Miragall, F. (1983). *Verh. Anat. Ges.* **77**: 741–743.

Briggs, M.H. and Duncan, R.B. (1961). *Nature* **191**: 1310–1311.

Briggs, R.B. and Duncan, M. (1962). *Nature* **195**: 1313–1314.

Brink, F. (1954). *Pharmacol. Rev.* **6**: 243–298.

Brink, F. and Posternak, J.W. (1948). *J. Cell Comp. Physiol.* **32**: 211–233.

Broadwell, R.D. (1975a). *J. Comp. Neurol.* **163**: 329–346.

Broadwell, R.D. (1975b). *J. Comp. Neurol.* **164**: 389–410.

Broadwell, R.D. and Jacobowitz, D.M. (1976). *J. Comp. Neurol.* **170**: 321–346.

Brodal. A. (1947a). *J. Comp. Neurol.* **87**: 1–16.

Brodal, A. (1947b). *Brain* **70**: 179–224.

Brodal, A. (1981). Neurological Anatomy in Relation to Clinical Medicine, 3rd ed. Oxford University Press, Oxford, New York.

Brodmann, K. (1909). Vergleichende Lokalisationslehre der Grosshirnrinde. Barth, J.A., Leipzig.

Brofeldt, S., Secher, C. and Mygind, N. (1983). *Eur. J. Resp. Dis.* **64 Suppl.** **128**: 436–440.

Broman, I. (1920). *Anat. Hefte* **58**: 143–191. Cited by Wysocki, C.J. (1979)

Bronshtein, A.A. (1977). Odour Receptors of Vertebrates, Monog. Academy of Sciences U.S.S.R. Institute of Comparative Physiology and Biochemistry, Leningrad.

Bronshtein, A.A. and Ivanov, V.P. (1965). *J. Evol. Biochem. Physiol.* **1**: 251–261. Cited by Graziadei, P. (1971)

Bronshtein, A.A. and Minor, A.V. (1977). *Tsitologiia*, **19**: 33–39.

Bronson, F.H. and Whitten, W.K. (1968). *J. Reprod. Fertil.* **15**: 131–134.

Brooksbank, B. (1970). *Experientia* **26**: 1012–1014.

Brooksbank, B.W.L., Brown, R. and Gustafsson, J.A. (1974). *Experientia* **30**: 864–865.

Brown, K.S. and Robinette, R.P. (1967). *Nature* **215**: 406–408.

Brown, H.E. and Beidler, L.M. (1966). *Fed. Proc.* **25**: 329.

Bruce, H.M. (1959). *Nature* **184**: 105.

Bruce, H.M. (1960a). *J. Reprod. Fertil.* **1**: 96–103.

Bruce, H.M. (1960b). *J. Reprod. Fertil.* **1**: 311–312.

Bruce, H.M. (1962). *Anim. Behav.* **10**: 3–4.

Bruce, H.M. (1963). *J. Reprod. Fertil.* **6**: 451–460.

Bruce, H.M. (1964). *Discovery* **25**: 19–22.

Bruce, H.M. (1965). *J. Reprod. Fertil.* **10**: 141–143.

Bruce, H.M. and Parrott, D.M.V. (1960). *Science* **131**: 1526.

Brunjes, P.C. (1985). *Brain Res. Bull.* **14**: 233–237.

Brunjes, P.C., Smith-Crafts, L.K. and McCarty, R. (1985). *Brain Res.* **354**: 1–6.

Brunn, A.V. (1892). *Arch Mikroskop. Anat.* **39**: 632–651. Cited by Moulton (1962, 1964).

Bullock, T.H. and Northcut, R.G. (1984). *Neurosci. Lett.* **44**: 155–160.

Burch, R.E., Sackin, D.A., Ursick, J.A., Jetton, M.M. and Sullivan, J.F. (1978). *Arch. Intern. Med.* **138**: 743–746.

Burghardt, G.M. (1970). Advances in Chemoreceptions, 1: Communication by Chemical Signals. ed. by Johnson, J.W., Moulton, D.R. and Trrk, A. Appleton-Century Crofts, New York, pp. 241–308.

Burns, S.M., Mitchell, J.A., Getchell, M.L. and Getchell, T.V. (1981). *Chem. Sens.* **6**: 307–315.

Byzov, A.L. and Flerova, G.Z. (1964). *Biofizika* **9**: 217–225.

Cagan, R.H. and Kare, M.R. (1981). Biochemisty of Taste and Olfaction. Acad. Press, New York, pp. 1–455.

Cagan, R.H. and Zeiger, W.N. (1978). *Proc. Natl. Acad. Sci. USA.* **75**: 4679–4683.

Cain, D.P. (1974b). *Psych. Bull.* **81**: 654–671.

Cain, D.P. and Bindra, D. (1972). *Expl Neurol.* **35**: 98–110.

Cain, W.S. (1974a). *Ann. N.Y. Acad. Sci.* **237**: 28–34.

Cain, W.S. (1982). *Yale J. Biol. Med.* **55**: 515–519.

Cain, W.S. and Gent, J.F. (1986). Clinical Measurement of Taste and Smell. ed. by Meiselman H.L. and Rivlin, R.S. Macmillan Publ. New York, pp. 170–186.

Cain, W.S. and Krause, R.J. (1979). *Neurol Res.* **1**: 1–9.

Cain, W.S. and Murphy, C.L. (1980). *Nature* **284**: 255–257.

Cain, W.S., Gent, J., Catalanotto, F.A. and Goodspeed, R.B. (1983). *Am. J. Otolaryngol.* **4**: 252–256.

Cajal, S. and Ramón, Y. (1911). Histologie du Systeme Nerveux. A. Maloine, Paris.

Cajal, S. and Ramón Y. (1955). Studies on the Cerebral Cortex (Limbic Structures). translated by Kraft, L.M., Year Book: Chicago.

Cajal, S. and Ramón, Y. (1972). The Structure of the Retina. translated by Thorpe, S.A. and Glickstein, M. Thomas Springfield, p. 58. Cited by Rafols, J.A. and Getchell, T.V. (1983).

Calleja, C. (1983). La Region Olfactoria del Cerebro. Moya, Madrid, pp.4–40. Cited by Fallon *et al.* (1978).

Callens, M. (1965). Peripheral and Central Regulatory Mechanisms of the Excitability in the Olfactory System. Arscia, Uitgaven, Brussels.

Cameron, D.J. and Wright, I.S. (1964). *Ann. Intern. Med.* **61**: 128–133.

Cancalon, P. (1978). *Chem. Sens. Flav.* **3**: 381–396.

Cancalon, P. (1980). Olfaction and Taste VII. ed. by Van der Starre, H. IRL Press London, p. 73.

Carson, J.L., Collier, A.M., Knowls, M.R., Boucher, R.C. and Rose, J.G. (1981). *Proc. Natl. Acad. Sci. USA.* **78**: 6996–6999.

Carsou, V., Hagan, J. and Manning, H. (1969). *Arch. Otolaryngol.* **90**: 500–503.

Catalanotto, F.A., Dore-Duffy, P. and Donaldson, J.O. (1986). Clinical Measurement of Taste and Smell. ed. by Meiselman, H.L. and Rivlin, R.S. Macmillan Publish. Co., New York, pp. 519–528.

Cattarelli, M. (1982). *Brain. Res.* **6**: 339–364.

Cauna, N. (1982). The Nose. Upper airway physiology and atmospheric environment. ed. by Proctor, D.F. and Andersen, I, Elsevier, Amsterdam, pp. 45–70.

Champion, R. (1966). *Br. J. Plast. Surg.* **19**: 182–185.

Chen, Z. and Lancet, D. (1984). *Proc. Natl. Acad. Sci. USA.* **81**: 1859–1863.

Chen, Z., Ophir, D. and Lancet, D. (1984). *Soc. Neurosci. Abstr.* **10**: 861.

Chen, Z., Pace, U., Ronen, D. and Lancet, D. (1986a). *J. Biol. Chem.* **261**: 1299–1305.

Chen, Z., Ophir, D. and Lancet, D. (1986b). *Brain Res.* **368**: 329–338.

Chi, C.C. and Flynn, J.P. (1971). *Science* **171**: 703–706.

Chitanondh, H. (1966). *Confin. Neurol.* **27**: 181–196.

Chuah, M.I. and Farbman, A.I. (1983). *J. Neurosci.* **3**: 2197–2205.

Church, A.C., Bunney, B.S. and Krieger, N.R. (1982). *Brain Res.* **234**: 369–376.

Ciges, M., Labella, T., Gayoso, M. and Sanchez, G. (1977). *Acta Otolaryngol.* **83**: 47–58.

Clancy, A.N., Coquelin, A., Macrides, F., Gorski, R.A. and Noble, E.P. (1984). *J. Neurosci.* **4**: 2222–2229.

Clancy, A.N., Schoenfeld, T.A. and Macrides, F. (1985). *Chem. Sens.* **10**: 399.

Clark, W.E. Le Gros (1951). *J. Neurol. Neurosurg. Psychiat.* **14**: 1–10.

Clark, W.E. Le Gros (1956). *Yale J. Biol. Med.* **29**: 83–95.

Clark, W.E. Le Gros (1957). *Proc. Roy. Soc. B.* **146**: 299–319.

Clark, W.E. Le Gros and Boggon, R.H. (1933). *J. Anat.* **67**: 215–226.

Clark, W.E. Le Gros and Warwick, R.T. (1946). *J. Neurol. Neurosurg. Psychiat.* **9**: 101–111.

Clark, W.E. Le Gros and Meyer, M. (1947). *Brain* **70**: 304–328.

Claus, R. and Alsing, W. (1976). *J. Endocr.* **68**: 483–484.

Colby, D.R. and Vandenberg, J.G. (1974). *Biol. Reprod.* **11**: 268–279.

Comfort, A. (1971). *Nature* **230**: 432–434.

Cone, T.E. Jr. (1968). *Pediatrics* **41**: 993–995.

Connoly, F.H. and Gittleson, N.L. (1971). *Br. J. Psychiat.* **119**: 443–444.

Conrad, L.C.A. and Pfaff, D.W. (1976). *J. Comp. Neurol.* **169**: 185–220.

Coombs, J.S., Eccles, J.C. and Fatt, P. (1955). *J. Physiol.* **130**: 291–325.

Cooper, J.G. and Bhatnager, K.P. (1976). *J. Anat.* **122**: 571–601.

Corrigal, W.A., Crain, S.M. and Bornstein, M.B. (1976). *J. Neurobiol.* **7**: 521–536.
Corwin, J., Serby, M., Conrad, P. and Rotrosen, J. (1985). *IRCS Med. Sci.* **13**: 260.
Costanzo, R.M. and Becker, D.P. (1986). Clinical Measurement of Taste and Smell. ed. by Meiselman, H.L. and Rivlin, R.S, pp. 565–578.
Costanzo, R.M. and Graziadei, P.P.C. (1983). *J. Comp. Neurol.* **215**: 370–381.
Costanzo, R.M. and Mozell, M.M. (1976). *J. Gen. Physiol.* **68**: 297–312.
Costanzo, R.M. and O'Connell, R.J. (1978). *Brain Res.* **139**: 327–332.
Costanzo, R.M. and O'Connell, R.J. (1980). *J. Gen. Physiol.* **76**: 53–63.
Costanzo, R.M., Shipley, M.T. and Van Octaghen, S. (1984). Cytochrome oxidase (CO) activity in the olfactory system: Ontogeny and response to peripheral deafferentation. Presented at Achem S VI, Sarasota, Fla.
Cragg, B.G. (1960). *Exp. Neurol.* **2**: 547–572.
Cragg, B.G. (1962). *Exp. Neurol.* **5**: 406–427.
Criag, A.D. Jr., Wiegand, S.T. and Price, J.L. (1982). *J. Comp. Neurol.* **206**: 28–48.
Criswell, E.W., McClure, F.L., Schafer, R. and Brower, K.R. (1980). *Science* **210**: 425–426.
Crosby, E.C. and Humphrey, T. (1939). *J. Comp. Neurol.* **71**: 121–213.
Crosby, E.C. and Humphrey, T. (1941). *J. Comp. Neurol.* **74**: 309–352.
Crosby, E.C. and Humphrey, T. (1944). *J. Comp. Neurol.* **81**: 285–305.
Curtis, R.F., Ballatine, J.A., Keverne, E.B. and Bonsall, R.P. (1971). *Nature* **232**: 396–398.
Cuschieri, A. (1974). *J. Anat.* **118**: 477–489.
Cuschieri, A. and Bannister, L.H. (1975a). *J. Anat.* **119**: 277–286.
Cuschieri, A. and Bannister, L.H. (1975b). *J. Anat.* **119**: 471–498.
Cuvier, M. (1811). *Ann. d. Mus. d'Hist. Nat.* **18**: 412–424. Cited by Wysocki (1979).
D'Ambrosio, A., Zivikovic, B. and Bartholini, G. (1982). *Brain Res.* **238**: 470–474.
Dahlström, A., Fuxe, K., Olson, L. and Ungerstedt, U. (1965). *Life Sci.* **4**: 2071–2074.
Danner, H. and Pfister, C. (1981). *J. Hirnforsch.* **22**: 685–696.
Daval, G. and Leveteau, J. (1974). *Brain Res.* **78**: 395–410.
Daval, G. and Leveteau, J. (1975). *J. Physiol.* (Paris) **71**: 275A.
Davidson, G.M. (1945). *Psychiat. Quart.* **19**: 692–696.
Davis, B.J. and Macrides, F. (1981). *J. Comp. Neurol.* **203**: 475–493.
Davis, B.J., Macrides, F., Youngs, W.M., Schneider, S.P. and Rosene, D.L. (1978). *Brain Res. Bull.* **3**: 59–72.
Davis, B.J., Burd, G.D. and Macrides, F. (1982). *J. Comp. Neurol.* **204**: 377–383.
De France, J.F. (1976). The Septal Nuclei, Advances in Behavioral Biology. Vol. 20. Plenum Press, New York. Cited by Brodal, A. (1981).
De Groot, J. (1966). Neuroendocrinology, Vol. 1. ed. by Martini L. and Ganong W.F. Acad. Press, New York, pp. 81–106.
Delaeu, J.C. and Holley, A. (1980). *Chem. Sens.* **5**: 205–218.
Delaleu, J.C. and Holley, A. (1983). *Neurosci. Lett.* **37**: 251–256.
De Lorenzo, A.J. (1957). *J. Biophys. Biochem. Cytol.* **3**: 839–850.
De Lorenzo, A.J. (1960). *Ann. Otol. Rhinol. Lar.* **69**: 410–420.
De Lorenzo, A.J.D. (1963). Olfaction and Taste I. ed. by Zotterman, Y. Macmillan Publish. Co., New York, pp. 5–18.
De Lorenzo, A.J.D. (1968). Medical Physiology. ed. by Mountcastle, V.B. The C. V. Mosby, St. Louis, pp. 1626–1642.
De Lorenzo, A.J.D. (1970). Taste and Smell in Vertebrates. ed. by Wolstenholme, G.E.W. and Knight, J. Churchill, London, pp. 151–173.
De Molina, A. Fernandez, and Ispizua, A. (1972). *Physiol. Behav.* **8**: 135–138.
De Molina, A. Fernandez and Ruiz-Marcos, A. (1967). *Trab-Inst.-Cajal-Invest. -Biol.* **59**: 137–151.
De Molina, A. Fernandez and Garcia-Sanchez, J.L. (1967). *Physiol. Behav.* **2**: 225–227.
De Morsier, G. (1963). *Acta Neuropath.* (Berlin) **1**: 433.

Dennis, B.J. (1985). *Proc. Austral. Physiol. Pharmacol. Soc.* **16**: 23–41.

Dennis, B.J. and Kerr, D.I.B. (1968). *Brain Res.* **11**: 373–369.

Dennis, B.J. and Kerr, D.I.B. (1975). *J. Comp. Neurol.* **159**: 129–148.

Dennis, B.J. and Kerr, D.I.B. (1976). *Brain Res.* **110**: 593–600.

Denny-Brown, D. (1974). Handbook of Neurological Examination and Case Recording. rev. ed. Cambridge: Harvard University Press. Cited by Cain, W.S. and Gent, J.F. (1986).

Den Otter, C.J. (1981). Sense Organs. ed. by Laverack, M.S. and Cosens, D.J. Blackie, Glasgow, p. 186.

De Olmos, J.S. (1972). The Neurology of the Amygdala. ed. by Eleftheriou, E. Plenum Press, New York. pp. 145–204.

De Olmos, J.S. and Ingram, W.R. (1972). *J. Comp. Neurol.* **146**: 303–334.

De Olmos, J., Hardy, H. and Heimer, L. (1978). *J. Comp. Neurol.* **166**: 31–48.

De Simone, J.A. and Getchell, T.V. (1986). *Chem. Sens.* **11**: 268; *JASTS* **19**: 46–48, 1985.

De Simone, J.A., Heck, G.L., Mierson, S., Getchell, M.L. and Getchell, T.V. (1985). *Chem. Sens.* **10**: 394.

Desmet, J.E. (1953). *J. Physiol.* **121**: 191–205.

Dethier, V.G. and Chadwick, L.E. (1950). *J. Gen. Physiol.* **33**: 589–599.

Devor, M. (1975). *Science* **190**: 998–1000.

Devor, M. (1976a). *J. Comp. Neurol.* **166**: 31–48.

Devor, M. (1976b). *J. Comp. Neurol.* **166**: 49–72.

Dickens, G. and Trethowan, W.H. (1971). *J. Psychosom. Res.* **15**: 259–268.

Dionne, V.E. (1986). *Chem. Sens.* **11**: 594.

Dmitriew, V.V., Tarasov, E.K. and Lukash, A.I. (1984). *Fiziol. Zh, SSSR* **70**: 311–315.

Dodd, J., Solter, D. and Jessel, T.M. (1984). *Nature* **314**: 469–472.

Dodd, J. and Jessel, T.M. (1985). *J. Neurosci.* **5**: 3278–3294.

Dodd, G. and Persaud, K. (1981). Biochemistry of Taste and Olfaction. ed. by Cagan, R.H. and Kare, M.R. Acad. Press, New York. pp. 333–357.

Dogiel, A. (1886). *Biol. Zbl.* **6**: 428–431.

Dominic, C.J. (1965). *J. Reprod. Fertil.* **10**: 469–472.

Doty, R.L. (1976). Mammalian Olfaction, Reproductive Processes and Behavior. ed. by Doty, R.L. Acad. Press, New York, pp. 295–321.

Doty, R.L. (1977). Chemical Signals in Vertebrates. ed. by Müller-Schwarze, D. and Mozell, M.M. Plenum Press, New York, pp. 273–286.

Doty, R.L. (1979). *Am. J. Otolaryngol.* **1**: 57–79.

Doty, R.L. (1981). *Chem. Sens.* **6**: 351–376.

Doty, R.L. (1986). Clinical Mesurement of Taste and Smell. ed. by Meislmen, H.L. and Rivlin, R.S. Macmillan Publ. Co., New York, pp. 377–413.

Doty, R.L. and Dunbar, I.A. (1974). *Physiol. Behav.* **35**: 729–731.

Doty, R.L., Ford, M., Preti, G. and Huggins, G. (1975). *Science* **190**: 1316–1318.

Doty, R.L., Brugger, W.E., Jurs, P.C., Orndorff, M.A., Snyder, P.J. and Lowry, L.D. (1978a). *Physiol. Behav.* **20**: 175–185.

Doty, R.L., Kligman, A., Leyden, J. and Orndorff, M.M. (1978b). *Behav. Biol.* **23**: 373–380.

Doty, R.L., Snyder, P.J., Huggins, G.R. and Lowry, L.D. (1981). *J. Comp. Physiol. Psychol.* **95**: 45–60.

Doty, R.L., Green, P.A., Ram, C. and Yankell, S.L. (1982a). *Horm. Behav.* **16**: 13–22.

Doty, R.L., Hall, J.W., Flickinger, G.L. and Sondheimer, S.J. (1982b). Olfaction and Endocrine Regulation. ed. by Breipohl, W. IRL Press, London, pp. 35–42.

Doty, R.L., Shaman, P.M. and Dann, M. (1984). *Physiol. Behav.* **32**: 489–502.

Doty, R.L., Applebaum, S.A., Settle, R.G. and Vollmecke, T. (1985a). *Chem. Sens.* **10**: 411.

Doty, R.L., Applebaum, S., Zusho, H. and Settle, R.G. (1985b). *Chem. Sens.* **10**: 414.

Douek, E. (1970). *J. Laryng.* **84**: 1185–1191.

Douek, E. (1974). The Sense of Smell and Its Abnormalities. Churchill Livingstone, Edinburgh, pp. 1–243.

Douek, E., Bannister, L.H. and Dodson, H.C. (1975). *Proc. Roy. Soc. Med.* **68**: 467–474.

Döving, K.B. (1964). *Acta Physiol. Scand.* **60**: 150–163.

Döving, K.B. (1966a). *J. Physiol.* **186**: 97–109.

Döving, K.B. (1966b). *Acta Physiol. Scand.* **68**: 404–418.

Doving, K.B. and Pinching, A. (1973). *Brain Res.* **52**: 115–129.

Dreesen, T.D. and Koch, R.B. (1982). *Biochem. J.* **203**: 69–75.

Dubois-Dauphin, M., Tribollet, E. and Dreifuss, J. J. (1981). *Brain Res.* **219**: 269–289.

Duchamp, A. (1982). *Chem. Sens.* **7**: 191–210.

Duchamp, A. and Sicard, G. (1984a). *Chem. Sens.* **9**: 1–14.

Duchamp, A. and Sicard, G. (1984b). *Chem. Sens.* **8**: 355–366.

Duchamp, A., Revial, M.F., Holley, A. and MacLeod, P. (1974). *Chem. Sens. Flav.* **1**: 213–233.

Dulzen, D.E. and Ramirez, V.D. (1983). *Horm. Behav.* **17**: 139–145.

Duncan, R.B. and Briggs, M. (1962). *Arch. Otolaryngol.* **75**: 116–124.

Dusser De Barenne, J.G. (1933). *Arch. Neurol. Psychiat.* **30**: 884–901.

Eccles, J.C. (1964). The Physiology of Synapses. Springer, Berlin, p. 304.

Eccles, R. (1982). The Nose. ed. by Proctor, D.F. and Anersen, I. Elsevier, Amsterdam, pp. 191–214.

Edwards, C.H. (1973). Neurology of Ear, Nose and Throat Diseases. Butterworths, London. Cited by Cain, W.S. and Gent, J.F. (1986).

Edwards, C. and Hagiwara, S. (1959). *J. Gen. Physiol.* **43**: 315–321.

Edwards, D.A. (1974). *Behav. Biol.* **11**: 287–302.

Eichenbaum, H., Shedlack, K.J. and Eckmman, K.W. (1980). *Brain Behav. Evol.* **17**: 255–275.

Eleftherieu, B.E. (ed.) (1972). The Neurobiology of the Amygdala. Plenum Press, New York.

Elsberg, C.A. (1935a). *Bull. Neurol. Inst. N.Y.,* **4**: 1–19.

Elsberg, C.A. (1935b). *Bull. Neurol. Inst. N.Y.* **4**: 496–500.

Elsberg, C.A. (1935c). *Bull. Neurol. Inst. N.Y.* **4**: 511–522.

Elsberg, C.A. (1935d). *Bull. Neurol. Inst. N.Y.* **4**: 535–543.

Elsberg, C.A. (1936). *Bull. Neurol. Inst. N.Y.* **4**: 535–543.

Elsberg, C.A. (1937). *Arch. Neurol. Psychiat.* **37**: 223.

Elsberg, C.A. and Levy, I. (1935). *Bull. Neurol. Inst. N.Y.* **4**: 5–19. Cited by Sumner, D. (1962).

Engen, T. (1962). *Rep. Psychol. Lab. Univ. Stockholm, Rept.* No. 127.

Engström, H. and Bloom, G. (1953). *Acta Oto-Laryngol.* **43**: 11–21.

Erickson, R.P. (1968). *Psychol. Rev.* **75**: 447–465.

Eskenazi, B., Cain, W.S., Novelly, R.A. and Mattson, R. (1986). *Neuropsychologia* **24**: 553–562.

Evans, H.E. and Christensen, G.C. (1979). Anatomy of the dog. Saunders Co., Philadelphia, London, Toronto.

Evarts, E.V. (1968a). *J. Neurophysiol.* **31**: 14–27.

Evarts, E.V. (1968b). *Electroence. Clin. Neurophysiol.* **245**: 83–86.

Exner, S. (1877). Fortgesetzte Studien Über die Endigungsweise des Geruchsnerven. S.-B. Wien Akad. **76**: Abt. 3. Cited by Takagi, S.F. (1971).

Fallon, J.H. and Moore, R.Y. (1979). *J. Comp. Neurol.* **170**: 533–544.

Fallon, J.H., Loughlin, S.E. and Ribak, C.E. (1983). *J. Comp. Neurol.* **218**: 91–120.

Farbman, A.I. (1977). *Anat. Rec.* **189**: 187–200.

Farbman, A.I. and Margolis, F.L. (1980). *Develop. Biol.* **74**: 205–215.

Farbman, A.I., Ritz, S.M. and Brunjes, P. (1985). *Chem. Sens.* **10**: 404.

Fatt, P. and Ginsborg, B.L. (1958). *J. Physiol.* **142**: 516–543.

Fatt, P. and Katz, B. (1953). *J. Physiol.* **121**: 374–389.

Fein, B.T., Kamin, P.B. and Fein, N.N. (1966). *Ann. Allergy* **24**: 278–283.

Felix, D. and McLennan, H. (1971). *Brain Res.* **25**: 661–664.

Ferrer, N.G. (1969). *J. Comp. Neurol.* **136**: 337–347.

Fesenko, E.E., Novoselov, V.I. and Krapivinskaya, L.D. (1979). *Biochim. Biophys. Acta* **587**: 424–433.

Fesenko, E.E., Kolesnikov, S.S. and Lyubarsky, A.L. (1985). *Nature* **313**: 310–313.

Finkenzeller, P. (1966). *Pflüger Arch. ges. Physiol.* **292**: 76–80.

Firestein, S. and Werblin, F. (1985). *Proc. Ann. Meet. Soc. Neurosci.* 15th, Dallas, TX, 970.

Firestein, S. and Werblin, F. (1986). *Chem. Sens.* **11**: 598.

Firestein, S. and Werblin, F. (1987). *Proc. Natl. Acad. Sci.* **84**: 6292–6296.

Fleming, A., Vaccarino, F., Tambosso, L. and Chee, P. (1979). *Science* **203**: 372–374.

Forschheimer, L. (1916). *Dermatol. Zbl.* **6**: 98–101. Cited by Bednár, M. and Langfelder, O. (1916).

Fox, C.A. (1940). *J. Comp. Neurol.* **72**: 1–62.

Fox, C.A. (1943). *J. Comp. Neurol.* **79**: 277–295.

Fox, C.A. (1949). *Anat. Rec.* **104**: 537–538.

Fox. C.A., McKinley, W.A. and Magoun, H.W. (1944). *J. Neurophysiol.* **7**: 1–16.

Frankenhaeuser, B. (1957). *J. Physiol.* **137**: 245–260.

Frankenhaeuser, B. and Hodgkin, A.L. (1957). *J. Physiol.* **137**: 218–244.

Freeman, W.J. (1959). *J. Neurophysiol.* **22**: 644–665.

Freeman, W.J. (1968a). *J. Neurophysiol.* **31**: 1–13.

Freeman, W.J. (1968b). *J. Neurophysiol.* **31**: 337–348.

Freeman, W.J. (1974a). *Electroence. Neurophysiol.* **36**: 33–45.

Freeman, W.J. (1974b). *Brain Res.* **65**: 91–107.

Freeman, W.J. and Watts, J.W. (1947). *J. Comp. Neurol.* **86**: 65–93.

Frisch, D. (1964). *Anat. Rec.* **148**: 283.

Frisch, D. (1967). *Am. J. Anat.* **121**: 87–119.

Erüwald, V. (1935). *Jpn. J. Med. Sci, III, Biophysics.* **5**: 22–23.

Fujita, S.C. and Obata, K. (1984). *Neurosci., Res.* **1**: 131–141.

Fujita, H., Oikawa, I., Ihara, H. and Takagi, S.F. (1964). *Jpn. J. Physiol.* **14**: 615–629.

Fujita, S.C., Mori, K., Imamura, K. and Obata, K. (1985). *Brain Res.* **326**: 192–196.

Fujiwara, M., Sasakawa, S., Itokawa, Y. and Ikeda, K. (1964). *J. Vitaminol.* **10**: 79–87.

Fukuda, H., Torii, S., Kanemotoi, H., Takino, R., Miyauti, S. and Hamauzu, Y. (1985). *JASTS* **19**: 65–68; *Chem. Sens.* **11**: 269.

Fukumoto, Y., Nakashima, H., Uetake, M., Matsuyama, A. and Yoshida, T. (1957). *Zinrui Idengaku Zassi* **2**: 7–16.

Fukushima, Y. and Oka, S. (1979). *J. Otorhinolaryng.* **17**: 90.

Fulton, J.E. (1949). The Physiology of the Nervous System. Oxford Univ. Press, New York.

Furstenberg, A.C., Crosby, E. and Farrior, B. (1943). *Arch. Otol.* **48**: 529–530.

Gamble, E.A. McC. (1898). *Am. J. Psychol.* **10**: 82–142. Cited by Wenzel, B.M. (1948).

Gandelman, R., Zarrow, M.X., Denenberg, V.H. and Meyers, M. (1971). *Science* **171**: 210–211.

Ganser, S. (1882). *Morphol. Jb.* **7**: 591–725. Cited by Wenzel, B.M. (1948)

Gasser, H.S. (1956). *J. Gen. Physiol.* **39**: 473–469.

Gasser, H.S. (1958). *Exp. Cell Res. Suppl.* **5**: 3–17.

Gattefosés, R.M. (1937). Aromatherapie. Gent, J.F., France.

Gennings, J.N., Gower, D.B. and Bannister, L.H. (1977). *Biochim. Biophys. Acta* **496**: 547–556.

Gent, J.F., Cain, W. and Bartoshuk, L.M. (1986). Clinical Measurement of Taste and Smell. ed. by Meiselman, H.L. and Rivlin, R.S. Macmillan Publ. Co., New York, pp. 107–116.

Gerebtzoff, M.A. and Phillippot, E. (1957). *Acta Oto-Rhino-Laryng. Belg.* **13**: 297–300.

Gerebtzoff, M.A. and Schkapenko, G. (1951). *C.R. Assoc. Anat.* **68**: 511–516.

Gesteland, R.C. (1964). *Ann. N.Y. Acad. Sci.* **116**: 440–447.

420 References

Gesteland, R.C. (1967). Olfaction and Taste II. ed. by Hayashi, T. Pergamon Press, Oxford, pp. 821–831.

Gesteland, R.C. (1971). Handbook of Sensory Physiology. IV. Chemical Senses. I. Olfaction. ed. by Beidler, L.M. Springer-Verlag, Berlin, pp. 133–150.

Gesteland, R.C. (1976a). Frog Neurobiology. ed. by Llinas, R. and Precht. W. Springer-Verlag, Heidelberg, FRG, pp. 234–250.

Gesteland, R.C. (1976b). Structure-Activity Relationships in Chemoreception. ed. by Berz, G. IRL Press, London.

Gesteland, R.C. (1982). *Environmental Progress* **1**: 94–97.

Gesteland, R.C. (1987). Olfactory Research: Issues and Answers. Text of a lecture at Takasago Symposium held on May 13, 1987, Tokyo.

Gesteland, R.C., Howland, B., Lettvin, J.Y. and Pitts, W.H. (1959). *Proc. Inst. Radio Engr.* **47**: 1856–1862.

Gesteland, R.C., Lettvin, J.Y., Pitts, W.H. and Rojas, H. (1963). Olfaction and Taste I. ed. by Zotterman, Y. Pergamon Press, Oxford, pp. 19–34.

Gesteland, R.C., Lettvin, J.Y. and Pitts, W.H. (1965). *J. Physiol.* **181**: 525–559.

Gesteland, R.C., Yancey, R.A., Mair, R.C., Adamek, G.D. and Farbman, A.I. (1980). Olfaction and Taste VII. ed. by van der Starre, H. IRL Press, London, pp. 143–148.

Gesteland, R.C., Yancey, R.A. and Farbman, A.I. (1982). *Neuroscience* **7**: 3127–3136.

Getchell, T.V. (1973). *J. Physiol.* **234**: 533–551.

Getchell, T.V. (1974a). *J. Neurophysiol.* **37**: 1115–1130.

Getchell, T.V. (1974b). *J. Gen. Physiol.* **64**: 241–261.

Getchell, T.V. (1977a). Food Intake and Chemical Senses. ed. by Katsuki, Y., Sato, M., Takagi, S.F. and Oomura, Y. University of Tokyo Press, Tokyo, pp. 3–13.

Getchell, T.V. (1977b). *Brain Res.* **123**: 275–286.

Getchell, T.V. (1986). *Physiol. Rev.* **66**: 722–817.

Getchell, M.L. and Gesteland, R.C. (1972). *Proc. Natl. Acad. Sci. USA.* **69**: 1494–1498.

Getchell, M.L. and Getchell, T.V. (1974). *Ann. N.Y. Acad. Sci.* **237**: 62–71.

Getchell, T.V. and Getchell, M.L. (1977a). *Chem. Sens. Flav.* **2**: 313–326.

Getchell, T.V. and Getchell, M.L. (1977b). Olfaction and Taste I. ed. by Le Magnen, J. and MacLeod, P. IRL Press, London, pp. 105–112.

Getchell, M.L. and Getchell, T.V. (1984). *J. Comp. Physiol. A*, **155**: 435–443.

Getchell, T.V. and Shepherd, G.M. (1975a). *J. Physiol.* **251**: 497–552.

Getchell, T.V. and Shepherd, G.M. (1975b). *J. Physiol.* **251**: 523–548.

Getchell, T.V. and Shepherd, G.M. (1978a). *J. Physiol.* **282**: 521–540.

Getchell, T.V. and Shepherd, G.M. (1978b). *J. Physiol.* **282**: 541–560.

Getchell, T.V., Heck, G.L., DeSimone, J.A. and Price, S. (1980). *Biophys. J.* **29**: 397–472.

Getchell, M.L., Rafols, J.A. and Getchell, T.V. (1984a). *Anat. Rec.* **208**: 553–567.

Getchell, T.V., Margolis, F.L. and Getchell, M.L. (1984b). *Prog. Neurobiol.* **23**: 317–345.

Getchell, M.L., Zielinski, B. and Getchell, T.V. (1985a). *Chem. Sens.* **10**: 398.

Getchell, M.L., Zielinski, B. and Getchell, T.V. (1985b). *Chem. Sens.* **10**: 398–399.

Getchell, M.L., Zielinski, B. and Getchell, T.V. (1986). Abstract of ISOT 9. ed. by Roper, S.D. P. 30.

Getchell, M.L., Zielinski, B., DeSimone, J.A. and Getchell, T.V. (1987). *J. Comp. Physiol. A.* **160**: 155–168.

Ghorbanian, S.N., Paradise, J.L. and Doty, R.L. (1978). *Pediatr. Res.* **12**: 371.

Giachetti, I. and MacLeod, P. (1975). Olfaction and Taste V. ed. by Denton, D.A. and Coghlan, J.P. Acad. Press, New York, pp. 303–307.

Giachetti, I. and MacLeod, P. (1977a). *Brain Res.* **125**: 166–169.

Giachetti, I. and MacLeod, P. (1977b). Food Intake and Chemical Senses. ed. by Katsuki, Y., Sato, M., Takagi, S.F. and Oomura, Y. Univ. Tokyo Press, Tokyo, pp. 45–49.

Giesen, M. (1971). *Arch. Klin. exp. Ohr. Nas. Kehlk. Heilk.* **196**: 337. Cited by Ohba, H. (1982).

Giesen, M. and Mrowinski, D. (1970). *Arch. Ohr. Nas.-Kehlk. Heilk.* **196**: 377–380. Cited by Plattig, K.-H., and Kobal, G. (1979).

Gilbert, A.N. (1985). *Bull. Clin. Neurosci.* **50**: 38.

Gilbert, A.N. (1986). *Neurobiology of Aging* **7**: 578–579.

Gilbert, A.N., Yamazaki, K., Beauchamp, G.K. and Thomas, L. (1986). *J. Comp. Psychol.* **100**: 262–265.

Gilman, A.G. (1984). *Cell* **36**: 577–579.

Girgis, M. (1969). *Acta Anat.* **72**: 502–519.

Gladysheva, O.S., Kukushkina, D.M., Marynova, G.I. and Safronov, V.P. (1982). Khim. Signaly Zhivton. ed. by Sokolov, V.E., Nauka, Moscow, pp. 178–197, Izd.

Gladysheva, O.S. and Martynova, G. (1982). *Gegenbauers Morphol. Jahresb.* **128**: 78–83. Cited by Getchell, M.L. *et al.* (1987).

Glaze, J.A. (1928). *Am. J. Psychol.* **40**: 569–575.

Glees, P. and Spoerri, P.E. (1978). *Cell Tiss. Res.* **188**: 149–152.

Goetzl, F.R. and Stone, F. (1948). *Gastroenterology* **10**: 708–713.

Goland, P.P. (1937). *Arch. Surg.* **35**: 1173–1182. Cited by Doty, R.L. (1979).

Goldberg, S.J., Turpin, J.A. and Price, S. (1979). *Chem. Sens. Flav.* **4**: 207–213.

Goldwyn, R.M. and Shore, S. (1968). *Plast. Reconstr. Surg.* **41**: 427–432.

Golgi, C., (1875). Opera omnia. Vol. 1, Istologia normale. Ulrico Hoepli, Milano, 1903. pp. 113–132.

Gompper, H.I. (1950). *Z. mikr.-anat. Forsch.* **56**: 102–128.

Good, P.R., Geary, N. and Engen, T. (1976). *Chem. Sens. Flav.* **2**: 45–50.

Goodspeed, R.B., Catalanotto, F.A., Gent, J.F., Cain, W.S., Bartoshuk, L.M., *et al.* (1986a). Clinical Measurement of Taste and Smell. ed. by Meiselman, H.L. and Rivlin, R.S. Macmillan Publ. Co., New York, pp. 451–466.

Goodspeed, R.B., Gent, J.F., Catalanotto, F.A. Cain, W.S. and Zagraniski, R.T. (1986b). Clinical Measurement of Taste and Smell. ed. by Meiselman, H.L. and Rivlin, R.S. Macmillan Publ. Co., New York, pp. 514–518.

Goodwin, T.W. (1952). The Comparative Biochemistry of the Carotenoids. Chapman and Hall, London. Cited by Moulton, D.G. (1962).

Gower, D.B. (1972). *J. Steroid Biochem.* **3**: 45–103.

Growers, W.R. (1881). Epilepsy and Other Chronic Convulsive Diseases. London, Cited by Douek, E. (1974).

Grafe, M.R. (1983). *J. Neurosci.* **3**: 617–630.

Graham, L.T. Jr. (1973). *Life Sci.* **12**: 443–447.

Graham, J. and Gerard, R.W. (1946). *F. Cell Comp. Physiol.* **28**: 99–117.

Graham, C.A. and McGrew, W.C. (1980). *Psychoneuroendocrinology* **5**: 245–252.

Grassi, V. and Castronuovo, A. (1989). *Arch Mikrosc. Anat.* **34**: 385–390. Cited by Rafols, J.A. and Getchell, T.V. (1983).

Graziadei, P.P.C. (1971). Handbook of Sensory Physiology, Vol. IV. Chemical Senses I, Olfaction. ed. by Beidler, L.M. Springer-Verlag, Berlin, Heidelberg, New York, pp. 27–58.

Graziadei, P.P.C. (1972). Olfaction and Taste IV. ed. by Schneider, D. Wissenschaftliche Verlagsgesellschaft, Stuttgart, pp. 13–19.

Graziadei, P.P.C. (1973a). The Ultrastructure of Sensory Organ. ed. by Friedmann, I. North-Holland/American Elsevier, Amsterdam, London, New York, pp. 267–284.

Graziadei, P.P.C. (1973b). *Tissue Cell* **5**: 113–131.

Graziadei, P.P.C. (1975). Methods in Olfactory Research. ed. by Moulton, D.G., Turk, A. and Johnston Jr., J.W. Acad. Press, London, pp. 191–240.

Graziadei, P.P.C. (1977). Chemical Signals in Vertebrates. ed. by Mueller-Schwarze, D. and Mozell, M.M. Plenum Press, New York and London, pp. 435–454.

Graziadei, P.P.C. and Bannister, L.H. (1967). *Z. Zellforsch.* **80**: 220–228.
Graziadei, P.P.C. and DeHan, R.S. (1973). *J. Cell Biol.* **59**: 525–530.
Graziadei, P.P.C. and Gagne, H.T. (1973). *Z. Zellforsch.* **138**: 315–326.
Graziadei, P.P.C. and Metcalf, J.F. (1971). *Z. Zellforsch.* **116**: 305–318.
Graziadei, P.P.C. and Monti Graziadei, G.A. (1976). *J. Neurocytol.* **5**: 11–32.
Graziadei, P.P.C. and Monti Graziadei, G.A. (1978a). Handbook of Sensory Physiology IX. ed. by Jacobson, M. Springer-Verlag, New York, Berlin, Heidelberg, pp. 55–83.
Graziadei, P.P.C. and Monti Graziadei, G.A. (1978b). Neuronal Plasticity. ed. by Cotman, C.W. Raven Press, New York, pp. 131–153.
Graziadei, P.P.C. and Monti Graziadei, G.A. (1979). *J. Neurocytol.* **8**: 1–18.
Graziadei, P.P.C. and Monti Graziadei, G.A. (1980). *J. Neurocytol.* **9**: 145–162.
Graziadei, P.P.C. and Monti Graziadei, G.A. (1986a). *J. Comp. Neurol.* **247**: 344–356.
Graziadei, P.P.C. and Monti Graziadei, G.A. (1986b). *Neuroscience* **19**: 1025–1035.
Graziadei, P.P.C. and Okano, M. (1978). *Acta Anat. Nippon* **53**: 47–48.
Graziadei, P.P.C. and Okano, M. (1979). *Acta Anat.* **104**: 220–236.
Graziadei, P.P.C., Levine, R.R. and Monti Graziadei, G.A. (1978). *Proc. Natl. Acad. Sci. USA.* **75**: 5230–5234.
Graziadei, P.P.C., Levine, R.R. and Monti Graziadei, G.A. (1979). *Neuroscience* **4**: 713–727.
Green, J.D., Mancia, M. and von Baumgarten, R. (1962). *J. Neurophysiol.* **25**: 467–488.
Greer, C.A. and Shepherd, G.M. (1982). *Brain Res.* **235**: 156–161.
Greer, C.A., Stewart, W.B., Kauer, J.S. and Shepherd, G.M. (1981). *Brain Res.* **217**: 279–293.
Greer, C.A., Stewart, W.B., Teicher, M.H. and Shepherd, G. (1982). *J. Neurosci.* **2** (12): 1744–1759.
Griffith, I.P. (1976). *Practioner* **217**: 907–913.
Griffiths, N.M. and Patterson, R.L.S. (1970). *J. Sci. Food Agric.* **21**: 4–6.
de Groot, J. (1966). Neuroendocrinology, Vol. 1. ed. by Martini, L. and Ganong, W.F. Acad. Press, New York, pp. 81–106.
Grossman, S. (1953). *N.Y.J. Med.* **53**: 1236.
Grossman, S. (1976). The Septal Nuclei. ed. by De France, J.F. Plenum Press, New York, pp. 361–422.
Guevara-Aguilar, R., Solano-Flores, L.P., Donatti-Albarran, O.A. and Aguilar-Baturoni, H.U. (1982). *Brain Res. Bull.* **8**: 711–719.
Guild, A.A. (1956). *J. Laryngol. Otol.* **70**: 408–414.
Guillery, R.W. (1957). *J. Anat.* **91**: 91–115
Guillery, R.W. (1959). *J. Anat.* **93**: 403–419.
Guillot, M. (1948a). *C.R. Soc. Biol.* **142**: 161–162.
Guillot, M. (1948b). *C.R. Acad. Sci.* **226**: 1307–1309.
Haberly, L.B. (1969). *Brain Res.* **12**: 481–484.
Haberly, L.B. (1973a). *J. Neurophysiol.* **36**: 762–774.
Haberly, L.B. (1973b). *J. Neurophysiol.* **36**: 775–788.
Haberly, L.B. (1983). *J. Comp. Neurol.* **215**: 163–187.
Haberly, L.B. (1985). *Chem. Sens.* **10**: 219–238.
Haberly, L.B. and Behan, M. (1983). *J. Comp. Neurol.* **219**: 448–460.
Haberly, L.B. and Bower, J.M. (1984). *J. Neurophysiol.* **51**: 90–112.
Haberly, L.B. and Price, J.L. (1977). *Brain Res.* **129**: 152–157.
Haberly, L.B. and Price, J.L. (1978a). *J. Comp. Neurol.* **178**: 711–740.
Haberly, L.B. and Price, J.L. (1978b). *J. Comp. Neurol.* **181**: 781–808.
Haberly, L.B. and Shepherd, G.M. (1973). *J. Neurophysiol.* **36**: 789–802.
Haberly, L.B., Hansen, D.J., Feig, S.L. and Presto, S. (1987). *J. Comp. Neurol.* **266**: 269–290.
Hagan, P.J. (1967). *Arch. Otolaryngol.* **85**: 85–89.
Hagiwara, S. (1975). *J. Gen. Physiol.* **65**: 617–644.
Hagiwara, S. and Naka, K. (1964). *J. Gen. Physiol.* **48**: 141–146.

Hagiwara, S. and Nakajima, S. (1966). *J. Gen. Physiol.* **49**: 807–818.

Hagiwara, S., Kusano, K. and Saito, S. (1960). *J. Neurophysiol.* **23**: 505–515.

Halász, N., and Shepherd, G.M. (1983). *Neuroscience* **10**: 579–619.

Halász, N., Ljungdahl, A. and Hökfelt, T. (1979). *Brain Res.* **167**: 221–240.

Hall, E. (1972). The Neurobiology of the Amygdala. ed. by Eleftheriou, B.E. Plenum Press, New York, pp. 95–121.

Hallman, B.L. and Hurst, J.W. (1953). *J. Am. Med. Assoc.* **52**: 322.

Halpern, M. (1980a). Animals and Environmental Fitness. ed. by Gilles, R. Pergamon Press, New York, pp. 263–282.

Halpern, M. (1980b). Comparative Neurology of the Telencephalon. ed. by Ebbesson, S.O.E. Plenum Press, New York, pp. 257–295.

Halpern, M. (1983). Advances in Vertebrate Neuro-ethology. ed. by Ewert, J.-P. Capranica, R.R. and Ingle, D.J. Plenum Press, New York, pp. 141–176.

Halpern, M. (1987). *Ann. Rev. Neurosci.* **10**: 325–362.

Halpern, M. and Kubie, J.L. (1984). *Trends Neurosci.* **7**: 472–477.

Hamill, O.P., Marty, A., Neher, E., Sakmann, B. and Sigworth, F.J. (1981). *Pfl. Arch.* **391**: 85–100.

Hamilton, C.R., Henkin, R.I., Weir, G. and Kliman, B. (1973). *Ann. Intern. Med.* **78**: 47–55.

Hansen, R. (1970). *Munch. Med. Wochenschr.* **112**: 2167–2169.

Hansen, R. and Glass, L. (1936). *Klin. Wochenschr.* **15**: 891–894.

Hara, T.J. and Gorbman, A. (1967). *Comp. Biochem. Physiol.* **21**: 185–200.

Harding, J.W., Graziadei, P.P.C., Monti Graziadei, G.A. and Margolis, F.L. (1977). *Brain Res.* **132**: 11–28.

Harding, J.W., Getchell, T.V. and Margolis, F.L. (1978). *Brain Res.* **140**: 271–285.

Harris, E.J. and Hutter, O.F. (1956). *J. Physiol.* **133**: 58–59.

Hartman, B.K. (1954). *Proc. Am. Soc. Hort. Sci.* **64**: 335–342.

Hartman, B.K. and Margolis, F.L. (1975). *Brain. Res.* **96**: 176–180.

Hartzell, H.C., Kuffler, S.W., Stickgold, R. and Yoshikami, D. (1977). *J. Physiol.* **271**: 817–846.

Hartzell, H.C. and Fambrough, D.M. (1972). *J. Gen. Physiol.* **60**: 248–262.

Hasama, B. (1934). *Pflüger. Arch.* **234**: 784–755.

Hasegawa, N. (1978). Olfactory Disorders——Olfactometry and -therapy. ed. by Toyota, B., Kitamura, T. and Takagi, S.F. Igaku-Shoin, Ltd., Tokyo, pp. 157–158.

Hasler, A.D. (1957). The Physiology of Fishes. ed. by Brown, M.E. Acad. Press, New York, Vol. 2, pp. 187–210.

Hatanaka, T., Matsuzaki, O. and Shibuya, T. (1982). *Zool. Mag.* **91**: 190–193.

Haug, F.-M. (1976). *Advan. Anat. Embryol. Cell. Biol.* **52**: Fasc. 4, 73.

Hawkes, R., Leclerc, N. and Colonnier, M. (1984). *Soc. Neurosci. Abst.* **10**: 44.

Hayashi, M., and Hinoki, M. (1987). Effects of a musk-like odor on olfactory vertigo, with reference to results of equilibrium tests for olfactory vertigo and those of equilibrium tests after adrenaline injection. (In preparation)

Haynes, L. and Yau, K.-W. (1985). *Nature* **317**: 61–64.

Heck, G.L. and Erickson, R.P. (1973). *Behav. Biol.* **8**: 687–712.

Heck, G.L., DeSimone, J.A. and Getchell, T.V. (1984). *Chem. Sens.* **9**: 273–283.

Hedgewig, R. (1980). *Gegenbaurs Morph., Jahrb. Leipzig* **126**: 676–722.

Heimer, L. (1968). *J. Anal.* **103**: 413–432.

Heimer, L. (1969). *Ann. N.Y. Acad. Sci.* **167**: 129–146.

Heimer, L. (1972). *Brain Behav. Evolut.* **6**: 484–523.

Heimer, L. and Kalil, R. (1978) *J. Camp. Neurol.* **178**: 559–610.

Heimer, L. and Larsson, K. (1967). *Physiol. Behav.* **2**: 207–209.

Heimer, L. and Nauta, W.J.H. (1969). *Brain Res.* **13**: 284–297.

Heimer, L., Robards, M. and Switzer, III. R.C. (1976). *Exp. Brain. Res. Suppl.* **1**: 141–147.

Heimer, L., Van Hoesen, G.W. and Rosene, D.L. (1977). *Int. J. Neurol.* **12**: 42–52.

Heindl, A. (1932). *Monatsschr. Ohrenheilk.* **66**: 931–941.

Heller, S.B. and Halpern, M. (1982). *J. Comp. Physiol. Psychol.* **96**: 967–983.

Hempstead, J.L. and Morgan, J.I. (1983a). *Chem. Sens.* **8**: 107–119.

Hempstead, J.L. and Morgan, J.I. (1983b). *Brain Res.* **288**: 289–295.

Hempstead, J.L. and Morgan, J.I. (1985). *J. Neurosci.* **5**: 438–449.

Hendry, S.H.C., Hockfield, S., Jones, E.G. and Mckay, R. (1984). *Nature* **307**: 267–269.

Henkin, R.I. (1966). *Life Sci.* **5**: 1031–1040.

Henkin, R.I. (1967a). Olfaction and Taste II. ed. by Hayashi, T. Pergamon Press, Oxford, London, New York, pp. 235–252.

Henkin, R.I. (1967b). *J. Clin. Endocrinol. Metab.* **27**: 1436–1440.

Henkin, R.I. (1968). *J. Clin. Endocrinol. Metab.* **28**: 624–628.

Henkin, R.I. (1974). Biorhythms and Human Reproduction. ed. by Ferin, M., Halberg, F., Richard, R.M. and van de Wiele, R.L. Wiley, New York, pp. 277–285.

Henkin, R.I. (1975). Handbook of Physiology, Section 7, Vol. 6. ed. by Blaschko, H., Smith, A.D. aad Sayers, G. American Physiological Society, Washington, D.C., pp. 209–330.

Henkin, R.I. and Bartter, F.C. (1964). *Clin. Res.* **112**: 270.

Henkin, R.I. and Bartter, F.C. (1966a). *J. Clin. Invest.* **45**: 1631–1639.

Henkin, R.I. and Hoye, R.C. (1966b). *Life Sci.* **5**: 331–341.

Henkin, R.I. and Kopin, I.J. (1964). *Life Sci.* **3**: 1319–1925.

Henkin, R.I. and Powell, G.F. (1962). *Science* **138**: 1107–1108.

Henkin, R.I. and Smith, F.R. (1971). *Lancet* **i**: 823–826.

Henkin, R.I., Gill, J.R., Jr. and Bartter, F.C. (1962). *Clin. Res.* **10**: 400.

Henkin, R.I., Christiansen, R.L. and Bosma, J.F. (1966). *Clin. Res.* **14**: 236.

Henkin, R.I., McGlone, R.E., Daly, R. and Bartter, F.C. (1967). *J. Clin. Invest.* **46**: 429–435.

Henkin, R.I., Hoye, R.C., Ketcham, A.S. and Gould, W.J. (1968). *Lancet* **ii**: 479–481.

Henkin, R.I., Talal, N., Larson, A.L. and Mattern, C.F.T. (1972). *Ann. Intern. Med.* **76**: 375–383.

Henkin, R.I., Larson, A.L. and Powell, R.D. (1975). *Ann. Otol.* **84**: 672–681.

Henkin, R.I., Schechter, P.J., Friedenwald, W.T., Desmets, D.L. and Raff, M.A. (1976). *Am. J. Med. Sci.* **272**: 285–299.

Henning, H. (1916). Der Geruch. Verlag von J.A. Barth, Leipzig

Herberhold, C. (1971). *Arch. Klin. Exp. Ohr. Nas. Kehlk. Heilk.* **202**: 394–397.

Herberhold, C. (1975). *Arch. Oto-Rhino-Laryng.* **210**: 67–164.

Herberhold, C., Genkin, H., Brändle, L.W., Leitner, H. and Wöllmer, W. (1982). Olfactory and Endocrine Regulation. ed. by Breipohl, W. IRL Press, London, pp. 343–351.

Herkenham, M. (1979). *J. Comp. Neurol.* **183**: 487–518.

Hernández-Peón, R., Lavin, A., Alcocer-Cuarón, C. and Marcelin, (1960). *Electroence. Neurophysiol.* **12**: 41–58.

Herrick, C.J. (1910). *J. Comp. Neurol.* **20**: 413–547.

Herrick, C.J. (1924). *J. Comp. Neurol.* **37**: 317–359.

Hertz, J., Cain, W.S., Bartoshuk, L.M. and Dolan, T.F., Jr. (1975). *Physiol. Behav.* **14**: 89–94.

Heywood, P.G. and Costanzo, R.M. (1985). *Chem. Sens.* **10**: 414.

Higashino, S. and Takagi, S.F. (1964). *J. Gen. Physiol.* **48**: 323–335.

Higashino, S., Takagi, S.F., and Yajima, M. (1961). *Jpn. J. Physiol.* **11**: 530–543.

Higashino, S., Takeuchi, H. and Amoore, J.E. (1969). Olfaction and Taste III. ed. by Pfaffmann, C. Rockefeller University Press, New York, pp. 192–211.

Hinds, J.W. (1968). *J. Comp. Neurol.* **134**: 287–304.

Hinds, J.W. (1972a). *J. Comp. Neurol.* **146**: 233–252.

Hinds, J.W. (1972b). *J. Comp. Neurol.* **146**: 253–276.

Hinds, J.W. and Hinds, P.L. (1976). *J. Comp. Neurol.* **169**: 41–62.

Hinoki, M. (1985). *Acta Otolaryngol.* (Stockh.) Suppl. **419**: 30–52.

Hinoki, M., Nakanishi, K., Kishimoto, S., Ushio, N., Kitamura, H., Higashitsuji, H., Hayashi, M., Tamaki, S., Uehara, N., Matsuura, K. and Saijo, H. (1981). *Agressologie* **22**: 45–57.

Hirata, Y. (1964). *Arch. Histol. Jpn.* **24**: 293–302.

Hirose, S. (1944). *Okayama Igakkai Zasshi* **56**: 791–798.

Hirsch, J.D. and Margolis, F.L. (1979). *Brain Res.* **161**: 227–291.

Hirsch, J.D. and Margolis, F.L. (1981). Biochemistry of Taste and Olfaction. ed by Cagan R.H. and Kare, M. R. Acad. Press, New York, pp. 311–332.

Hjorth-Simonsen, A. (1972). *J. Comp. Neurol.* **146**: 219–232.

Hodgkin, A.L. and Huxley, A.F. (1952). *J. Physiol.* **117**: 500–544.

Hoffmann, C.K. (1867). Ueber die Membrana Olfactaria. Inaug.-Diss., Amsterdam. Cited by Takagi, S.F. (1971).

Holley, A. (1986). A paper read in ECRO VII held at Davos in September 22–26.

Holley, A. and Döving, K.B. (1977). Olfaction and Taste VI. ed. by LeMagnen, J. and MacLeod, P. IRL Press, London, pp. 113–123.

Holley, A. and MacLeod, P. (1977). *J. Physiol.* (Paris) **73**: 725–828.

Holms, G. (1960). An Introduction to Clinical Neurology. Edinburgh. Cited by Sumner, D. (1962).

Hopfield, J.J. (1982). *Proc. Natl. Acad. Sci. USA.* **79**: 2554–2558.

Hopkins, A.E. (1926). *J. Comp. Neurol.* **41**: 253–289.

Horiguchi, S., Ueno, R., Hyodo, M. and Hayaishi, O. (1986). *Eur. J. Pharmacol.* **122**: 173–179.

Hornung, D.E. and Mozell, M.M. (1977). *Brain Res.* **128**: 158–163.

Hornung, D.E. and Mozell, M.M. (1981). Biochemistry of Taste and Olfaction. ed. by Cagan, R.H. and Kare, M.R. Acad. Press, New York, pp. 33–45.

Hosoya, Y. and Yoshida, H. (1937). *Jpn. J. Med. Sci. III. Biophys.* **5**: 2223.

Hotchkiss, W.T. (1956). *Arch. Otolaryngol.* **64**: 478–479.

Hoye, R.C., Ketcham, A.S. and Henkin, R.I. (1970). *Am. J. Surg.* **120**: 485–491.

Huber, G.C., Kappers, C.V.A. and Crosby, E.C. (1962). Correlative Anatomy of the Nervous System. Macmillan Publ. New York, p. 419.

Huff, F.J., Corkin, S. and Growdon, J.H. (1986). *Brain Lang.* **28**: 235–249.

Huggins, G.R. and Preti, G. (1976). *Am. J. Obstet. Gynec.* **126**: 129–136.

Humphrey, T. (1940). *J. Comp. Neurol.* **73**: 431–468.

Humphrey, T. (1967). *J. Hirnforsch.* **9**: 437–469.

Humphrey, T. and Crosby, E.C. (1938). *Univ. Michigan Univ. Hosp. Bull.* **4**: 61–62.

Hunter, A.J., Fleming, D. and Dixson, A.F. (1984). *J. Anat.* **138**: 217–225.

Ichihara M., Miyao, T., Komatsu, A., Kamio, T., Taira, K., Shirakura, K., Sakuma, Y., Kanke, Y., Murayama, T., Shiokawa, H., Kōda, T., Ichihara, H., Asaka, H., Oda, R. and Ogawa, T. (1961). *J. Otolaryngol. Jpn.* **64**: 1498–1505.

Ichihara, M. (1963). Clinic of Olfaction. Kanahara Publ. Co., Tokyo.

Ichihara, M., Yokokawa, R., Miyao, T., Komatsu, A., Kamio, T., Shirakura, K., Takeuchi, S., Muramatsu, Y., Sakuma, Y., Kobayashi, T., Watanabe, K. and Seki, S. (1959). *J. Otorhinolaryng. Jpn.* **62**: 955–958.

Ichihara, M., Miyao, T., Komatsu, A., Kamio, T., Ichihara, H. *et al.* (1960). *Jpn. J. Otol. Tokyo* **63**: 1460–1469.

Ichihara, M., Miyao, T., Komatsu, A., Kamio, T., Taira, K. *et al.* (1961). *Jpn. J. Otol. Tokyo* **64**: 1498–1505.

Iino, M. and Takagi, S.F. (1978). *Jpn J. Physiol.* **28**: 149–157.

Iino, M. and Takagi, S.F. (1983). Unpublished data.

Iizumi, O. and Kitamura, T. (1978). Olfactory Disorders——Olfactometry and -therapy. ed. by Toyota, B., Kitamura, T. and Takagi, S.F. Igaku-Shoin, Ltd., Tokyo, pp. 17–20.

Iizumi, O., Kitamura, T., Kaneko, T., Uchida, K. and Suzuki, H. (1975). *Practica Otologica* **68**: 879–889.

Iizumi, O., Kitamura, T., Kaneko, T., Naito, J., Uchida, K. and Suzuki, H. (1976). *J. Otolaryng. Jpn.* **79**: 70–75.

Iizumi, O., Uchida, K., Suzuki, H., Saruta, T., Kaneko, T. and Kitamura, T. (1979). *J. Jpn. Rhinolog. Soc.* **17**: 88–89p.

Ikeda, R. (1961). *Minzoku Eisei* **27**: 304–308.

Imamura, K. and Onoda, N. (1983). *Jpn. J. Physiol.* **33**: 135–138.

Imamura, K., Onoda, N., Obata, E., Iino, M. and Takagi, S.F. (1980). *JASTS* **14**: 17–20.

Imamura, K., Onoda, N. and Takagi, S.F. (1984a). *Jpn. J. Physiol.* **34**: 55–73.

Imamura, K., Mori, K., Onoda, N. and Takagi, S.F. (1984b). *Neurosci. Lett.* (*Suppl.*) **17**: S135.

Imamura, K., Mori, K., Fujita, S.C. and Obata, K. (1985a). *Brain Res.* **328**: 362–366.

Imamura, K., Mori, K., Fujita, S.C. and Obata, K. (1985b). *J. Physiol. Soc. Jpn* **47**: 505.

Ingersoll, D.W. (1981). *Dissert. Abstr.* **41B**: 3215.

Inoue, S. and Hogg, J.C. (1977). *J. Ultrastruct. Res.* **61**: 89–99.

Ishibashi, Y., Ohba, H., Ooshiro, K., Yoneno, K., Tokuda, K. and Umeda, R. (1978). *J. Jpn. Rhinolog. Soc.* **17**: 85–86.

Ishikawa, A. (1938). *J. Otolaryngol. Jpn.* **44**: 724–737.

Ito, K. (1968). *Kitakanto Med. J.* **18**: 405–417.

Ito, M., Kostyuk, P.G. and Oshima, T. (1962). *J. Physiol.* **164**: 150–156.

Jackowski, A., Parnevalas, J.G. and Lieberman, A.R. (1978). *Brain Res.* **195**: 17–28.

Jackson, Hughlings (1888). Cited by Douek, E. (1974).

Jackson, R.T. (1959). *The Physiologist* **2** (3): 63–64.

Jackson, R.T. (1960). *J. Cell Comp. Physiol.* **55**: 143–147.

Jackson, R.T. and Benjamin, R.M. (1974). *Brain Res.* **75**: 193–201.

Jafek, B.W. (1983). *Laryngoscope* **93**: 1576–1599.

Jagadowski, K.P. (1901). *Anat. Anz.* **XIX**: 257–267.

Jahnke, K. (1976). *Acta Oto-Laryng. Suppl.* **336**: 5–40.

Jahr, C.E. and Nicoll, R.A. (1980). *Science* **207**: 1473–1475.

Jahr, C.E. and Nicoll, R.A. (1982). *J. Physiol.* **326**: 213–234.

Jahr, C.E. and Nicoll, R.A. (1982). *Nature* **297**: 227–229.

Jan, L.Y. and Revel, J.P. (1974). *J. Cell Biol.* **62**: 257–273.

Jancsar, S.M. and Leonard, B.E. (1984). *Prog. Neuropsychopharmacol. Biol. Psychiat.* **8**: 263–269.

Jastreboff, P.J., Pedersen, P.E., Greer, C.A., Stewart, W.B. Kauer, J.S., Benson, T.E. and Shepherd, G.M. (1984). *Proc. Natl. Acad. Sci. USA.* **81**: 5250–5254.

Jennes, L. and Stumpf, W.E. (1980). *Cell Tissue Res.* **209**: 239–256.

Joachims, H.Z., Altman, M.M. and Mayer S.W. (1975). *J. Laryngol. Otol.* **89**: 335–343.

Johns, M.A. (1980). Cited by Beauchamp *et al.* (1980).

Johnston, J.W., Moulton, d.G. and Turk, A. (1970). Advances in Chemoreception I: Communication by Chemical Signals. Appleton-Century-Croft, New York, p. 416.

Johnston, R.E. (1983). Pheromones and Reproduction in Mammals. ed. by Vandenbergh, J.G. Acad. Press, New York, pp. 3–37.

Johnston, R.E. (1985). Taste, Olfaction and the Central Nervous System. ed. by Pfaff, D.W. Rockefeller University Press, New York, pp. 322–346.

Johnston, R.E. and Rasmussen, K. (1984). *Physiol. Behav.* **33**: 95–104.

Jones, F.N. (1954). *Am. J. Psychol.* **67**: 147. Cited by Sumner, D. (1962).

Jones, B.A., Moskowitz, H.R. and Butters, N. (1975) *Neuropsychologica* **13**: 173–179.

Jordan, U. (1972). *Folia Morph.* **31**: 417–432.

Jourdan, F. (1982). *Brain Res.* **240**: 341–344.

Jourdan, F., Duveau, A., Astic, L. and Holley, A. (1980). *Brain Res.* **188**: 139–154.

Jorgensen, M.B., and Buch, N.H. (1961). *Acta Otolaryngol.* (Stockh.) **53**: 539–545.

Joyner, R.E. (1963). *J. Occup. Med.* **5**: 37–42.

Juge, A., Holley, A. and Delaleu (1979a). *J. Physiol.* (Paris) **75**: 919–927.

Juge, A., Holley, A. and Rajon, D. (1979b). *J. Physiol.* (Paris) **75**: 929–938.

Kaada, B.R. (1951). *Acta Physiol. Scand* **24**: Suppl. 83.

Kaada, B.R. (1960).Handbook of Physiology, Nurophysiology. Sect. 1. Vol. II. Chapt. 55, Am Phisiol. Soc., Washington, D.C., pp. 1345–1372.

Kaada, B.R., Pribram, K.H. and Epstein, J.A. (1949). *J. Neurophysiol.* **12**: 347–356.

Kahn, R.J. (1965). Doctoral dissertation, Yeshiva University, 1965. Dissertation Abstracts International, 1965, 26, 3478A. (University Microfilms, No. 65–11, 980)

Kaise, H. (1969). *Kitakanto Med. J.* **19**: 396–408.

Kaissling, K.-E. (1971). Handbook of Sensory Physiology, Vol. IV. ed. by Beidler, L. M., Springer Verlag, Berlin, pp. 351–431.

Kaissling, K.-E. (1987). R.H. Wright Lectures on Insect Olfaction. ed. by Colbow, K. Simon Frazer Univ. Press, Burnaby, B.C. Canada.

Kalil, K. and Reh, T. (1979). *Science* **205**: 1158–1161.

Kallmann, F.J., Schoenfeld, W.A. and Barrera, S.E. (1944). *Am. J. Ment. Defic.* **48**: 203–236.

Kalmus, H. (1957). Molecular Structure and Organoleptic Perception. Soc. Chem. Ind. London, pp. 13–27.

Kamata, H., Tomizawa, K., Kamei, T. and Makino, S. (1985). *J. Jpn. Rhinol. Soc.* **24**: 141.

Kamei, T. and Makino, S. (1987). *Nihon Rinsho* **45**: Spring Suppl. 1293.

Kamio, T. (1960). *Jpn. J. Otol. Tokyo.* **66**: 469–485.

Kamiyama, K. (1983). Wonder of the Forest. Iwanami-Shinsho 242, Iwanami Shoten, Ltd., Tokyo.

Kamiyama, K. and Suzuki, F. (1986). *Arch. Kyoritsu Women's College, Tokyo* **32**: 90–100.

Kamiyama, K. and Suzuki, F. (1987). *Arch. Kyoritsu Women's College, Tokyo* **33**: 1–8.

Kamiyama, K. and Tokin, B. (1980). Mysterious Power of Trees—Phytoncide. Blue Backs series No. B424. Kohdansha, Ltd., Tokyo.

Kamo, N., Miyake, M., Kurihara, K. and Kobatake, Y. (1974). *Biochim. Biophys. Acta* **367**: 11–23.

Kamo, N., Kashiwagura, T., Kurihara, K. and Kobatake, Y. (1980). *J. Thear. Biol.* **83**: 111–130.

Kanai, T. (1940). *Okajima Folia Anat. Jpn.* **19**: 199–213.

Kanda, T., Kitamura, T., Kaneko, T. and Iizumi, O. (1973). *Jpn. J. Otolarygol.* (Tokyo) **76**: 739–743

Kaneko, N., Debski, E.A., Wilson, M.C. and Whitten, W.K. (1980). *Biol. Reprod.* **22**: 873–878.

Karja, J., Njuutinen, J. and Karfalainen, P. (1982). *Arch. Otolar.* **108**: 99–101.

Kashiwayanagi, M. and Kurihara, K. (1984). *Brain Res.* **293**: 251–258.

Kashiwayanagi, M. and Kurihara, K. (1985). *Brain Res.* **359**: 97–103.

Kashiwayanagi, M., Sai, K. and Kurihara, K. (1987). *J. Gen. Physiol.* **89**: 443–457.

Kasuga, S., Okano, M. and Sugawa, Y. (1978). Cell differentiation in the olfactory pit of the chick embryo. Ninth Intern. Congr. on Electron Microscopy, Toronto, **2**: 622–623.

Kasuga, S., Inoue, Y. and Okano, M. (1980). *Bull. Coll. Agr. Vet. Med. Nihon Univ.* **37**: 122–135.

Katz, B. and Miledi, R. (1969). *J. Physiol.* **195**: 481–492.

Kauer, J.S. (1980). Olfaction and Taste VII, ed. by van der Starre, H., pp. 227–236.

Kauer, J.S. and Moulton, D.G. (1974). *J. Physiol.* **243**: 717–737.

Kauer, J.S. and Shepherd, G.M. (1975). *Brain Res.* **85**: 108–113.

Kauer, J.S. and Shepherd, G.M. (1977). *J. Physiol.* **272**: 495–516.

Keller, A. and Margolis, F.L. (1975). *J. Neurochem.* **24**: 1101–1106.

Keller, A. and Margolis, F.L. (1976). *J. Biol. Chem.* **251**: 6232–6237.

Kennedy, C., Des Rosiers, M.H., Jehle, J.W. *et al.* (1975). *Science* **187**: 850–853.

Kerekovic, M. (1972). *Acta Otorhinolaryngol. Belg.* **26**: 518–523.

Kerjanscki, D. (1977). Olfaction and Taste VI. ed. by Le Magnen, J. and MacLeod, P. IRL Press, London, pp. 75–85.

Kerjaschki, D. and Hörandner, H. (1976). *J. Ultrastruct. Res.* **54**: 420–444.

Kerkut, G.A. and Thomas, R.C. (1963). *J. Physiol.* **168**: 23.

Kerkut, G.A. and Thomas, R.C. (1964). *Comp. Biochem. Physiol.* **11**: 199–213.

Kerr, D.I.B. (1977). Olfaction and Taste VI, ed. by Le Magnen, J. and MacLeod, P. IRL Press, London, pp. 97–103.

Kerr, D.I.B. (1985). *Proceedings of the Australian Physiological and Pharmacological Society* **16**: 1–3.

Kerr, D.I.B. and Hagbarth, K.E. (1955). *J. Neurophysiol.* **18**: 362–374.

Keverne, E.B. (1978). *Trends Neurosci.* **1**: 32–34.

Keverne, E.B. (1979). Chemical Ecology: Odour Communication in Animals. ed. by Ritter, F.J. Elsevier/North-Holland Biomed. Press, Amsterdam, pp. 75–83.

Keverne, E.B. (1982). Olfaction and Endocrine Regulation. ed. by Breipohl, W. IRL Press, London, pp. 127–140.

Keverne, E.B. and Michael, R.P. (1971). *J. Endocr.* **51**: 313–322.

Keverne, E.B., Murphy, C.L., Silver, W.L., Wysocki, C.J. and Meredith, M. (1986). *Chem. Sens.* **11**: 119–133.

Kevetter, G.A. and Winans, S.S. (1981). *J. Comp. Neurol.* **197**: 81–98.

Keyaki, Y. (1964). *Jpn. J. Otol.* **67**: 1320–1337.

Kievit, J. and Kuypers, H.G.J.M. (1977). *Expl. Brain Res.* **29**: 299–322.

Kimura, K. (1961). *Kumamoto Med. J.* **14**: 37–46.

King, M. and Vires, N. (1979). *J. Appl. Physiol.* **47**: 26–31.

Kirk, R.L. and Stenhouse, N.S. (1953). *Nature* **171**: 698–699.

Kishi, K., Mori, K., Tazawa, Y. and Ojima, H. (1982a). *Neurosci. Lett. Suppl.* **9**: S118.

Kishi, K., Mori, K. and Tazawa, Y. (1982b). *Neurosci. Lett.* **28**: 127–132.

Kishi, K., Mori, K. and Ojima, H. (1984). *J. Comp. Neurol.* **225**: 511–526.

Kitamura, T. and Iizumi, O. (1978). Olfactory Disorders—Olfactometry and therapy. ed. by Toyota, B., Kitamura, T. and Takagi, S.F. Igaku Shoin, Ltd., Tokyo, pp. 38–41.

Kiyohara, S. and Tucker, D. (1978). *Physiol. Behav.* **21**: 987–994.

Klawans, H.L.Jr. (1973). The Pharmacology of Extrapyramidal Movement Disorders. Karger, Basel.

Kleene, S.J. and Gesteland, R.C. (1981). *Brain Res.* **229**: 536–540.

Kleene, S.J. and Gesteland, R.C. (1983). *J. Neurosci. Methods* **9**: 173–183.

Kloek, J. (1961). *Psychiat. Neurol. Neurochir.* **64**: 309–344.

Kobal, G. and Plattig, K.H. (1978). *Z. EEG-EMG* **9**: 135–145.

Kobal, G., Hummel, Th. and Van Toller, C. (1987). *Chem. Sens.* **12**: 183p.

Kobayashi, H., Negishi, M., Yamamoto, A. and Zusho, H. (1984). *JASTS* **18**: 101–104; *Chem. Sens* **10** (1): 136–137, 1985.

Kogure. S., Onoda, N. and Takagi, S.F. (1978). *Proc. Jpn. Acad.* **54** (B): 478–483.

Kogure, S., Onoda, N. and Takagi, S.F. (1980). Advances in Physiological Sciences. Vol. **16**, Sensory Functions. ed. by Grastyán, E. and Molnár, P. Akademiai Kiado, Budapest, Pergamon Press, Oxford, pp. 367–375.

Koelega, H.S. and Köster, E.P. (1974). *Ann. N.Y. Acad. Sci.* **237**: 234–246.

Kohonen, T., Oja, E. and Lehtio, P. (1981). Parallel Methods of Associative Memory. ed. by Hinton, G.E. and Anderson, J.A. Erlbaum Assoc., Hillsdale, pp. 105–144.

Kohonen, T. (1977). Associative Memory—A system—Theoretic Approach. Springer, New York.

Koikegami, H., Fuse, S., Yokoyama, T., Watanabe, T. and Watanabe, H. (1955). *Folia Psychiat. Neurol. Jpn.* **8**: 336–370.

Koikegami, H. (1957). *Acta Med. Biol.* (*Niigata*) **5**: 21–72.

Koketsu, K. (1965). *Proc. 23rd Intern. Congr. Physiol. Sci. Tokyo*, p. 521.

Kolmer, W. (1927). Moellendorff's Handbuch der Mikroskopischen Anatomie des Menschen. Springer-Verlag, Berlin, pp. 192–249.

Kolnberger, I. (1971). *Z. Zellforsch.* **122**: 53–67.

Kolnberger, I. and Altner, H. (1971). *Z. Zellforsch.* **118**: 254–262.

Komisaruk, B.R. and Beyer, C. (1972). *Brain Res.* **36**: 153–170.

Konishi, J. (1966). *J. Gen. Physiol.* **49**: 1241–1264.

Kopsch, Fr. (1912). Rauber's Lehrbuch der Anatomie des Menschen. Abteilung 6: Sinnesorgane. Verlag von Georg Thieme, Leipzig, pp. 84–85.

Kosaka, T., Hataguchi, Y., Hama, K., Nagatsu, I. and Wu, J.-Y. (1985). *Brain Res.* **343**: 166–171.

Köster, E.P. (1965). *Inst. Rhinol.* **3**: 57–64.

Köster, E.P. (1968). *Olfactologia* **1**: 43–51.

Kratzing, J.E. (1971a). *Aust. J. Biol. Sci.* **24**: 787–796.

Kratzing, J.E. (1971b). *J. Anat.* **108**: 247–260.

Kratzing, J.E. (1978). *J. Anat.* **125**: 601–613.

Kraupa-Runk, M. (1916). *Muenchener Medizinische Wochenschrift.* **63**: 46. Cited by Bednár, M. and Langfelder, O. (1930).

Kreiner, J. (1961). *J. Comp. Neurol.* **116**: 117–133.

Krettek, J.E. and Price, J.L. (1977a). *J. Comp. Neurol.* **171**: 157–192.

Krettek, J.E. and Price, J.L. (1977b). *J. Comp. Neurol.* **172**: 687–722.

Krettek, J.E. and Price, J.L. (1977c). *J. Comp. Neurol.* **172**: 723–752.

Kretted, J.E. and Price, J.L. (1978a). *J. Comp. Neurol.* **178**: 225–254.

Krettek, J.E. and Price, J.L. (1978b). *J. Comp. Neurol.* **178**: 255–280.

Krieger, N.R. (1981). Biochemistry of Taste and Olfaction. ed. by Cagan, R.H. and Kare, M.R. Acad. Press, New York, pp. 417–441.

Krieger, N.R. and Heller, J.S. (1979). *J. Neurochem.* **33**: 299–302.

Krieger, N.R. Kauer, J.S., Shepherd, G.M. and Greengard, P. (1977). *Brain Res.* **131**: 303–312.

Krieger, N.R., Megill, J.R. and Sterling, P. (1983). *J. Comp. Neurol.* **215**: 465–471.

Kubie, J.L., Vagvolgyi, A. and Halpern, M. (1978). *J. Comp. Physiol. Psychol.* **92**: 627–641.

Kubie, J.L., Mackay-Sim, A. and Moulton, D.G. (1980). Olfaction and Taste VII. ed. by van de Starre, H. IRL Press, London, pp. 163–166.

Kuffler, S.W. and Eyzaguirre, C. (1955). *J. Gen. Physiol.* **39**: 155–184.

Kuffler, S.W. and Nicholls, J.G. (1966). *Ergeb. Physiol.* **57**: 1–90.

Kurahashi, T. and Shibuya, T. (1985). *Animal Physiol.* **2**: 167.

Kurahashi, T. and Shibuya, T. (1986a). *JASTS* **20**: 33–36.

Kurahashi, T. and Shibuya, T. (1986b). *Animal Physiol. (Jpn.)* **3**: 184.

Kurahashi, T. and Shibuya, T. (1986c). *Zool. Sci.* **3**: 982

Kurihara, K. (1988). The Beidler Symposium on Taste and Smell, ed. by Miller, I.J.Jr. Book Service Assoc., Inc. Washington-Salem, N.C., pp. 65–74.

Kurihara, K. and Koyama, K. (1972). *Biochem. Biophys. Res. Commun.* **48**: 30–34.

Kurihara, K., Kano, K. and Kobatake, Y. (1978). *Adv. Biophys.* **10**: 27–95.

Kurihara, K., Miyake, M. and Yoshii, K. (1981). Biochemistry of Taste and Olfaction. ed. by Cagan, R.H. and Kare, M.R. Acad. Press, New York, pp. 249–286.

Kuroishi, T. and Zusho, H. (1979). *J. Otorhinolaryng. Jpn.* **17**: 84–85.

Labows, J.N., Jr. (1979). *Perfum. Flavor* **4**: 12–17.

Labows, J.N., Preti, G., Hoelzle, E., Leyden J. and Kligman, A. (1979). *Steroids* **34**: 249–258.

Labows, J.N., McGineley, K.J. and Klegman, A.M. (1982). *J. Soc. Cosmet. Chem.* **34**: 193–202.

Laing, D.G. (1982). *Perception* **11**: 221–230.

Laing, D.G. (1983). *Perception* **12**: 99–117.

Laing, D.G. (1985). *Chem. Sens.* **10**: 412.

Lammers, H.J. and Lohman, A.H.M. (1974). *Prog. Brain Res.* **41**: 61–78.

Lancet, D. (1984). *Trends Neurosci.* **7**: 35–38.

Lancet, D. (1986). *Annu. Rev. Neurosci.* **9**: 329–356.

Lancet, D., Greer, C.A., Kauer, J.S. and Shepherd, G.M. (1981). *Soc. Neurosci. Abstr.* **7**: 661.

Lancet, D. and Pace, U. (1984). *Soc. Neurosci. Abstr.* **10**: 655.

Lancet, D. Greer, C.A., Kauer, J.S. and Shepherd, G.M. (1982). *Proc. Natl. Acad. Sci. USA.* **79**: 670–674.

Land, L.J. (1973). *Brain Res.* **63**: 153–166.

Land, L.J. and Shepherd, G.M. (1974). *Brain Res.* **70**: 506–510.

Land, L.J., Eager, R.P. and Shepherd, G.M. (1970). *Brain Res.* **23**: 250–254.

Langworthy, O.R. (1932). *J. Comp. Neurol.* **54**: 437–499.

Lavin, A., Alcocer-Cuarón, C. and Hernàndes-Peòn, R. (1959). *Science* **129**: 332–333.

Lawson, S.N., Harper, A.A., Harper, E.L., Garson, J.A. and Anderton, B.H. (1984). *J. Comp. Neurol.* **228**: 263–272.

Leigh, A.D. (1943). *Lancet* **i**: 38–40.

Le Magnen, J. (1948). *C.R. Acad. Sci. Paris* **226**: 753–754.

Le Magnen, J. (1949). Odeurs et Parfums. Que Sais-Je? Presses Universitaire de France.

Le Magnen, J. (1950). *C.R. Acad. Sci. Paris* **230**: 1103.

Le Magnen, J. (1952). *Arch. des Sci. Physiol.* **6**: 125–160.

Le Magnen, J. (1956). *C.R. Soc. Biol. Paris* **150**: 32–35.

Le Magnen, J. and Rapaport, A. (1951). *C.R. Soc. Biol. Paris* **145**: 800–803.

Leichnetz, G.R. and Astruc, J. (1976). *Brain Res.* **109**: 455–472.

Leigh, A.D. (1943). *Lancet* **i**: 38–40.

Lenhardt, E. and Rollin, H. (1969). *H.N.O.* (Berlin) **17**: 104–106.

Leon, M. and Moltz, H. (1972). *Physiol. Behav.* **8**: 683–686.

Leonard, C.M. (1969). *Brain Res.* **12**: 321–343.

Leonard, C.M. (1972). *Brain Behav. Evolut.* **6**: 524–541.

Leonard, C.M. and Scott, J.W. (1971). *J. Com. Neurol.* **141**: 313–330.

Lettvin, J.Y. and Gesteland, R.C. (1965). *Cold Spring Harb. Symp. Quan. Biol.* **30**: 217–225.

Leveteau, J. and MacLeod, P. (1966a). *Science* **153**: 175–176.

Leveteau, J. and MacLeod, P. (1966b). *J. Physiol.* (Paris) **58**: 717–729.

Leveteau, J. and MacLeod, P. (1969a). Olfaction and Taste III. ed. by Pfaffmann, C. Rockefeller University Press, New York, pp. 212–215.

Leveteau, J. and MacLeod, P. (1969b). *J. Physiol.* (Paris) **61**: 5–16.

Leveteau, J., MacLeod, P. and Daval, G. (1969). *Physiol. Behav.* **4**: 479–482.

Leveteau, J., Daval, G. and MacLeod, P. (1972). Olfaction and Taste IV. ed. by Schneider, D. Wissenschaftl. Verlagsgesell. MBH, Stuttgart, pp. 135–141.

Leys, D. (1945). *Lancet* **i**: 461–464.

Libet, B. (1970). *Fed. Proc.* **29**: 1945–1956.

Libet, B. and Gerard, R.W. (1938). *Proc. Soc. exp. Biol. N.Y.* **38**: 886–888.

Libet, B. and Gerard, R.W. (1939). *J. Neurophysiol.* **2**: 153.

Libet, B. and Gerard, R.W. (1941). *J. Neurophysiol.* **42**: 438–455.

Lidow, M.S. and Menco, B. Ph. M. (1984). *J. Ultrastract. Res.* **86**: 18–30.

Lidow, M.S., Kleene, S.T. and Gesteland, R.C. (1985). *AChemS* VII. No. 120.

Lidow, M.S., Kleene S.J. and Gesteland R.C. (1986). *Dev. Brain Res.* **28**: 145–162.

Lidow, M.S., Gesteland, R.C., Shipley, M.T. and Kleene, S.T. (1987). *Dev. Brain Res.* **31**: 243–258.

Lindstrom, C.G. and Lindstrom, D.W. (1975). *Acta Otolaryngol.* **80**: 447–451.

Lindvall, T. (1977). Olfaction and Taste VI. ed. by LeMagnen, J. and MacLeod, P. IRL Press, London, pp. 449–457.

Ling, G. and Gerard, R.W. (1949a). *J. Cell Comp. Physiol.* **34**: 383–396.

Ling, G. and Gerard, R.W. (1949b). *J. Cell Comp. Physiol.* **34**: 413–438.

Linás, R. and Sugimori, M. (1980). *J. Physiol.* **305**: 197–213.

Liss, L. and Gomez, F. (1958). *Arch. Otolaryngol.* **67**: 167–171.

Lohman, A.H.M. (1963). *Acta Anat.* **53** (Suppl. 49): 1–109.

Lohman, A.H.M. and Lammers, H.J. (1967). *Prog. Brain Res.* **233**: 65–82.

Long, J.A. and Jones, A.L. (1967). *Lab. Invest.* **16**: 355–370. Cited by Yamamoto. M. (1976).

Loo, S.K. (1977a). *J. Anat.* **123**: 135–145.

Loo, S.K. (1977b). *Acta Anat.* **98**: 221–223.

Loo, S.K. and Kanagasuntheram, R. (1972). *J. Anat.* **112**: 165–172.

Löhner, L. (1924). *Pflügers Arch. Ges. Physiol.* **202**: 24–25.

Lovell, M.A., Jafek, B.W., Moran, D.T. and Rowley, III C. (1982). *Arch. Otolaryngol.* **108**: 247–249.

Lundberg, A. (1957a). *Acta Physiol. Scand.* **40**: 21–34.

Lundberg, A. (1957b). *Acta Physiol. Scand.* **40**: 35–58.

Luskin, M.B. and Price, J.L. (1982). *J. Comp. Neurol.* **209**: 249–263.

Luskin, M.B. and Price, J.L. (1983). *J. Comp. Neurol.* **216**: 292–302.

Lusting, A (1924). Die Degeneration des Epithels der Riechnschleimhaut des Kanninchens nach Zerstörung der Riechlappen desselben. S.-B. Wien. Akad. **89**: Abst. 2. Cited by Nagahara (1940).

Luvara, A. and Murizi, M. (1961). *Bollettino delle Malattie dell'Orecchio, della Gola. del Naso* **79**: 367–375. Cited by Doty, R.L. (1986).

Lygonis, C.S. (1969). *Heredity* **61**: 413–416.

Ma, W.C. (1982). Perception of Behavioral Chemicals, ed. by Norris, D.M. Elsevier Biomedical Press, Amsterdam, p. 267.

MacDowell, R.J.S. (1948). Handbook of Physiology and Biochemistry, pp. 712–715. Blakiston. Philadelphia, Pa.

MacFarlane, A. (1975). The Human Neonate in Parent-Infant Interaction. ed. by Porter, R. and O'Connor, M. Elsevier, Amsterdam, pp. 103–117.

Machne, X. and Segundo, J.P. (1956). *J. Neurophysiol.* **19**: 232–240.

Mackay-Sim, A. and Nathan, M.F. (1984). *Anat. Embryol.* **170**: 93–97.

Mackay-Sim, A. and Shaman, P. (1984). *Brain Res.* **297**: 207–216.

Mackay-Sim, A., Shaman, P. and Moulton, D.G. (1982). *J. Neurophysiol.* **48**: 584–596.

MacLeod, P. (1959). *J. Physiol.* (Paris) **51**: 85–92.

MacLeod, P. (1971). Handbook of Sensory Physiology, Vol. IV, Chemical Senses, Part 1, Olfaction. ed. by Beidler, L.M. Springer-Verlag, Berlin, pp. 182–207.

MacLeod, N.K. and Reinhardt, W. (1983). *Neuroscience* **10**: 119–129.

MacLeod, N.K., Reinhardt, W. and Ellendorf, F. (1979). *Brain Res.* **164**: 323–327.

Macrides, F. and Davis, B.J. (1983). Chemical Neuroanatomy. ed. by Emson, P.C. Raven Press, New York, pp. 391–426.

Macrides, F. and Schneider, S.P. (1982). *J. Comp. Neurol.* **208**: 419–430.

Macrides, F., Davis, B.J., Youngs, W.M., Nadi, N.S. and Margolis, F.L. (1981). *J. Comp. Neurol.* **203**: 495–514.

Macrides, F., Davis, B.J. and Burd, G.D. (1982). Olfaction and Endocrine Regulation. ed. by Breipohl W. IRL Press, London, pp. 83–91.

Macrides, F., Schoenfeld, A., Marchand, J.E. and Clancy, A.N. (1985). *Chem. Sens.* **10**: 175–202.

Madison, D.M. (1977). Chemical Signals in Vertebrates. ed. by Muller-Schwarze, D. and Mozell, M.M. Plenum Press, New York, pp. 135–168.

Maesaka, A. (1964). *Juzen Igaku Zassi of the Kanazawa University* **70**: 198–221.

Maier, W. (1980). Evolutionary Biology of the New World Monkeys and Continental Drift. ed. by Ciochon, R.L. and Chiarelli, A.B. Plenum Press, New York, pp. 219–241.

Mair, R.G. (1986). *Experientia,* **42**: 213–223.

Mair, R.G. and Engen, T. (1976). *Sens. Proc.* **1**: 33–39.

Mair, R.G. and McEntee, W.J. (1986). Clinical Measurement of Taste and Smell. ed. by Meiselman, H.L. and Rivlin, S.R. Macmillan Publ. Co., New York, pp. 550–564.

Mair, R.G., Bouffard, J.A., Engen, T. and Morton, T.H. (1978). *Sens. Proc.* **2**: 90–98.

Mair, F.G., Gesteland, R.C. and Blank, D.L. (1982). *Neuroscience* **7**: 3091–3103.

Makino, S. (1977). *Igakuno-Ayumi* **100**: 196.

Makino, S. (1985). Unpublished data (by personal communication).

Makino, S. and Endo, S. (1978). *J. Jpn. Rhinolog. Soc.* **17**: 89–90.

Makino, S., and Ishii, H. (1978). Olfactory Disorders—Olfactometry and -therapy. ed. by Toyota, B., Kitamura, T. and Takagi, S.F. Igaku-Shoin, Ltd., Tokyo, pp. 54–56.

Males, J.L. and Schneider, R.A. (1972). *Acta Endocrinol.* **71**: 7–12.

Malm, L., Wihl, J.A., Lamm, C.J. and Lindqvist, N. (1981). *Allergy* **36**: 209–214.

Malpighi, M. (1681). Cited by Shepherd, G.M. (1981).

Manaker, S., Winokur, A., Rostene, W.H. and Rainbow, T.C. (1985). *J. Neurosci.* **5**: 167–174.

Mancia, M., Green, J.D. and von Baumgarten, R. (1962). *Arch. Ital. Biol.* **100**: 463–475.

Mantyh, P.W. and Hunt, S.P. (1985). *J. Neurosci.* **5**: 551–561.

Margolis, F.L. (1972). *Proc. Natl. Acad. Sci. USA.* **69**: 1221–1224.

Margolis, F.L. (1974). *Science* **184**: 909–911.

Margolis, F.L. (1975). *Soc. Neurosci. Symp.* **3**: 167–188.

Margolis, F.L. (1977). Aspects of Behavioral Neurobiology, Society of Neuroscience Symposia, Vol. III. ed. by Ferrendelli, J.A. Society for Neuroscience, Bethesda, MD, U.S.A. pp. 167–188.

Margolis, F.L. (1980a). Proteins in the Nervous System. ed. by Schneider, D. Raven Press, New York. pp. 59–84.

Margolis F.L. (1980b). The Role of Peptides in Neuronal Functions. ed. by Barker, J.L. and Smith, T. Dekker, New York, pp. 545–572.

Margolis, F.L. (1981). Biochemistry of Taste and Olfaction. ed. by Cagan, R.H. and Kare, M.R. Acad. Press, New York, pp. 369–394.

Margolis, F.L. (1982). *Scand. J. Immunol.* (Suppl.) **9**: 181–199.

Margolise, F.L. (1985). *Trends. Neurosci.* **8**: 542–546.

Margolis, F.L. and Tarnoff, J. (1973). *J. Biol. Chem.* **248**: 451–455.

Margolis, F.L., Roberts, N., Ferriero, D. and Feldman, J. (1974). *Brain Res.* **81**: 469–483.

Margolis, F.L., Grannot-Reisfeld, N., Grillo, M. and Farbman, A. (1983). *Biochim. Biophys. Acta* **744**: 237–248.

Marques, D.M. (1979). *Neural Biol.* **26**: 311–329.

Marsden, H.M. and Bronson, F.A. (1964). *Science* **144**: 1469.

Marshall, T.M. (1966). *J. Kent. Med. Assoc.* **64**: 322–327.

Marshall, J.R. and Henkin, R.I. (1971). *Ann. Intern. Med.* **75**: 207–211.

Marshall, D.A. and Maruniak, J.A. (1986). *Brain Res.* **366**: 329–332.

Masson, C., Kouprach, S., Giachtti, I. and MacLeod, P. (1977): Olfaction and Taste VI. ed. by Le Magnen, J. and MacLeod. IRL Press, London, p. 195.

Masukawa, L.M., Hedlund, B. and Shepherd, G.M. (1985a). *J. Neurosci.* **5**: 128–135.

Masukawa, L., Hedlund, B. and Shepherd, G. (1985b). *J. Neurosci.* **5**: 136–141.

Masukawa, L., Hedlund, B. and Shepherd, G. (1986). *Chem. Sens.* **11**: 634.

Masukawa, L., Kauer, J.S. and Shepherd, G.M. (1983). *Neurosci. Lett.* **35**: 59–64.

Mateson, J.F. (1954). *Ann. N.Y. Acad. Sci.* **58**: 83–95.

Mathews, D.F. (1972a). *J. Gen. Physiol.* **60**: 166–180.

Mathews, D.F. (1972b). *Brain Res.* **47**: 389–400.

Matsuda, R. (1938). Olfactory sensation during menstruation. Cited by Ichihara, M. (1963).

Matsuda, Y., Yoshida, S. and Yonezawa, T. (1978). *Brain Res.* **154**: 69–82
Matsuzaki, M. (1961). *Minzoku-Eisei Tokyo.* **27**: 308–310.
Matsuzaki, O., Shibuya, T. and Hatanaka, T. (1980a). *JASTS* **14**: 173–176.
Matsuzaki, O., Shibuya, T. and Hatanaka, T. (1980b). *Zool. Mag.* (Tokyo) **89**: 192–195.
Matsuzaki, O., Shibuya, T. and Hatanaka, T. (1982). *Zool. Mag.* (Tokyo) **91**: 221–229.
Matthes, E. (1927). *Z. Wiss. Biol. Abt.* **C5**: 83–166.
Maue, R.A. and Dionne, V.E. (1987a). *Pflügers Arch.* **409**: 244–250.
Maue, R.A. and Dionne, V.E. (1987b). *J. Gen. Physiol.* **90**: 95–125.
Mayo Clinic (1964). Clinical Examinations in Neurology. 2nd ed. by Saunders, W.B. Philadelphia. Cited by Cain, W.S. and Gent, J.F. (1986).
McClintock, M.K. (1971). *Nature* **229**: 244–245.
McConnell, R.J., Menendez, C.E., Smith, F.R., Henkin, R.I. and Rivlin, R.S. (1975). *Am. J. Med.* **59**: 354–364.
McCormack, L.J. and Harris, H.E. (1955). *J. Am. Med. Assoc.* **157**: 318–321.
McCotter, R.E. (1912). *Anat. Rec.* **6**: 299–318.
McLennan, H. (1971). *Brain Res.* **29**: 177–187.
Meibach, R.C. and Siegel, A. (1977). *Brain Res.* **119**: 1–20.
Meisami, E. (1976). *Brain Res.* **107**: 437–444.
Meisami, E. and Manoochehri, S. (1977). *Brain Res.* **128**: 170–175.
Meixner, C.H. (1955). Changes in olfactory sensitivity during the menstrual cycle. Unpublished master's thesis, Brown University.
Menco, B. Ph. M. (1977). Communications Agricultural Univ. Wageningen: 77–13, 1–157.
Menco, B. Ph. M. (1980a). *Cell Tissue Res.* **207**: 183–209.
Menco, B. Ph. M. (1980b). *Cell Tissue Res.* **211**: 5–29.
Menco, B. Ph. M. (1983). Nasal Tumors in Animals and Man, Vol. I. Anatomy, Physiology, and Epidemiology. ed. by Reznik, G. and Stinson, S.F. CRC Press, Inc., Boca Raton, Florida, pp. 45–102.
Menco, B. Ph. M. (1984). *Cell Tissue Res.* **235**: 225–241.
Menco, B. Ph. M. (1988). Personal communication.
Menco, B. Ph. M., Dodd, G.H., Davey, M. and Bannister, L.H. (1976). *Nature* **263**: 597–599.
Menco, B. Ph. M., Leunissen, J.L.M., Bannister, L.H. and Dodd, G.H. (1978). *Cell Tissue Res.* **193**: 503–524.
Menco, B. Ph. M., Laing, D.G. and Panhuber, H. (1980). Olfaction and Taste VII. ed. by van der Starre, H. IRL Press, London, Washington, D.C., p. 93.
Menco, B. Ph. M. and Van der Wolk, F.M. (1980). Electron Microscopy 1980, Vol. 2, Biology. ed. by Brederoo, P. and De Priester, W. 7th Eur. Congr. Electron Microscopy Found., Leiden, The Netherlands, pp. 26–27.
Menco, B. Ph. M. and van der Wolk, F.M. (1982). *Cell Tissue Res.* **223**: 1–27
Menevse, A., Dodd, G. and Poynder, T.M. (1977b). *Biochem. Biophys. Res Commun.* **77**: 671–677.
Menevse, A., Dodd, G., Poynder, T.M. and Squirrel, D. (1977a). *Biochem. Soc. Trans.* **5**: 191–194.
Menevse, A., Dodd, G. and Poynder, T.M. (1978). *Biochem. J.* **176**: 845–854.
Meredith, M. (1980). Chemical Signals in Vertebrates and Aquatic Invertebrates. ed. by Muller-Schwarze, D. and Silverstein, R. Plenum Press, New York, pp. 303–326.
Meredith, M. (1982). Olfaction and Endocrine Regulation. ed. by Breipohl, W. IRL Press, London, pp. 223–236.
Meredith, M. (1983). Pheromones and Reproduction in Mammals. ed. by Vandenbergh, J.G. Acad. Press, New York, pp. 199–252.
Meredith, M. (1986). *Physiol. Behav.* **36**: 737–743.
Meredith, M. and Burghardt, G.M. (1978). *Physiol. Behav.* **21**: 1001–1008.

Meredith, M. and O'Connell, R.J. (1979). *J. Physiol.* **286**: 301–316.

Meyer, A. (1957). Modern Trends in Neurology. ed. by Williams, D. Butterworth, London, pp. 301–306.

Meynert, (1878). Cited by Shepherd, G.M. (1981).

Michael, R.P. and Keverne, E.B. (1968). *Nature* **218**: 746–749.

Michael, R.P. and Keverne, E.B. (1970). *Nature* **225**: 84–85.

Michael, R.P., Keverne, E.B. and Bonsall, R.W. (1971). *Science* **172**: 964–966.

Michael, R.P., Bonsall, R.W. and Warner, P. (1974). *Science* **186**: 1217–1219.

Michael, R.P., Bonsall, R.W. and Kutner, M. (1975). *Psychoneuroendocrinology* **1**: 153–163.

Milas, N.A., Postman, W.M. and Heggie, R. (1939). *J. Am. Chem. Soc.* **61**: 1929–1930. Cited by Moulton, D.G. (1962).

Miller, R.F. and Dacheux, R. (1976). *Brain Res.* **104**: 157–162.

Millhouse, O.E. (1969). *Brain Res.* **15**: 341–363.

Miller, W.H. and Nicol, G.D. (1979). *Nature* **280**: 64–66.

Miller, W.H., Gorman, R.E. and Bitensky, M.W. (1971). *Science* **174**: 259–297.

Minor, A.V. (1980). Sensory Systems. Olfaction and Taste. Leningrad, pp. 3–18.

Minor, A.C. and Sakina, N.L. (1973). *Neirofiziologiya* **5**: 415–422.

Miragall, F. (1983). *J. Neurocytol.* **12**: 567–576.

Miragall, F., Breipohl, W. and Bhatnager, K.P. (1979). *Cell Tissue Res.* **200**: 397–408.

Miyake, M. and Kurihara, K. (1983a). *Biochim. Biophys. Acta* **762**: 248–255.

Miyake, M. and Kurihara, K. (1983b). *Biochim. Biophys. Acta* **762**: 256–264.

Mohun, M. (1943). *Arch. Otolaryng.* **37**: 699–709.

Moncrieff, R.W. (1951). The Chemical Senses. 2nd Edition. Leonard Hill, London.

Moncrieff, R.W. (1957). *Am. Perf. Aroma* **69**: 40–43.

Monrad-Krohn, G.H. (1958). The Clinical Examination of the Nervous System. London. Cited by Sumner, D. (1962),

Monti Graziadei, G.A. (1983). *Brain Res.* **262**: 303–308.

Monti Graziadei, G.A., Margolis, F.L., Harding, J.W. and Graziadei, P.P.C. (1977). *J. Histochem. Cytochem.* **25**: 1311–1316.

Moore, B.W. and McGregor, D. (1965). *J. Biol. Chem.* **240**: 1647–1653.

Moore, R.Y. and Halaris, A.E. (1975). *J. Neurol.* **164**: 171–183.

Moran, D.T., Rowley, J.C. III, and Jafek, B.W. (1982a). *Brain Res.* **253**: 39–46.

Moran, D.T., Rowley, J.C. III, Jafek, B.W. and Lovell, M.A. (1982b). *J. Neurocytology* **11**: 721–746.

Moran, D.T., Jafek, B.W. and Rowley, J.C. III. (1985). *Chem. Sens.* **10**: 420.

Morgane, P.J. and Panksepp, J. (1979). Handbook of the Hypothalamus. Vol. 1, Anatomy of the Hypothalamus. Marcel and Dekker, Inc., New York, Basel.

Mori, K. (1983). *Soc. Neurosci. Abstr.* **9**: 1020.

Mori, K. (1987a). *Prog. Neurobiol.* **29**: 275–320.

Mori, K. (1987b). *Adv. in Neurol. Soc.* **31**: 320–329; *Brain Res.* **408**: 215–221.

Mori, K. and Kishi, K. (1982). *Brain Res.* **247**: 129–133.

Mori, K. and Shepherd, G.M. (1979). *Brain Res.* **172**: 155–159.

Mori, K. and Takagi, S.F. (1975). *Brain Res.* **100**: 685–689.

Mori, K. and Takagi, S.F. (1977a). Food Intake and Chemical Senses. ed. by Katsuki, Y., Sato, M., Takagi, S.F. and Oomura, Y. University of Tokyo Press, Tokyo, pp. 33–43.

Mori, K. and Takagi, S.F. (1977b). *Proc. Int. Union Physiol. Sci. XXVII*, Paris, p. 527

Mori, K. and Takagi, S.F. (1977c). Olfaction and Taste VI. ed. by LeMagnen, J. and MacLeod, P. IRL Press, London, p. 201.

Mori, K. and Takagi, S.F. (1978a). *J. Physiol.* **279**: 569–588.

Mori, K. and Takagi, S.F. (1978b). *J. Physiol.* **279**: 589–604.

Mori, K., Kogure, S. and Takagi, S.F. (1977). *Brain Res.* **133**: 150–155.

Mori, K., Satou, M. and Takagi, S.F. (1978). *Proc. Jpn. Acad.* **54**: 484–489.

Mori, K., Satou, M. and Takagi, S.F. (1979). *Exp. Neurol.* **64**: 295–305.

Mori, K., Nowycky, M.C. and Shepherd, G.M. (1981a). *J. Physiol.* **314**: 281–294.

Mori, K., Nowycky, M.C. and Shepherd, G.M. (1981b). *J. Physiol.* **314**: 295–309.

Mori, K., Nowycky, M.C. and Shepherd, G.M. (1981c). *J. Physiol.* **314**: 311–320.

Mori, K., Nowycky, M.C. and Shepherd, G.M. (1981d). *Neurosci. Lett. Suppl.* **6**: S103.

Mori, K., Nowycky, M.C. and Shepherd, G.M. (1981e). *J. Physiol.* **320**: 89.

Mori, K., Kishi, K., Tazawa, Y. and Ojima, H. (1982a). *Neurosci. Lett. Suppl.* **9**: S118.

Mori, K., Nowycky, M.C. and Shepherd, G.M. (1982b). *J. Neurosci.* **2**: 497–502.

Mori, K., Kishi, K. and Ojima, H. (1983a). *J. Physiol. Soc. Jpn.* **45**: 500.

Mori, K., Kishi, K. and Ojima, H. (1983b). *J. Comp. Neurol.* **219**: 339–355.

Mori, K., Nowycky, M.C. and Shephered, G.M. (1984). *J. Neurosci.* **4**: 2291–2296.

Mori, K., Fujita, S.C., Imamura, K. and Obata, K. (1985). *J. Comp. Neurol.* **242**: 214–229.

Mori, K., Imamura, K., Fujita, S.C. and Obata, K. (1987). *Neuroscience* **20**: 259–278.

Morris, N. and Udry, J.R. (1978). *J. Biosoc. Sci.* **10**: 147–157.

De Morsier, G. (1962): *Acta Neuropath.* (Berlin) **1**: 433.

Moruzzi, G. and Magoun, H.W. (1949). *Electroence. Neurophysiol.* **1**: 455–473.

Mortimer, H., Wright, R.P. and Collip, J.B. (1937). *Can. Med. Ass. J.* **35**: 615–621.

Motokizawa, F. (1974a). *Brain Res.* **67**: 334–337.

Motokizawa, F. (1974b). *J. Physiol. Soc. Jpn.* **36**: 312.

Motokizawa, F. (1976). Electrobiology of Nerve, Synapse, and Muscle. ed. by Reuben, J.P., Purpura, D.P., Bennett, L.V.P. and Kandel, E.L. Raven Press, New York, pp. 235–241.

Motokizawa, F. (1985). *Chem. Sens.* **10**: 129.

Motokizawa, F. (1986). *Chem. Sens.* **11**: 272.

Motokizawa, F. and Ino, Y. (1981a). *Neuroscience* **6**: 39–46.

Motokizawa, F. and Ino, Y. (1981b). *Neurosci. Lett. Suppl.* **6**: S102.

Motokizawa, F., Ino, Y., Tsujimoto, Y. and Yasuda, T. (1977). Olfaction and Taste VI. ed. by Le Magnen, L. and MacLeod, P. IRL Press, London, p. 202.

Motokizawa, F., Ino, Y. and Ohta, N. (1982). *J. Physiol. Soc. Jpn.* **46**: 394.

Moulton, D.G. (1958). Ph.D. Thesis. Univ. Birmingham.

Moulton, D.G. (1962). *Nature* **195**: 1312–1313.

Moulton, D.G. (1963). Olfaction and Taste I. ed. by Zotterman, Y. Pergamon Press, Oxford, England, pp. 71–84.

Moulton, D.G. (1965). *Cold Spring Harb. Symp. Quant. Biol. Sensory Receptors* **30**: 201–206.

Moulton, D.G. (1967). Olfaction and Taste II. ed. by Hayashi, T. Pergamon Press, New York, pp. 109–116.

Moulton, D.G. (1971). Handbook of Sensory Physiology IV, Chemoreception. ed. by Beidler, L.M. Springer Verlag, Berlin, pp. 59–74.

Moulton, D.G. (1974). *Ann. NY Acad. Sci.* **237**: 52–61.

Moulton, D.G. (1975). Olfaction and Taste V. ed. by Denton, D.A. and Coghlan, J.P. Acad. Press, New York, pp. 111–114.

Moulton, D.G. (1976). *Physiol. Rev.* **56**: 578–593.

Moulton, D.G. and Beidler, L.M. (1967). *Physiol. Rev.* **47**: 1–52.

Moulton, D.G. and Fink, R.P. (1972). Olfaction and Taste IV. ed. by Schneider, D. Stuttgart Wissenschaftliche Verlagsgessellschaft, MbH, pp. 20–26.

Mourant, A.E. (1950). *Cold Spring Harbor Symp.* **15**: 242–245. Cited by Brown, K.S. and Robinette, R.R. (1967).

Mozell, M.M. (1962). *Am. J. Physiol.* **203**: 353–358.

Mozell, M.M. (1964). *Nature* **208**: 1180–1183.

Mozell, M.M. (1966). *J. Gen. Physiol.* **50**: 25–41.

Mozell, M.M. (1967). Olfaction and Taste II. ed. by Hayashi, T. Pergamon Press, Oxford, pp. 117–124.

Mozell, M.M. (1970). *J. Gen. Physiol.* **56**: 46–63.

Mozell, M.M. (1971). Handbook of Sensory Physiology, Vol. 4, Chemical Senses. ed. by Beidler, L.M. Springer-Verlag, Berlin, pp. 205–215.

Mozell, M.M. and Hornung, D.E. (1981). *Chem. Sens.* **6**: 267–276.

Mozell, M.M. and Phaffmann, C. (1954). *Ann. N.Y. Acad. Sci.* **58**: 96–108.

Mugford, R.A. and Nowell, N.W. (1970). *Nature* **226**: 967–968.

Mugnami, E., Oertel, W.H. and Wouterloo, F.F. (1984). *Neurosci. Lett.* **47**: 221–226.

Müller, W. (1971). *Z. Vergl. Physiologie* **72**: 370–385.

Muller, J.F. and Marc, R.E. (1984). *J. Comp. Neurol.* **222**: 482–495.

Mullins, L.J. (1955) *Ann. N.Y. Acad. Sci.* **62**: 249–276.

Murakami, M. and Shimoda, Y. (1977). *J. Physiol.* **264**: 801–818.

Murphy, M.P. (1980). *Behav. Neural Biol.* **30**: 323–340.

Murphy, J.P. and Gellhorn, E. (1945). *J. Neurophysiol.* **8**: 431–447.

Mustaparta, H. (1971). *Acta Physiol. Scand.* **82**: 154–166.

Naessen, R. (1970). *Acta Otolaryng.* (Stockh.) **70**: 1–51.

Naessen, R. (1971a). *Acta Otolaryng.* (Stockh.) **71**: 49–62.

Naessen, R. (1971b). *Acta Otolaryng.* **71**: 335–348.

Nagahara, Y. (1940). *Jpn. J. Med. Sci.* (Sect. 5, Path.) **5**: 165–199.

Nagano, H. (1965). *Gunma Symp. Endocrinol.* **2**: 19–28.

Nagano, H. (1977). *J. Otorhinolaryng. Jpn.* **80**: 25–32.

Nagano, H., Soda, T., Shiraishi, K., Hamazaki, J., Eura, S. *et al.* (1976). *J. Jpn. Rhiol. Soc.* **118**: 63.

Naguro, T. and Breipohl, W. (1982). *Cell Tissue Res.* **227**: 519–534.

Naito, J., Kawamura, K. and Takagi, S.F. (1984). *Neurosci. Res.* **1**: 19–33.

Nakamura, Y. (1916). *Dainihon Jiji Ih.* 22 (1916). Cited by Nagahara, Y. (1940).

Nakamura, T. and Gold, G.H. (1987). *Nature* **325**: 442–444.

Nakanishi, K. and Hinoki, M. (1981). *Equilibrium Res.* **40**: 208–216.

Nakashima, M., Mori, K. and Takagi, S.F. (1978). *Brain Res.* **154**: 301–316.

Nakashima, M., Kanenae S. and Yoshida, M. (1979). *J. Otolaryng. Jpn.* **82**: 1216.

Nakashima, M., Kimmelman, C.P. and Snow, J.B. Jr. (1984). *Arch. Otolaryngol.* **110**: 641–646.

Narumiya, S., Ogorochi, T., Nakao, K. and Hayashi, O. (1982). *Life Sci.* **31**: 2093–2103.

Nauta, W.J.H. (1958). *Brain* **81**: 319–340.

Nauta, W.J.H. (1961). *J. Anat.* **95**: 515–531.

Nauta, W.J.H. (1962). *Brain* **85**: 505–522.

Nauta, W.J.H. (1964). The Frontal Granular Cortex and Behavior. ed. by Warren, J.M. and Akert, K. McGraw Hill, New York, pp. 397–409.

Nauta, W.J.H. (1971). *J. Psychiat. Res.* **8**: 167–187.

Nauta, W.J.H. (1972). *Acta Neurobiol. Expl.* **32**: 125–140.

Negishi, M., Watanabe, N., Nagai, D. and Zusho, H. (1982). *JASTS* **16**: 21–24.

Negus, V. (1958). The Comparative Anatomy and Physiology of the Nose and Paranasal Sinuses. E & S Livingstone, Edinburgh.

Neher, E., Sakmann, B. and Steinbach, J.H. (1978). *Pflüg. Arch.* **375**: 219–228.

Neutra, M. and Leblond, C.P. (1969). *Scient. Am.* **220**: 100–107.

Nicoll, R.A. (1969). *Brain Res.* **14**: 157–172.

Nicoll, R.A. (1970a). *Brain Res.* **19**: 491–493.

Nicoll, R.A. (1970b). *Nature* **227**: 623–625.

Nicoll, R.A. (1971a). *Brain Res.* **35**: 137–149.

Nicoll, R.A. (1971b). *Science* **171**: 824–825.

Nicoll, R.A. (1972). *Exp. Brain Res.* **14**: 185–197.

Nishi, S. (1974). The Peripheral Nervous System. ed. by Hubbard, J.I. Plenum Press, New York, pp. 225–255.

Nishida, H. (1971). *J. Otorhinolaryng. Jpn.* **74**: 1631–1652.

Nishida, H. and Maeda, G. (1971). *Otolayngology* **43**: 425–431.

Nishida, H., Kumagami, H. and Jinnai, H. (1973). *J. Otorhinolaryng. Jpn.* **76**: 1449–1458.

Noferi, G. and Giudizi, S. (1946). *Rivista di Clinica Medica* 5, Supple. **1**: 89–100. Cited by Doty, R.L. (1986).

Noirot, E. (1969). *Animal Behav.* **17**: 547–550.

Nomura, T. and Kurihara, K. (1987a). *Biochemistry* **26**: 6135–6140.

Nomura, T. and Kurihara, K. (1987b). *Biochemistry* **26**: 6141–6145.

Nowycki, M.C., Waldow, U. and Shepherd, G.M. (1978). *Soc. Neurosci. Abstr.* **4**: 583.

Nowycky, M.C., Mori, K. and Shepherd, G.M. (1980). *Neurosci. Abstr.* **6**: 181.

Nowycky, M.C., Mori, K. and Shepherd, G.M. (1981a). *J. Neurophysiol.* **46**: 639–648.

Nowycky, M.C., Mori, K. and Shepherd, G.M. (1981b). *J. Neurophysiol.* **46**: 649–658.

Nygren, L.G., Olson, L. and Seiger, A. (1971). *Histochem.* **28**: 1–15.

Ochi, J. (1963). *Jpn. J. Physiol.* **13**: 113–128.

Ogata, S., Nishimura, K., Indo, M., Kawasaki, M. and Torii, S. (1986). *JASTS* **20**: 149–152; *Chem. Sens.* **12**: 517–518, 1987.

Ogorochi, T., Narumiya, S., Mizuno, N., Yamashita, K., Miyazaki, H. and Hayaishi, O. (1984). *J. Neurochem.* **43**: 71–82.

Ohba, H. (1982). *Practica Otologia (Tokyo)* **75**: 1081–1104.

Ohkado, T. (1984). *Oto-Rhino-Laryngol. (Tokyo)* **27**: 139–170.

Ohomori, H. (1985). *J. Physiol.* **359**: 189–217.

Ojima, H., Mori, K. and Kishi, K. (1984). *J. Comp. Neurol.* **230**: 77–87.

Okano, M. (1965). *Arch. Histol. Jpn.* **26**: 169–185.

Okano, M. (1973). *J. Jpn. Med. Assoc.* **69**: 1356–1365.

Okano, M. (1981). *Adv. Neurol. Sci. (Tokyo)* **25**: 249–326

Okano, M. and Takagi, S.F. (1974). *J. Physiol.* **242**: 353–370.

Okano, M., Weber, A.F. and Frommes, S.F. (1967). *J. Ultrastruct. Res.* **17**: 487–502.

O'Leary, J.L. (1973). *J. Comp. Neurol.* **150**: 217–238.

Olds, J. (1956). *J. Comp. Physiol. Psychol.* **49**: 281–285.

Oley, N., DeHan, R.S., Tucker, D., Smith, J.C. and Graziadei, P.P.C. (1975). *J. Comp. Physiol. Psychol.* **88**: 477–495.

de Olmos, J. and Ingram, W.R. (1972). *J. Comp. Physiol.* **146**: 303–333.

de Olmos, J., Hardy, H. and Heimer, L. (1978). *J. Comp. Neurol.* **181**: 213–244.

Olsen, K.D. and Desanto, L.W. (1983). *Arch. Otolaryngol.* **109**: 797–802.

Olszewski, (1952). The Thalamus of the *Macaca mulatta*: An Atlas for Use with the Stereotoxic Instrument. Karger, Basel.

Onagawa, (1957). *J. Physiol. Soc. Jpn.* **19**: 189–193.

Onoda, N. (1988a). *Neuroscience* **2**: 1003–1012.

Onoda, N. (1988b). *Neuroscience* **26**: 1013–1022.

Onoda, N. and Fujita, S.C. (1988). *Neuroscience* **26**: 993–1002.

Onoda, N. and Iino, M. (1980). *Proc. Jpn. Acad.* **56**: 300–305.

Onoda, N. and Imamura, K. (1984). *Neurosci. Res.* **1**: 457–461.

Onoda, N. and Kogure, S. (1980). Brain Mechanisms of Sensation. ed. by Katsuki, Y., Norgren, R. and Sato, M. John Wiley and Sons, Tokyo, Chichester, Brisbane, Toronto, pp. 227–240.

Onoda, N. and Mori, K. (1980). *J. Neurophysiol.* **44**: 29–39.

Onoda, N., Imamura, K., Ariki, T. and Iino, M. (1981a). *Proc. Jpn. Acad.* **57** (B): 355–358.

Onoda, N., Imamura, K., Obata, E. and Iino, M. (1981b). *Neurosci. Lett. Suppl.* **6**: S104.

Onoda, N., Imamura, K., Ariki, T. and Iino, M. (1982a). *Neurosci. Lett. Suppl.* **9**: S119.

Onoda, N., Ariki, T., Imamura, K. and Iino, M. (1982b). *Proc. Jpn. Acad.* **58** (B): 222–225.

Onoda, N., Imamura. K., Obata, E. and Iino, M. (1984). *J. Neurophysiol.* **52**: 638–652.

Oomura, Y. (1976). Hunger: Basic Mechanism and Clinical Implications. ed. by Novin, D., Wyrwicka, W. and Bray, G.A. Raven Press, New York, pp. 145–157.

Oomura, Y., Shimizu, H., Kita, H., Ishizuka, S., Aou, S., Yoshimatsu, H. and Yamabe, K. (1980). *Brain Res. Bull.* Suppl. 4,5: 151–161.

Ooshima, K. and Takagi, S.F. (1973). *J. Jpn. Med. Ass.* **69**: 1392–1404.

Orci, L., Perrelet, A. and Dumant, Y. (1974). *Proc. Natl. Acad. Sci. USA.* **71**: 307–310.

Orona, E., Scott, J.W. and Rainer, E.C. (1983). *J. Comp. Neurol.* **217**: 227–237.

Orona, E., Rainer, E.C. and Scott, J.W. (1984). *J. Comp. Neurol.* **226**: 346–356.

Orrego, (1961). *Arch. Ital. Biol.* **99**: 446–465.

Osterhammel, P., Terkildsen, K. and Zilstorff, K. (1969). *J. Laryngol.* **83**: 731–733.

Osumi, T. (1984). *Oto-Rhino-Laryngol. (Tokyo)* **27**: 139–170.

Ottoson, D. (1954). *Acta Physiol. Scand.* **32**: 384–386.

Ottoson, D. (1956). *Acta Physiol. Scand.* **35**, Suppl. 122: 1–83.

Ottoson, D. (1958) *Acta Physiol. Scand.* **43**: 167–181.

Ottoson, D. (1959a). *Acta Physiol. Scand.* **47**: 136–148.

Ottoson, D. (1959b). *Acta Physiol. Scand.* **47**: 149–159.

Ottoson, D. (1959c). *Acta Physiol. Scand.* **47**: 160–172.

Ottoson, D. (1963a). Olfaction and Taste I. ed. by Zotterman, Y. Pergamon Press, Oxford, pp. 149–159.

Ottoson, D. (1963b). *Pharmacol. Rev.* **15**: 1–42.

Ottoson, D. (1970). Ciba Symp. on Taste and Smell in Vertebrates. ed. by Wolstenlholme, G.E.W. and Knight, J. Churchill, London, pp. 343–356.

Ottoson, D. (1971). Handbook of Sensory Physiology, 4: 1. ed. by Beidler, L.M. Springer Verlag, NY, pp. 95–131.

Ottoson, D. and Shepherd, G.M. (1967). Progress in Brain Research. Sensory Mechanisms. Vol. 23, ed. by Zotterman, Y. Elsevier, Amsterdam, pp. 83–138.

Pace, U., Hanski, E., Salomon, Y. and Lancet, D. (1985). *Nature* **316**: 255–258.

Pager, J., Giachetti, I., Holley, A. and Le Magnen, J. (1972). *Physiol. Behav.* **9**: 573–579.

Pager, J. (1974). *Physiol. Behav.* **13**: 523–526.

Pangborn, R.M. (1967). The Chemical Senses and Nutrition. ed. by Kare, M.R. and Maller, O. Johns Hopkins Univ. Press, Baltimore, p. 149.

Panhuber, H., Mackay-Sim, A. and Laing, D.G. (1987). *Brain Res.* **428**: 307–311.

Papez, J.W. (1937). *Arch. Neurol. Psychiat. (Chic.)*, **38**: 725–743.

Parkes, A.S. and Bruce, H.M. (1962). *J. Reprod. Fertil.* **4**: 303–308.

Parsons, T.S. (1970). *Forma et Functio* **3**: 105–111.

Parsons, T.S. (1979). Evolution of the Vomeronasal Organ. Paper presented at Symposium on the Neural and Behavioral Basis of the Dual Olfactory Hypothesis. At 50th Annual Meeting of Eastern Psychologial Association, Philadelphia, PA.

Pasquier, D.A. and Reinoso-Suarez, F. (1976). *Brain Res.* **108**: 165–169.

Patterson, P.M. and Lauder, B.A. (1948). *J. Heredity* **39**: 295–297.

Patterson, R.L.S. (1968). *J. Sci. Food Agric.* **19**: 31–38.

Peabody, C.A. and Tinklenberg, J.R. (1985). *Am. J. Psychiat.* **142**: 524–525.

Pearlman, S.J. (1934). *Ann. Oto. Rhinol. Lar.* **43**: 739–768.

Pearson, A.A. (1941a). *J. Comp. Neurol.* **75**: 39–66.

Pearson, A.A. (1941b). *J. Comp. Neurol.* **75**: 199–217.

Pearson, A.A. (1942). *Ann. Oto. Rhinol. Lar.* **51**: 317–333.

Pearson, K. (1976). Simpler Networks and Behavior. ed. by Fentress, J.C. Sinauer Associates, Sunderland, MA, pp. 99–110.

Pearson, K., Stein, R. and Malhotra, S. (1970). *J. Exptl. Biol.* **53**: 299–316.

Pedersen, P.E. and Benson, T.E. (1986). *J. Comp. Neurol.* **252**: 555–562.

Pedersen, P.E., Jastreboff, P.J., Stewart, W.B. and Shepherd, G.M. (1986a). *J. Comp. Neurol.* **250**: 93–108.

Pedersen, P.E., Shepherd, G.M. and Greer, C.A. (1986b). *Chem. Sens.* **11**: 649–650.

Peele, T.L. (1977). The Neuroanatomic Basis for Clinical Neurology. McGraw Hill, New York, pp. 538–544.

Pelosi, P. (1977). Olfaction and Taste VI. ed. by. LeMagnen, J. and MacLeod, P. IRL Press, London, p. 70.

Penfield, W.G. and Erickson, T.C. (1941). Epilepsy and Cerebral Localization. Charles C. Thomas. Springfield, Ill, p. 623.

Persaud, K.C., Wood, P.H., Squirrel, D.J. and Dodd, G.H. (1981). *Biochem. Soc. Trans.* **9**: 107–108.

Peters, G. (1958). Handbuch der Speziellen Pathologischen Anatomie und Histologie. ed. by Lubarsch, O., Henke, F. and Roessle, R. Vol. XIII. Erkrankungen des zentralen Nerven Systems. II. Bandteil A. IV. Springer, Berlin, pp. 525–590.

Pfaff, D.W. and Gregory, E. (1971). *J. Neurophsiol.* **34**: 208–216.

Pfaff, D.W. and Pfaffmann, C. (1969a). *Brain Res.* **15**: 137–156.

Pfaff, D.W. and Pfaffmann, C. (1969b). Olfaction and Taste III. ed. by Pfaffmann, C. Rockefeller Univ. Press, New York, pp. 258–267.

Phillips, C.G., Powell, T.P.S. and Shepherd, G.M. (1963). *J. Physiol.* **168**: 65–88.

Phillips, H.S., Hostetter, G., Kerdelhue, B. and Kozlowski, G.P. (1980). *Brain Res.* **193**: 574–579.

Phillips, H.S., Ho, B.T. and Linner, J.G. (1982). *Brain Res.* **246**: 193–204.

Pickel, V.M., Segal, M. and Bloom, F.E. (1974). *J. Comp. Neurol.* **155**: 15–42.

Pinching, A.J. (1972). Olfaction and Taste IV. ed. by Schneider, D. Wissen Schaftliche Verlagsgesellschaft MBH Stuttgart. pp. 40–47.

Pinching, A.J. (1977). *Brain.* **100**: 377–388.

Pinching, A.J. and Döving, K.B. (1974). *Brain Res.* **82**: 195–204.

Pinching, A.J. and Powell, T.P.S. (1971a). *J. Cell Sci.* **9**: 305–345.

Pinching, A.J. and Powell, T.P.S. (1971b). *J. Cell Sci.* **9**: 347–377.

Pinching, A.J. and Powell, T.P.S. (1971c). *J. Cell Sci.* **9**: 379–409.

Pinching, A.J. and Powell, T.P.S. (1972). *J. Cell Sci.* **10**: 585–619.

Plattig, K.-H., and Kobal, G. (1976). Cited by Plattig, K.-H., and Kobal, G. (1977b).

Plattig, K.-H. and Kobal, G. (1977a). Olfaction and Taste VI. ed. by Le Magnen, J. and MacLeod, P. IRL Press, London, p. 203.

Plattig, K.-H. and Kobal, G. (1977b). Food Intake and Chemical Senses. ed. by Katsuki, Y., Sato, M., Takagi, S.F., and Oomura, Y. Univ. of Tokyo Press, Tokyo, pp. 51–70.

Plattig, K.-H. and Kobal, G. (1979). Human Evoked Potentials. ed. by Lehmann, D. and Callaway, E. Plenum Press, pp. 285–301.

Pollack, M.S., Wysocki, C.J., Beauchamp, G.K., Braun, D. Jr., Callaway, C. and Dupont, B. (1982). *Immunogenetics* **15**: 579–589.

Polyzonis, B.M., Kafandaris, P.M., Gigis, P.I. and Demetriou, T. (1979). *J. Anat.* **128**: 77–83.

Porrino, L.T., Crane, A.M. and Goldman-Rakic, P.S. (1981). *J. Comp. Neurol.* **198**: 121–136.

Porter, K.R. and Bonneville, M.A. (1964). An Introduction to the Fine Structure of Cells and Tissues. 2nd ed. Leeand Febinger, Philadelphia. Cited by Yamamoto, M.(1976).

Potiquet, M. (1891). *Revue Lar.* **2**: 737–753. Cited by Pearlman, S.J. (1934).

Potter, H. and Nauta, W.J.H. (1979). *Neuroscience* **4**: 361–367.

Powell, T.P.S., Cowan, W.M. and Raisman, G. (1965). *J. Anat.* **99**: 791–813.

Powers, J.B. and Winans, S.S. (1975). *Science N.Y.* **187**: 961–963.

Preti, G. and Huggins, G.R. (1975). *J. Chem. Ecol.* **1**: 361–376.

Preti, G. and Huggins, G.R. (1978). The Human Vagina. ed. by Hafex, E.S.E. and Evans, T.N. Elsevier, North-Holland Biomedical Press, Amsterdam, pp. 151–166.

Preti, G., Huggins, G.R. and Silverberg, G.D. (1979). *Fertil. Steril.* **32**: 47–54.

Pribram, K.H., Chow, K.L. and Semmes, K.J. (1953). *J. Comp. Neurol. Physiol.* **98**: 433–448.

Price, J.L. (1968). *Brain Res.* **7**: 483–486.

Price, J.L. (1969). *Brain Res.* **14**: 542–545.

Price, J.L. (1973). *J. Comp. Neurol.* **150**: 87–108.

Price, J.L. (1985). *Chem. Sens.* **10**: 239–258.
Price, J.L. and Powell, T.P.S. (1970a). *J. Cell Sci.* **7**: 91–123.
Price, J.L. and Powell, T.P.S. (1970b). *J. Cell Sci.* **7**: 125–155.
Price, J.L. and Powell, T.P.S. (1970c). *J. Cell Sci.* **7**: 157–187.
Price, J.L. and Powell, T.P.S. (1970d). *J. Cell Sci.* **7**: 631–651.
Price, J.L. and Powell, T.P.S. (1970e). *J. Anat.* **107**: 215–237.
Price, J.L. and Powell, T.P.S. (1971). *J. Anat.* **110**: 105–126.
Price, J.L. and Slotnick, B.M. (1983). *J. Comp. Neurol.* **215**: 63–77.
Price, J.L., Sasaki, L.S. and Slotnick, B.M. (1984). *AChemS. Abstr.* **6**: 107.
Price, S. (1977). *Neurosci. lett.* **4**: 49–50.
Price, S. (1981). Biochemistry of Taste and Olfaction. ed. by Cagan, R.H. and Kane, M.R. Acad. Press, New York, pp. 69–84.
Price, S. (1984). *Chem. Sens.* **8**: 341–354.
Price, S. (1986a). *Ann. Rev. Neurosci,* **9**: 329–355.
Price, S. (1986b). *JASTS* **20**: 8–10.
Price, S., and Turpin, J. (1980). Olfaction and Taste VII. ed. by van der Starre, H. IRL Press, London, pp. 65–68.
Price, S. and Willey, A. (1987). *Ann. N.Y. Acad. Sci.* **510**: 561–564.
Pumphrey, R.J. (1935). *J. Cell Comp. Physiol.* **6**: 457–467.
Punter, P.H., Menco, B. Ph. M. and Boelens, H. (1981). Odor Quality and Molecular Structure. ed. by Moskowitz, H.R. and Warren, C.B. ACS Symp. Ser. 148, American Chemical Society, Washington, D.C., pp. 93–108. Cited by Menco, B. Ph. M. (1984).
Quadagno, D.M., Shubeita, H.E., Deck, J. and Francoeur, D. (1981). *Psychoneuroendocrinology* **6**: 239–244.
Quastler, H. and Weingarten, H. (1930). *Arch. Entwickl.-Mech. Org.* **122**: 763–769.
Rafols, J.A. and Getchell, T.V. (1983). *Anat. Rec.* **206**: 87–101.
Raisman, G. (1972). *Expl. Brain Res.* **14**: 395–408.
Raisman, G. (1985). *Neuroscience* **14**: 237–254.
Rall, W. (1969). *Biophys. J.* **9**: 1483–1508.
Rall, W. and Shepherd, G.M. (1968). *J. Neurophysiol.* **31**: 884–915.
Rall, W., Shepherd, G.M., Reese, T.S. and Brightman, M.W. (1966). *Exp. Neurol.* **14**: 44–56.
Rausch, R., and Serafetinides, E.A. (1975a). *Nature* **255**: 557–558.
Rausch, R. and Serafetinides, E.A. (1975b). Olfaction and Taste V. ed. by Denton, D.A. and Coghlan, J.P. Acad. Press, New York, pp. 321–324.
Read, E.A. (1908). *Am. J. Anat.* **8**: 17–47.
Reep, R.L. and Winas, S.S. (1982). *Neuroscience* **7**: 1265–1288.
Reese, T.S. (1965). *J. Cell Biol.* **25**: 209–230.
Reese, T.S. and Shepherd, G.M. (1972). Structure and Function of Synapses. ed. by Pappas, G.D. and Purpura, D.P. Raven Press, New York, pp. 121–136.
Reinhardt, W., MacLeod, N.K., Ladewig, J. and Ellendorff, F. (1983). *Neuroscience* **10**: 131–139.
Revial, M.F., Duchamp, A. and Holley, A. (1978). *Chem. Sens. Flav.* **3**: 7–21.
Revial, M.F., Sicard, G. Duchamp, A. and Holley, A. (1982). *Chem. Sens.* **7**: 175–190.
Retzius, G. (1892). *Biol. Untersuch. N.F.* **4**: 62–64. Cited by Rafols, J.A. and Getchell, T.V. (1983).
Reynolds, J. and Keverne, E.B. (1979). *J. Reprod. Fert.* **57**: 31–35.
Rhein, L.D. and Cagan, R.H. (1980). *Proc. Natl. Acad. Sci. USA.* **77**: 4412–4416.
Rhein, L.D. and Cagan, R.H. (1981). Biochemistry of Taste and Olfaction. ed. by Cagan, R.H. and Kare, M.R. Acad. Press, New York, pp. 47–68.
Ribak, C.E., Vaughn, J.E., Saito, K., Barber, R. and Roberts, E. (1977). *Brain Res.* **126**: 1–18.
Richards, C.D. and Sercombe, R. (1968). *J. Physiol.* **197**: 667–683.
Ries, E.A. and Langworthy, O.R. (1937). *J. Comp. Neurol.* **68**: 1–47.

Ritter, F.N. (1964). *Arch. Otolaryngol.* **79**: 169–171.

Robecchi, M.G. (1972). *Minerva Otorinolaryn.* **22**: 196–204. Cited by Kasuga, S., Inoue, Y. and Okano, M. (1980).

Roberts, A. and Bush, B.M.H. eds. (1981). Neurons without Impulses. Society for Experimental Biology Seminar Series, Vol. 6. Cambridge University Press, Cambridge, England, p. 261. Cited by Mori, K. *et al.* (1982b).

Robinson, G.A., Butcher, R.W. and Sutherland, E.W. (1971). Cyclic AMP. Acad. Press, New York, London, p. 531.

Rodolfo-Masera, T. (1943). *Arch. Ital. Anat. Embriol.* **48**: 157–212. Cited by Wysocki, C.J. (1979).

Rogers, K.E., Grillo, M., Sydor, W., Poonian, M. and Margolis, F.L. (1985). *Proc. Natl. Acad. Sci. USA.* **82**: 5218–5222.

Rohrbaugh, J.W., Syndulko, K. and Lindsley, D.B. (1976). *Science* **191**: 1055–1057.

Ropartz, P. (1968). *Animal Behav.* **16**: 97–100.

Rose, S. (1927). *J. Psychol. Neurol. (Lpz)* **34**: 250–255.

Rose, S. (1929). *J. Psychol. Neurol. (Lpz)* **40**: 1–51.

Rose, J.E. and Woolsey, C.N. (1948). *Res. Publs. Ass. Res. Nerv. Ment. Dis.* **27**: 210–232.

Rosenberg, B., Misra, T.N. and Switzer, R. (1968). *Nature* **217**: 423–427.

Rosenblatt, J.S. (1972). *Sci. Am.* **227**: 18–25.

Rosene, D.L. and Heimer, L. (1977). *Anat. Rec.* **187**: 698.

Rossberg, G., Schaupp, H. and Schmidt, W.Z. (1966). *Z. Laryngol. Rhinol.* **45**: 571–590.

Rossman, C.M., Lee, R.M.K.W., Forrest, J.B. and Newhouse, M.T. (1984). *Am. Rev. Respir. Dis.* **129**: 161–167.

Rothstein, A and Meier, R. (1951). *J. Cell Comp. Physiol.* **38**: 245–270.

Routtenberg, A. (1971). *J. Comp. Physiol. Psychol.* **75**: 269–276.

Rubert, S.L., Hollender, M.H. and Mehrhof, E.G. (1961). *Arch. Gen. Psychiat.* **5**: 313–318.

Rundles, R.W. (1946). *Blood* **1**: 209–219.

Russell, M.J. (1976). *Nature* **260**: 520–522.

Russell, M.J., Switz, G.M. and Thompson, K. (1980). *Pharmacol. Biochem. Behav.* **13**: 737–738.

Ryan, R.E. Sr. and Ryan, R.E. Jr. (1974). *Postgrad. Med.* **56**: 159–162.

Sachs, E. Jr., Brendler, S.J. and Fultion, J.F. (1949). *Brain* **72**: 227–240.

Saini, K.D. and Breipohl, W. (1976). *Am. J. Anat.* **147**: 433–446.

Saito, Y. (1966). *Jpn. J. Otol.* **69**: 1893–1912.

Samanen, D.W. and Forbes, W.B. (1984). *J. Comp. Neurol.* **225**: 201–211.

Sanada, S. and Asaka, E. (1979). *J. Otorhinolarng.* **17**: 86–87.

Sandberg, M., Bradford, H.F. and Richards, C.D. (1984). *J. Neurochem.* **43**: 276–279.

Sanders-Woudstra, J.A.R. (1971). Cited by Nauta, W.J.H.J.

Santen, R.J. and Paulsen, C.A. (1973). *J. Clin. Endocrinol. Metab.* **36**: 47–54.

Sapolsky, R.M. and Eichenbaum, H. (1980). *Brain Behav. Evolut.* **17**: 276–290.

Satir, P. (1977). Mammalian Cell Membrane, Vol. 2. ed. by Jamieson, G. and Robinson D.M. Butterworth, London, pp. 323.

Sato, S. (1978). Olfactory Disorders—Olfactometry and -therapy. ed. by Toyota, B., Kitamura, T. and Takagi, S.F. Igaku-Shoin, Ltd., Tokyo, pp. 125–139.

Sato, S. and Kurosumi. K. (1965). *Gunma Symp. Endocrinol.* **2**: 61–69.

Sato, T., Shimizu, M. and Watanabe, H. (1956). *Otolaryngology* (Tokyo) **28**: 585–586.

Satou, M., Mori, K., Tazawa, Y. and Takagi, S.F. (1980). Integrative Control Functions of the Brain, Vol. 3. ed. by Ito, M. Kodansha Scientific, Tokyo, and Elseviei Scientific Co, Amsterdam. pp. 91–93.

Satou, M., Mori, K., Tazawa, Y. and Takagi, S.F. (1982a). *J. Neurophysiol.* **48**: 1142–1156.

Satou, M., Mori, K., Tazawa, Y. and Takagi, S.F. (1982b). *J. Neurophysiol.* **48**: 1157–1163.

Satou, M., Mori, K., Tazawa, Y. and Takagi, S.F. (1983a). *J. Neurophysiol.* **50**: 74–88.

Satou, M., Mori, K., Tazawa, Y. and Takagi, S.F. (1983b). *J. Neurophysiol.* **50**: 89–101.
Satou, M., Mori, K., Tazawa, Y. and Takagi, S.F. (1983c). *Exp. Neurol.* **81**: 571–585.
Scalia, F. (1966). *J. Comp. Neurol.* **126**: 285–310.
Scalia, F. (1968). *Brain Behav. Evol.* **1**: 101–123.
Scalia, F. and Winans, S.S. (1975). *J. Comp. Neurol.* **161**: 31–56.
Scalia, F. and Winans, S.S. (1976). Mamalian Olfaction Reproductive Processes and Behaviour. ed. by Doty, R.L. Acad. Press, New York, pp. 8–28.
Schaal, B., Montagner, H., Hertling, E., Bolzoni, D., Moyse, A. and Quinchon, A. (1980). *Reprod. Nutr. Develop.* **20**: 843–858.
Schaupp, H. (1967). *Z. Laryngol. Rhinol.* **46**: 629–634.
Schaupp, H., and Seilz, J. (1969). *Arch. Klin. Exp. Ohr Nas. Kehlkopfheilk.* **195**: 179–191.
Schechter, P.J. and Henkin, R.I. (1974). *J. Neurol. Neurosurg. Psychiat.* **37**: 802–810.
Scherman, A.H., Amoore, J.E. and Weigel, V. (1979). *Otolaryng. Head Neck Surg.* **87**: 717–733.
Schiffman, S.S. (1977). *J. Gerontol.* **32**: 586–592.
Schiffman, S.S. (1979). Sensory Systems and Communication in the Elderly. ed. by Ordy, J.M. and Brizzee, K.R. Raven Press, New York, pp. 227–246.
Schiffman, S.S. (1983). *New Engl. J. Med.* **308** (21): 1275–1279; **308** (22):1337–1343.
Schiffman, S.S., Mose, J. and Erickson, R.P. (1976). *Exp. Aging Res.* **2**: 389–398.
Schiffman, S.S., Nash, M.L. and Dackis, C. (1978). *Physiol. Behav.* **21**: 239–242.
Schiffman, S.S., Orlandi, M.L. and Erickson, R.P. (1979). Sensory Systems and Communication in the Elderly. ed. by Orly, L.M. and Bizzee, K.R. Raven Press, New York, pp. 247–268.
Schleidt, M. and Hold, B. (1982). Olfaction and Endocrine Regulation. ed. by Breipohl., W. IRL Press, London, pp. 181–194.
Schleidt, M., Hold, B. and Attili, G. (1981). *J. Chem. Ecol.* **7**: 19–31.
Schmidt, H. (1925). *Klin. Wochenschr.* **4**: 1967–1968.
Schmitt, F.O. and Worden, F.G., eds. (1979). The Neurosciences: Fourth Study Program. MIT Press, Cambridge, Mass.
Schneeberg, N.G. (1952). *J. Am. Med. Assoc.* **149**: 1091–1093.
Schneider, R.A. (1967). *New Engl. J. Med.* **227**: 299–303.
Schneider, R.A. (1972). *Ann. Otol.* **81**: 272–277.
Schneider, R.A. and Macrides, F. (1978). *Brain Res. Bull.* **3**: 73–82.
Schneider, R.A. and Schmidt, C.E. (1967). *Physiol. Behav.* **2**: 305–309.
Schneider, S.P. and Scott, J.W. (1983). *J. Neurophysiol.* **50**: 358–378.
Schneider, R.A. and Wolf, S. (1955). *J. Appl. Physiol.* **8**: 337–342.
Schneider, R.A., Costiloe, J.P., Howard, R.P. and Wolf, S. (1958). *J. Clin. Endocr. Metab.* **18**: 379–390.
Schoenfeld, T.A. and Macrides, F. (1984). *J. Comp. Neurol.* **227**: 121–135.
Scholfield, C.N. (1978a). *J. Physiol.* **275**: 535–546.
Scholfield, C.N. (1978b). *J. Physiol.* **279**: 547–557.
Schramm, M. and Selinger, Z. (1984). *Science* **225**: 1350–1356.
Schroffner, W.G. and Furth, E.D. (1970). *J. Clin. Endocrinol.* **31**: 267–270.
Schultz, E.W. (1941). *Proc. Soc. Exp. Biol.* **46**: 41–43.
Schultz, E.W. (1960). *Amer. J. Pathol.* **37**: 1–19.
Schultze, M. (1856). *Monatsber. Deut. Akad. Wiss., Berlin* **21**: 504–515.
Schultze, M. (1862). *Abh. Naturforsch. Ges. Hall.* **7**: 1–100. Cited by Dogiel, A. 1886, and by Jagadowski, K.P., 1901.
Schurr, P.H. (1975). *Proc. Roy. Soc. Med.* **68**: 470–472.
Schwarzkroin, P.A. and Slawsky, M. (1977). *Brain Res.* **135**: 157–161.
Scollo-Lavizzari, G. and Akert, K. (1963). *J. Comp. Neurol.* **121**: 259–270.
Scott, J.W. (1981). *J. Neurophysiol.* **46**: 918–931.
Scott, J.W. and Chafin, B.R. (1975). *Brain Res.* **88**: 64–68.
Scott, J.W. and Leonard, C.M. (1971). *J. Comp. Neurol.* **141**: 331–344.

Scott, J.W. and Pfaff, D.W. (1970). *Physiol. Behav.* **5**: 407–411.

Scott, J.W. and Pfaffmann, C.P. (1967). *Science* **158**: 1592–1594.

Scott, J.W. and Pfaffmann, C.P. (1972). *Brain Res.* **48**: 251–264.

Scott, J.W., McBridge, R.L. and Schneider, S.P. (1980). *J. Comp. Neurol.* **194**: 519–534.

Scott, J.W., Ranier, E.C., Pemberton, J.L., Orona, E. and Mouradian, L.E. (1985). *J. Comp. Neurol.* **242**: 415–424.

Segal, M. (1977). *Exp. Neurol.* **57**: 750–765.

Segal, M. (1979). *Brain Res.* **162**: 137–141.

Segal, M. and Landis, S. (1974). *Brain Res.* **78**: 1–15.

Segovia, S., Orensanz, L.M., Valencia, A. and Guillamon, A. (1984a). *Devel. Brain Res.* **16**: 312–314.

Segovia, S., Paniagua, R., Nistal, M. and Guillamon, A. (1984b). *Devel. Brain Res.* **14**: 289–291.

Seifert, K. (1968). *Archiv Klin. exp. Ohren-, Nasen- und Kehlkopfheilk.* **192**: 182–213.

Seifert, K. (1970). Normale und Pathologische Anatomie, Heft 21, ed. by Bargmann, W. and Doerr, W. Georg Thieme Verlag, Stuttgart. pp. 1–99.

Seifert, K. (1971a). *Arch. Klin. Exp. Ohren- Nasen- Kehlkopfheilk.* **200**: 223–251.

Seifert, K. (1971b). *Arch. Klin. Exp. Ohren-Nasen-Kehlkopfheilk.* **200**: 252–274.

Seifert, K. (1972). *Acta Oto-Rhino-Laryngol. Belg.* **26**: 463–490.

Seifert, K. and Ule, G. (1967). *Z. Zellforsch.* **76**: 147–169.

Sem-Jacobsen, C.W., Bickford, G., Dodge, H.J. and Petersen, C. (1953). *Proc. Staff. Meet. Mayo. Clin.* **28**: 166.

Sem-Jacobsen, C.W., Petersen, M.C., Dodge, H.W. Jr., Jacks, Q.D. and Lazarte J.A. (1956). *Am. J. Med. Sci.* **232**: 243–251.

Senf W., Menco, B.P.M., Punter, P.H. and Duyvesteyn, P. (1980). *Experientia* **36**: 213–215.

Sen Gupta, P. (1967). Olfaction and Taste II. ed. by Haysahi, T. Pergamon Press, London, New York, pp. 193–201.

Serby, M., Corwin, J., Conrad, P. and Rotrosen, J. (1985). *Am. J. Psychiat.* **142**: 781–782.

Settle, R.G. (1986). Clinical Measurement of Taste and Smell. ed. by Meiselman, H. L. and Rivlin, R.S. Macmillan Publish. Co., New York, pp. 487–513.

Seydell, E.M. and McKnight, W.P. (1948). *Arch. Otolaryngol.* **47**: 465–470.

Shafa, F. (1978). *Experientia* **35**: 880–881.

Shafa, F. and Meisami, E. (1977). *Brain Res.* **136**: 355–359.

Shammah-Lagnado, S.J. Negrao, N. (1981). *J. Comp. Neurol.* **201**: 51–63.

Sharp, F.G., Kauer, J.S. and Shepherd, G.M. (1975). *Brain Res.* **98**: 596–600.

Sharp, F.R., Kauer, J.S. and Shepherd, G.M. (1977). *J. Neurophysiol.* **40**: 800–813.

Shepherd, G.M. (1971). *Brain Res.* **32**: 212–217.

Shepherd, G.M. (1972). *Physiol. Rev.* **52**: 864–917.

Shepherd, G.M. (1977). Handbook of Physiology. The Nervous System Celluar Biology of Neurons. ed. by Kandel, E.R., Vol. I, Part. 2, Am. Physiol. Soc., Bethesda, MD, pp. 945–968.

Shepherd, G.M. (1979). The Synaptic Organization of the Brain. Oxford Univ. Press, Oxford.

Shepherd, G.M. (1981). Brain Mechanisms of Sensation. ed. by Katsuki, Y., Norgren, R. and Sato, M. Wiley, New York, pp. 209–223.

Shepherd, G.M. (1985). Taste, Olfaction, and the Central Nervous System. ed. by Pfaff, D.W. Rockefeller Univ. Press, New York, pp. 307–321.

Shepherd, G.M. and Brayton, R.K. (1979). *Brain Res.* **175**: 377–382.

Shepherd, G.M., Nowycky, M.C., Greer, C.A. and Mori, K. (1981). Advances in Physiological Sciences. Vol. 30. Neural Communication and Control. ed. by Szekely, G.Y., Labos, E. and Damjanovich, S., pp. 263–278.

Sherman, A.H., Amoore, J.E. and Weigel, V. (1979). *Otolaryngol. Head Neck Surg.* **87**: 717–733.

Shibuya, T. (1960). *Jpn. J. Physiol.* **10**: 317–326.

Shibuya, T. (1964). *Science* **143**: 1338–1340.

Shibuya, T. (1969). Olfaction and Taste III. ed. by Pfaffmann, C. Rockefeller University Press, New York, pp. 109–116.

Shibuya, T. and Shibuya, S. (1963). *Science* **140**: 495–496.

Shibuya, T. and Takagi, S.F. (1962). *Gunma J. Mec. Sci.* **11**: 63–68.

Shibuya, T. and Takagi, S.F. (1963). *J. Gen. Physiol.* **47**: 71–82.

Shibuya, T. and Tonosaki, K. (1972). Olfaction and Taste V. ed. by Schneider, D. Wissenschaftliche Verlagsgesellschaft MBH, Stuttgart, FRG, pp. 102–108.

Shibuya, T. and Tucker, D. (1967). Olfaction and Taste II. ed. by Hayashi, T. Pergamon Press, Oxford, pp. 219–233.

Shibuya, T., Ai, N. and Takagi, S.F. (1962). *Proc. Jpn. Acad.* **38**: 231–233.

Shimizu, T., Yamamoto, S. and Hayaishi, O. (1979). *J. Biol. Chem.* **254**: 5222–5228.

Shimizu, T., Yamashita, A. and Hayaishi, O. (1982). *J. Biol. Chem.* **257**: 13570–13573.

Shinomiya, M., Tange, S. and Okuno, K. (1978). *J. Jpn. Rhinol. Soc.* **17**: 85.

Shipley, M.T., Halloran, F.J. and De La Torre, J. (1985). *Brain Res.* **329**: 294–299.

Shirakura, M., Asaka, H. and Okamoto, M. (1984). *JASTS* **18**: 73–76.

Shirakura, M., Asaka, H. and Okamoto, M. (1985a). *Chem. Sens.* **10**: 134.

Shirakura, M., Asaka, H. and Okamoto, M. (1985b). *JASTS* **19**:30.

Shirakura, M., Asaka, H. and Okamoto, M. (1986). *Chem. Sens.* **11**: 271.

Shirley, S., Polak, E. and Dodd, G.H. (1981). *Biochem. Soc. Trans.* **9**: 108.

Sicard, G. (1985). *Brain Res.* **326**: 203–212.

Sicard, G. and Holley, A. (1984). *Brain Res.* **292**: 283–296.

Siegel, A., Edinger, H. and Troiano, R. (1973). *Exp. Neurol.* **38**: 202–217.

Siegel, A., Fukushima, T., Meibach, R., Burke, L., Edinger, H. and Weiner, S. (1977). *Brain Res.* **135**: 11–23.

Silver, W.L. and Moulton, D.G. (1982). *Physiol. Behav.* **28**: 927–931.

Silver, W.L., Mason, J.R., Marshall, D.A. and Maruniak, J.A. (1985). *Brain Res.* **333**: 45–54.

Simmons, P.A. and Getchell, T.V. (1981a). *J. Neurophysiol.* **45**: 516–528.

Simmons, P.A. and Getchell, T.V. (1981b). *J. Neurophysiol.* **45**: 529–549.

Simmons, P.A., Rafols, J.A. and Getchell, T.V. (1981). *J. Comp. Neurol.* **197**: 237–257.

Singh, N., Grewal, M.S. and Austin, J.H. (1970). *Arch. Neurol.* **22**: 40–44.

Singh, P.B., Brown, R.E. and Roser, B. (1987). *Nature* **327**: 161–164.

Skeen, L.C. (1977). *Brain Res.* **124**: 147–153.

Skeen, L.C. and Hall, W.C. (1977). *J. Comp. Neurol.* **172**: 1–36.

Skeen, L.C., Due, B.R. and Douglas, F.E. (1985). *Neurosci. Lett.* **54**: 301–306.

Skeen, L.C., Due, B.R. and Douglas, F.E. (1986). *Neurosci. Lett.* **63**: 5–10.

Sklar, P.B., Anhot, R.R.H. and Snyder, S.H. (1968). *Chem. Sens.* **11**: 665.

Skouby, A.P. and Zilstorff-Pedersen, K. (1954). *Acta Physiol. Scand.* **323**: 252–258.

Slotnick, B.M. and Kaneko, N. (1981). *Science* **214**: 91–92.

Small, R. (1977). *Anat. Rec.* **187**: 716–717.

Small, R.K. and Leonard, C.M. (1983). *J. Comp. Neurol.* **214**: 353–369.

Smart, I.H.M. (1971). *J. Anat.* **109**: 243–251.

Smith, C.G. (1938). *Can. Med. J.* **39**: 138–140.

Smith, C.G. (1942). *J. Comp. Neurol.* **77**: 589–594.

Smith, C.G. (1951). *Anat. Rec.* **109**: 661–671.

Smith, D.B., Allison, T., Goff, W.R. and Princitato, J.J. (1971). *Electroence. Neurophysiol.* **30**: 313–317.

Snider, R.S. and Lee, J.C. (1961). A Stereotaxic Atlas of the Monkey Brain (*Macaca mulatta*). Univ. of Chicago Press, Chicago.

Snow, J.B. (1969). *Laryngoscope* **79**: 1485–1469.

Snyder, S.H., Sklar, P.B. and Pevsner, J. (1988). *J. Biol. Chem.* **263**: 13971–13974.

Sokoloff, L. (1977). *J. Neurochem.* **27**: 13–26.

Solano-Flores, L.P., Aguilar-Baturoni, H.U. and Guevara-Aguilar, R. (1980). *Brain Res. Bull.* **5**: 383–389.

Solano-Flores, L.P., Aguilar-Baturoni, H.U. and Guevara-Aguilar, R. (1981). *Brain Res. Bull.* **7**: 655–660.

Somjen, G.G. (1975). *Ann. Rev. Physiol.* **37**: 163–190.

Sparkes, R.S., Simpson, R.W. and Paulsen, A. (1968). *Arch. Intern. Med.* **121**: 534–538.

Spencer, W.G. (1894). *Phil. Trans. R. Soc. Ser.* B **185**: 609–657. Cited by Kaada, B.R. (1960).

Spencer, W.G. and Kandel, E.R. (1961). *J. Neurophysiol.* **24**: 272–285.

Standbygaerd, E. (1954). *Arch. Otolaryngol.* **59**: 485–490.

Steegman, A.T. (1970). Examination of the Nervous System: A Student's Guide. 3rd ed. Chicago: Year Book Medical Publishers. Cited by Cain, W.S. and Gent, J.F. (1986).

Steel, E. (1984). *Anim. Behav.* **32**: 597–608.

Steel E. and Keverne, E.B. (1985). *Physiol. Behav.* **35**: 195–200.

Stein, M. and Ottenberg, P. (1958). *Psychosom. Med.* **20**: 60–65.

Steinbrecht, R.A. (1969). Olfaction and Taste III. ed. by Pfaffmann, C. Rockefeller University Press, New York, pp. 3–33.

Steinbrecht, R.A. (1980a). *Tissue Cell* **12**: 73–100.

Steinbrecht, R.A. (1980b). Olfaction and Taste VII. ed. by van der Starre, H. IRL Press, London, pp. 135–138.

Stephan, H. (1965). *Acta Anat.* **62**: 215–253.

Stephan, H. (1975). Allocortex. Handbuch der mikroskopischen Anatomie des Menschen, ed. by Bargmann, W., Vol. 4: Nervensystem, Part. 9. Springer-Verlag, Berlin.

Stephan, H. and Andy, O.J. (1962). *J. Hirnforsch.* **5**: 229–244.

Sternberg, H. (1931). *Mschr. Ohr. Hk.* **65**: 171–172.

Stevens, C.F. (1969). *J. Neurophysiol.* **32**: 184–192.

Stevens, J.C. and Cain, W.S.G. (1985a). *Chem. Sens.* **10**: 517–529.

Stevens, J.C., Cain, W.S. and Meyers, T.D. (1985b). *Chem. Sens.* **10**: 419.

Steward, O. (1976). *J. Comp. Neurol.* **167**: 285–314.

Stewart, W.B., Kauer, J.S. and Shepherd, G.M. (1979). *J. Comp. Neurol.* **185**: 715–734.

Stewart, W.B., Pedersen, P.E., Greer, C.A. and Shepherd, G.M. (1985). *Chem. Sens.* **10**: 400.

Stone, H. (1963). *J. Appl. Physiol.* **18**: 746–751.

Stone, H. and Pryer, G. (1967). *Perc. and Psychophys.* **2**: 516–519.

Stranbygärd, E. (1954). *A.M.A. Archives of Otolaryngol.* **59**: 485–491.

Struble, R.G. and Walters, C.P. (1982). *Brain Res.* **236**: 237–251.

Struble, W.B., Kauer, S.J. and Shepherd, G.M. (1979). *J. Comp. Neurol.* **185**: 715–734.

Stryer, L. (1986). *Ann. Rev. Neurosci.* **9**: 87–119.

Stuiver, M. (1960). *Acta Otolaryng.* (Stock.) **51**: 135–141. Cited by Sumner, D. (1962).

Suga, B. and Nakashima, M. (1972). *Otologia* **18**: 371–382.

Suga, B. and Nakashima, M. (1973). *Otorhinology* **19**: 644–647.

Suga, B. and Nakashima, M. (1978). Olfactory Disorders—Olfactometry and -therapy. ed. by Toyota, B., Kitamura, T. and Takagi, S.F. Igaku-Shoin, Ltd., Tokyo, pp. 81–85.

Sumner, D. (1962). *Lancet* **ii**: 895–897.

Sumner, D. (1964). *Brain* **87**: 107–120.

Suzuki, H. (1984). *J. Otolaryng. Jpn.* **87**: 1–15.

Suzuki, N. (1967). *J. Fac. of Science, Hokkaido University* Ser. VI, Zool. **16**: 174–185.

Suzuki, N. (1977). Food Intake and Chemical Senses. ed. by Katsuki, T., Sato, M., Takagi, S.F. and Oomura, Y. Univ. of Tokyo Press, Tokyo, pp. 13–22.

Suzuki, N. (1978a). *Comp. Biochem. Physiol.* **61**A: 461–467.

Suzuki, N. (1978b). *JASTS* **12**: 9–12.

Suzuki, N. (1980). *JASTS* **14**: 33–36.

Suzuki, N. (1982). Chemoreception in Fishes. ed. by Hara, T.J. Elsevier Scientific Publ. Co., Amsterdam, pp. 93–108.

Suzuki, N. (1984). *Brain Res.* **311**: 181–185.

Suzuki, N. (1986). *Chem. Sens.* **11**: 669.

Suzuki, N. (1987). *Zoo. Sci.* **4**: 964.

Suzuki, N. (1988). The Beidler Symposium on Taste and Smell. ed. by Miller, I.J., Jr. Book Service Assoc., Inc., Washington-Salem, N.C. pp. 173–181.

Swanson, S.L., Santen, S.J. and Smith, D.W. (1971). *J. Pediatr.* **78**: 1037–1039.

Swanson, L.W. (1978). Functions of the Septo-Hyppocampal System. Ciba Foundation Symposium 58 (new sereis). Elsevier Excepta Medica, Amsterdam, pp. 25–43.

Swanson, L.W. and Cowan, W.M. (1977). *J. Comp. Neurol.* **172**: 49–84.

Switzer, R.C. III, Johnson, J.I.Jr. (1977). *Acta Anat.* **99**: 36–42.

Switzer, R.C., de Olmos, J. and Heimer, L. (1985). The Rat Nervous System. ed. by Paxinos, G. Acad. Press, Sydney.

Szentagothai, J. (1970). The Neurosciences; Second Study Program. ed. by Schmitt, F.O. Rockefeller University Press, New York, pp. 427–443.

Tagatz, G.T., Fialkow, P.J., Smith, D. and Spadoni, L. (1970). *New Eng. J. Med.* **283**: 1326–1329.

Takagi, S.F. (1962). *Jpn. J. Physiol.* **12**: 365–382.

Takagi, S.F. (1967). Olfaction and Taste II. ed. by Hayashi, T. Pergamon Press, Oxford, pp. 167–179.

Takagi, S.F. (1968). Theories of Odor and Odor Measurement. ed. by Tanyolac, N. Technivision, Maidenhead, England, pp. 509–521.

Takagi, S.F. (1969). Olfaction and Taste III. ed. by Pfaffman, C. Rockefeller University Press, New York, pp. 71–91.

Takagi, S.F. (1971). Handbook of Sensory Physiology, Vol. VI, Chemical Senses I, Olfaction. ed. by Beidler, L.M., Springer-Verlag, Berlin, Heidelberg, New York, pp. 75–94.

Takagi, S.F. (1974). Stories on Olfaction. Iwanami-Shoten, Ltd., Tokyo.

Takagi, S.F. (1976). *Proc. Sixth Intern. Tobacco Scientif. Congr.* pp. 64–70.

Takagi, S.F. (1978). Handbook of Perception. Vol. VIA: Tasting and Smelling. ed. by Carterette, E.C. and Friedman, M.P. Acad. Press, New York, pp. 233–243.

Takagi, S.F. (1979a). Integrative Control Functions of the Brain, Vol. 2. ed. by Ito, M. Kodansha Scientific, Tokyo and Elsevier Scientific Co., Amsterdam, pp. 51–67.

Takagi, S.F. (1979b). *Trends Neurosci.* **2**: 313–315.

Takagi, S.F. (1980). Olfaction and Taste VII. ed. by van der Starre, H. IRL Press, Oxford, pp. 275–278.

Takagi, S.F. (1981). *Chem. Sens.* **6**: 329–333.

Takagi, S.F. (1983). *Proc. Int. Union Physiol. Sci. XXIX*, Sydney. Vol. XV, 446, *Chem. Sens.* **8**: 267 (1984).

Takagi, S.F. (1984a). *Seitai no Kagaku* **35**: 67–79.

Takagi, S.F. (1984b). *Jpn. J. Physiol.* **34**: 561–573.

Takagi, S.F. (1984c). *Chem. Sens.* **8**: 267.

Takagi, S.F. (1986a). *Progr. Neurobi.* **27**: 195–249.

Takagi, S.F. (1986b). Emotions—Neuronal and Chemical Control. ed. by Oomura, Y. Jpn. Scientif. Soc. Press, Tokyo, and Karger, Basel, Munchen, Paris, London, New York, Tokyo, pp. 157–165.

Takagi, S.F. (1986c). *Chem. Sens.* **11**: 574.

Takagi, S.F. (1987). Olfaction and Taste IX. ed. by Roper, S.D. and Atema, J. *Ann. N.Y. Acad. Sci.* **510** (30): 113–118.

Takagi, S.F. (1989). *Chem. Sens.* **14**: 25–46.

Takagi, S.F. and Okano, M. (1971). *Proc. Int. Union Physiol. Sci. XXV*, Munich, p. 553.

Takagi, S.F. and Okano, M. (1974). *Gunma Symp. Endocrin.* **11**: 2–14.

Takagi, S.F. and Omura, K. (1963). *Proc. Jpn. Acad.* **39**: 253–255.

Takagi, S.F. and Shibuya, T. (1959). *Nature* **184**: 60.

Takagi, S.F. and Shibuya, T. (1960a). *Jpn. J. Physiol.* **10**: 99–105.

Takagi, S.F. and Shibuya, T. (1960b). *Nature* **186**: 724.

Takagi, S.F. and Shibuya, T. (1960c). *Jpn. J. Physiol.* **10**: 385–395.

Takagi, S.F. and Shibuya, T. (1960d). *Jpn. J. Physiol.* **10**: 499–508.

Takagi, S.F. and Shibuya, T. (1960e). Electrical Activity of Single Cells. ed. by Katsuki, Y. Igaku-Shoin, Tokyo, pp. 1–10.

Takagi, S.F. and Shibuya, T. (1961). *Jpn. J. Physiol.* **11**: 23–37.

Takagi, S.F. and Wyse, G.A. (1965). *Proc. Int. Union Physiol. Sci. XXIII*, Tokyo, p.379.

Takagi, S.F. and Yajima, T. (1964). *Nature* **202**: 1220.

Takagi, S.F. and Yajima, T.Y. (1965). *J. Gen. Physiol.* **48**: 559–569.

Takagi, S.F., Shibuya, T., Higashino, S. and Arai, T. (1960). *Jpn. J. Physiol.* **10**: 571–584.

Takagi, S.F., Wyse, G.A. and Yajima, T. (1966). *J. Gen. Physiol.* **50**: 473–489.

Takagi, S.F., Wyse, G.A., Kitamura, H. and Ito, K. (1968a). *J. Gen. Physiol.* **51**: 552–578.

Takagi, S.F., Wyse, G.A., Yajima, T., Kitamura, H. and Ito, K. (1968b). *Proc. Int. Union. Physiol. Sci. XXIV* Abstr. volunteer papers, Washington. p. 1280.

Takagi, S.F., Kitamura, H., Imai, K. and Takeuchi, H. (1969a). *J. Gen. Physiol.* **53**: 115–130.

Takagi, S.F., Aoki, K., Iino, M. and Yajima, T. (1969b). Olfaction and Taste III. ed. by Pfaffmann, C. Rockefeller Univ. Press, New York, pp. 92–108.

Takagi, S.F., Iino, M. and Yarita, H. (1978a). *Jpn. J. Physiol.* **28**: 109–128.

Takagi, S.F., Iino, M., Yarita, H. and Mori, K. (1978b). *Jpn. J. Physiol.* **28**: 129–148.

Takagi, T. and Shibuya, T. (1967). *Jpn. J. Exptl. Morph.* **21**: 483.

Takahashi, R. (1979). Stories on Nose. Iwanami Shoten, Tokyo.

Takenaka, B., Hayashido, H., Makino, I., Uematsu, H. and Oibe, S. (1958). *Otolaryngology* (Jpn) **30**: 1035–1038.

Takasaka, T., Sato, M. and Onodera, A. (1980). *Ann. Otol. Rhinol. Laryngol.* **89**: 37–45.

Takata, N. (1929). *Arch. Ohren-Nasen-u. Kehlkopfheilk* **121**: 43–78.

Takeuchi, N. (1963a). *J. Physiol.* **167**: 128–140.

Takeuchi, N. (1963b). *J. Physiol.* **167**: 141–155.

Takeuchi, A. and Takeuchi, N. (1960). *J. Physiol.* **154**: 52–67.

Takimonto, T., Sakamoto, E. and Mitsui, Y. (1962). *Tokushima J. Exp. Med.* **9**: 8–23.

Tanabe, T., Iino, M., Ooshima, Y. and Takagi, S.F. (1973). *J. Physiol. Soc. Jpn.* **35**: 550.

Tanabe, T., Iino, M., Ooshima, Y. and Takagi, S.F. (1974). *Brain Res.* **80**: 127–130.

Tanabe, T., Iino, M., Ooshima, Y. and Takagi, S.F. (1975a). Olfaction and Taste V. ed. by Denton, D.G. and Coghlan, J.P. Acad. Press, New York, pp. 309–312.

Tanabe, T., Iino, M. and Takagi, S.F. (1975b). *J. Neurophysiol.* **38**: 1284–1296.

Tanabe, T., Yarita, H., Iino, M., Ooshima, Y. and Takagi, S.F. (1975c). *J. Neurophysiol.* **38**: 1269–1283.

Tanabe, H., Kurihara, K. and Kobatake, Y. (1980). *Biochemistry* **19**: 5339–5344.

Tanaka, D. Jr. (1976). *Brain Res.* **110**: 21–38.

Taniguchi, K. and Mikami, S. (1985). *Cell Tissue Res.* **240**: 41–48.

Taniguchi, K. and Mochizuki, K. (1982). *Jpn. J. Vet. Sci.* **44**: 419–426.

Taniguchi, K., Taniguchi, K. and Mochizuki, K. (1982a). *Jpn. J. Vet. Sci.* **44**: 709–716.

Taniguchi, K., Taniguchi, K. and Mochizuki, K. (1982b). *Jpn. J. Vet. Sci.* **44**: 881–890.

Taniguchi, K., Taniguchi, K. and Mikani, S. (1986). Ontogeny of Olfaction. ed. by Breipohl, W. Springer-Verlag, Berlin, Heidelberg, pp. 83–94.

Tazawa, Y. (1983). *Kitakanto Med. J.* **33**: 43–56.

Tazawa, Y., Mori, K. and Takagi, S.F. (1981). *J. Physiol. Soc. Jpn.* **43**: 320.

Tazawa, Y., Mori, K. and Takagi, S.F. (1982a). *J. Physiol. Soc. Jpn.* **44**: 388.

Tazawa, Y., Mori, K. and Takagi, S.F. (1982b). *Neurosci. Lett. Suppl.* **9**: S119.

Tazawa, Y., Onoda, N. and Takagi, S.F. (1983). *Neurosci. Lett. Suppl.* **13**: S62.

Tazawa, Y., Onoda, N. and Takagi, S.F. (1987). *Neurosci. Res.* **4**: 357–375.

Tecce, J.J. and Scheff, N.M. (1969). *Science* **164**: 331–333.

Teicher, M.H. and Blass, E.M. (1976). *Science* **193**: 422–424.

Teicher, M.H., Stewart, W.B., Kauer, J.S. and Shepherd, G.M. (1980). *Brain Res.* **194**: 530–535.

Teicher, M.H., Shaywitz, B.A. and Lumia, A.P. (1984). *Devel. Brain Res.* **12**: 97–110.

Thamke, B., Schulz, E. and Schönheit, B. (1973). *J. Hirnforschung* **14**: 435–449.

Thomas, L. (1975). Fourth International Convocation of Immunology. ed. by Neter, E. and Milgrom, F. Karger, Basel, Switzerland, pp. 2–11.

Thommesen, G. (1978). *Acta Physiol. Scand.* **102**: 205–217.

Thommesen, G. and Doving, K.B. (1977). *Acta Physiol. Scand.* **99**: 270–280.

Thornhill, R.A. (1967). *J. Cell Sci.* **2**: 591–602.

Thornhill, R.A. (1970). *Z. Zellforsch. Mikiroskop. Anat.* **109**: 147–157.

Tidsall, F.F., Brown, A. and DeFries, R.D. (1938). *J. Pediat.* **13**: 60–62.

Tisserand, M. (1985). Aromatherapy for Women. Thorsons Publishing Group, Ltd., Wellingborough, Northamptonshire, England.

Tisserand, R. (1977). The Art of Aromatherapy. Daniel, C.W. Pub. Co., Ltd., Sofron Walden, Essex, England. Cited by Tisserand, M. (1985).

Tobias, T.J. (1975). *Brain Res.* **83**: 191–212.

Tokuda, K., Yoneno, K., Oshiro, T., Ishibashi, Y., Ohba, H. and Umeda, R. (1978). *J. Otorhinolaryng. Jpn.* **17**: 83–84.

Tokumoto, H., Watanabe, K., Fukushima, D., Shimizu, T. and Hayaishi, O. (1982). *J. Biol. Chem.* **257**: 13576–13580.

Tonoike, M. and Kurioka, Y. (1982). *Bull. Electotechnical Lab.* **46**: 62–73.

Tonoike, M., Yamanaka, T., Sobagaki, H., Nishimoto, A., Takebayashi, M. and Kurioka, Y. (1979). *Jpn. J. Ergonomics,* **15**: 251–258.

Torii, S. (1986). *Fragrance J.* **77**: 16–20.

Torii, S. (1987). Psychology and Biology of the Fragrance. ed. by Van Toller, S. and Dodd, G.H. Chapman Hall, London. pp. 107–120.

Toyota, T. (1962). *Juzen Igaku Zasshi, Kanazawa Univ.* **68**: 124–148.

Toyota, B., Kitamura, T. and Takagi, S.F. (1978). Olfactory Disorders—Olfactometry and -therapy. Igaku-Shoin, Ltd., Tokyo.

Trautwein, W. and Dudel, J. (1958). *Arch. Ges. Physiol.* **266**: 324–334.

Trotier, D. (1986). *Pflügers Arch.* **407**: 589–595.

Trotier, D. and MacLeod, P. (1983). *Brain Res.* **268**: 225–237.

Tryjilloo-Cenöz, O. (1961). *Z. Zellforsch. Mikroskop. Anat.* **54**: 654–676.

Tucker, D. (1960). *Physiologist* **3**: 167.

Tucker, D. (1963a). *J. Gen. Physiol.* **46**: 453–489.

Tucker, D. (1963b). Olfaction and Taste I. ed. by Zottermann, Y. Pergamon Press, Oxford, pp. 45–69.

Tucker, D. (1967). *Fed. Proc.* **26**: 544.

Tucker, D. (1971). Handbook of Sensory Physiology, Vol. 4, Chemical Senses, Part 1, Olfaction. ed. by Beidler, L.M. Springer-Verlag, Berlin, pp. 151–181.

Tucker, D. (1978). Encyclopedia of Food Science. ed. by Peterson, M.S. and Johnson, A.H. Arie Publishing Co., Westport, Conn., pp. 581–585.

Tucker, D. and Kiyohara, S. (1978). Neuroscience Abstract, Vol. 4, November.

Tucker, D. and Shibuya, T. (1965). *Cold Spr. Harb. Symp. Quant. Biol.* **30**: 207–215.

Turner, P. (1968). Drugs and Sensory Function. ed. by Herxheimer, A.W. Churchill, London.

Turner, P. and Patterson, D.S. (1966). *Acta Otolaryngol.* **62**: 149–156.

Turner, B.H., Cupta, K.C. and Mishkin, M. (1978). *J. Comp. Neurol.* **177**: 381–396.

Ueki, S. and Domino, E.F. (1961). *J. Neurophysiol.* **24**: 12–25.

Ueno, R., Narumiya, S., Ogorochi, T., Nakayama, T., Ishikawa, Y. and Hayaishi, O. (1982). *Proc. Natl. Acad. Sci. USA.* **79**: 6093–6097.

Ueno, R., Honda, K., Inoue, S. and Hayaishi, O. (1983). *Proc. Natl. Acad. Sci. USA.* **80**: 1735–1737.

Umeda, R. (1972). *Clin. Physiol. Jpn.* **2**: 586–591.

Umeda, R. (1981a). On the Diagnosis and Therapy of Olfactory Disorders. An assigned report to the 82nd general assembly of the Otorhinolaryngol. Soc. of Japan in Kyoto Dept. Otorhinolarnygol. School of Medicine, Kanazawa Univ, Kanazawa.

Umeda, R. (1981b). *Jpn. J. Otol.* (Tokyo) **84**: 1362–1370.

Ungerstedt, U. (1971). *Acta Physiol. Scand. Suppl.* **367**: 1–48.

Urbach, E. (1941/42). *J. Allergy* **13**: 387–396.

Ussing, H.H. and Zerahn, K. (1951). *Acta Physiol. Scand.* **23**: 110–127.

Usukura, J. and Yamada, E. (1975). *Zool. Mag.* **87**: 339.

Usukura, J. and Yamada, E. (1978). *Cell Tiss. Res.* **188**: 83–98.

Usukura, J. and Yamada, E. (1979). *Adv. in Neurol. Sci.* **23**: 783–791.

Vaccarezza, O.L., Sepich, L.N. and Tramezzani, J.H. (1981). *J. Anat.* **132**: 167–185.

Valencia, A., Segovia, S. and Guillamon, A. (1986). *Devel. Brain Res.* **24**: 287–290.

Valverde-Garcia, F. (1965). Studies on the Piriform Lobe. Harvard University Press, Cambridge, Mass.

Van Allen, M.W. (1969). Pictorial Manual of Neurologic Tests. Chicago: Year Book Medical Publishers. Cited by Cain, W.S. and Gent, J.F. (1986).

Van As, W., Kauer, J.S., Menco, B.P.M. and Köster, E.P. (1985). *Chem. Sens.* **10**: 1–22.

Van Boxtel, A. and Köster, E.P. (1978). *Chem. Sens. Flav.* **3**: 39–44.

Van Den Eeckhaut, J. (1978). L'Odorat Clinique Office Internat. et Librair. Cited by Amoore, J.E. and Ollman, B.G. (1983).

Van der Lee, S. and Boot, L.M. (1956). *Acta Physiol. Pharmacol.* (Neerl.) **5**: 213–215a.

Van der Wolk, F.M. and Menco, B. Ph. M. (1980). Olfaction and Taste VII. ed. by van der Starre, H. IRL Press, London, pp. 49–52.

Van Drongelen, W. (1978). *J. Physiol.* **277**: 423–435.

Van Drongelen, W., Holley, A. and Döeving, K.B. (1978). *J. Theoret. Biol.* **71**: 39–48.

Van Gehuchten, A. (1890). *La Cellule* **6**: 393–407.

Van Hoesen, G.W. and Pandya, D.N. (1975a). *Brain Res.* **95**: 1–24.

Van Hoesen, G.W. and Pandya, D.N. (1975b). *Brain Res.* **95**: 39–59.

Van Hoesen, G.W., Pandya, D.N. and Butters, N. (1972). *Science* **175**: 1471–1473.

Van Hoesen, G.W., Pandya, D.N. and Butters, N. (1975). *Brain Res.* **95**: 25–38.

Van Hoesen, G.W., Mesulam, M.M. and Haaxma, R. (1976). *Brain Res.* **109**: 375–381.

Vandenbergh, J.G., Whitsett, J.M. and Lombardi, J.R. (1975). *J. Reprod. Fert.* **43**: 515–523.

Venström, D. and Amoore, J.E. (1968). *J. Foods Sci.* **33**: 264–265.

Verkleij, A.J. and Ververgaert, P.H.J. Th. (1978). *Biochim. Biophys. Acta* **515**: 303–327.

Verkleij, A.J., Mombers, C., Leunissen-Bijvelt, J. and Ververgaert, P.H.J. Th. (1979). *Nature* **279**: 162–163.

Vierling, J.S. and Rock, J. (1967). *J. Appl. Physiol.* **22**: 311–315.

Vincent, S.R., McIntosh, C.H.S., Buchan, A.M.J. and Brown, J.C. (1985). *J. Comp. Neurol.* **238**: 169–186.

Vinnikow, J.A. and Titow, L.K. (1949). *C.R. Acad. Sci. URSS* **65**: 903–906. Cited by Moulton, D.G. (1962).

Vollmecke, T.A. and Doty, R.L. (1985). *Chem. Sens.* **10**: 413.

Von Bekésy, G. (1964). *J. Appl. Physiol.* **19**: 369–373.

Von Skramlik, E. (1925). *Z. Sinnesphysiol.* **56**: 69–140.

Von Skramlik, E. (1926). Handbuch der Physiologie der neideren Sinne Band I. Die Physiologie des Geruchs- und Geschmackssinnes. Thieme, Leipzig, pp. 281–305.

Walker, A.E. (1940). *J. Comp. Neurol.* **73**: 59–86.

Wallace, P. (1977). *Physiol. Behav.* **19**: 577–579.

Walsh, R.R. (1956). *Am. J. Physiol.* **186**: 255–257.

Walter, W.G., Cooper, R., Aldridge, V.J., McCallum, W.C. and Winter, A.L. (1964). *Nature* **203**: 380–384.

Wang, R.T. and Halpern, M. (1980a). *J. Morphol.* **164**: 47–67.

Wang, R.T. and Halpern, M. (1980b). *Am. J. Anat.* **157**: 399–428.

Wang, R.T. and Halpern, M. (1982a). *Brain Res.* **237**: 23–39.

Wang, R.T. and Halpern, M. (1982b). *Brain Res.* **237**: 41–59.

Wang, R.T., Vagvolgyi, A., Mendelsohn, B. and Halpern, M. (1979). *Neurosci. Abstr.* **5**: 103.

Ward, C.D., Hess, W.A. and Calne, D.B. (1983). *Neurology* Abstr. **5**: 103.

Warner, M.D., Peabody, C.A., Flattery, J.J. and Tinklenberg, J.R. (1986). *Psychiat.* **21**: 116–118.

Watanabe, K., Shimizu, T., Iguchi, S., Wakatsuki, H., Hayashi, M. and Hayaishi, O. (1980). *J. Biol. Chem.* **255**: 1779–1782.

Watanabe, Y., Watanabe, Y. (Yumiko), Kaneko, T. and Hayaishi, O. (1985). Advances in Prostaglandin, Thromboxane, and Leukotriene Research. ed. by Hayaishi, O. and Yamamoto, S. Vol. 15. Raven Press, New York, pp. 553–554.

Watanabe, Y., Mori, K., Imamura, K., Takagi, S.F. and Hayaishi, O. (1986). *Brain Res.* **378**: 216–222.

Watkin, J.W. (1985). *Am. J. Primat.* **8**: 309–355.

Watkin, J.W. and Silverman, A.J. (1983). *J. Comp. Neurol.* **218**: 426–432.

Webline, F.S. (1977). *J. Physiol.* **264**: 767–785.

Wedgwood, M. (1974). *J. Physiol.* **239**: 88–89.

Weffenbach, J.M. and McCarthy, V.P. (1984). *Chem. Sens.* **9**: 193–199.

Weidmann, S. (1956). Electrophysiologie der Herzmuskelfaser. H. Huber Press, Bern.

Weight, F.F. (1974). Synaptic Transmission and Neuronal Interaction. ed. by Bennett M.V.L. Raven Press, New York, pp. 141–152.

Wender, M. and Szmeja, Z. (1971). *Neurol. Neurochir. Pol.* **64**: 767–772.

Wenzel, B.M. (1948). *Psychol. Bull.* **45**: 231–247.

Wenzel, B.M. and Salzman, A. (1968). *Exptl. Neurol.* **22**: 472–479.

Wesolowski, H. (1967). *Folia Biol.* **15**: 303–324.

White, E.L. Jr. (1965). *Anat. Rec.* **152**: 465–480.

White, E.L. (1972). *Brain Res.* **37**: 69–80.

White, E.L. (1973). *Brain Res.* **60**: 299–313.

White, J.L. and Meredith, M. (1985). Cited by Kevern *et al.* (1986). *Chem. Sens.* **11**: 119–133.

Whitten, W.K. (1956). *J. Endocrinol.* **13**: 399–404.

Whitten, W.K. (1957). *Nature* **180**: 1436.

Whitten, W.K. (1958). *J. Endocrinol.* **17**: 303–313.

Wiegand, S.J. and Price, J.L. (1980). *Neurosci. Abstr.* **6**: 307.

Willey, T.J. (1973). *J. Comp. Neurol.* **152**: 211–232.

Wilson, K.C.P. and Raisman, G. (1980). *Brain Res.* **185**: 103–113.

Wilson, J.A.F. and Westerman, R.A. (1967). *Z. Zellforsch. Mikroskop. Anat.* **83**: 196–206.

Winans, S.S. and Powers, J.B. (1977). *Brain Res.* **126**: 325–344.

Winans, S.S. and Scalia, F. (1970). *Science* **170**: 330–332.

Witkin, J.W. (1985). *Am. J. Primat.* **8**: 309–315.

Witkin, J.W. and Silverman, A.-J. (1983). *J. Comp. Neurol.* **218**: 426–432.

Wolf. G. and Sutin, J. (1966). *J. Comp. Neurol.* **127**: 137–155.

Wollard, H.H. (1925). *Proc. Zool. Soc. Lond.* **3**: 1071–1184.

Wright, R.H. (1961). *Nature* **190**: 1101–1102.

Wright, W.H. (1964). The Science of Smell. George Allen and Unwin, Ltd., London.

Wright, R.H. (1978). The Sense of Smell. ed. by Wright, R.H. Boca Rotan, Fl., pp. 213–215.

Wright, R.H., Reid, C. and Evans, H.G.V. (1956). *Ind. Chem.* 973. Cited by Ottoson, D. (1958).

Wysocki, C.J. (1979). *Neurosci. Biobehav. Rev.* **3**: 301–341.

Wysocki, C.J., Katz, Y. and Bernhard, R. (1983). *Biol. Reprod.* **28**: 917–922.

Wysocki, C.J., Beauchamp, G.K., Wellington, L. and Greeley, S. (1984). Cited by Pedersen, P.E. *et al.* (1986).

Wysocki, A.J., Wysocki, L.M., Mittelberg, R. and Beauchamp, G.K. (1985). *Chem. Sens.* **10**: 420.

Wysocki, C.J., Bean, N.J. and Beauchamp, G.K.(1986). Chemical Signals in Vertebrates: Ecology, Evolution, and Comparative Biology. ed. by Duvall, D., Muller Schwarze, D. and Silverstein, R.M. Plenum Press, New York.

Yamada, E. (1965). *Gunma Symp. Endocrinol.* **2**: 1–17.

Yamada, S., Watanabe, Y., Hirano, Y. and Naito, J. (1978). Olfactory Disorders— Olfactometry and -therapy, ed. by Toyota, B., Kitamura, T. and Takagi, S.F. Igaku-Shoin, Tokyo, pp. 41–43.

Yamamoto, M. (1976). *Arch. Histol. Jpn.* **38**: 359–412.

Yamamoto, C. and Iwama, K. (1960). *Proc. Jpn. Acad.* **36**: 295–298.

Yamamoto, C. and Iwama, K. (1962). *Jpn. J. Physiol.* **38**: 63–67.

Yamamoto, T. and Kawamura, Y. (1975). *Brain Res.* **94**: 447–463.

Yamamoto C. and Matsui S. (1976). *J. Neurochem.* **26**: 487–491.

Yamamoto, T., Tonosaki, A. and Kurosawa, T. (1965). *Acta Anat. Nippon.* **40**: 342–353.

Yamamoto, C., Yamamoto, T. and Iwama, K. (1963). *J. Neurophysiol.* **26**: 403–415.

Yamashita, A., Watanabe, Y. and Hayaishi, O. (1983). *Proc. Natl. Acad. Sci. USA.* **80**: 6114–6118.

Yamazaki, K., Boyse, E.A., Mike, V., Thaler, H.T., Mathieson, B.J., Abbott, J. *et al.* (1976). *J. Exp. Med.* **144**: 1324–1335.

Yamazaki, K., Yamaguchi, M., Andrews, P.W., Peake, B. and Boyes, E.A. (1978). *Immunogenetics* **6**: 253–259.

Yamazaki, K., Yamaguchi, M., Baranoski, L., Bard, J., Boyse, E.A. and Thomas, L. (1979). *J. Exp. Med.* **150**: 755–760.

Yamazaki, K., Beauchamp, G.K., Bard, J., Thomas, L. and Boyse, E.A. (1982). *Proc. Natl. Acad. Sci. USA.* **79**: 7828–7831.

Yamazaki, K., Beauchamp, G.K., Wysocki, C.J., Bard, J., Thomas, L. and Boyse, E.A. (1983). *Science* **221**: 186–188.

Yamazaki, K., Beauchamp, G.K., Thomas, L. and Boyse, E.A. (1984). *J. Mol. Cell Immunol.* **1**: 79–82.

Yamazaki, K., Beauchamp, G.K., Thomas, L. and Boyse, E.A. (1985). *J. Exp. Med.* **162**: 1377–1380.

Yamazaki, K., Beauchamp, G.K., Matsuzaki, O., Bard, J., Thomas, L. and Boyse, E.A. (1986a). *Proc. Natl. Acad. Sci. USA.* **83**: 4438–4440.

Yamazaki, K., Beauchamp, G.K., Matsuzaki, O., Kupniewski, D., Bard, J., Thomas, L. and Boyse, E.A. (1986b). *Proc. Natl. Acad. Sci. USA.* **83**: 740–741.

Yarita, H. and Takagi, S.F. (1978). *Proc. Jpn. Acad.* **54**: 30–34.

Yarita, H., Kogure, S., Iino, M. and Takagi, S.F. (1977). Olfaction and Taste VI. ed. by Le Magnen, J. and MacLeod, P. IRL Press, London, p. 212.

Yarita, H., Iino, M., Tanabe, T., Kogure, S. and Takagi, S.F. (1980). *J. Neurophysiol.* **43**: 69–85.

Yasuda, K., Nishida, Y. and Ikeda, Y. (1973). *Otologia, Fukuoka Jibi to Rinsho* **19**: 35–38.

Yfantis, C. (1980). Cited by Schleidt, M. and Hold, B. (1982).

Yokota, T., Reeves, A.G. and MacLean, P.D. (1970). *J. Neurophysiol.* **33**: 96–107.

Yoshida, H. (1950). *Hokkaido J. Med. Sci.* **25**: 454–458.

Yoshii, K. and Kurihara, K. (1983a). *Brain Res.* **274**: 239–248.

Yoshii, K. and Kurihara, K. (1983b). *Brain Res.* **279**: 185–191.

Yoshii, K. and Kurihara, K. (1983c). *Brain Res.* **280**: 63–67.

Zarrow, M.X., Gandelman, R. and Denenberg, V.H. (1971). *Hormones Behav.* **2**: 227–238.

Zilstorff-Pedersen, K. (1955). *J. Laryngol.* **69**: 602–607.

Zilstorff-Pedersen, K. (1965). *Arch. Otolaryngol.* **82**: 53–55.

Zilstorff-Pedersen, K. (1972). *O.R.L. Digest*, 37–43. Cited by Doty, R.L. (1979).

Zimmerman, H.M. and Netsky, M.G. (1950). *Res. Publs. Ass. Res. Nerv. Ment. Dis.* **28**: 271–312.

Zotterman, Y. (1949). *Acta Physiol. Scand.* **18**: 181–189.

Zusho, H. (1978). *J. Jpn. Rhinolog.* **17**: 81.

Zusho, H. (1979). *JASTS* **13**: 147–150.

Zusho, H. (1982). *Arch. Otolaryngol.* **108**: 90–92.

Zusho, H., Asaka, H. and Okamoto, M. (1981). *Auris Nasus Larynx (Tokyo)*, **8**: 19–26.

Zusho, H., Yamamoto, K., Negishi, M., Kobayashi, K., Doty, R.L. and Appelbaum, S. (1983a). *JASTS* **17**: 145–148; *Chem. Sens.* **9**: 84, 1984.

Zusho, H., Kobayashi, H., Yamamoto, A., Negishi, M., Yamamoto, K. and Amoore, J.E. (1983b). *JASTS* **17**: 149–152; *Chem. Sens.* **9**: 84–85, 1984.

Zusho, H., Kobayashi, H. and Yamamoto, K. (1984). *JASTS* **18**: 97–100; *Chem. Senses*, **10**: 136, 1985.

Zusho, H., Kobayashi, K., Yamamoto, K. and Fujii, S. (1985). *JASTS* **19**: 105–109; *Chem. Sens.* **11**: 273, 1986.

Zwaardemaker, H. (1895). Die Physiologie des Geruchs. Englemann, Leipzig.

Zyo, K., Omukai, F. and Ban, T. (1963). *Med. J. Osaka Univ.* **13**: 193–239.

Appendix: The Art of Smell

A. The Japanese Cult of Incense, kōdō

1. A description of an incense gathering

The time is early spring. The place is a quiet room inside a sequestered building called "Rosei-tei" in the grounds of the famous Ginkakuji (Silver Pavillion) located at the foot of Higashiyama in the ancient Japanese capital of Kyoto. A dozen people are sitting in quiet repose against the walls of this 10-mat room surrounding an elderly man in uniform. The old man leans forward toward a small brazier in front of him, so small that it can easily be placed on his palm. At first, he buries a glowing round charcoal inside the ash of the brazier, covering the charcoal nearly entirely, leaving only a small hole in the center of the ash. One can see only a part of the red glowing charcoal through the hole. After he trims the ash and makes folds toward the hole from the surrounding ash with an ash trimmer, he places a small mica plate on the ash surrounding the hole, and then puts a tiny piece of incense wood on the center of the plate, so that the wood may be warmed by heat rising through the hole.

Then he returns to his upright position, takes the brazier on the palm of his left hand, and, half covering the top of the brazier with his right hand, brings it slowly to his nose. Not a sound can be heard within the room as time flows at a snail's pace through this peculiar silence. This is a typical scene from an incense gathering held in Rosei-tei twice every month.

The old man is Hachiya Soyu, the present and 19th master of the Hachiya family, a continuous line of incense masters of the Shino school, which can trace its origins back to the age of Ashikaga Yoshimasa (1436–1490). The master passes the brazier with its heated incense to the person on his left, who, in turn, passes it to her left as the incense makes its way around the room. The deportment of each person as she receives the incense brazier from her neighbor closely resembles that of the tea ceremony, with its unique manners and customs and an elegance all its own.

With the passing of time as the incense makes its way around the room, each participant imitates the actions of the master by holding the incense brazier in her left hand, partially covering it with her right hand and "listening" for the aroma, memorizing exactly what she "hears." At first three aromas are passed around the room from hand to hand. These are known as "tameshi" (initial contacts). After this "tameshi," 10 pieces of incense are passed around, and the participants try to decide which of the 10 pieces are similar to each of the "tameshi." The 10 pieces include three pieces each of the three scents used in the "tameshi" plus one piece that is different from the other three and has yet to be sniffed. This piece is called "kyaku" (the guest). These four distinct kinds of incense are then wrapped in individual papers, placed together and thoroughly rearranged. The challenge is to determine the order of appearance of each aroma as it was originally passed around, a nearly impossible task.

Figure is a scene from a typical gathering for the cult of incense. When the small incense brazier is placed gently on the palm of the left hand, partially covered by the right hand, (see the woman at the left in Fig. 1) and brought close to the nose, the moment of the first sniff draws a faint aroma of the Orient into the nose. This instant offers a fragrance as full of colors as a firework display in the sky on a warm summer night, which disappears almost as quickly as it first appears. The cult of incense includes the teaching of the "nana soku" (seven breaths), a lesson based on the principle that after "hearing" the aromas seven times, any further analysis or memorization becomes futile, regardless of the number of times the incense is smelled. It is a well-known fact that our sense of smell tires easily. In effect, after seven sniffings—or, for that matter, even three, if your decisions have yet to be made and committed to memory, any further attempts become increasingly more difficult.

Most of the participants are women, yet on the day of the gathering there is hardly any fragrance of cosmetics. Neither are any words exchanged among the members of the group. In order for the judgement, analysis, memorization of aromas to take place, a total concentration of mind and body is necessary. Any slight distraction would serve to render the decision-making process impossible. All evil intentions need to be expelled from the mind and all physical relationships with the outside world cut to allow for the necessary mental concentration. A neat, uncluttered room provides the most conducive atmosphere. After "hearing" the aroma, each incense brazier should be passed quietly to the next person. Decisions are recorded on the designated form before the next brazier is accepted. After all of the incense pieces have made their rounds, the incense is disposed of according to the traditions of the particular school of kōdō, and the forms are collected. The decisions of each participant are then checked for accuracy of order.

The cult of incense has evolved from a long Japanese cultural tradition. It is the enjoyment of the elegance of an aroma, a sensation that exceeds mere enjoyment. It is thought of as a means of character development. Just what are the origins of this cult of incense?

2. What is kōdō?

The origin of incense burning can be traced to India through China. Incense in China was used primarily in connection with Buddhist worship. When Buddhism was introduced into Japan from China, the custom of incense burning naturally followed. There is a well-known story about the introduction of scented wood. Around 630 A.D., a large piece of wood drifted ashore on the Japanese island of Awaji in the Inland Sea. When the islanders burned it, they found that it emitted a wonderful odor. Hearing of this event, Prince Shotoku ordered them to present this driftwood to the Emperor. After that time, scented woods were imported directly from China, and indirectly through Korea. They were used in Buddhist ceremonies. In the Heian era (700–1150 A.D.) incense came to be used in the daily life of the aristocrats. The peers used incense to scent their guest rooms, clothes, and furnishings. They also came up with a game that can be called "incense matching" or "incense comparing." The nobles competed with each other to see who could judge correctly two identical odors among ten pieces of incense.

In the succeeding period, the Ashikaga era (1350–1500), the use of incense in Buddhist ceremonies declined, but the aesthetic factor of incense burning came to be more appreciated. It is said that *kōdō*, the incense ceremony or the cult of incense, was born in Japan around 1450 A.D. when the shogun Ashikaga Yoshimasa was governing Japan on behalf of the emperor. The shogun enjoyed aesthetic recreations so much that he wanted to formulate a code of etiquette to govern incense burning. The peer Sanjonishi Sanetaka and the warrior Shino Soshin came up with a number of rules for the incense ceremony with the aid of a number of tea ceremony masters. Hence, the ceremonies came to resemble each other in many points. Nowadays, there remain only three sects. The first is an aristocratic sect named "Oieryu" which was founded by Sanjonishi and is popular among the members of the Imperial Household to this day in the Tokyo area. The twenty-second master, Sanjonishi Sanetori, holds an incense ceremony once a month in a room on the 34th floor of a modern 36-story building in Tokyo. The second sect is the Shino sect, which was popular among the feudal lords, warriors, and the general public, and today is popular among the people in the Kyoto and Nagoya areas. The twentieth master in descent of the Shino sect, Hachiya Soyu, offers incense ceremony lessons twice a month in an attractive old hut located on the grounds of the famous Ginkakuji in Kyoto. The name of the hut, "Rosei-tei," means "a hut in which to enjoy cleanliness." The third sect is the Kannonji sect, which is said to be popular in Kyoto.

Scented woods have been classified into six categories according to the country of their origin. Their names are the ancient Japanese titles of these countries or islands, all of which are located in the southern part of Asia: Kyara and Manaka (India), Rakoku (Thailand), Manaban (Malacca), Sumotara (Sumatra), and Sassori (an unknown location). The exact places of origin are still unknown since the merchants refused to share their sources with their customers. It is said that the wood from the trees in these locations was first covered with honey and later buried in the ground or exposed to the open air for a long time. During this time the easily decaying portions disappeared, leaving only the resinous parts of the wood. Since *kōdō* took place in a literary atmosphere, these scented woods were given poetic names taken from classic Japanese literature, including the famous *Tale of Genji*.

One of the most fascinating things about *kōdō* is its very name, which means

"the hearing of incense" or "listening to incense." One never speaks of smelling or sniffing incense in the ceremony. Nothing can better describe the degree of concentration which is demanded for differentiating odors than the metaphor of hearing. Any perfumer will quickly admit that noise is the greatest enemy of odor discrimination.

Before any further explanation of the incense ceremony, a listing of the various utensils is in order. First on the list is the censer, or a small brazier, which is usually a valuable piece of porcelain. It is half-filled with a specially made non-odorous ash. A round piece of charcoal, also made to burn without odor specifically for the ceremony, is ignited. A pair of tongs is used to bury the charcoal in the ash. Then an ash trimmer is used to trim the ash over the charcoal, leaving a small hole above the center of the charcoal. The hole serves as an exit for heat from the glowing charcoal. Then a mica plate pincer is used to put a mica plate on the hole. Finally, a feather sweeper is used to clean the sides of the censer.

Now, with the preparations for incense heating completed, a tiny fragment of incense wood is taken from its wrapper and put on the heated mica plate. Soon the incense can be "heard," and the incense ceremony commences. Figure is a typical scene from the ceremony previously described.

From the original game, a number of more or less complicated formats have been developed. One is called *genjiko*. In this game, there are five pieces each of five different kinds of incense, totalling 25 pieces. These are mixed up and, without a pre-test, they are submitted to the members of the party for identification in sets of five. A special system of notation employing five vertical rods was devised for this game.

When the first and third pieces of incense are judged to be identical, the tops of the first and third vertical bars from the right are connected by a top bar. If the first and second, and fourth and fifth are connected, this means that the first and the last two are of the same kind, while the third one is a different variety. Five independent rods without a connecting bar means that all of the five odors are different. Thus, 52 different combinations can be made by means of these vertical bars.

Another, and perhaps a more fascinating, version of *kōdō* is the game known

---

Figure. An incense gathering
The man sitting in black kimono on the right is Sanjonishi Sanetori, the twentieth master of the "Oieryu" sect. The woman on the extreme right is in charge of incense heating for this gathering. The women on the left is about to "hear the incense." She has a small brazier in her left hand and, half covering the brazier with her right hand, is raising it to her nose. The woman in black kimono in the back center is passing a brazier to her neighbor, while other guests are silently waiting for the next brazier to be passed.
A white slip of paper is seen in front of each participant. After "hearing" each incense, the participant writes a judgement on this paper. In the end, these papers are gathered and the results are compared by the woman in the center foreground, who is in charge of registering the names of all participants and their results on the large sheet of white paper before her.

as *keiba-ko*, which means "incense horse race." In this game participants are divided into two teams, each with its own horse symbol, one red and one white. These are moved along a beautifully laid-out race track scoreboard as correct answers are received. If one team loses more than 5 points—in other words, if a cavalier falls more than 5 points behind his rival—he has to dismount until he can overtake his rival. But *keiba-ko*, unlike the incense ceremony, is more a game than a serious ritual.

For an incense party, the guest room should be tastefully decorated as on other social occasions. There is, however, a noticeable difference in the decorations. No fragrant flowers are used. Moreover, to avoid unnecessary confusion, participants in the party are required not to carry incense bags, to wear fragrant clothes or to bring with them handkerchiefs that have been sprinkled with perfume. In order to obtain the best results, a participant should neither have completed a hearty meal nor be in the process of fasting.

A brief explanation of the *dō* of *kōdō* and of the spirit of the incense ceremony becomes necessary at this point. Very probably, readers are already familiar with *sadō* (tea ceremony), *kadō* (flower arrangement), and *judō* (Japanese-style wrestling). There are many other *dō* in Japan: for instance, *shōdō* is the art of calligraphy or writing with a brush and Chinese ink; *kendō* refers to Japanese swordsmanship. All of these words commonly employ the word *dō* ("way" or "path") in their names. Tea ceremony and flower arrangement are thought of by the Japanese as aesthetic means to enhance our personality and deepen our philosophy of life. Through the very elegant acts of making tea or arranging flowers, we are able to achieve total concentration. Thus, we can segregate oursouls from the daily lives in which we are constantly disturbed by difficult problems and unpleasant experiences.

The incense ceremony is not merely a form of play or a practical technique. Rather, it has been elevated to an aesthetic medium for seeking human perfection. The scented woods used in Japan do not have the fragrance or sweetness found in European perfumes. They do have their own elegant simplicity and profundity, however, and are capable of taking us into sublime mental states. Moreover, the differences in the odors among these woods are small and unclear in many cases. Consequently, in order to perceive such small differences, it is essential to stay in a very quiet atmosphere, to remove all distractions from our minds, and to concentrate our attention wholly on the sound of incense that we need to hear. All other things need to be forgotten. Through the act of incense hearing we are able to retrieve our subjectiveness and enjoy meditating on the past, present and future. During the centuries since the introduction of Buddhism into Japan, all of our techniques and means of entertainment have come to contain a philosophical meaning characterized in the Japanese language by the use of the suffix-*dō*. *Dō* by itself has the meaning of the "way" or "road" on which a person walks, but as a suffix it means a way of living which a spiritual person must seek in order to complete his or her own being. This has been a characteristic element of the Japanese mind. Very probably, most readers are familiar with the sect of Buddhism known as Zen. In the tea ceremony, flower arrangement, and the incense ceremony, the influence of Zen has been very great indeed.

During World War II, most cities and towns in Japan were burnt to the ground. Even after the war, for many years people suffered severely from housing and food shortages. The common desire of the Japanese during those years was to relieve their hunger, to restore Japan economically, and then to create a new democratic country. During the 40 years since the end of the war, the economy of Japan has recovered miraculously, and it continues to advance. At the same time, the traditional spirit of Japan, although once devastated by the war, has also been reviving gradually.

In the 1980s and the years to come, it is a most important challenge for Japan to deal with the spiritual poverty which exists alongside the material richness of the country. The revival of *kōdō* will be an aid in this process.

3. Why has there been no development of a perfume culture in Japan?

Since ancient times, the Japanese have been an agricultural people. For centuries they continued to center their agricultural efforts around the raising of rice and wheat, and lived in a poverty peculiar to farmers, with hunting being restricted to a few members of the warrior class. With the importation and rise in popularity of Buddhism, meat-eating nearly disappeared. A light diet composed primarily of grains and plants evolved, and the use of spices was quite limited.

At the same time, as Japan is a relatively hot country, the Japanese developed a fondness for bathing, with a daily washing and cleansing of the body becoming almost a ritual. The Japanese as a group are known to have only faint body odors, due to a major extent to the diet that evolved over the years. The Japanese have fine, soft, delicate skin that is easily influenced by various sources of stimulation. Therefore the application of substances that would serve to cover or change a strong body odor has never become a habit. At the very most, they have limited their use of aromas to the scenting of their clothes and, on occasion, their rooms, in order to please visitors.

Moreover, due to the location of Japan in a temperate climate, there are few tropical plants and flowers which emit intense fragrances. There has been no particular call for an interest in powerful fragrances, and no real development of a perfume culture either in Japan or in China, a major source of traditional Japanese culture.

God has given us our wonderful sense of smell. How have we taken advantage of this gift? Our use of our sense of smell seems to depend to a large extent upon our cultural heritage.

The cult of incense was created in Japan and has developed over the course of more than 500 years. It is a world of art and character formation that has evolved and been passed along from one generation to the next. When fragrance is thought of as a spiritual world, the cult of incense becomes the highest response of the Japanese to the gift of smell.

How do the people of the West take advantage of this gift?

(B) Perfumery in the West

1. Chanel No. 5

The great French perfumer Ernest Beaux was born in Moscow in 1883. He entered the Rallet Company, where his elder brother was the general manager, in 1898, and was able to study the skills of perfumery with the renowned perfumer Lemercier.

With the outbreak of World War I, Beaux fled Russia for France, where he joined the army and was stationed in Northern Europe. In this part of Europe on summer nights, when the sun never falls below the horizon, he would leave his barracks and go off to the shores of a nearby lake. From there Beaux could see a number of other lakes with their shores covered in a carpet of beautiful flowers. Amidst these flowers and the fragrant aromas that they would emit, he could spend his nights in fragrant fantasy. The image of this transient aroma became engraved upon his mind.

With the end of the war, Beaux decided to attempt to express through perfumes his memory of the transient aromas of Northern Europe. He used jasmine, rose, and various other abundant natural refined flower oils freely. He succeeded, through the skillful arrangement and combination of a few kinds of aldehydes, in creating new synthetic aromas with the distinctive features of each major fragrance (basic aroma) that up until this time had been nonexistent.

It was at this time, with Paris leading the world of fashion and "haute couture" and serving as a center for all kinds of splendid activities, that Coco Chanel first met with Beaux, who had already developed a name for himself as a perfumer. It was an event that should be recorded as extremely significant in the history of perfumery. Beaux received a commission from Chanel to create perfumes, and was able to come up with the two series of perfumes we now know as numbers 1 through 5 and numbers 20 through 24. Coco selected perfume No. 5 to lead the way in her perfume enterprise.

When Beaux first asked the name to be selected for this perfume, she replied instantly, "Actually, it is to be presented as a part of my fifth fashion show, which is being held on the fifth of May. The number 5 is quite a lucky number for me, so I will call it 'Chanel No. 5'." This is the story behind the birth of the most famous perfume in the world.

In the years to come, a number of interesting episodes about Chanel No. 5 would gain a fame all their own. When Marilyn Monroe was asked by a reporter what she wears to bed, she responded by saying, "I put on a little Chanel No. 5." This line is still fresh in our minds. In this manner, the reputation of Chanel No. 5 continued to grow.

Beaux wrote in his memoirs: "Perfume is art, and the perfumist a creative artist. With the progress of science, the birth of new, synthetic aromas is preeminent. The perfumist has at his command new fragrances and must continue to create a sense of originality through an inundation of new aromas. This is the real work, and the perogative of the perfumer."

Indeed, perfumery is an art that has been improved and refined over the ages. The perfumer deserves to be considered a great artist.

It becomes necessary at this point to delve into the background of the development of the culture of perfume as it arose in the West.

2. Perfumes and the ancient Egyptians

We know that the current use of perfumes began as a luxury with the ancient Egyptians. They had three primary uses for perfumes: in religious ceremonies, in the process of mummification, and in daily life.

Incense was burned on an enormous scale in the large shrines. The air in the rooms of private homes was also full of these aromas. The women of ancient Egypt carried something similar to a spice box, which emitted a strong and pleasant fragrance.

The oldest method for making incense and perfumes can be found in Exodus **30**: 34–37:

> The Lord said to Moses, "Take fragrant spices: gum resin, aromatic shell, galbanum; add pure frankincense to the spieces in equal proportions. Make it into incense, perfume made by the perfumer's craft, salted and pure, a holy thing. Pound some of it into fine powder, and put it in front of the Tokens in the Tent of the Presence, where I shall meet you; you shall treat it as most holy. The incense prepared according to this prescription you shall not make for your own use."

The ingredients for incense must first be emulsified into a fine powder and mixed before being burned.

These fragrant spices (stacte) are now called styrax. According to the first-century Roman writer Plineus, myrrh is the sap that exudes from a spiny shrub.

In Egypt, an appreciation for aromatic essences began to develop in the first century B.C., with Cleopatra. There are a number of reports about her relationship with perfumes. She was an extremely attractive woman, and she made it a regular practice to use perfumes, which led to their popularity at the time. Her use of perfumes and fragrances on the ship in which she met with Anthony is a famous story that is well known even 2000 years after the event. The numerous perfumes available now are, in effect, the creations and the arrangements of those of the early Egyptians.

In recent years, a number of beautifully curved alabaster perfume oil containers from the tomb of Tutankahmen, which date back to 1350 B.C., have been found. Some of the essences have histories of over 3000 years. Regardless of their age, however, their fragrance remains. In Luxor, too, numerous beautiful pots full of perfumed fragrant plasters have been found just as they were when they were first sealed up in the tombs.

It is surprising to discover that the formulas for perfume compounds which were in use 3000 years ago are quite similar to the mixed perfumes of today. However these formulas are limited to those of the simpler perfumes.

3. *Frankincense and myrrh*

Frankincense and myrrh appear quite frequently throughout history. But the explanations of these essences are not all that easy to understand. In Matthew 2: 11 it is stated as follows:

> At the sight of the star they were overjoyed. Entering the house, they saw the child with Mary his mother, and bowed to the ground in homage to him; then they opened their treasures and offered him gifts: gold, frankincense, and myrrh.

"Myrrh" appears many times in the Bible. Myrrh can be found in southern Arabia, Abyssinia, Somaliland, India, and the East Indies as a product of the *boswellia carterii* tree. When the bark of this tree is cut, sap oozes out from the cuts. The moisture in this sap evaporates in the sunlight, and what remains is "frankincense." Nowadays, this is called olibanum oil.

Frankincense was used primarily as a sacred fragrance offered to the gods in Egypt and other ancient cultures. The aroma of frankincense is extremely refreshing. Smelling this aroma exhilarates the heart and soul. It is the fragrance of the forest. Its sacred aroma cleanses the human soul and serves to please the gods as well. Frankincense was not used as an offering to the dead or for the spirits of ancestors. Nor was it found in the presence of mummies.

The scientific name for the plant which produces the oil of myrrh is "oleum myrrhae commiphora myrrha." It is found in Arabia, Abyssinia, and along the coast of Somalia in east Africa. Myrrh is made from the sap of the tree that oozes from scars in the bark and has been dried in the sunlight. The fragrance of myrrh excites the heart and soul. It has a warmth of its own, that comforts the heart. It is also a powerful germicide. The fragrance serves as a kind of panacea. It is an appropriate aroma for cleansing the stench of death, and was used in the preservation of corpses. It was burned in the temples of ancient Egypt and was used in the mummification process. Between the 25th century B.C. and the period around the birth of Christ, the Egyptians mummified their dead. It is known that during this particular period, myrrh and cinnamon were used in this process. The techniques employed in mummifying the dead were a highly guarded secret, and only special persons were engaged in the process.

4. *The Jews and perfume*

The Jews learned of the use of perfumes and incense from the people of the ancient world. Influenced by the Egyptians, they were quick to pick up the use of fragrant essences in their religious rights and ceremonies. It is said that they also learned to apply perfumes to their bodies. Jewish women, like their Egyptian counterparts, used perfumes as comsmetics. One Jewish woman in particular, Judith, stands out as a heroine. She cleansed herself with perfume and covered her body with the best oils, dressed herself in refined clothes, and tempted the commander of the Assyrian army that was besieging her city town. When he became drunk, she lay him on her bed and killed him.

5. The Greeks and perfume

The use of perfumes among the women of ancient Greece appears to have been quite common. Perfumes were especially popular among prostitutes, who took baths in perfume. Afterwards they would put the "fragrant oils from the East" all over their bodies. These same prostitutes would put fragrant liquids in their mouths in order to sweeten their breath. At the time, there were hundreds of perfumists who had opened perfume shops in the city of Athens, where they sold many different types of perfumes. Some of these perfume merchants had attached their own names to their products. For example, a record still exists of the shop of the perfumist Peron.

It is thought likely that the Greeks inherited their knowledge regarding perfume and their habits of applying perfumes from the Egyptians.

6. The Romans and perfume

The fancy of the Romans for perfume and essences, particularly during the age of the Roman Empire, appears quite similar to that of the Greeks. They adopted the practice of applying different perfumes to the various parts of their bodies, which was a custom with the Greeks. In the age of Christ, at the zenith of Rome, the number of perfume shops in the city was equal to that of Athens. The women of Rome kept three kinds of perfumes on their cosmetic tables: salve-like solid ointments, liquid ointments mixed with oils, and powders. The more popular among these perfumes were used by the women of Rome after their baths. In the Roman Empire, the use of perfumed oils, ointments, and pomade was extremely popular.

7. The Arabs and perfume——The discovery of alcohol and the origins of perfume

With the downfall of the Roman Empire, the center of influence moved to the Eastern Roman Empire. The aromas of all of the fragrant essences from the Orient could be found in the halls of the Byzantine empire. The port cities served as centers for the trade in perfumes. The Arabs came to be the perfumists of the world in Constantinople. The most important discovery in the history of perfumery was made by the Arabs at this time. This was the discovery and the use of alcohol.

The great scientist, physician, and philosopher Ibn Sina Avicenna, who lived between 980 and 1036, through the discovery of the skills of distillation, discovered a way to extract and maintain the fragrances of plants, and made the production of the perfumes we know today possible. In his first experiment, he chose the favorite flower of the Arabs, the rose. He was successful in extracting its fragrance for use in a perfume. This became what we call rose water.

The use of rose water spread throughout Europe in the days of the Crusades. The Crusaders were quite attracted to things Oriental, and, on their return to their homes, brought with them Eau de Chypre, the perfume of Oriental harems, which became quite fashionable throughout Europe. It seems to have been a perfume made from a single flower. This was the beginning of the second introduction of perfume into Europe, the largest in scope after the introduction of perfumes during the age of the Romans.

Most of these perfumes were imported from the Middle East, but by the end of the 12th century perfumes were being produced by the Europeans themselves and distributed among the women of France. These perfumes, however, were simply imitations of the originals from the Middle East.

8. Europeans and perfumes after the age of the Romans

The first original European perfume to be produced after the age of the Romans seems to have been what we call "lavender water." According to tradition, the Benedictine nun, St. Hildegard, who lived in the 12th century, discovered the means of distilling it.

Lavender water was soon being produced in England, Germany, and France, and is still quite popular today.

There is another cosmetic water that has enjoyed an equally long history. It was called "Hungarian water," and it first appeared as a popular cosmetic among the noble classes around 1370. It is said to have been produced by Queen Elizabeth of Hungary.

The great age of perfumery in England began with Queen Elizabeth I. Queen Elizabeth liked to use a variety of perfumes. Among her favorites were musk and rose, and she also favored a perfume known as "compound water," a perfume which we have yet to understand. The queen's expenditures for perfumes were exorbitant, and her palace was said to literally reek from the aroma perfumes.

At the same time in France, a perfume culture was still in the early stages of development. In the process of marrying King Henry II, Catherine de Medici (1519–1589) was said to have brought her personal perfumist from Florence to France. This perfumist was Reni, who eventually opened his own shop in Paris. Catherine, in effect, should be considered the founder of the industry in France as it is today.

Lady Pompadour, the lover of Louis XV, used perfumes lavishly. She was fond of flowers, especially those with a sweet aroma. The wife of Louis XVI, Queen Marie Antoinette (1755–1793), despised the spicy fragrances of the perfumes from the Orient as well as strong animal odors. Her response was to further popularize the light, pleasant fragrances that could be distilled from roses and violets.

The popularity of perfumes was able to survive the upheaval brought on by the French Revolution. Josephine, the wife of Napoleon, continued in the tradition of Marie Antoinette, favoring simple, natural fragrances. For example, eau de cologne was still quite new in the time of Josephine. At the beginning of the 18th century, Paul de Feminis, an Italian born in the German city of Köln, came up with a cosmetic water that would eventually work its way up to the Number

Two spot on a list of popular perfumes which were being marketed at the time, displacing Hungarian water. It was a refined oil made from lemons, bergamot, and other refined citrus oils, with lavender as its base. It was called "l'eau admirable," and retained its popularity for nearly 100 years.

Johann Maria Farina, a descendant of de Feminis, added rose water to this mixture and sold his new product under the name "eau de cologne." It became an immediate success that was to be imitated by other perfumists, and served as the original mild cosmetic water. It is still popular today.

In Paris, Pierre Guerrine made his own lavender-based cosmetic water, which found favor with the empress Impératrice Eugenie and among the upper classes in Paris. Guerlain's name became quite well-known all over Europe; his perfume was destined to become associated with the empress Eugenie who would give it the name, "Eau de Cologne Imperiale."

9. *The creation of perfumes*

The original inhabitants of Europe were horse-riding nomads. They were said to have been meat-eaters. It can be assumed that the use of spices at that time would have been quite popular. It can also be assumed that their diet would have given them a fairly strong body odor.

The Egyptians, with the development of their concept of hygene, believed that cleanliness was next to godliness. The Romans inherited this same line of thought. A complicated ritual form of bathing originated with the Egyptians. Egyptian women took baths more frequently, and seemed to enjoy their bathing more, than their Roman counterparts. After their baths, wealthy Egyptian women would lie on their beds unclothed and submit to massages in which ointments were rubbed all over their bodies by slaves.

But this particular style of bathing was soon to disappear. During the reign of Queen Elizabeth I, bathing was thought to have been a rather uncommon practice. There are records of this period about the lavish use of perfumes:

Names such as "sweet (floral) water" and "ultimately sweet exquiste aroma" were given to those perfumes which were favored at the time. At first, these names sound extremely elegant but in effect, the overall scent of the palace of Queen Elizabeth is thought to have been so offensive because these perfumes were meant to cover the horrible smells of unwashed bodies clothed in unwashed clothes. Such perfumes were necessary at this particular point in history.

The perfumes of the era were frighteningly powerful, and not only in England. In 1580, one of the first books directed towards the general public on the subject of cosmetics was published in France. *Les secrets de Maistre Alerys le Piemn Tois* recommended that women use a cosmetic water called "Eternal Beauty." But an analysis shows that this "Eternal Beauty" was a frightening product, less a cosmetic product of science than a magical witches' brew. This kind of alchemical mixture was made not so much to enhance the sexual charms of women but to simply overpower the strong odor of an unwashed body. This product with its stimulating smell is said to have somehow managed to survive into the 18th century.

With the exceptions of ancient Egypt and Rome, it seems apparent that the custom of bathing has died out and body odors continued to grow ever more overpowering. To make life more bearable, people seemed to develop a serious concern about the use and application of perfumes to their bodies, using skills that developed over a period of 4,000 years.

In this century, we have seen the creation of such wonderful perfumes as Chanel No. 5, No. 19, Soir de Paris (Chanel), Mitsouko, Ma Griffe, Vol de Nuit (Guerlain), and so on. They have been cultivated over the years and represent the balance of the "art of perfumery." Perfumes are the ultimate response of Europeans to the gift of smell.

10. Perfumes and the cult of incense

Perfumes are the epitome of the European's expression of a certain kind of beauty. With the application of perfume to the body, we can emphasize our own individuality.

The cult of incense, likewise, is an art form that enables the Japanese to enhance their own character and create an abundant spiritual life through the medium of aroma. With the mastery of these skills, it becomes possible for the Japanese to refine and express themselves in a way that is entirely different from that of the people from the West.

The Japanese of today have been deeply impressed by the wonders of perfumes from the Western world, and have made the application of perfumes an almost daily routine. So perfumes have made a significant contribution to the understanding of Western culture by the Japanese, who have long had their own cult of incense.

Finally, it should be pointed out that the skills of perfume-making have been imported, and under the scrutiny of Japanese perfumers, a number of perfumes considered to be "Oriental" have been successfully produced and are marketed all over the world under the names "Zen," "Koto," "Mai," and others.

References
Katada, M. (1986), Attractiveness and wonder of smell. Nihon Kogyo Shinbun-Sha, Tokyo. (In Japanese)
Moroe, T. (1986). The path of incense. Kohfu-Sha, Publishing Co., Tokyo. (In Japanese)
Takagi, S. F. (1974). Stories on olfaction. Iwanami Shinsho, Iwanami Shoten. Tokyo, (In Japanese)

Postscript: An Answer to Dr. Gerard

Elucidation of the generative mechanism of the regular 6/sec wave which is elicited by nicotine solution in the frog olfactory bulb was the study theme Dr. Gerard presented to me in Chicago in 1954. This problem has remained in a corner of my mind ever since then.

A decade later, in 1964, Hirata discovered a very interesting reciprocal synapse in the olfactory bulb. Precise electro-physiological studies in our laboratory (Mori and Takagi, 1978 a, b) revealed the remarkable activity of this dendrodendritic synapse: when nicotine solution was dripped onto and spread over the frog olfactory bulb, a great many or most of the mitral cells are activated. As a consequence, depolarization occurs not only in the cell bodies but also in the primary and secondary dendrites. This depolarization activates synapses from the mitral secondary dendrite to the peripheral process of the granule cells and depolarizes the cells. Excitation of the granule cells in turn hyperpolarizes the mitral cells and many other cells through the dendrodendritic synapses from the peripheral processes to the mitral secondary dendrites.

Since the secondary dendrites extend extensively (see Fig. VII-8) and form dendrodendritic synapses with a large number of granule cells (see Fig. VII-3, 7), it is very likely that a great many or most of the mitral cells in the olfactory bulb are connected with each other via granule cells. Thus, once mitral cells are excited by nicotine solution, elicited depolarization of the cells is always accompanied by the hyperpolarization of the same and many adjacent mitral cells. When hyperpolarization subsides, the remaining nicotine solution again elicits depolarization of mitral cells. In this way, a great many mitral cells coincidentally repeat depolarization and hyperpolarization and evoke alpha-wave like rhythmical waves in the olfactory bulb. Understanding of the function of this reciprocal synapse leads to an explanation of the generative mechanism of the rhythmic nicotine wave discovered by Drs. Libet and Gerard.

I finally found an answer to the question that Dr. Gerard had put to me in Chicago, but, sadly, not until four years after his death. A deep regret that I could not convey my answer directly to him still lingers in my heart.

Index

aspartate (Asp), 278, 279
aspiration (of brain), 337, 338
axillary odor, 15, 17, 21; axillary bacteria, 16
axonal conduction velocity of AON neuron, 298
  of AOB mitral/tufted cell, 404
  of MOB mitral cell, 404

bad odors, 24; definition of, 25
basal cell, 129, 397
  mitotic division of, 122
basal forebrain, 294
  cholinergic cells in magnocellular groups, 303
  islands of Calleja complex of, 309
basal ganglion of substantia innominata, 327
basal lamina (basement membrane) of olfactory mucosa, 108, 126, 131
basal nucleus of amygdala, *see* amygdaloid nuclei
basal nucleus of Meynert, 347
basement membrane, basal lamina, 108, 126, 131
basolateral nuclei of amygdala, *see* amygdaloid nuclei
bed nucleus of stria terminalis, 405
behavioral studies on monkey, 355–358
betamethasone disodium phosphate, Rinderon, 94, 95
bicuculine, a GABAergic antagonist, 260, 272
biogenic amines, 178, 278
birhinal olfactory stimulation, 296
birhinal olfactory test, 57, 88
blastema cell, Blastema Zellen, 123, 129
blood-brain barrier, 124
body odors, 15, 18, 24, 35 (*see* axillary, hand, foot, breast, hand, other odors, 17–19)
Bowman's gland, 134, 396
breast odor, 15, 19
breath odor, 19, 21
Bruce effect, 26, 406
butyric acid, 18, 19

Ca²⁺-spike, 263
carbohydrate expressions of receptor cells and their axons (globo-series or lacto-series carbohydrates), 401
carnosine (β-alanyl-L-histidine), 123, 278, 279, 283
catecholamines, 278
central segment of MD, 319, 321, *see*
mediodorsal thalamic nuclei
centrifugal fibers to OB, 288–291, 295, 405;— influence on OB, 290, 291; — input, 295, 302, 327
centroposterior area of orbitofrontal cortex (CPOF), *see* orbitofrontal cortex
chemoinvestigatory behaviors (sniffiing/licking, flehmen), 407
chemosensory dysfunction, 71
chemosensory recognition of genetic individuality, 407
chloride ion (Cl⁻) conductance, 297, 304;—mediated IPSP, 303
cholecystokinin (CCK), 278
cholinergic cell, *see* basal forebrain, 303
chronic unanesthetized animal, 360, 361; experiments on—, 358–361
classification of abnormal olfaction, 79
classification of olfactory disorders, 72–75
collision test, 262, 298, 345
conditioned reflex, 5
content-addressable memory (CAM), 308
contingent negative variation (CNV), 31, 32, 33
contralateral OB (COB), 295, 297, 303, 311
convergence of olfactory and gustatory sensations, 385
copulins, 21, 26
cortico-hypothalamic pathway, 324
corticomedial nuclei of amygdala, *see* amygdaloid nuclei
corticothalamic inhibitory fiber, 317
CPOF, *see* orbitofrontal cortex
cumulative sum analysis test (cusum test), 378
cyclic nucleotides, 199, 218
  adenylate cyclase, 217
  cAMP, 199, 217, 218
  cGMP, 199, 218
  cyclic nucleotide-gated conductance, 199, 200, 201, 217;—sensitive channel, 218
  odorant-sensitive adenylate cyclase, 216
  role in EOG, 184, in olfactory reception, 199–201, 217
  experiments on—, 217, 218
l-cysteine ethylester hydrochloride, Cystanine, 96, 97
cytochrome oxidase (CO) staining, 107, 108

(VMA), 24;—nucleus (VMH), 322

identification of odor, 41, test for—, 85
immunohistochemical studies on olfactory epithelium, 139
  on vomeronasal nerve pathway, 397
IMP, *see* intramembranous particles
impedance change of olfactory epithelium during EOG, 159
imprinting, 27
incense cult of Japan, 453
individual body odor, 15, 20
induced wave of OB, 256
information processing, olfactory, 273, 274, 276, 372
infundibular region, 322
inhibition: autogenetic type, 258; Renshaw type, 258; recurrent, 258; lateral type, 268
inhibitory interneuron, 305, 306
inhibitory postsynaptic potential, *see* IPSP
insulin, 278
intensity of odor, 7
interbulbar lateral inhibition, 296
internal segment of MD, *see* mediodorsal thalamic nuclei
internal tufted cell (iT), 238
interspecies chemosensory recognition of individuals, 408
intracellular HRP labeling, 247, 248, 251, 252, 286, 298
intracellular recording of MDmc neuron, 317, 318
  of HPC neuron, 330
  of olfactory cell, 201–207
  of supporting cell, 208–210
  of VNN cell, 404
intramembranous particles (IMPs), 115, 116, 117, 118, 119, 126
intrinsic neurons of PPC, 301, 302
intrinsic wave of OB, 256
ion substitution experiment, 168, 174
ionic stimulation of olfactory epithelium, 223–225
ipsilateral OB (IOB), 296, 297, 298
IPSP (fast, slow), 243, 246, 258–260, 265, 270–272, 290, 296, 297, 303, 304, 305, 306, 330, 404; prolonged —, 404
islands of Calleja complex of rat basal forebrain, 309
isovaleric acid, 16, 37, 41, 42, 43, 44, 51, 52
IS-spike of mitral cell, 261

Jacobson's organ (=vomeronasal organ), 103, 106, 140, 391, 393, 395

kakosmia, 75, 79
Kneipp therapy, 29
Kreiner's classification of orbital gyrus (ORG), 381

lamina propria of olfactory mucosa, 101, 108, 121, 131
laminar structures (patterns) of AOB, 401
  of AON, 295
  of OB, 237–242
  of OT, 308
  of PPC, 300
latency of EOG, 157
lateral hypothalamic area (LHA), *see* hypothalamus
  cellular (neuronal) responses of LHA to odors, 370
lateral olfactory tract (LOT), 235, 237, 260, 264–272, 286, 290, 297, 298, 300, 302, 303, 305, 306, 307, 328, 354, 355, 372, 373, 387, 388
lateral orbital area (LO), 388
lateral preoptic area, 290
lateroposterior area of orbitofrontal cortex (LPOF), 307, 308, 312, 316, 327, 334, 335, 343, 349, 353, 357
  cellular responses to odors in LPOF, 361–363
Lee-Boot effect, 25, 406
lesion studies by means of aspiration or electrocoagulation, 337
lesions of OFC, 382
limbic system, 293, 294, 316;—areas, 328, 375;—limbic circuit, 294
liposomes, 211
local axon collaterals of principal cells in OB, 247, 268, 269
locus coeruleus nucleus (adrenergic) (LC), 302, 309, 310, 314, 330, 405
long-lasting synaptic excitation, Es, 272, 273
long-lasting synaptic inhibition, ls, 270, 271, 272
LPOF, *see* orbitofrontal cortex
luteinizing hormone-releasing hormone (LHRH), 278, 279, 404

M-spike of mitral cell, 261
MA, *see* medial portion of amygdala
MAb, *see* monoclonal antibodies
Macaca irus, *see* monkey